Lost Stars

Lost Stars

Lost, Missing, and Troublesome Stars from the Catalogues of Johannes Bayer, Nicholas-Louis de Lacaille, John Flamsteed, and Sundry Others

Morton Wagman

The McDonald & Woodward Publishing Company
Blacksburg, Virginia

The McDonald & Woodward Publishing Company
Blacksburg, Virginia and Granville, Ohio
www.mwpubco.com

Lost Stars : Lost, Missing, and Troublesome Stars from the Catalogues of Johannes Bayer, Nicholas-Louis de Lacaille, John Flamsteed, and Sundry Others

Text © 2003 by Morton Wagman

All rights reserved
Printed in the United States of America
by McNaughton & Gunn, Inc., Saline, Michigan

11 10 09 08 07 06 05 04 03   10 9 8 7 6 5 4 3 2 1

First printing September 2003

Library of Congress Cataloging-in-Publication Data

Wagman, Morton, 1933-
  Lost stars : lost, missing, and troublesome stars from the catalogues of Johannes Bayer, Nicholas-Louis de Lacaille, John Flamsteed, and sundry others / Morton Wagman.
     p. cm.
Includes bibliographical references and index.
  ISBN 0-939923-78-5 (hardcover : alk. paper)
  1. Constellations — Names. 2. Stars — Names. I. Title.
QB802.W34 2003
523.8'02'16 — dc22

2003014985

Reproduction or translation of any part of this work, except for short excerpts used in reviews, without the written permission of the copyright owner is unlawful. Requests for permission to reproduce parts of this work, or for additional information, should be addressed to the publisher.

Contents

Preface vii
Introduction 1
Part One : Lettered Stars 19
Part Two : Numbered Stars 331
Glossary 497
Appendix I: Locating Bayer's Stars on His Charts 501
Appendix II: Lost Constellations 519
Select Bibliography 521
Index 535

To

Jeanette Lindenbaum Wagman

and

Etary Wagman

Preface

I first became interested in astronomy and star lore many years ago after reading William Tyler Olcott's *A Field Book of the Stars* and *Star Lore of All Ages.* However, I was puzzled to discover that, although astronomers identified certain stars, usually the brighter ones, with letters and numbers, in many constellations some letters and numbers were missing. In looking over numerous star charts of the constellation Scorpius, for instance, I noticed that stars were assigned letters from Alpha to Omega and beyond, but no stars bearing the letters Gamma and Phi were to be found. No matter how carefully I examined the star charts, using even a magnifying glass, I could not locate any stars with these two letters. Similarly, in Telescopium, there were stars identified as Alpha and Delta, but there was neither a Beta nor a Gamma. A similar situation prevailed with numbered stars. The stars in the constellation Canes Venatici were numbered from 1 to 25, but there was no star numbered 22. Why?

I was puzzled by yet another aspect of astronomy and star lore. Why were the constellations named the way they were? Why, for instance, was a ram or a bull or a lion placed among the stars? And why was each placed in a particular sector of the heavens? Was there any relationship between a constellation's name and its location in the sky? Or was it all just coincidence or happenstance?

In this book I have attempted to answer these questions of my youth. In Part One: Lettered Stars, I examine each of the eighty-eight constellations recognized by modern astronomers and explain in detail why some lettered stars appear to be missing from stellar atlases and catalogues. In the second section of this work, Part Two: Numbered Stars, I have done the same for those stars assigned numbers. In addition, I have sought to seek out and explain, wherever possible, the origin of the names of constellations now in use in the west. This search led me all the way to the ancient Middle East, the cradle of Western Civilization and the birthplace of astronomy.

Throughout this work I have made frequent use of words such as erroneous, mistaken, incorrect, and the like. At first, I was elated to think that a humble student like myself would find errors in the works of the giants of science. But as time passed, I became more appreciative of their labor and the painstaking way in which they checked and rechecked their calculations, checked and rechecked their work. I finally came to realize that their mistakes in no way detract either from their greatness or their enormous contribution to human knowledge. Besides, we would all do well to bear in mind the old adage: *errare humanum est,* to err is human.

Part Two of this work, Numbered Stars, appeared in an abbreviated form in the August 1987 issue of the *Journal for the History of Astronomy,* and the account of Hercules in Part One, Lettered Stars, appeared in a somewhat shortened version in the May 1992 issue of the same journal.

The research for this work was supported in part by a grant from The City University of New York PSC-CUNY Research Award Program.

I owe many debts to those who have helped me with this work, debts that I can never fully repay. I am forever grateful to my teachers at Erasmus Hall High School, Brooklyn College, and Columbia University for guiding me in my search for knowledge. I appreciate the hard work of Carol A. Boone of McDonald & Woodward for helping me prepare this work for publication. Thanks are also in order for the assistance provided to me by the librarians at the New York Public Library, the United States Naval Observatory, the Library of Congress, Brown University, Harvard University, the Bodleian Library of Oxford University, University of Illinois at Urbana, and Princeton University. I am especially grateful to the hard-working librarians at Kingsborough Community College who, one way or another, succeeded in securing for me many hard-to-find dusty old tomes from all over the country. I am heavily in debt to the many scientists and scholars cited in the bibliography; they were the pathfinders, and their help was invaluable. But I am especially indebted to two very special individuals without whose assistance this book could never have been written: my mother, Jeanette Lindenbaum Wagman, and my sister, Etary Wagman. My debt to them is incalculable, my affection for them is immeasurable. To them I dedicate this book.

Morton Wagman
History Department
Kingsborough Community College
The City University of New York
Brooklyn, New York

Introduction

Introduction

Four hundred years ago, Nicholas Copernicus, along with Tycho Brahe, Johannes Kepler, and Galileo Galilei, ignited a revolution in Europe — a scientific revolution that shook the very foundations of intellectual thought and freed Western man, once and for all, from the restraints and stifling influence of the past. Before the sixteenth century, European scholars rarely, if ever, dared to challenge the writings of the classical Greek and Roman scientists. The writings of Aristotle on natural science, Celsus on medicine, and Claudius Ptolemy on astronomy were considered sacred and sacrosanct. What European living in the Dark Ages following the collapse of the civilization of the ancient world would even dream of matching the genius of these intellectual giants? But all of this changed as a new spirit of inquiry, intellectual curiosity, and increased self-confidence — the Renaissance — began to pervade Europe. Based on their own observations and research, Europeans of the sixteenth century not only challenged the writings of the classical scholars but also expounded their own theories and findings, which often directly contradicted the teachings of the ancients.

Nicholas Copernicus (1473-1543), a relatively insignificant Polish priest, radically postulated that the sun, not Earth, stood at the center of our universe. Tycho Brahe (1546-1601), a Dane, compiled the first new star catalogue since that of Claudius Ptolemy, which had been created some fifteen hundred years earlier. Galileo Galilei (1564-1642), an Italian, using a telescope, established the fact, once and for all, that the sun was at the center of our planetary system. He did this by observing that Venus, like our Moon, went through various phases, and this could only occur if the sun was the central body of our planetary system. Based on the observational notes of his teacher, Tycho Brahe, Johannes Kepler (1571-1630), a German, developed his theory of the elliptical journey of the planets around the sun. In describing Kepler's achievement, one modern historian has noted:

> *Thus for the first time, and forever, the principle that for a thousand years had been accepted by all astronomers as the basis of astronomy has been destroyed, that the circle was the natural and true orbit of heavenly bodies (*Pannekok, *History of Astronomy*, p. 241*).*

Ranking with these giants of science was their German contemporary Johannes Bayer (1572-1625), the father of stellar cartography and author of what is generally recognized as the first modern atlas of the heavens.

Bayer's *Uranometria* of 1603, a combination star atlas and catalogue, introduced innovations in astronomy that were so significant and far reaching that his work has served as a model for all subsequent sky charts and stellar catalogues. His atlas contains fifty-one stellar charts: one for each of the classic forty-eight constellations cited in Claudius Ptolemy's second-century work, the *Almagest;* one depicting the twelve new southern constellations devised in 1603 by Pieter Dircksen Keyzer and Frederick de Houtman; and two planispheres of the northern and southern hemispheres. Bayer's charts were prepared with an eye to the needs of the practical astronomer. Based largely on the catalogues of Tycho Brahe, the charts are remarkably accurate, at least for observations made with the naked eye. Not only are these charts large, 11½ by 8½ inches, and beautifully drawn, but historically they were among the first to show the stars as they actually appeared in the heavens to the earth-bound observer — not backward as they appeared on celestial globes and as they were drawn on most earlier charts. Bayer's charts mark off ecliptical latitude and longitude in one-degree units along their borders, thus enabling astronomers to determine easily and quickly stellar positions with a straight-edge ruler. Accompanying each constellation chart is a list of stars down to the 6th magnitude, the limit of visibility with the naked eye. To facilitate stellar identification, each chart also includes stars from neighboring constellations. And most important of all, his catalogue broke the rigid system of Ptolemaic identification and enumeration.

For over fifteen hundred years, astronomers had faithfully patterned their works after that of Claudius Ptolemy, going so far as to copy his catalogue verbatim and to include the very same stars in the very same order as he had done. But Bayer was among the first to change this. He not only added hundreds of new stars — stars that were plainly visible but which Ptolemy and his followers neglected — but he arranged his catalogue systematically. He grouped stars by magnitude class and listed them in order of decreasing brightness, not haphazardly or according to allegorical pictograms, methods that had characterized the work of Ptolemy and his followers. He also organized his constellations more logically than did Ptolemy. The latter, for example, had considered Arcturus, although a 1st-magnitude star and one of the brightest in the heavens, an *informis*, or unformed star, outside the border of the constellation Bootes, the Herdsman. He treated Hamal, a 2nd-magnitude star, the same way — as an unformed star outside the border of Aries, the Ram. What is truly amazing is that astronomers after Ptolemy — Al-Sufi, the editors of the Alphonsine Tables, Ulugh Beg, and countless others — did exactly the same thing. Bayer was among the first to include Arcturus within Bootes and Hamal within Aries. But of all his innovations, the most far-reaching and most important was his method of stellar identification. He identified stars with a systematic pattern of Greek and Roman letters.

Bayer's innovations were accepted and, as a result, his *Uranometria* became the most popular and widely used atlas among astronomers. It was reprinted many times, occasionally without its star lists, and became the standard by which all other star atlases were judged. Later astronomers copied Bayer's constellation figures and adopted his method of enumeration. In the nineteenth century, eminent astronomers, such as Frederick W. A. Argelander, author of the *Bonner Durchmusterung,* and Eduard Heis, reproduced Bayer's drawings in their own atlases. And as recently as 1938, the prominent Czech astronomer Josef Klepesta edited a somewhat revised version of the *Uranometria* and copied the exact same constellation figures devised by Bayer. So influential and famous was Bayer's work that when new constellations were formed, mostly in the southern skies, astronomers emulated his methods, especially his system of stellar enumeration.

Bayer's method of stellar identification proved both practical and simple. He designated stars in a constellation with lower-case letters from the Greek alphabet, from Alpha to Omega, in order of the stars' decreasing magnitude class. After exhausting these twenty-four letters, he turned to the lower-case Roman alphabet. He started, however, with a capital A instead of a lower-case Roman a to avoid confusing it with its Greek look-alike Alpha. He continued lettering from b to z, but omitted j and v since in the Roman alphabet i and j and u and v are written the same way and would be indistinguishable. In only one constellation, Hercules, did Bayer ever reach the letter z.

Bayer frequently identified several stars with the same letter, a practice that eventually led to the use of indices, superscripts, or in the case of Benjamin A. Gould, the author of *Uranometria Argentina,* subscripts. Bayer used the same letter for stars that were nearly adjacent to each other, classifying them as *duplex* or double stars; for example, Alpha Capricorni (two adjacent stars in the Goat's head), Omega Cygni (two very close stars in the Swan's claw), Nu Sagittarii (two nearly adjacent stars in the Archer's chin), and Theta Tauri (two adjacent stars under the Bull's eye). Later astronomers would label them Alpha^1 Cap and Alpha^2 Cap, Omega^1 Cyg and Omega^2 Cyg, and so forth. He also used the same letter for those stars he called *binae* or pairs of stars; for example, d Cancri (a pair in the lower part of the Crab's shell), Pi Cygni (a pair in the Swan's tail), c Scorpii (a close pair in the body of the Scorpion), and A Virginis (two close stars in the Virgin's head). These were usually closely located stars of the same magnitude, and they too would later be assigned indices. Finally, Bayer used the same letter to describe several stars or a group of stars of the same brightness — usually those fainter than 3rd magnitude — which were in close proximity and somehow associated with each other in the constellation's mythology. In Orion, for instance, he assigned Pi to a series of six stars that he considered 4th magnitude and that were located in the lion skin that the Giant Hunter bore in his right hand. Later astronomers would designate them Pi^1, Pi^2, Pi^3, Pi^4, Pi^5, and Pi^6 Orionis. When assigning the same letter to several stars, Bayer usually placed a single letter in the midst of the group, although occasionally he would mark individual stars with the same letter, as, for example, in Virgo, where he assigned the letter A to each star of the pair in the Virgin's head.

Bayer was not the first to identify stars with letters or numbers. Several earlier astronomers, notably

Alessandro Piccolomini, had done the same, but Bayer's work is markedly superior to the others. As noted earlier, his method of stellar identification is systematic, not arbitrary or haphazard. His *Uranometria,* moreover, is an exceptionally fine example of the printer's art. In fact, when first examining it, one is instantly struck by its almost modern appearance, the exceptionally fine quality of its plates, and its stellar designations. Each gradation of magnitude is carefully and clearly delineated, as are double stars and nebulous objects. *Uranometria* far surpasses anything that preceded it. Consequently, Bayer's system of stellar nomenclature set the pattern for others to follow and has prevailed through the ages.

Bayer's practice of lettering stars was adopted by other cartographers. In the late seventeenth and early eighteenth century, Augustin Royer in France and John Senex in England were the first two cartographers to prepare celestial charts that included many of Bayer's lettered stars, but the first astronomer to systematically attempt to trace all of Bayer's stars was John Flamsteed, Britain's first Astronomer Royal. While compiling his own catalogue of over 2,900 stars, he made certain to note whenever he thought one of the stars he observed was synonymous with one of Bayer's. As he carefully and meticulously observed the heavens and reduced his observations to the year 1690, the epoch, or date, of his catalogue, he began to expand and to elaborate on Bayer's system of stellar enumeration. If he observed several stars in close proximity to what he thought was a Bayer star, as, for example, Sigma Ursae Majoris, he designated them "1 ad [to] Sigma, 2 ad Sigma" in order of right ascension or celestial longitude. He did this because the stars were close to each other, or because he was not precisely sure which star Bayer meant as Sigma, or because he felt Bayer's Sigma may have included the combined light of both stars. As with Bayer's duplicate letters, this method of stellar enumeration gave rise eventually to the use of superscripts.

Flamsteed's use of indices unfortunately caused considerable confusion. When he came to a group of stars to which Bayer had assigned the same letter, he was faced with a serious problem. Since he worked with a copy of the *Uranometria* that included only the star charts and not the star lists, he sometimes could not determine which letter applied to which stars. In the case of Pi Orionis, for example, where Bayer assigned Pi to six stars, Flamsteed assigned Pi to only two stars — the two that were nearest the letter Pi on Bayer's star chart. Flamsteed made a more glaring error in Auriga, the Charioteer. Bayer listed Psi in Auriga in his catalogue as a group of ten 6th-magnitude stars, but he inadvertently omitted the letter from his chart. As a result, Flamsteed never realized that Psi existed in the constellation and so left all ten stars unlettered in his catalogue.

Flamsteed's famous catalogue — "Catalogus Britannicus," which was included in the third volume of his *Historia Coelestis Britannica* — was published in 1725, six years after his death. Based on almost forty-five years of work and thousands of observations, this catalogue was the most accurate star catalogue of its day. Since Flamsteed made extensive use of the telescope, the pendulum clock, and the micrometer for determining stellar positions, his was, for all intents and purposes, the first catalogue to take advantage of the new scientific technology that was then emerging in Western Europe. It was also, up to that time, the largest catalogue that had ever been compiled, double the size of its nearest competitor, Johannes Hevelius's *Prodromus Astronomiae* of 1690. Because of its size and because it was not printed until after Flamsteed's death in 1725, "Catalogus Britannicus" is not entirely free from errors, especially those involving stellar identifications.

In 1712, a dozen or so years before the authorized version of Flamsteed's "Catalogus Britannicus" appeared, Edmund Halley arranged for the publication of a manuscript copy of the catalogue that had been loaned to Isaac Newton and the Royal Society. Flamsteed was so furious with Halley — and Newton, who had encouraged Halley in this enterprise — that he later obtained all the unsold copies, about three hundred of the four hundred printed, and burned them, as he said, as a sacrifice to Heavenly Truth. Flamsteed claimed that Halley did not properly edit the manuscript and that he had no say in its publication. He took exception, for instance, to the way Halley described the stars, to the way Halley organized the constellations, to the way Halley grouped the stars within the constellations, and to the way Halley numbered the stars — Flamsteed preferred leaving the stars unnumbered. No wonder he called the 1712 catalogue the "corrupted Catalogue." Surprisingly though, the pirated edition contains some useful information on stellar identification that is not found in the 1725 edition. Halley's edition not only properly equates Flamsteed's stars with Bayer's stars but also corrects errors in identification that somehow appear in the later edition. Consequently, I have used the pirated edition, the "corrupted Catalogue," extensively in the preparation of the present work since it is an important source of stellar identification.

A quarter of a century later, John Bevis, using both editions of the "Catalogus Britannicus," sought to reconcile the two catalogues and to correct as many of Flamsteed's errors as he could find. He was especially concerned with properly identifying Bayer's lettered stars and correlating them correctly with Flamsteed's stars. Bevis's atlas and catalogue, *Uranographia Britannica* (1747-1749) — or as it is usually but erroneously called, *Atlas Celeste* — was never formally published. Only about twelve copies in various stages of completion were prepared from plates. Consequently, one can question its importance since it had virtually no impact on later astronomical studies. Nevertheless, Bevis's work is included here largely because it anticipates many of the corrections and adjustments to Flamsteed's catalogue that were made by later scholars.

In 1835, Francis Baily, apparently unaware of Bevis's *Uranographia,* prepared a revised and corrected edition of Flamsteed's catalogue. It was based, to a certain extent, on the work of earlier astronomers like James Bradley and Caroline and William Herschel, who also had sought to correct Flamsteed's catalogue. However, Baily's revision was based most importantly on new source material that he himself had found. He discovered at the Greenwich Observatory seventy volumes of Flamsteed's manuscripts, including his computation book and notes of his daily observations. After meticulously studying these papers, Baily produced, for all intents and purposes, a totally new catalogue, which included several hundred stars that Flamsteed had observed but omitted, for one reason or another, from his printed catalogue. Among these are several of Bayer's stars, such as A Bootis and b Persei.

At about the same time in Germany, F. W. A. Argelander prepared an atlas and catalogue of naked-eye stars visible from Central Europe. Argelander copied very closely the figures of the constellations in Bayer's star charts and entitled the atlas, appropriately enough, *Uranometria Nova* (1843). In both the atlas and catalogue, which were published separately, Argelander sought to reconcile Bayer's star lists with Flamsteed's catalogue. He was concerned with removing letters from the dimmer stars that Flamsteed had included in his catalogue since he felt Bayer could not possibly have observed them with the naked eye.

In the next generation, Eduard Heis published *Atlas Coelestis Novus* (1872), his own atlas and catalogue of stars visible to the naked eye. Like Argelander, Heis based the figures in his atlas on Bayer's drawings. Heis, who not only had an exceptionally sharp eye but used a special instrument to block out extraneous light, succeeded in cataloguing over 5,000 "naked-eye" stars, 2,000 more than Argelander. Based on his own observations, he proposed various changes in stellar letter designations that had appeared earlier in the works of Baily and Argelander. As a result, differences arose among astronomers over which stars merited Bayer letters and which indices or superscripts applied to which stars.

While astronomers like Flamsteed, Baily, Argelander, and Heis were expanding and refining Bayer's system of stellar identification, others were involved in extending it to the newly discovered stars in the southern hemisphere. The first to do so was Nicholas-Louis de Lacaille, whom Benjamin Gould dubbed "the true Columbus of the southern sky." Before Lacaille, only three other people had tried to chart and catalogue systematically the southern skies: Pieter Dircksen Keyzer and Frederick de Houtman, whose catalogue of 304 stars was published in 1603 as an appendix to Houtman's dictionary of the Malay and Madagascan languages, and Edmund Halley, who, after spending about two years at Saint Helena scanning the southern skies, published his chart and catalogue of 341 stars in 1679. Lacaille, by contrast, observed at Cape Town almost 10,000 stars in the eleven-month period from August 6, 1751, to July 18, 1752.

In preparing his chart and catalogue, Lacaille noted that Bayer's work on the southern constellations was badly in need of revision. In the *Uranometria,* Bayer had included six southern Ptolemaic constellations whose actual appearance in the heavens bore little or no resemblance to the stars in his charts. Bayer had not actually observed them since they were too far south to be seen from Central Europe; instead, he had relied upon the works of Ptolemy and others. Lacaille decided to redesign, reletter, and rechart the constellations Ara, Centaurus, Corona Australis, Lupus, Piscis Austrinus, and Argo Navis, which he divided into four separate constellations — Carina, Puppis, Pyxis, and Vela. Lacaille also separated Crux from Centaurus, setting it up as a separate constellation with its own stellar letter designations. And since he was in a far better position to observe southern skies than either Ptolemy or Bayer, he added new stars and additional letters to several of the more northerly of the southern constellations such as Canis Major, Eridanus, Hydra, Sagittarius, and Scorpius.

In addition to revising and relettering Ptolemy's southern constellations in Bayer's catalogue, Lacaille also proposed to revise the twelve new constellations in the southern hemisphere that Bayer had included on a single chart in his *Uranometria*. As mentioned, these constellations had been devised by the Dutch explorers and seamen Keyzer and Houtman, whose catalogue of southern stars appeared in the same year as Bayer's *Uranometria*. Unlike the stars in the Ptolemaic constellations in the *Uranometria,* these stars were left unlettered, without any accompanying description, and, Lacaille discovered, inaccurately mapped. He undertook, therefore, to letter and chart the stars in the twelve new constellations of Apus, Chamaeleon, Dorado, Grus, Hydrus, Indus, Musca (which Bayer had originally called Apis, the Bee), Pavo, Phoenix, Triangulum Australe, Tucana, and Volans.

Finally, noting that there were vast areas in the southern firmament still largely uncharted, Lacaille set about the herculean task of filling in the gaps. He mapped and designed fourteen new constellations, all but one of which he named in honor of the various instruments that symbolized for him the Age of Reason, the enlightenment that was spreading learning and understanding throughout Western Europe. He memorialized the artist's easel and palette (Pictor), chemical furnace (Fornax), compass dividers (Circinus), engraver's burins (Caelum), mariner's compass (Pyxis), microscope (Microscopium), pendulum clock (Horologium), pneumatic pump (Antlia), reflecting octant (Octant), rhomboidal reticle (Reticulum), sculptor's studio (Sculptor), square and ruler (Norma), and telescope (Telescopium). Table Mountain (Mensa), where he set up his small, eight-power telescope overlooking the Cape of Good Hope, was the other of his new constellations. He also decided to extend to these new constellations — and to Columba, which had been devised half a century earlier by Petrus Plancius — Bayer's system of stellar designations.

Lacaille basically followed Bayer's system of lettering, but he did make some significant changes. After exhausting the Greek alphabet, Lacaille used lower-case Roman letters starting with lower-case a instead of capital A as Bayer had done. He adopted this practice since, after using all the lower-case Roman letters, he introduced as his third alphabet capital Roman letters from A to Z. Like Bayer, Lacaille omitted some Roman letters. He did not use lower-case j or capital J; he did not use capital U, but he did use lower-case u; he did not use lower-case v, but he did use capital V. By and large, he kept his lettering to a minimum. He used only Greek letters in the new southern constellations and added Roman letters only to the larger Ptolemaic constellations, like Centaurus and Lupus, and to the the newly devised components of the former Argo Navis. In only two constellations, Puppis and Vela, did he ever get as far as Z. Like Bayer, Lacaille assigned the same letter to closely situated stars. Unlike Flamsteed, he did not use indices; he merely repeated the same letter. Other astronomers would later assign indices, superscripts, or subscripts to these stars. Usually Lacaille lettered stars only of the 6th magnitude or brighter, but occasionally he lettered dimmer ones.

Using his small telescope with a focal length of only 26¼ French inches (71 centimeters), with its ½-inch aperture, and working under the most trying of conditions, Lacaille observed almost 10,000 stars in a relatively short period. He had no time to recheck his observations, nor did he have an opportunity to return to the Cape of Good Hope. Considering the scope of his work, it is truly amazing how few errors he made. Most of these occurred in determining magnitude. As he himself admitted in his "Notations to the Catalogue" of 1763, if he had any doubt about a star's magnitude, he would assign it the fainter rather than the brighter magnitude. He further admitted that it was most difficult to determine gradations of magnitude when dealing with faint stars, those of 6th magnitude and beyond, since their magnitude was determined at the moment of their passage across the lens of his telescope; moreover, because of their dimness, they were especially effected by the brighter stars that appeared in his field of vision, by the moon, and by any number of unfavorable viewing conditions that may have played a role in distorting their light. Lacaille was obviously more concerned with fixing stellar positions and mapping uncharted regions of the heavens than with arranging stars, particularly the fainter ones, in order of descending magnitude. In fact, for his first catalogue, published in 1756, he had time to reduce the positions of only 1,935 stars approximately to the epoch 1752; he enlarged this to 1,942 in his second catalogue, which was published posthumously in 1763. In both of these works, he lettered only about half the stars listed. All of his observations were eventually reduced and published under the direction of Francis Baily almost a century after they had been made.

When Baily edited *Lacaille's Catalogue* (1847) and the *British Association Catalogue* (1845), which includes most of Lacaille's stars, he proposed several changes in lettering. He suggested that stars fainter than 5th magnitude should not be distinguished by letter and that Lacaille's Roman letters should be dropped for all constellations except the subdivisions of Argo Navis. It was not a firm rule, however, and Baily retained some of Lacaille's letters for stars fainter than magnitude 6 while removing letters from some stars brighter than magnitude 5. He removed most of Lacaille's Roman letters, but not all. There is, moreover, no consistent pattern in the lettering between *Lacaille's Catalogue* and the *British Association Catalogue,* probably because Baily did not live to see them completed. He died in August 1844, while the galley proofs were being prepared. As a result, some stars are lettered in one catalogue and not lettered in the other.

A generation later, Benjamin A. Gould revised many of Lacaille's constellations in his *Uranometria Argentina* of 1879, a catalogue of stars in the Southern Hemisphere visible to the naked eye. He established boundaries for the southern constellations that were adopted virtually without change by the International Astronomical Union in 1930, and he dropped or relettered many of Lacaille's letters for stars dimmer than 6th magnitude. He relettered or removed completely the capital letters from Lacaille's stars that were lettered from R to Z to avoid confusing them with variable stars that Argelander had begun to letter starting with R. For example, in Centaurus Gould removed Lacaille's R, S, T, X, and Y (Lacaille did not use U and W) and left these stars unlettered since he considered them too faint to merit a letter designation; he considered Lacaille's V and Z bright enough to be relettered v and J. In addition, he added letters to various bright stars that Lacaille had overlooked and to some faint ones within twenty-five degrees of the southern celestial pole. He also extended Bayer's lettering system to two "modern," or post-Ptolemaic, constellations devised by Hevelius — Scutum and Sextans — that had escaped lettering.

As Lacaille and Gould extended Bayer's lettering system to the southern skies, Francis Baily was doing the same for the north. In his introduction to the *British Association Catalogue,* Baily announced his intention of assigning letters to those stars in the nine modern constellations included in the catalogue of Hevelius that had also been included in the catalogue of Flamsteed. These constellations included Camelopardalis, Canes Venatici, Coma Berenices, Lacerta, Leo Minor, Lynx, Monoceros, Sextans, and Vulpecula. He omitted Hevelius's Scutum since it had been excluded by Flamsteed. Baily assigned letters to the stars in all of these constellations except Sextans, where, he asserted, there was no star bright enough to merit a letter. In fact, to avoid what he considered the profusion and confusion of letters and alphabets that Lacaille had used, he limited his lettering only to the Greek alphabet and only to stars brighter than magnitude 4½. Actually, Baily's criticism of Lacaille was unfounded. As Gould later pointed out, Baily mistakenly accused Lacaille of using duplicate letters and duplicate alphabets when, as a matter of fact, Lacaille was very careful in lettering stars and committed neither of the practices for which he was faulted by Baily.

Although Flamsteed, Lacaille, Baily, and Gould were primarily responsible for extending Bayer's lettering system, other astronomers also added Greek and Roman letters. John Bevis, for instance, in his *Uranographia* of 1747-1749 added numerous lower-case Gothic letters, but since very few copies of his work were ever published, his letter designations were largely unknown and never adopted by other astronomers. On the other hand, Johann Bode, in 1801, prepared an extensive catalogue with a series of large, carefully drawn stellar charts that were widely used throughout the nineteenth century. Bode not only identified his stars with Bayer and Lacaille's letters but, in some cases, continued lettering where Bayer left off. In Ursa Minor, for example, Bayer ended with Theta, but Bode added other letters including Lambda, Pi^1, and Pi^2, which have become standard designations in almost every modern atlas and catalogue. He was also responsible for correcting numerous errors — many of them copying or typographical errors — in Flamsteed's catalogue and for correctly identifying many of Bayer's stars. Oddly enough, Baily later made many of the very same corrections, but for some reason failed either to mention Bode or to acknowledge his assistance.

In addition to Bode, other astronomers have proposed adding letters, but, with a few exceptions, their letter designations have not survived in the scientific literature while those of Bayer, Flamsteed, Lacaille, Bode, Baily, and Gould have withstood the test of time.

Over the centuries, many of Bayer's letters, as well as those of Flamsteed and Lacaille, have, for one reason or another, been lost, dropped, or misplaced, causing considerable confusion and resulting in stellar discordances between various atlases and catalogues. As a consequence, some astronomers have gone so far as to suggest the elimination of letters, especially Roman letters. In fact, the latest edition of *The Bright Star Catalogue,* as well as *Sky Catalogue 2000.0,* has done just that — all references to Roman letters in these two catalogues have been eliminated. This is patently unfair and smacks of prejudice. Astronomers should not discriminate against stars with Roman letters. They are just as important, if somewhat dimmer, as those with Greek letters and should be afforded equal treatment and consideration. Discrimination should have no place in the scientific community.

It is the primary purpose of this study to identify wherever possible those lettered stars that have been lost, missed, or dropped; those that have generated confusion; and those that have been misidentified. An effort has been made not merely to list these stars but to analyze and discuss why and how the discordances or problems first arose. Particular attention has been directed to those stars with indices or superscripts since these have caused extensive and widespread confusion — Flamsteed initiating indices; Baily correcting Flamsteed; Argelander correcting Baily and Flamsteed; and Heis correcting Argelander, Baily, and Flamsteed. Consequently, all stars with indices have been carefully scrutinized to determine how their indices originated.

A secondary purpose of this study is to assemble information to help explain the origins of the constellations. While most peoples of the world have seen figures in the night sky and developed their own constellations, why ancient people grouped together certain stars into certain constellations is more difficult to explain. Some constellations like Scorpius, Leo, and Taurus do indeed bear some resemblance to their namesakes. Some like Libra, Aries, Orion, and Gemini were probably grouped together because certain religious or agricultural events occurred when the group appeared with the rising or setting sun. The emergence of agriculture required the development of an accurate calendar to alert farmers when to prepare their fields for planting and nothing was more accurate for the ancients than the annual movement of the sun across the sky. Other groupings are inexplicable, at least at the present time. They bear no resemblance to their names and seem to have no connection to any event, although to the ancients they must have been important, important enough to have been commemorated in the heavens. More recently in the eighteenth century, Lacaille devised fourteen constellations that helped to honor the Age of Reason.

The present work deals exclusively with the eighty-eight constellations recognized by the International Astronomical Union (IAU). No effort has been made to discuss other constellations unless their history has a bearing on one of the eighty-eight. Detailed information about astral legends around the world can be found in such works as Allen's *Star Names,* Lum's *The Stars in Our Heaven,* Ridpath's *Star Tales,* Staal's *New Patterns in the Sky,* and Krupp's *Beyond the Blue Horizon,* all of which discuss in considerable detail stellar mythology and star lore of other peoples and cultures.

This work is divided into two parts. Part One discusses lost, missing, and troublesome *lettered* stars, primarily from the catalogues of Bayer, Flamsteed, and Lacaille, while Part Two discusses lost, missing, and troublesome *numbered* stars from Flamsteed's "Catalogus Britannicus" of 1725. In both parts, the constellations are arranged in alphabetical order.

Each constellation is introduced with its Latin name, its three-letter official IAU abbreviation, a translation of its name, and/or the popular version of its name. A brief note follows indicating whether it is of Ptolemaic or more recent origin, who devised it, who lettered it, what letters it contains in the Greek and Roman alphabets, and, in Part Two, how many numbers Flamsteed assigned to it.

Part One includes, wherever possible, a description of the origin of each constellation.

Following the introduction of each constellation in Part One is a synopsis of its lettered stars. The synopsis is divided into nine columns. The first column lists the star's letters. The letters in this column, arranged in alphabetical order, are those assigned by Bayer, Lacaille, or other prominent astronomers. In the case of Lacaille, however, his capital Roman letters beginning with R are not included since they are currently used to identify only variable stars; however, they are included in the section that follows which explains in detail the various lost, missing, and troublesome stars. A letter followed by two superscripts, such as Gamma^{1,2}, indicates a closely located double star. The second column is the star's magnitude as described by Bayer in his *Uranometria.* The third column is for southern stars. It notes the star's magnitude as described by Lacaille in his *Coelum Australe* of 1763. The fourth column is the star's V, or visual magnitude — how its brightness would appear to the human eye as measured by a photoelectric photometer. The sources for this column are the fourth edition of Hoffleit's *The Bright Star Catalogue,* Hoffleit *et al.*'s *A Supplement to the Bright Star Catalogue,* and Hirshfeld and Sinnott's *Sky Catalogue 2000.0.* The letter c following a star's visual magnitude in this column indicates that this is a double- or multiple-star system and that the magnitude represents the combined light of the entire system. In some instances, the figure used for combined magnitude is from Pickering's *The Revised Harvard Photometry.* The fifth column lists the star's Flamsteed number, if it has one. The sixth column, for southern stars, lists the star's number in Lacaille's *Coelum Australe.* The seventh column is for stars whose letters are derived from catalogues other than those of Bayer, Flamsteed, or Lacaille. The eighth column is the star's HR number, its designation in Pickering's *The Revised Harvard Photometry* and its successors, the various editions of *The Bright Star Catalogue.* Finally, the ninth column is the star's HD number, its designation in the *Henry Draper Catalogue.* An examination of a constellation's synopsis will enable the reader to determine at a glance how many lettered stars there are in the constellation and who first lettered, catalogued, and identified its stars.

Following each constellation's synopsis in Part One is a detailed account of lost, missing, or troublesome stars in alphabetical order starting with the Greek alphabet, Roman lower-case alphabet, and Roman capital alphabet, the format adopted by Bayer and Lacaille. Although some of the stars have since been shifted to other constellations, they are listed here in the constellation in which they were first lettered, with a notation indicating their current location if a change has occurred. In addition, included here are all of Bayer's and Lacaille's Roman-lettered stars since many catalogues are no longer lettering them. These stars are not lost yet, but they are in imminent danger of falling into obscurity. Also included here are stars not listed in the synopsis. These are stars with temporary or transient letters that have been removed for one reason or another from most modern atlases and catalogues, such as the letters assigned by Flamsteed and removed by later astronomers. Also in this category are Lacaille's capital Roman letters beginning with R; these letters were removed in the nineteenth century to avoid confusing these stars with variable stars. Following a star's letter is its Flamsteed number, if it has one. The star is next identified by its Revised Harvard Photometry or HR number in *The Bright Star Catalogue*. If it is not listed in *The Bright Star Catalogue*, its Henry Draper or HD number is cited. Since Bayer and several other astronomers mentioned in this work did not use telescopes, the star's V, or visual magnitude, has been included as a helpful guide to its visibility. (For the possibility that Bayer may indeed have used a telescope, see Chi and h Persei.)

If a star is properly identified and lettered in Tirion's *Sky Atlas 2000.0*, *The Bright Star Catalogue*, *A Supplement to the Bright Star Catalogue*, and *The Revised Harvard Photometry*, the notations *SA, BS, BS Suppl.,* and *HRP* have been added as appropriate. *The Revised Harvard Photometry* has been used as a source because it is the forerunner of *The Bright Star Catalogue* and because of its copious and informative notes on stellar nomenclature that have largely been omitted from the latter work. Finally, the star's listings in other catalogues have been cited. No effort has been made to list all the catalogues consulted, only those that have been especially helpful in establishing a star's identity. If an asterisk (*) appears following a catalogue citation, this indicates that the referenced catalogue has properly identified the star with its letter. See Explanation and Interpretation of Citation Lines included in this Introduction for an explanation and example of these citations.

After the star's identification has been established, a detailed account explains why this star has caused problems for astronomers and stellar cartographers. In some instances, the explanation is simply a sentence or two. In others, the explanation can extend to several paragraphs and even several pages, as with Chi and h Persei and Upsilon Eridani.

The constellation synopsis in Part Two lists all of Flamsteed's numbered stars and is divided into six columns. The first column is the star's Flamsteed number. The second column is the star's letter, if it has one. The third column is its magnitude as described in the "Catalogus Britannicus." It should be noted that Flamsteed did not use fractions to refine magnitude gradations. Instead, he used a system of integers and decimal points. For example, when he determined a star's magnitude as brighter than 4 but less than 3, he would write 3.4, which Baily later interpreted to mean 3½. Similarly, when Flamsteed described a star as brighter than magnitude 4 but considerably less than 3, he would write 4.3. Baily converted this to 3¾. This system, incidentally, was employed by astronomers such as Argelander and Heis well into the nineteenth century. In the present work, Baily's suggested magnitude fractions as found in his revised edition of Flamsteed's catalogue have been adopted. In some instances, the "Catalogus Britannicus" omits stellar magnitude altogether, or notes "obs" [*obscura,* indistinct or obscure] or "tel" [*telescopica,* telescopic]. The fourth column in the synopsis is the star's V, or visual magnitude. The fifth and sixth columns indicate the star's HR and HD numbers respectively.

In Part Two, the missing or troublesome stars are listed in numerical order as they appear in Flamsteed's "Catalogus Britannicus" of 1725. Otherwise, the format in Part Two follows that used in Part One. That is, if a star is properly identified in *Sky Atlas 2000.0, The Bright Star Catalogue, A Supplement to the Bright Star Catalogue,* and *The Revised Harvard Photometry,* the notations *SA, BS, BS Suppl.,* and *HRP* have been added as appropriate. The star's listings in other major catalogues have been cited, and if an asterisk follows a catalogue citation, this indicates that the catalogue has identified the star with its proper Flamsteed number.

Lost, missing, or troublesome stars in both Part One and Part Two are identified in the synopsis with a plus (+) sign. This indicates that information about them will be found in the pages following the synopsis. Also included here are those stars from Flamsteed's catalogue that are fainter than magnitude 6.5, hence they are not listed in *The Bright Star Catalogue,* which excludes stars dimmer than that magnitude. They are included in this section, although without a plus sign in the synopsis, since they are not lost as yet but are in danger of losing their Flamsteed numbers. In effect, the current work is an annotated and updated list of all the stars in the catalogues of Bayer and Flamsteed, as well as all the newly lettered stars in Lacaille's *Coelum Australe* of 1763.

In cases where no modern or twentieth-century source could be found to verify stellar identity, the doubtful star's position was carefully analyzed in relation to neighboring stars and its precession calculated to minimize the possibility of error or misidentification. The coordinates of these doubtful stars were reduced from either 2000.0 or 1950.0 to 1850.0, the epoch of the *British Association Catalogue,* which was found to contain almost all the missing stars. In some cases, the coordinates of doubtful stars were reduced back to 1690.0, the epoch of Flamsteed's "Catalogus Britannicus." This was done primarily with Flamsteed's numbered stars, especially those fainter ones, those no longer in their original constellation because of boundary changes made by the IAU in 1930, and those with incorrect or missing coordinates in Flamsteed's catalogue. This exercise in stellar precession over a period of three centuries brought new respect and admiration for the accuracy, observational skill, and mathematical ability of the First Astronomer Royal, who worked in an age before the aberration of light, stellar proper motion, and the nutation — the slight "nodding" — of Earth's axis were fully known or understood.

The final determining factor in classifying a star as missing was whether it appeared in the fourth edition of *The Bright Star Catalogue* with its proper nomenclature; that is, with its proper letter and its proper Flamsteed number. *The Bright Star Catalogue* was selected because its widespread use and availability made it both an ideal and practical starting point in the search for missing stars.

Table 1. Explanation and interpretation of citation lines

EXAMPLE FOR STARS IN PART ONE: LETTERED STARS

Hypothetical star Alpha in hypothetical constellation Constella:

Alpha, 99, HR 9999, 5.55V, *SA, HRP, BS,* CA 8888*, BR 7777.

Explanation:

Alpha, 99: In addition to the star's letter designation as Alpha, it is also Flamsteed's 99 in the constellation Constella.

HR 9999: It is listed as star number 9999 in Pickering's *Revised Harvard Photometry,* the forerunner of *The Bright Star Catalogue.*

5.55V: Its visual apparent magnitude is 5.55.

SA: It is designated as Alpha in Tirion's *Sky Atlas 2000.0.*

HRP: It is designated as Alpha in Pickering's *Revised Harvard Photometry.*

BS: It is designated as Alpha in Hoffleit's *The Bright Star Catalogue.*

CA 8888*: It is designated as Alpha and numbered 8888 in Lacaille's *Coelum Australe.*

BR 7777: It is star number 7777 in Brisbane's *Catalogue of 7385 Stars* but without letter designation.

EXAMPLE FOR STARS IN PART TWO: NUMBERED STARS

Hypothetical star 66 in hypothetical constellation Astrum:

66, HD 555555, 6.66V, *SA, BS Suppl.,* BAC 4444, GC 3333*.

Explanation:

66: Flamsteed numbered this star 66 in the constellation Astrum.

HD 555555: It is star number 555555 in the Henry Draper catalogue.

6.66V: Its visual apparent magnitude is 6.66.

SA: It is designated as 66 in Tirion's *Sky Atlas 2000.0.*

BS Suppl.: It is designated as 66 in Hoffleit's *A Supplement to the Bright Star Catalogue.*

BAC 4444: It is star number 4444 in Baily's *British Association Catalogue* but without Flamsteed's number designation.

GC 3333*: It is designated as 66 and numbered 3333 in Boss's *General Catalogue.*

Notes: Although stars in both Part One and Part Two are referred to by their catalogue numbers, it should be understood that citations apply not only to the body of the catalogues but also to the accompanying notes. For an explanation of the catalogue and atlas abbreviations cited above, see Table 2. Catalogue and atlas abbreviations.

Table 2. Catalogue and atlas abbreviations

ABBREVIATION	CATALOGUE
ADS	Aitken, *New General Catalogue of Double Stars*
B	Bode, *Allgemeine Beschreibung...der Gestirne*
BAC	Baily, *Catalogue...of the British Association*
BD	Bonner Durchmusterung
BF	Baily's edition of Flamsteed's catalogue in *An Account of the Revd. John Flamsteed*
BKH	Backhouse, *Catalogue of 9842 Stars*
BR	Brisbane, *Catalogue of 7385 Stars*
BS	Hoffleit, *The Bright Star Catalogue*
BS Suppl.	Hoffleit, *A Supplement to the Bright Star Catalogue*
BV	Bevis, *Atlas Celeste [Uranographia Britannica]*
CA	Lacaille, *Coelum Australe Stelliferum*
CA(P)	Lacaille, "Table des Ascensions Droites," Lacaille's preliminary catalogue of 1756
CAP	Maclear, *Catalogue of 4810 Stars,* the Cape Catalogue of 1850
G	Gould, *Uranometria Argentina*
GC	Boss, *General Catalogue*
H	Heis, *Atlas Coelestis Novus*
HA	Baily's edition of Halley's southern catalogue in "Catalogues of Ptolemy *et al.*"
HD	Cannon and Pickering, *The Henry Draper Catalogue*
HF	Halley's 1712 edition of Flamsteed's catalogue in *Historiae Coelestis*
HEV	Baily's edition of Hevelius's catalogue in "Catalogues of Ptolemy *et al.*"
HR	Stellar designations in Pickering's *The Revised Harvard Photometry* and Hoffleit's *The Bright Star Catalogue*
HRP	Pickering, *The Revised Harvard Photometry.*
L	Lacaille, *Catalogue of 9766 Star*
LL	Lalande, *Catalogue of . . . Stars in the Histoire Céleste*
M	Messier's nebulous objects in *The Messsier Album*
NGC	Dreyer, *New General Catalogue*
P	Piazzi, *Praecipuarum Stellarum*
R	Rumker, *Preliminary Catalogue*
RAS	Baily, "General Catalogue of the Principal Stars" for the Royal Astronomical Society.
SA	Tirion, *Sky Atlas 2000.0*
Sky Cat.	Hirshfeld and Sinnott, *Sky Catalogue 2000.0*
T	Taylor, *Taylor's General Catalogue*
T(P)	Taylor, *A General Catalogue of the Principal Fixed Stars,* Taylor's first or preliminary catalogue
UN	Argelander, *Uranometria Nova*
W	Werner and Schmeidler, *Synopsis*

Notes: For additional information on these works, see the annotated bibliography. Abbreviations in italics refer to books; whereas, the others refer to specific catalogue numbers. BAC 432, for example, refers to star number 432 in Baily's *Catalogue . . . of the British Association.*

Table 3. The constellations

ABBREVIATION	LATIN NAME	TRANSLATION OR POPULAR NAME
And	Andromeda	Andromeda, Chained Lady
Ant	Antlia	Air Pump
Aps	Apus	Bird of Paradise
Aqr	Aquarius	Water Bearer
Aql	Aquila	Eagle
Ara	Ara	Altar
Ari	Aries	Ram
Aur	Auriga	Charioteer
Boo	Bootes	Herdsman, Bear Keeper, Bear Driver
Cae	Caelum	Sculptor's or Engraver's Burin or Chisel
Cam	Camelopardalis	Giraffe
Cnc	Cancer	Crab
CVn	Canes Venatici	Hunting Dogs
CMa	Canis Major	Greater, Larger, or Big Dog
CMi	Canis Minor	Lesser, Smaller, or Little Dog
Cap	Capricornus	Goat, Sea Goat
Car	Carina	Ship's Keel
Cas	Cassiopeia	Cassiopeia, Queen, Lady in the Chair
Cen	Centaurus	Centaur
Cep	Cepheus	Cepheus, King
Cet	Cetus	Sea Monster, Whale
Cha	Chamaeleon	Chameleon
Cir	Circinus	Pair of Compass Dividers
Col	Columba	Dove, Noah's Dove
Com	Coma Berenices	Berenice's Hair
CrA	Corona Australis	Southern Crown
CrB	Corona Borealis	Northern Crown
Crv	Corvus	Raven, Crow
Crt	Crater	Cup, Bowl
Cru	Crux	Cross, Southern Cross
Cyg	Cygnus	Swan
Del	Delphinus	Dolphin
Dor	Dorado	Swordfish, Goldfish
Dra	Draco	Dragon
Equ	Equuleus	Little Horse
Eri	Eridanus	River Eridanus
For	Fornax	Chemical Furnace
Gem	Gemini	Twins
Gru	Grus	Crane
Her	Hercules	Hercules, Kneeler
Hor	Horologium	Pendulum Clock
Hya	Hydra	Female Water Snake, Large Water Snake
Hyi	Hydrus	Male Water Snake, Small Water Snake
Ind	Indus	Indian
Lac	Lacerta	Lizard
Leo	Leo	Lion

(table continued on next page)

Table 3. The constellations *(continued)*

ABBREVIATION	LATIN NAME	TRANSLATION OR POPULAR NAME
LMi	Leo Minor	Lesser, Smaller, or Little Lion
Lep	Lepus	Hare
Lib	Libra	Scales, Balance
Lup	Lupus	Wolf
Lyn	Lynx	Lynx or Tiger
Lyr	Lyra	Lyre, Harp
Men	Mensa	Table, Table Mountain
Mic	Microscopium	Microscope
Mon	Monoceros	Unicorn
Mus	Musca	Fly
Nor	Norma	Square, Architect's Square
Oct	Octans	Octant
Oph	Ophiuchus	Ophiuchus, Serpent Bearer
Ori	Orion	Orion, Giant Hunter
Pav	Pavo	Peacock
Peg	Pegasus	Pegasus, Flying Horse
Per	Perseus	Perseus, Champion
Phe	Phoenix	Phoenix
Pic	Pictor	Painter's Easel
Psc	Pisces	Fishes
PsA	Piscis Austrinus	Southern Fish
Pup	Puppis	Ship's Stern
Pyx	Pyxis	Ship's Compass
Ret	Reticulum	Reticle
Sge	Sagitta	Arrow
Sgr	Sagittarius	Archer
Sco	Scorpius	Scorpion
Scl	Sculptor	Sculptor, Sculptor's Implements
Sct	Scutum	Shield
Ser	Serpens	Serpent
Sex	Sextant	Astronomical Sextant
Tau	Taurus	Bull
Tel	Telescopium	Telescope
Tri	Triangulum	Triangle
TrA	Triangulum Australe	Southern Triangle
Tuc	Tucana	Toucan
UMa	Ursa Major	Larger, Greater, or Big Bear
UMi	Ursa Minor	Lesser, Smaller, or Little Bear
Vel	Vela	Ship's Sails
Vir	Virgo	Virgin
Vol	Volans	Flying Fish
Vul	Vulpecula	Little Fox

Table 4. Lower-case Greek alphabet

α Alpha	ε Epsilon	ι Iota	ν Nu	ρ Rho	φ Phi
β Beta	ζ Zeta	κ Kappa	ξ Xi	σ Sigma	χ Chi
γ Gamma	η Eta	λ Lambda	ο Omicron	τ Tau	ψ Psi
δ Delta	θ Theta	μ Mu	π Pi	υ Upsilon	ω Omega

Table 5. The changing position of the vernal equinox through the ages

	AGE OF TAURUS	AGE OF ARIES	AGE OF PISCES	AGE OF AQUARIUS
	4200 B.C. TO 2000 B.C.	2000 B.C. TO A.D. 200	A.D. 200 TO A.D. 2400	A.D. 2400 TO A.D. 4600
Mar 21 - Apr 19	Taurus	Aries	Pisces	Aquarius
Apr 20 - May 20	Gemini	Taurus	Aries	Pisces
May 21 - Jun 21	Cancer	Gemini	Taurus	Aries
Jun 22 - Jul 22	Leo	Cancer	Gemini	Taurus
Jul 23 - Aug 22	Virgo	Leo	Cancer	Gemini
Aug 23 - Sep 22	Libra	Virgo	Leo	Cancer
Sep 23 - Oct 23	Scorpius	Libra	Virgo	Leo
Oct 24 - Nov 21	Sagittarius	Scorpius	Libra	Virgo
Nov 22 - Dec 21	Capricornus	Sagittarius	Scorpius	Libra
Dec 22 - Jan 19	Aquarius	Capricornus	Sagittarius	Scorpius
Jan 20 - Feb 18	Pisces	Aquarius	Capricornus	Sagittarius
Feb 19 - Mar 20	Aries	Pisces	Aquarius	Capricornus

Notes: For an explanation of why the vernal equinox appears to move, see **Stellar Coordinates** below. Dates are approximate and adopted from Gleadow's *Origin of the Zodiac*.

Star Charts

Among the illustrations included in this work are all forty-nine constellation charts from Bayer's *Uranometria* of 1603; his last chart, Tabula XLIX, depicts a planisphere of the southern heavens showing the twelve new constellations devised by Keyzer and Houtman, the Dutch travelers to the East Indies. Also included is Lacaille's planisphere of the southern heavens from his *Coelum Australe* of 1763, which illustrates the constellations he devised and revised; and Flamsteed's charts, all twenty-seven, from his posthumously published *Atlas Coelestis* of 1729. In addition, there are eighteen charts — from a total of fifty-six — from Hevelius's *Prodromus Astronomiae* of 1690, also published posthumously. Hevelius's charts are included to illustrate the twelve constellations he devised as well as those "modern" constellations that were probably created by the Dutch clergyman, Petrus Plancius. As noted earlier, the figures in Hevelius's atlas are drawn backward, as they would appear on a celestial globe.

The photograph of Lacaille's planisphere is from his *Coelum Australe Stelliferum*. The chart of Canes Venatici with Cor Caroli is from Lalande and Méchain's *Atlas Céleste de Flamstéed,* third edition. Both of these works are at the Science, Industry, and Business Library, The New York Public Library, Astor, Lenox, and Tilden Foundations. Flamsteed's star charts are reproduced from his *Atlas Coelestis* at the United States Naval Observatory with the exception of the complete chart of Hydra, which is from the History of Science Collection, the John Hay Library, Brown University. The charts from the works of Bayer and Hevelius are from the author's own collection.

Stellar Coordinates

Bayer, like Hevelius, designated stellar positions on his charts using ecliptic longitude and latitude, both measured in degrees (°), minutes (1/60 of a degree, represented by the symbol '), and seconds (1/60 of a minute, represented by the symbol "). Ecliptic **latitude** is measured in degrees north or south of the ecliptic (0°) to a maximum of 90° north to the north ecliptic pole in Draco and 90° south to the south ecliptic pole in Dorado. Ecliptic **longitude** is measured eastward along the ecliptic from its origin at the First Point of Aries, ♈, 0°, to a maximum of 360°. On seventeenth-century stellar charts, ecliptic longitude is marked off in twelve units of 30° each, with each unit designated by a sign of the zodiac. As seen from Earth, the sun seems to move eastward along the ecliptic about 1° each day, passing through the twelve constellations of the zodiac at a rate of about one constellation a month. The sun completes its circumnavigation around the celestial globe in 365.24219 days. The signs of the zodiac can be seen either at the top or bottom — and occasionally on both the top and bottom — of the charts of both Bayer and Hevelius (see, for example, Figure 1). Flamsteed also included ecliptic longitude and latitude on his charts, but, in addition, he used the modern format of right ascension and declination, thus cluttering his charts with lines going seemingly in all directions (see, for example, Figure 81). The twelve signs of the zodiac — tiny figures on the charts of Bayer, Hevelius, and Flamsteed located every 30° of ecliptic longitude — are as follows:

♈	Aries
♉	Taurus
♊	Gemini
♋	Cancer
♌	Leo
♍	Virgo
♎	Libra
♏	Scorpius
♐	Sagittarius
♑	Capricornus
♒	Aquarius
♓	Pisces

Modern charts employ equatorial coordinates using right ascension, or celestial longitude, and declination, or celestial latitude, in determining a star's position in the heavens. The celestial sphere is divided into twenty-four hours (^{h}) of right ascension, with each hour consisting of sixty minutes (^{m}) and each minute consisting of sixty seconds (^{s}). Right ascension begins at 0^h, the First Point of Aries, ♈. Occasionally, although this method is quite rare in modern astronomy, astronomers may use arc degrees, from 0° to 360°, instead of units of time to measure right ascension around the celestial sphere. Declination is measured in degrees from the celestial equator, 0°, north or south of the equator to a maximum of 90° north to the north celestial pole and 90° south to the south celestial pole. In astronomical works, a plus (+) sign is sometimes substituted for north and a minus (-) sign for south. For example, the Pole Star, Polaris, in the Little Bear (Ursa Minor), is 89° north, or simply +89°, while Sirius, the Dog Star, in the Big Dog (Canis Major), is 16° south, or simply -16°.

The First Point of Aries, ♈, identifies the point where ecliptic longitude and right ascension originate (0°, 0^h). It is the point where the ecliptic, the celestial equator, and the vernal equinoctial colure intersect. For the period from about A.D. 200 to about A.D. 2400, the First Point of Aries has been and will be in the constellation Pisces, but it is gradually moving westward because of a phenomenon called precession — the changing position of the stars as seen from Earth as a result of a slight wobble of Earth's axis of rotation. Each complete cycle of the wobble takes 25,800 years. At about A.D. 2400, the First Point of Aries will pass into Aquarius.

The change in the view of the heavens as seen from Earth as a result of precession occurs gradually, very gradually — about 50.2" a year or 1.4° every hundred years. For example, Alpha Ursae Minoris, currently the northern Pole Star, was not always the Pole Star. About three thousand years ago, that position was held by Thuban, Alpha Draconis, in the Dragon. And about twelve thousand years from now, Vega, Alpha Lyrae in the Lyre, will become the northern Pole Star. In order to take this gradual apparent movement of the stars into account, stellar cartographers and catalogers must always adjust their observations to a specific period or epoch. Thus, for example, Flamsteed began his systematic observation of the heavens in 1676 and continued virtually uninterrupted until his death in 1719. Since his observations extended over such a long period of time, almost forty-five years, the position, or coordinates, he plotted for a particular star one day would not be exactly the same even on the following day. To take this difference into account, astronomers like Flamsteed have established certain years or epochs to mark their catalogues or atlases and have "reduced" their observations to that particular base year or epoch.

Flamsteed chose the epoch 1690.0. That is, all the stars in his catalogue and atlas were reduced to the position they had been on January 1, 1690. This purely mathematical work of reduction was done either by the astronomer himself or, more likely, by his assistants. The latter were often called "computers."

Stellar Magnitude

Stellar magnitude describes a star's brightness as seen from Earth. Until the seventeenth century, there were basically six grades of magnitude that ranged from 1 to 6 in decreasing order of brightness, with stars of magnitude 6 being the limit visible to the human eye. The invention of the telescope dramatically increased the range of stellar visibility. Flamsteed, for example, one of the first astronomers to make use of the telescope in preparing a stellar catalogue, recorded stars of magnitude 7 and 8. The largest earth-bound telescope can detect stars of about magnitude 25. With the deployment of the Hubble Space Telescope (HST), astronomers are able to view stars as dim as magnitude 28 and perhaps beyond. In the nineteenth century, the gradation between magnitudes was established at approximately 2½ times; that is, a star of the 3rd magnitude is approximately 2½ times brighter than a star of the 4th magnitude. Based on this formula, some very bright stars have been classified with negative magnitude. The five brightest stars in the night sky are:

Sirius, Alpha Canis Majoris, in the Big Dog	-1.46
Canopus, Alpha Carinae, in the Ship's Keel	-0.72
Rigel Kent, Alpha Centauri, a double star in the Centaur with a combined magnitude of	-0.27
Arcturus, Alpha Bootis, in the Herdsman	-0.04
Vega, Alpha Lyrae, in the Lyre	0.03

Part One: Lettered Stars

ANDROMEDA, AND
Andromeda, Chained Lady

Andromeda (Figure 1) is a Ptolemaic constellation that Bayer lettered from Alpha to Omega and from A to c.

The mythology surrounding Andromeda is closely linked with at least four other constellations: Cepheus, Cassiopeia, Perseus, and Cetus, all of which, with the exception of Cetus, are grouped close together in the northern sky. There are several variations to the myth, but the basic story starts when Cassiopeia, wife of King Cepheus of Ethiopia, brags about her beauty to all who will listen. Ultimately, word of the Queen's boasting reaches the Nereids, a group of sea nymphs. The Nereids are deeply offended when they learn that Cassiopeia, a mere mortal, is claiming to be even more beautiful than they and they appeal to the sea god Poseidon to punish the Queen. Heeding the appeal of the sea nymphs, Poseidon sends a storm with huge waves crashing down along the coast of Ethiopia and, in addition, a ravenous sea monster, Cetus, to wreak havoc among the people. Hoping to find an end to the turmoil besetting his kingdom, Cepheus consults an oracle, who tells him that the only way to end the troubles is to sacrifice his daughter, Andromeda, to the sea monster. Accordingly, Andromeda is chained to a rock along the Mediterranean coast near Jaffa (Tel Aviv). Just as Cetus rises from the sea to devour Andromeda, the Greek hero Perseus arrives on the scene, kills Cetus, and frees Andromeda. The two marry, move to Argos in Greece, and live happily ever after. Upon Andromeda's death, the goddess Athena honors her by placing her image in the heavens.

Although basically a Greek myth embellished by the Romans, the story of Andromeda and her family was widely known throughout the ancient world. For example, the Jewish Historian Josephus (A.D. 37?-A.D. 100?), in *The Jewish War* (III, 9:419-20), noted:

> *Now Joppa [Jaffa] is not naturally an haven, for it ends in a rough shore where all the rest of it is straight, but the two ends bend towards each other, where there are deep precipices, and great stones that jut out into the sea, and where the chains wherewith Andromeda was bound have left marks, which attests to the antiquity of the fable.*

The story's setting has been shifted to all corners of the ancient world — from Ethiopia, to the Levantine coast, and ultimately to Greece. Some scholars have suggested that the story may have originated in the Middle East and that Perseus's battle with Cetus may be somehow related to the Mesopotamian account of Marduk's battle with the sea goddess Tiamat, while others have suggested that the name Andromeda is derived from the Phoenician name Adamath. But as far as is presently known, there is no relationship between Middle Eastern star lore and the Andromeda story. In Andromeda's place in the heavens, the Mesopotamians saw several constellations including the Stag, the Rainbow, the Deleter, the Field, and the Plow.

See Ridpath, *Star Tales,* pp. 21-23; Staal, *New Patterns in the Sky,* pp. 7-10; Schwab, *Gods & Heroes,* pp. 66-68; Allen, *Star Names,* pp. 31-35; Condos, *Star Myths,* pp. 27-28; Hunger and Pingree, *Mul.Apin,* pp. 137-38. For the battle between Marduk and Tiamat and her allies, see Canis Major and Hydra.

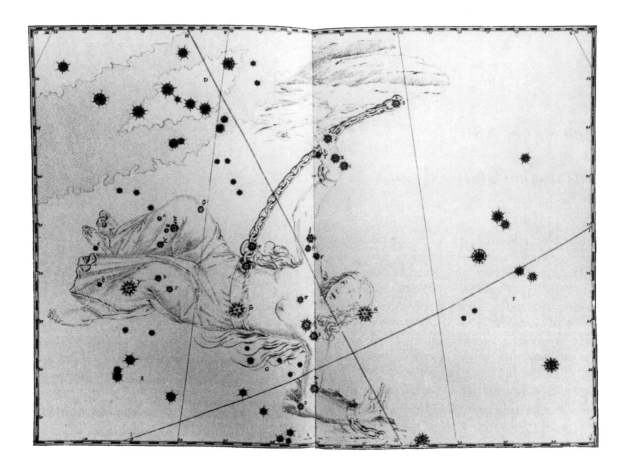

Figure 1. Bayer's Andromeda. The shaded portion of the chart in the upper left corner is the Milky Way. The line H, running from the top to the bottom of the chart, is the equinoctial colure. It is an imaginary great circle that runs completely around the celestial sphere from the celestial north pole through the equator to the south pole and around the other side past the equator and up to the north pole. Where the equinoctial colure meets the equator in the constellation Pisces is the First Point of Aries, ♈; where it meets the equator on the other side of the celestial globe in the constellation Virgo is the First Point of Libra, ♎. When the sun reaches these points on its yearly journey around the celestial sphere, it is the time of either the vernal (March, ♈) or autumnal (September, ♎) equinox. The line I is the celestial Tropic of Cancer, which is 23½° north of the celestial equator. When the sun reaches this point in the sky in June in the northern hemisphere, it is the time of the summer solstice, the highest point in the sky reached by the sun at noon. The Tropic of Capricorn, on the other hand, is 23½° south of the celestial equator. When the sun reaches this point in December, it marks the sun's lowest point in the sky at noon. It is the time of the winter solstice. The letter D is the constellation Cassiopeia; E is Triangulum; F is Pegasus; and G is Pisces.

The four vertical lines at the top of the chart, from the small one in the upper right corner to the small one in the upper left corner, each mark off 30° of ecliptic longitude. The celestial globe is divided into twelve sectors of 30° for each of the twelve signs of the zodiac. As the small signs at the top of the page and to the left of each line indicate, Andromeda passes through more than 90° of longitude: starting at Pisces (♓), to Aries (♈), to Taurus (♉), and finally ending at Gemini (♊). These lines extend from the north ecliptic pole located in the constellation Draco, 90° south to the ecliptic, and from there another 90° south to the south ecliptic pole in Dorado.

Table 6. The lettered stars of Andromeda

	MAGNITUDES			CATALOGUE NUMBERS				
Letter	Bayer	Lacaille	Visual	Flamsteed	Lacaille	Other	HR	HD
Alpha+	2		2.06	21			15	358
Beta	2		2.06	43			337	6860
Gamma[1,2]+	2		2.20c	57			603/4	12533/4
Delta	3		3.27	31			165	3627
Epsilon	4		4.37	30			163	3546
Zeta	4		4.06	34			215	4502
Eta	4		4.42	38			271	5516
Theta	4		4.61	24			63	1280
Iota	4		4.29	17			8965	222173
Kappa	4		4.14	19			8976	222439
Lambda	4		3.82	16			8961	222107
Mu	4		3.87	37			269	5448
Nu	4		4.53	35			226	4727
Xi+	4		4.88	46			390	8207
Omicron	4		3.62	1			8762	217675
Pi	5		4.36	29			154	3369
Rho	5		5.18	27			82	1671
Sigma	5		4.52	25			68	1404
Tau	5		4.94	53			477	10205
Upsilon+	5		4.95	50?			483	10307
Phi	5		4.25	42			335	6811
Chi+	5		4.98	52			469	10072
Psi	5		4.95	20			9003	223047
Omega+	6		4.83	48			417	8799
A+	6		5.27	49			430	9057
b	6		4.83	60			643	13520
c	6		5.30	62			670	14212

The Lost, Missing, or Troublesome Stars of Andromeda

Alpha, 21, HR 15, 2.06V, *SA, HRP, BS.*

Bayer noted that Alpha And was the same star as Delta Peg. He included five duplicates in his *Uranometria:*
Alpha And/Delta Peg
Gamma Aur/Beta Tau
Nu Boo/Psi Her
Xi Ari/Psi Cet
Kappa Cet/g Tau

The first three he intentionally duplicated and specifically identified as duplicate stars since these three had also been cited by Ptolemy as duplicates, but Ptolemy did not actually duplicate Alpha And/Delta Peg. He mentioned that these two stars were the same, but he included only Delta Peg in his catalogue, whereas Bayer catalogued both Delta Peg and Alpha And. Bayer was unaware of the last two pairs — Xi Ari/Psi Cet and Kappa Cet/g Tau — which were mistakenly duplicated. He corrected them some twenty-five years later in Julius Schiller's *Coelum Stellatum Christianum*, which some authorities consider a revised edition of the *Uranometria* since Bayer played an important role in its preparation. See Psi and Kappa Cet.

Ptolemy's catalogue lists 1,028 stars, of which three are duplicates: the two cited above (Gamma Aur/Beta Tau and Nu Boo/Psi Her) and Fomalhaut, which is included in both Piscis Austrinus and Aquarius. Bayer, on the other hand, included Fomalhaut only in the Southern Fish, where it has remained to the present day.

Bayer's charts, by his own count, include 1,725 stars, but only 1,275 are lettered and catalogued, including the five duplicates noted above. The rest can be found among the 119 stars that comprise the twelve new southern constellations devised by Keyzer and Houtman, all of which are depicted unlettered and uncatalogued on one chart, Tabula XLIX; and among the 325 or so *informes* or unformed stars outside the traditional forty-eight Ptolemaic constellations.

Gamma¹, Gamma², 57, HR 603/4, 2.26V, 4.84V, comb. mag. 2.20V, *SA, HRP* as Gamma, *BS,* ADS 1630.

Bayer described Gamma as a single 2nd-magnitude star. Later astronomers added the indices to identify the components of this double star.

Xi, 46, HR 390, 4.88V, *SA, HRP, BS,* HF 63*, BV 51*, B 200*, BF 159*.

In Halley's 1712 pirated edition of Flamsteed's catalogue, Xi is properly equated with 46. In the authorized edition of Flamsteed's catalogue, the "Catalogus Britannicus" published posthumously in 1725, Xi is erroneously equated with 49. Bevis, noting that 49 did not conform to Bayer's chart, shifted Xi to 46 (Tabula XX).

Upsilon, 50, HR 458, 4.09V, *SA, HRP, BS,* BV 52 as a, BF 184, BAC 480, W 37.

Bayer described a 5th-magnitude star in Andromeda's knee as Upsilon. He did not mean this star but HR 483 (4.95V, BV 55 as Upsilon, BF 199 as Upsilon, BAC 510, W 41 as Upsilon). HR 483 is Tycho's 19 And, Hevelius's 20 And, and Piazzi's I, 142. The error originated with Flamsteed, who noted that his 50 was synonymous with Bayer's Upsilon. It is not. It is synonymous with Ptolemy's 18, Hevelius's 40, and Piazzi's I, 119, but not with any of Bayer's stars. Bevis was among the first to catch this error. He noted that "the Character [Upsilon] is wrongly placed before the 50th [of Flamsteed] or Ptolemy's 18th." He correctly assigned Bayer's Upsilon to HR 483 and assigned his own letter, a, to Flamsteed's 50, HR 458 (Tabula XX). Baily also caught the mistake. In his revised edition of Flamsteed's catalogue, he removed Upsilon from 50 and assigned it to HR 483, noting that this "is the star designated by the letter Upsilon in Bayer's map; which I have here retained (*BF,* p. 515)." Argelander thought otherwise. He argued that the error was a result of a mistake of several degrees in Tycho's catalogue which misled Bayer. Nevertheless, based on Tycho's actual observations, not his catalogue, Argelander felt that Upsilon was indeed equivalent to Flamsteed's 50 (*Fide Uran.*, pp. 13-14). Baily, though, remained unconvinced and again affirmed in the *BAC* that "50 Andromedae is not Upsilon (p. 64)." In editing and reducing the stars in Lalande's *Histoire Céleste Française,* he assigned Bayer's Upsilon to LL 3054, which is synonymous with HR 483. Despite Baily's repeated efforts to set matters right, most astronomers have persisted down to the present day in designating 50 as Upsilon. Among the few who have noted Flamsteed's error are Werner and Schmeidler in their *Synopsis* (p. 54). Oddly enough, Flamsteed observed HR 483, but it was one of 458 stars that was inadvertently omitted from his printed catalogue (*BF,* p. 392; Baily, "Catalogue of 564 Stars," No. 32).

Chi, 52, HR 469, 4.98V, *SA, HRP, BS,* HF 69*, BV 57*, B 225*, BF 193*.

Flamsteed's catalogue of 1725 labels 52 as Lambda, a typographical error for Chi that Bevis later corrected (Tabula XX). Flamsteed consistently referred to 52 as Chi in his manuscripts and in Halley's pirated edition of his catalogue. The error in the catalogue arose because of the similarity in appearance of these two letters, χ and λ.

Omega, 48, HR 417, 4.83V, *SA, HRP, BS,* HF 65*, BV 53*, B 210*, BF 168*.

Although 48 is properly identified as Bayer's Omega in Halley's edition of Flamsteed's catalogue, Flamsteed's 1725 edition leaves 48 unlettered. Bevis relabeled it Omega (Tabula XX).

A, 49, HR 430, 5.27V, *SA, HRP,* HF 66*, BV 56*, B 216*, BF 174*.

Although 49 is properly equated with Bayer's A in Halley's edition of Flamseed's catalogue, the authorized catalogue erroneously equates 49 with Xi. Bevis corrected the discrepancy. See Xi.

b, 60, HR 643, 4.83V, *SA, HRP.*

c, 62, HR 670, 5.30V, *SA, HRP.*

d, 41, HR 324, 5.03V, BV 49, BF 114.

Bayer did not letter this star. The letter was assigned by Flamsteed but removed by Bevis and Baily as superfluous. See i Aql.

ANTLIA, ANT

Air Pump

Lacaille devised and lettered Antlia (Figure 2) from Alpha to Theta. On the chart that accompanied his first or preliminary catalogue published in 1756, he called it la Machine Pneumatique and noted that it represented experimental physics *(la Physique expérimentale)*. On the chart accompanying his later catalogue of 1763, he labeled it in Latin Antlia Pneumatica, the Pneumatic Pump. Later, Sir John Herschel suggested that this, like Lacaille's other constellations, be reduced to one word. Lacaille's own interpretation of the meaning of the fourteen new constellations he devised can be found in his *Remarques*, which he included at the end of his preliminary catalogue.

The constellation figures in the planispheres that accompany each catalogue are exactly the same. There are, however, several differences between the two charts. Lacaille used French in the preliminary chart and divided it into hours of right ascension. In the later planisphere of 1763, he used Latin instead of French and degrees rather than hours. The use of right ascension on the first chart and degrees on the second reflects the different system of coordinate determination utilized by Lacaille in the creation of the two catalogues. In addition, on the chart of 1763, Lacaille designated individual stars with Greek or Latin letters. There is also a difference in the number of stars between the two catalogues. His first lists 1,935 stellar objects while the later edition lists 1,942. In addition, the earlier chart is approximately based on the epoch 1752 while the latter is reduced precisely to 1750.

Lacaille's decision to name a group of southern stars in honor of the discovery and development of the air pump reflected the importance of this instrument to the scientific community of Western Europe. It commemorated especially the work of Otto von Guericke, who is perhaps best remembered for the "Magdeburg Hemispheres Experiment" in which two large bronze hemispheres were fitted snugly together; the air between them pumped out; and despite the efforts of two teams of eight horses, each team pulling in the opposite direction, the vacuum inside the two hemispheres held them together. Thanks to the experiments of von Guericke and the work of Robert Boyle, a new field of physics dealing with air, vacuums, and gases was opened to scientists. The Air Pump, along with the Microscope, the Pendulum Clock, the Mariner's Octant, and the Telescope with its Reticle Micrometer were all discoveries that contributed to the Scientific Revolution of the seventeenth and eighteenth centuries, and Lacaille honored these instruments with a place among the stars.

Numerous writers have criticized Lacaille for devising these and other new constellations in the southern skies since their stars do not by the farthest stretch of the imagination resemble any of the figures for which they are named. Lacaille, though, never intended the stars to resemble his figures. They were devised to commemorate instruments of modern science and the Age of Reason. In fact, very few constellations actually resemble their namesakes. In devising new constellations, Lacaille, like the ancients before him, intended to honor certain objects, individuals, or events by placing them in appropriate sectors of the heavens. The stars making up the constellation Libra, for example, do not resemble a scale but, instead, commemorate a time some 3,500 years ago when the sun passed through this sector of the heavens and the hours of daylight and darkness were in balance — the autumnal equinox. The ancients honored the event by naming this heavenly area Libra or Balance. Similarly, the stars in Aries do not resemble a Ram but rather signify a time when the sun was in this sector of the heavens 3,500 years ago during the vernal equinox, the new year, when the people of Mesopotamia celebrated the coming of spring by sacrificing a male sheep or ram to the gods. Consequently, they thought it appropriate to call this part of the sky Aries.

See Baily, *BAC,* p. 63; Lacaille, "Table des Ascensions," pp. 588-89; Warner, *Sky Explored,* pp. 142-43; Wolf, *History of Science,* I, 99-109. See Aries and Libra for more detailed information about the origin of these two constellations.

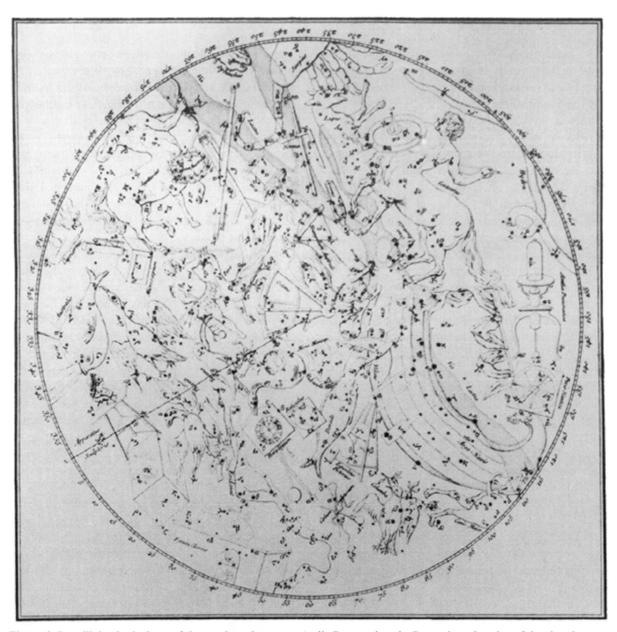

Figure 2. Lacaille's planisphere of the southern heavens. Antlia Pneumatica, the Pump, is at the edge of the chart between 140° and 170° right ascension. Enlarged quadrants of Lacaille's planisphere are shown as figures 2a, 2b, 2c, and 2d on the following pages.

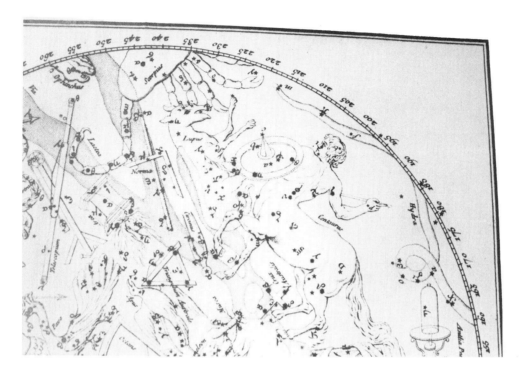

Figure 2a. An enlarged view of Lacaille's planisphere from 153° to 265° right ascension. This section of the chart shows his Antlia Pneumatica, Circinus, Norma, and Telescopium. It also depicts Keyzer and Houtman's Apus, Musca, Pavo, and Triangulum Australe; all or parts of Ptolemy's Ara, Centaurus, Hydra, Lupus, and Scorpius; and the Southern Cross, Crux Australis.

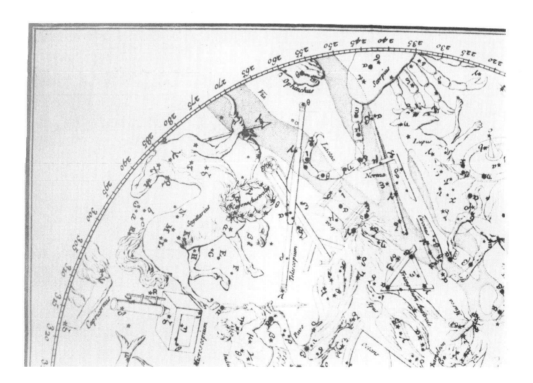

Figure 2b. An enlarged view of Lacaille's planisphere from 218° to 325° right ascension. This section of the chart shows his Circinus, Microscopium, Norma, Telescopium, and part of Octans. It also depicts all or parts of Keyzer and Houtman's Apus, Chamaeleon (his head), Indus, Musca, Pavo, and Triangulum Australe; and all or parts of Ptolemy's Ara, Centaurus, Corona Australis, Lupus, Sagittarius, and Scorpius.

Lost Stars

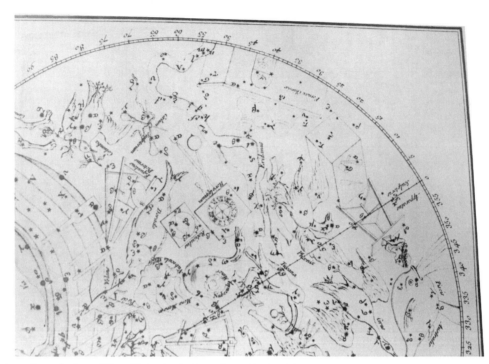

Figure 2c. An enlarged view of Lacaille's planisphere from 323° to 0° and from 0° to 94° right ascension. This section of the chart shows his Apparatus Sculptoris, Caelum Scalptorium, Equuleus Pictorius, Fornax Chimiae, Horologium, Mons Mensae, Octans, and Reticulum. It also depicts all or parts of Keyzer and Houtman's Chamaeleon, Dorado, Grus, Hydrus (unlabeled, with his head pointing into the Clock), Indus, Phoenix, Piscis Volans, and Tucana; all or parts of Ptolemy's Eridanus and Piscis Austrinus; and Plancius's Columba.

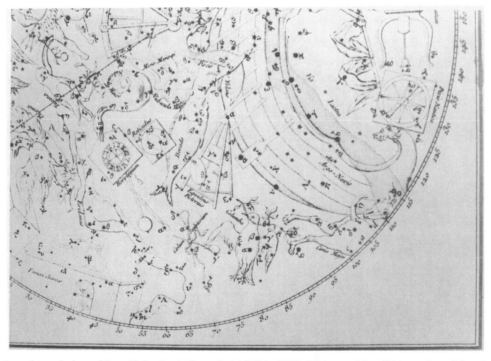

Figure 2d. An enlarged view of Lacaille's planisphere from 25° to 152° right ascension. This section of the chart shows his Caelum Scalptorium, Equuleus Pictorius, Fornax Chimiae, Horologium, Mons Mensae, Octans, Pixis Nautica, Reticulum, and the lower half of Antlia Pneumatica. It also depicts all or parts of Keyzer and Houtman's Chamaeleon (lower part), Dorado, Indus, Hydrus (unlabeled, with his head pointing into the Clock), Phoenix, Piscis Volans, and Tucana; parts of Ptolemy's Argo Navis, Canis Major, and Eridanus; and Plancius's Columba.

Table 7. The lettered stars of Antlia

		MAGNITUDES			CATALOGUE NUMBERS			
Letter	Bayer	Lacaille	Visual	Flamsteed	Lacaille	Other	HR	HD
Alpha		5	4.25		931		4104	90610
Beta+		5	5.81		996		4339	97023
Gamma+		6	6.95		927			90156
Delta		6	5.56		933		4118	90972
Epsilon		6	4.51		859		3765	82150
Zeta^1+		6	5.74c		861		3780/1	82383/4
Zeta^2		6	5.93		863		3789	82513
Eta		6	5.23		903		3947	86629
Theta		6	4.79		884		3871	84367
Iota+			4.60			G 84	4273	94890

The Lost, Missing, or Troublesome Stars of Antlia

Beta, in Hydra, HR 4339, 5.81V, CA 996 erroneously as Eta, BR 3430 erroneously as Eta, L 4623, BAC 3822, G 272 in Hydra.

Beta is misprinted as Eta in Lacaille's catalogues of 1756 and 1763 and on his chart accompanying the latter catalogue. The chart accompanying his catalogue of 1756 is unlettered. Baily included Beta unlettered in Hydra. See Gould, *Uran. Arg.*, p. 91.

Gamma, in Hydra, HD 90156, 6.95V, CA 927*, BR 2981*, L 4277*, BAC 3558*.

Gould intended to keep Gamma in Antlia. He drew Antlia's northern border with Hydra as a broad curve from -23° to -35° (1875.0) with Gamma inside the curve. On behalf of the International Astronomical Union in 1930, Eugène Delporte (*Atlas Céleste*) changed this to a series of steps that shifted Gamma to Hydra. This marked one of the very few instances in which the IAU disagreed with the constellation boundaries proposed by Gould. Gould excluded Gamma from his *Uran. Arg.* since he considered it less than 7th magnitude, which was the limiting magnitude for his catalogue (p. 296).

Zeta^1, HR 3780/1, 7.21V, 6.35V, comb. mag. 5.74V, *SA, HRP, BS,* CA 861 as Zeta, BR 2515*, L 3880*, BAC 3254*, G 8/9*.
Zeta^2, HR 3789, 5.93V, *SA, HRP, BS,* CA 863 as Zeta, BR 2521*, L 3884*, BAC 3262*, G 10*.

Lacaille designated these two nearly adjacent stars Zeta.

Iota, HR 4273, 4.60V, *SA, HRP, BS,* CA 980, BR 3293, L 4527, BAC 3755, G 84*.

Lacaille did not letter this star. Although he observed and catalogued it, he left it unlettered; he considered it 6th magnitude. Gould designated it Iota since it is brighter than 5th magnitude and the third brightest star in the constellation.

Apus, Aps

Bird of Paradise

Pieter Dircksen Keyzer and Frederick de Houtman, Dutchmen who traveled to the East Indies in 1595, devised Apus (Figure 3). In his catalogue of 1603, Houtman described it as De Paradijs Voghel, Bird of Paradise. It appeared as Paradysvogel Apis Indica on Petrus Plancius's Globe of 1598, which Bayer may have seen since he also called it Apis Indica and depicted it without any feet on his chart of the Southern Hemisphere — a reference to the supposed footless condition of the Bird of Paradise. Jacob Bartsch, in his *Usus Astronomicus* of 1624, referred to it as Apous, Apis, or Avis Indica, while his father-in-law, Johannes Kepler, in the Rudolphine Tables of 1627, called it Apus, Avis Indica. Apis is Latin for bee, but its use here may have resulted from a typographical error for avis, bird, or for *apous,* απους, Greek for footless. Lacaille called the constellation Apus and lettered its stars from Alpha to Kappa.

The different names for this constellation as well as for several others devised by Keyzer and Houtman arose because both men, although working together for a time, apparently prepared two separate catalogues; only Houtman's has survived. A manuscript catalogue, probably Keyzer's, was returned to the Netherlands in 1697, and a copy sent to Plancius, who may have passed it on to Bayer.

When Keyzer and Houtman catalogued the stars in the southern skies, they purposely organized the stars into twelve new constellations. Since ancient times, the number twelve has been considered especially important, signifying completion, probably because twelve months make up a lunar year, which served as the basis for man's first calendar before he turned to agriculture and developed the more sophisticated solar calendar. The importance attached to the number twelve can be seen in the twelve signs of the zodiac, the twelve tablets of Gilgamesh, the twelve tribes of Israel, the twelve major Olympian gods, the twelve labors of Hercules, the twelve disciples of Christ, and the like. Classical astronomers like Hipparchus and Ptolemy catalogued the stars of the forty-eight (a multiple of twelve) ancient constellations. Although they recognized other asterisms or star groups, such as Coma Berenices and Antinous, they rigidly adhered to the original forty-eight. Consequently, when Keyzer and Houtman mapped the southern skies, it was not at all surprising that they devised twelve constellations, no more and no less. Ninety years later, when Johannes Hevelius filled in the gaps in the northern skies, he too devised twelve constellations.

See Warner, *Sky Explored,* pp. 30, 203-4; Bartsch, *Usus Astronomicus,* p. 66; Kepler, *Gesammalte Werke,* X, 139; Allen, *Star Names,* p. 43; Houtman, *Spraeckende Woordboeck,* Appendix; Knobel, "On Frederick de Houtman's Catalogue," pp. 417-20. See also Camelopardalis, Canes Venatici, Dorado, Grus, Musca, and Tucana.

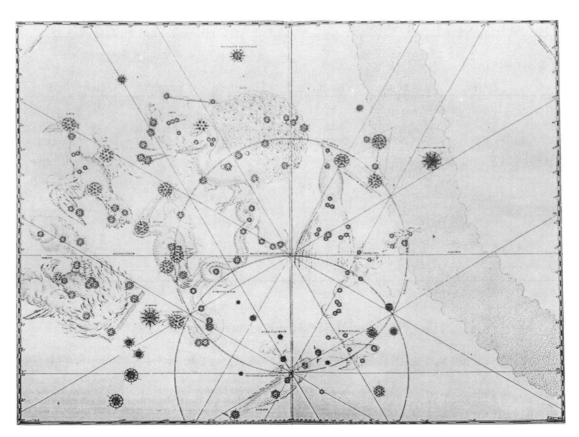

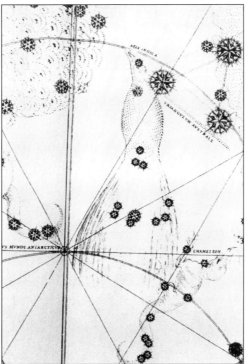

Figure 3. Top: Bayer's Tabula XLIX, his planisphere of the southern heavens. Apus is between the Peacock and the Southern Triangle. In the center of the chart between the tails of Hydrus and Apus is the southern celestial pole while below it, in the head of Dorado, is the southern ecliptic pole. For the difference between the celestial and ecliptic poles, see Draco, Figure 36. **Bottom: Bayer's Apus.** An enlarged view of Apus from Bayer's Tabula XLIX. Apus is referred to as Apis Indica and depicted as legless.

Table 8. The lettered stars of Apus

	MAGNITUDES			CATALOGUE NUMBERS				
Letter	Bayer	Lacaille	Visual	Flamsteed	Lacaille	Other	HR	HD
Alpha		5	3.83		1218		5470	129078
Beta		5	4.24		1361		6163	149324
Gamma		5	3.89		1346		6102	147675
Delta^1+		6	4.68		1327		6020	145366
Delta^2		6	5.27		1328		6021	145388
Epsilon		6	5.06		1196		5336	124771
Zeta		6	4.78		1415		6417	156277
Eta		6	4.91		1188		5303	123998
Theta		6	4.79		1181		3871	84367
Iota		6	5.41		1414		6411	156190
Kappa^1+		6	5.49		1282		5730	137387
Kappa^2		6	5.65		1291		5782	138800

The Lost, Missing, or Troublesome Stars of Apus

Delta^1, Delta^2, HR 6020/1, 4.68V, 5.27V, *SA, HRP, BS,* CA 1327/8 as Delta, BR 5584*, BR 5586*, L 6623, L 6628, BAC 5339/40*, G 40/1*.

Lacaille designated each component of this double star Delta.

Kappa^1, HR 5730, 5.49V, *SA, HRP, BS,* CA 1282 as Kappa, BR 5302*, L 6323, BAC 5068*, BAC 5068*, G 29*.
Kappa^2, HR 5782, 5.65V, *SA, HRP, BS,* CA 1291 as Kappa, BR 5373*, L 6390, BAC 5108*, G 33*.

Lacaille designated these two neighboring stars Kappa.

Aquarius, AQR

Water Bearer

Aquarius (Figure 4) is a Ptolemaic constellation that Bayer lettered from Alpha to Omega and from A to i. It is among the oldest and most important constellations in the heavens, dating back at least six thousand years. Devised by the Sumerians in Mesopotamia, Aquarius, together with Taurus, Leo, and Scorpius, marked the four turning points of the year for ancient man: the vernal and autumnal equinoxes — Taurus and Scorpius respectively — and the summer and winter solstices — Leo and Aquarius. The Sumerians noted that these cardinal points in their calendar occurred when the sun moved through the stars of these four constellations. Actually, the constellations preceded these four points in 4000 B.C. by 10°-25°; but because of the intensity of the sun's rays and the lack of sophisticated instruments, the Sumerians could not precisely determine the exact position of the sun against the background of the stars. They based their observations on the apparent position of the sun that *seemed* to pass through these four constellations at the four turning points of the solar year.

Aquarius was originally depicted in ancient Mesopotamian astronomical texts as an enormous ibex, which extended from Capricornus north to Pegasus. Like the other cardinal constellations of the ancient Near East — Taurus, Leo, and Scorpius — it bore a striking resemblance to the figure it represented. The ibex, with its huge curved horns, was the sacred animal of Enki (Akkadian Ea), the Sumerian god of wisdom, intelligence, skill, the earth, and especially of sweet or fresh water. He moved about the marshes, rivers, and canals of Mesopotamia on his boat *Dara-abzuk,* the Ibex of the Deep. Enki was also called En-uru, Lord Reed Bundle, and this perhaps explains his association with the ibex. Like their descendants, the modern Iraqis, the ancient Mesopotamians constructed their huts of long reeds that grew along the marshes of the lower reaches of the Tigris and Euphrates. And to many of these people, the reeds projecting above the surface of the water resembled the horns of the ibex.

The Sumerian cuneiform sign for this constellation in several astronomical texts is GU.LA, which can mean either great or the goddess of healing. Scholars have puzzled over this since neither term seems directly related to Enki. In the Mesopotamian pantheon, however, Enki was one of the four great gods, the others being An, who ruled the heavens; Enlil, who ruled the air and wind; and Ninhursaga or Ninmah, the Mother Goddess. To the Mesopotamians, Enki was the god who helped create man, was most friendly to humans, and was the savior of mankind. According to legend, when the gods decide to destroy all life on Earth, it is Enki (Ea) who warns Utnapishtim, the Mesopotamian Noah, of the coming deluge and advises him to build a huge ship to save himself, his family, and a sampling of all living things. As the god of sweet and fresh water — rain, dew, streams, and water that wells up from the ground — Enki, whose name in Sumerian translates as Lord of the Soil, was worshipped for bringing life-giving moisture to the parched land. His fresh waters not only fertilized the land but cleansed and purified mankind. He was also invested with the power to exorcise evil spirits. He was usually depicted holding a vase with water streaming out. In Sumerian mythology, he is entrusted by the gods with the water of life, *me balati,* which could cleanse man of sickness and disease and even restore life to the dead. In short, not only did Enki rank among the chief gods of Mesopotamia, but he also symbolized the numinous force that controlled and directed the power of life.

The Sumerians appropriately placed Enki's emblem, the ibex, at that point in the heavens that marked the winter solstice, when cooling rains (see below) broke the draught and deadly heat of summer and autumn and restored life to the parched, dying earth. Later generations would split the constellation into two parts: Aquarius, or Enki, carrying his water jug with the water of life, and Capricornus, the Goat-Fish, also an emblem of Enki.

The Greeks believed that the constellation represented Ganymede, the son of King Tros of Troy. According to one account, Zeus sees the boy and falls deeply in love with him. In the guise of an eagle (Aquila), Zeus swoops down and carries Ganymede off to become his catamite and cup bearer.

See Hartner, "The Earliest History of the Constellations in the Near East," especially note 44 for a discussion of the meaning of GU.LA; van der Waerden, "Babylonian Astronomy. II. The Thirty-Six Stars"; van der Waerden, "History of the Zodiac"; Jacobsen, *Toward the Image of Tammuz,* pp. 7, 21-22, 33; Kramer, *The Sumerians,* chaps. 4 and 5; Kramer, *Sumerian Mythology,* pp. 90, 94-95; Kramer, *Mythologies of the Ancient World,* p. 99; Marcus, "Enki"; Jacobsen, "Mesopotamian Religions"; Pritchard, *Anc. Near East. Texts,* pp. 68, 93, 99-100; Langdon, *The Babylonian Epic of Creation,* pp. 169-70; Langdon, *Semitic Mythology,* pp. 86-108, 182 (these latter two works by Langdon should be used with great care since later research has modified and corrected several of his conclusions); O'Neal, *Time and the Calendars,* pp. 59-60; Condos, *Star Myths,* pp. 29-31; Ridpath, *Star Tales,* pp. 25-26. See also Aquila, Capricornus, Leo, Scorpius, and Taurus.

The average monthly rainfall in modern Iraq, with winter maxima and summer minima, is as follows:

October	0.1"
November	0.8"
December	1.2"
January	1.2"
February	1.3"
March	1.3"
April	0.9"
May	0.2"
June	0.0"
July	0.0"
August	0.0"
September	0.0"

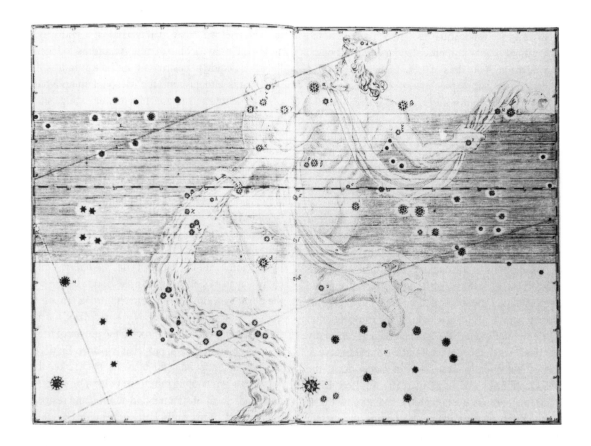

Figure 4. Bayer's Aquarius. The shaded section on the chart is the zodiac, the apparent path of the sun, moon, and planets across the heavens. Bayer's charts depict it as extending 8° north and south of the ecliptic, the path of the sun as seen from Earth. The lines P and R are, respectively, the celestial equator and the Tropic of Capricorn. Line Q is the equinoctial colure. The letter K is Capricornus; L is Pisces; M is Cetus; N is Piscis Austrinus; and O is the 1st-magnitude star Fomalhaut, Alpha PsA.

Table 9. The lettered stars of Aquarius

	MAGNITUDES			CATALOGUE NUMBERS				
Letter	Bayer	Lacaille	Visual	Flamsteed	Lacaille	Other	HR	HD
Alpha	3		2.96	34			8414	209750
Beta	3		2.91	22			8232	204867
Gamma	3		3.84	48			8518	212061
Delta	3		3.27	76			8709	216627
Epsilon	4		3.77	2			7950	198001
Zeta^{1,2}+	4		3.65c	55			8558/9	213051/2
Eta	4		4.02	62			8597	213998
Theta	4		4.16	43			8499	211391
Iota	4		4.27	33			8418	209819
Kappa	4		5.03	63			8610	214376
Lambda	4		3.74	73			8698	216386
Mu	5		4.73	6			7990	198743
Nu	5		4.51	13			8093	201381
Xi	5		4.69	23			8264	205767
Omicron	5		4.69	31			8402	209409
Pi	5		4.66	52			8539	212571
Rho	5		5.37	46			8512	211838
Sigma	5		4.82	57			8573	213320
Tau^1+			5.66	69			8673	215766
Tau^2	5		4.01	71			8679	216032
Upsilon	5		5.20	59			8592	213845
Phi	5		4.22	90			8834	219215
Chi	5		5.06	92			8850	219576
Psi^1+	5		4.21	91			8841	219449
Psi^2	5		4.39	93			8858	219688
Psi^3	5		4.98	95			8865	219832
Omega^1+	5		5.00	102			8968	222345
Omega^2	5		4.49	105			8988	222661
A^1+			5.34	103			8980	222547
A^2	5		4.82	104			8982	222574
b^1+	5		3.97	98			8892	220321
b^2	5		4.39	99			8906	220704
b^3	5		4.71	101			8939	221565
c^1+	5		4.47	86			8789	218240
c^2	5		3.66	88			8812	218594
c^3	5		4.69	89			8817	218640
d+	6		5.10	25			8277	206067
e+	6		5.46	38			8452	210424
f	6		5.56c	53			8544/5	212697/8
g^1+	6		4.69	66			8649	215167
g^2			5.26	68			8670	215721
h+	6		5.43	83			8782	218060
i^1+	6		5.24	106			8998	222847
i^2			5.29	107			9002	223024
i^3	6		5.18	108			9031	223640
k+			4.42	3		B 9	7951	198026

The Lost, Missing, or Troublesome Stars of Aquarius

Zeta¹, Zeta², 55, HR 8558/9, 4.59V, 4.42V, comb. mag. 3.65V, *SA, HRP* as Zeta, *BS*, ADS 15971.

Bayer described Zeta as a single 4th-magnitude star. Later astronomers added the indices to identify the components of this double star.

Tau¹, 69, HR 8673, 5.66V, *SA, BS*, BF 3120*, H 86.
Tau², 71, HR 8679, 4.01V, *SA, HRP* as Tau, *BS*, HF 66 as Tau, BF 3123*, H 88 as Tau.

Bayer described only one 5th-magnitude star as Tau. Halley's pirated edition of Flamsteed's catalogue equates 71 with Tau. Flamsteed's 1725 catalogue, however, labels these two neighboring stars, 69 and 71, as Tau-1 and Tau-2. Although Baily accepted these designations, Argelander felt otherwise. He agreed with the Halley edition that only the brighter star, 71, was synonymous with Bayer's Tau (*UN*, p. 114).

Psi¹, 91, HR 8841, 4.21V, *SA, HRP, BS*.
Psi², 93, HR 8858, 4.39V, *SA, HRP, BS*.
Psi³, 95, HR 8865, 4.98V, *SA, HRP, BS*.

Since Bayer noted that Psi was a group of three very closely located 5th-magnitude stars, Flamsteed designated his 91, 93, and 95 as Psi-1, Psi-2, and Psi-3.

Omega¹, 102, HR 8968, 5.00V, *SA, HRP, BS*, HF 90 as Omega, BV 111*, B 360*, BF 3239*.
Omega², 105, HR 8988, 4.49V, *SA, HRP, BS*, HF 93 as Omega, BV 112*, B 369*, BF 3247*.

Since Bayer noted that Omega was a couple of adjacent 5th-magnitude stars, Flamsteed designated his 102 and 105 as Omega-1 and Omega-2. Although the stars are properly labeled in Halley's pirated edition of Flamsteed's catalogue, the letter Omega is somehow omitted from 102 in Flamsteed's 1725 catalogue, probably as the result of an error in editing. Bevis and Bode relabeled 102 as Omega-1.

A¹, 103, HR 8980, 5.34V, *SA, HRP*, BF 3244*, G 256*.
A², 104, HR 8982, 4.82V, *SA, HRP*, BF 3245*, H 134 as A, G 257*.

Bayer noted that A was a single 5th-magnitude star, the most northerly of a group of three stars. He placed the letter A next to it on his chart. He designated the other two stars i. Since Flamsteed had only Bayer's charts and not his star lists, he was confused about these three stars. Adding to his confusion were several other factors: (1) The letter i was the last letter in the constellation; there were no additional letters to alert Flamsteed to search for i. (2) It was printed very small on the star chart. (3) It was placed in the stream flowing from the water jug and seemed to blend with the lines in the picture. Consequently, Flamsteed never realized that the letter i existed in Aquarius. He assigned A-1 to his 103 and A-2 to 104 since they were virtually adjacent. He then proceeded to assign A-3, A-4, and A-5 to his 106, 107, and 108 since they were alongside 103 and 104, although they, or at least 106 and 107, should have been designated i (see i, below). Baily corrected this by retaining only A¹ for 103 and A² for 104. Argelander and Heis, however, designated only 104, the brighter star, as A since they could not see 103 with the naked eye (*UN*, p. 115). But Gould saw both stars and restored their letters. For Flamsteed's use of an incomplete edition of Bayer's *Uranometria*, see *BF*, pp. 399-400.

b¹, 98, HR 8892, 3.97V, *SA, HRP*, BV 87*, BF 3205*, H 121*, G 231*.
b², 99, HR 8906, 4.39V, *SA, HRP*, BV 89*, BF 3211*, H 122*, G 235*.
b³, 101, HR 8939, 4.71V, *SA, HRP*, BV 98*, BF 3227 as b⁴, H 129*, G 244*.

Bayer noted that b was a group of three 5th-magnitude stars. Flamsteed assigned b-1, b-2, b-3, and b-4 to his 98, 99, 100 (HR 8932, 6.29V, BV 95, BF 3223 as b³, H 127, G 242), and 101 since he was not sure whether 100 or 101 was among the stars Bayer meant to include as b. Baily was also uncertain about 100 and 101, so he kept Flamsteed's designations. Bevis, though, felt only 98, 99, and 101 were synonymous with Bayer's b's; consequently, he relettered them as in the heading above. Argelander, who could not see 100 with his naked eye, agreed (*UN*, p. 114). Heis and Gould were able to observe 100, but they left it unlettered.

c¹, 86, HR 8789, 4.47V, *SA, HRP*.
c², 88, HR 8812, 3.66V, *SA, HRP*.
c³, 89, HR 8817, 4.69V, *SA, HRP*.

Since Bayer described c as a group of three 5th-magnitude stars, Flamsteed designated his 86, 88, and 89 as c-1, c-2, and c-3.

d, 25, HR 8277, 5.10V, *SA, HRP.*

This star is one of Flamsteed's twenty-two duplicate stars. It is the same as 6 Peg in Aquarius.

e^1, 37, HD 210422, 6.70V, BV 40, BF 3026*, BAC 7719*, G 116.

e^2, 38, HR 8452, 5.46V, *SA* as e, *HRP* as e, HF 38 as e, BV 38 as e, BF 3027*, BAC 7722*, H 49 as e, G 117 as e.

Bayer described e as a 6th-magnitude star, and Flamsteed designated his 38 as e. Bevis agreed, but Baily felt this was erroneous inasmuch as Bayer's chart shows e north of the ecliptic and 38 is below. Although he was certain 37, not 38, was Bayer's e, he kept Flamsteed's designation by assigning e^1 to 37 and e^2 to 38. Argelander, unable to see 37 with his naked eye, designated only 38 as e (*UN*, p. 113). Gould, high up in the rarified atmosphere of the Andes, observed 37 but left it unlettered. Most astronomers have felt that since 37's magnitude is too dim for naked-eye observation, Bayer could not possibly have seen it. But, for the possibility that he may have had telescopic assistance, see Chi and h Per.

f, 53, HR 8544/5, 6.57V, 6.35V, comb. mag. 5.56V, *SA, HRP.*

g^1, 66, HR 8649, 4.69V, *SA, HRP* as g, HF 61 as g, BV 60 as g, BF 3107*, H 84*, G 176 as g.

g^2, 68, HR 8670, 5.26V, *SA,* HF 63, BV 63, BF 3118*, H 85*, G 179.

Bayer's *Uranometria* describes only one 6th-magnitude star as g, and the Halley edition of Flamsteed's catalogue labels 66 as g. Flamsteed's 1725 catalogue, though, designates these two neighboring stars, 66 and 68, as g-1 and g-2. Baily felt that Bayer must have meant the brighter star, 66, but he kept Flamsteed's designations. Bevis, Argelander (*UN*, p. 114), and Gould, however, labeled only 66 as g. Heis, on the other hand, agreed with the authorized version of Flamsteed's catalogue and restored g to both stars.

h, 83, HR 8782, 5.43V, *SA, HRP,* BV 92*, H 103*, G 208*.

Bayer described h as a single 6th-magnitude star, but Flamsteed was not certain which of four closely located stars Bayer meant, or whether Bayer referred to the combined light of all of them. Consequently, he assigned h to each of them, and Baily agreed with his designations:

      h-1, 83, HR 8782, 5.43V, BV 92 as h, BF 3158 as h^1, H 103 as h.
      h-2, 84, HD 218081, 7.5V, BV 93, BF 3159 as h^2.
      h-3, 85, HD 218173, 6.9V, BV 94, BF 3161 as h^3.
      h-4, 87, HD 218331, 7.4V, BV 96, BF 3164 as h^4.

Bevis, though, felt that just 83, the brightest, was synonymous with Bayer's h. Argelander agreed because he could see only 83 with his naked eye (*UN*, p. 114). Gould observed both 83 and 84, noting that they appeared as a single object to the naked eye, but he specifically assigned h to 83.

i^1, 106, HR 8998, 5.24V, *SA, HRP,* BV 109, BF 3251*, H 137*, G 261*.

i^2, 107, HR 9002, 5.29V, *SA, HRP,* BV 110 as i-1, B 372 as i-1, BF 3253*, H 138, G 262/3*, ADS 16979.

i^3, 108, HR 9031, 5.18V, *SA, HRP,* BV 113 as i-2, B 385 as i-2, BF 3265, H 145 as i^2, G 270*.

Bayer noted that i was a couple of 6th-magnitude stars. Flamsteed, working without Bayer's star list, overlooked i and did not designate any star with this letter in Aquarius (see A, above). Bevis and Bode corrected this by designating his 107 as i-1 and his 108 as i-2. Baily, also aware of Flamsteed's omission, assigned i^1 to 106 and i^2 to 107. Argelander disagreed with Baily. After carefully examining the pair's position in the sky, he felt that while 106 was i^1, 108 and not 107 was equivalent to i^2 since it conformed better to Bayer's chart (*UN*, p. 115). Gould, however, concluded that Bayer must have seen 106 and 107 as one star since they are very close and 108 as the second of the i pair. Consequently, he designated these three stars i^1, i^2, and i^3 (*Uran. Arg.*, p. 94).

j

This letter is not used in Aquarius.

k, 3, HR 7951, 4.42V, *SA, HRP,* B 9*, H 3, G 9*.

Bayer did not letter this star. The letter was first assigned by Bode, who added numerous Roman capital and lower-case letters to stars in Aquarius, none of which has remained in the literature except k. It owes its survival to Gould, who felt a star of its brightness merited a letter, so he retained Bode's designation.

Aquila, Aql

Eagle

Aquila (Figure 5) is a Ptolemaic constellation that Bayer lettered from Alpha to Omega and from A to h. Like Ptolemy, Bayer included the asterism Antinous in Aquila.

The origin of Aquila can be traced back at least four thousand years to ancient Mesopotamia, where the constellation was known as Nasru, or Eagle. According to an old Akkadian myth, an eagle agrees to carry on his back Etana, King of Kish, to the heavenly abode of the gods where he hopes to obtain the plant of life that would perpetuate his dynasty. This myth may have been the source of the various classical legends of eagles carrying humans to the sky. Alexander the Great, for example, is reputed to have yoked together two giant eagles in order

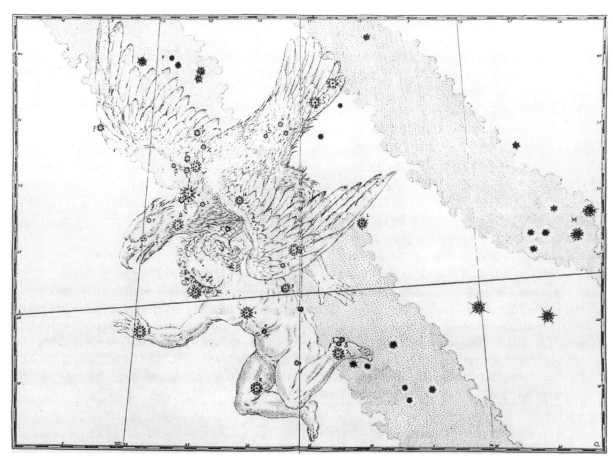

Figure 5. Bayer's Aquila. In his introduction to Aquila, Bayer described the Eagle as either kidnapping Ganymede (Ganymedis raptrix) or delivering Antinous to the heavens (Servans Antinoum). The line I is the celestial equator. The letter K is Sagitta; L is the tail of Serpens; and M is Ophiuchus's arm.

Table 10. The lettered stars of Aquila

	Magnitudes			Catalogue Numbers				
Letter	Bayer	Lacaille	Visual	Flamsteed	Lacaille	Other	HR	HD
Alpha	2		0.77	53			7557	187642
Beta	3		3.71	60			7602	188512
Gamma	3		2.72	50			7525	186791
Delta	3		3.36	30			7377	182640
Epsilon	3		4.02	13			7176	176411
Zeta	3		2.99	17			7235	177724
Eta	3		3.90	55			7570	187929
Theta	3		3.23	65			7710	191692
Iota	3		4.36	41			7447	184930
Kappa	3		4.95	39			7446	184915
Lambda	3		3.44	16			7236	177756
Mu	4		4.45	38			7429	184406
Nu	4		4.66	32			7387	182835
Xi	5		4.71	59			7595	188310
Omicron	5		5.11	54			7560	187691
Pi	5		5.72	52			7544	187259
Rho	5		4.95	67			7724	192425
Sigma	5		5.17	44			7474	185507
Tau	6		5.52	63			7669	190327
Upsilon	6		5.91	49			7519	186689
Phi	6		5.28	61			7610	188728
Chi	6		5.27	47			7497	186203
Psi	6		6.26	48			7511	186547
Omega[1]+	6		5.28	25			7315	180868
Omega[2]			6.02	29			7332	181383
A	6		5.53	28			7331	181333
b	6		5.16	31			7373	182572
c	6		5.80	35			7400	183324
d	6		5.49	27			7336	181440
e	6		5.03	36			7414	183630
f	6		5.01	26			7333	181391
g	6		5.42	14			7209	176984
h	6		5.42	15			7225	177463
i+			4.02	12			7193	176678
k+			5.12	37			7430	184492
l+			4.32	71		G 146	7884	196574

to view the world from above. In another legend, Zeus, in the guise of an eagle, abducts Ganymede, the beautiful son of the king of Troy, to serve as his cupbearer and catamite — the word itself deriving from the Etruscan-Roman form of Ganymede: Catmite, Catamitus. The Emperor Hadrian's familiarity with this legend undoubtedly played a role in his decision to commemorate his young homosexual lover, Antinous of Bythnia, with a place in the heavens directly below Aquila. Ptolemy included Antinous in his catalogue, and so too did Bayer, who pictured him on his chart of Aquila as a naked young boy being carried skyward in the talons of the Eagle.

As noted, the Greeks believed that Aquila represented Zeus in the guise of an eagle or the Eagle as the messenger and bird of Zeus swooping down to capture Ganymede. Another account relates that while Zeus was making a sacrifice at the Altar (Ara) just before his battle with the Titans, an eagle flew down to his side and henceforth became his favorite animal.

See Langdon, *The Legend of Etana and the Eagle;* Langdon, *Semitic Mythology,* chap. III; Langdon, *Babylonian Menologies,* p. 8; Pritchard, *Anc. Near East. Texts,* pp. 114-18, 517; van der Waerden, "Babylonian Astronomy. II. The Thirty-Six Stars," pp. 11, 13-14; Toomer, *Ptolemy's Almagest,* pp. 356-57; Ridpath, *Star Tales,* pp. 137-38; Condos, *Star Myths,* pp. 33-36.

The Lost, Missing, or Troublesome Stars of Aquila

Omega¹, 25, HR 7315, 5.28V, *SA, HRP, BS,* HF 27 as Omega, BV 29 as Omega, BF 2600 as Omega, H 30*.
Omega², 29, HR 7332, 6.02V, *SA, HRP, BS,* BF 2608, H 35*.

Bayer described only one 6th-magnitude star as Omega, and the 1712 pirated edition of Flamsteed's catalogue designates 25 as Omega. Flamsteed's authorized catalogue of 1725, however, labels these two neighboring stars, 25 and 29, as Omega-1 and Omega-2. Bevis and Baily, though, felt that only the brighter star, 25, was synonymous with Bayer's Omega. Argelander, unable to see 29 with his naked eye, agreed with them (*UN,* p. 75). Heis, seeing both stars, reinstated 29 as Omega.

A, 28, HR 7331, 5.53V, *SA, HRP.*

b, 31, HR 7373, 5.16V, *SA, HRP.*

c, 35, HR 7400, 5.80V, *SA, HRP.*

d, 27, HR 7336, 5.49V, *SA, HRP.*

e, 36, HR 7414, 5.03V, *SA, HRP.*

f, 26, HR 7333, 5.01V, *SA, HRP.*

g, 14, HR 7209, 5.42V, *SA, HRP.*

h, 15, HR 7225, 5.42V, *SA, HRP.*

i, 12, HR 7193, 4.02V, *SA, HRP,* BV 12, BF 2561, H 9, G 9*.

Bayer did not assign this star a letter. Although he observed this star and included it in his atlas, he considered it an *informis* or *sparsilis* star, an unformed or scattered star outside the borders of the constellation, so he left it unlettered. All in all, Bayer's charts included about 325 of these *informes*. Baily noted that i was added by Flamsteed, who used a considerable number of non-Bayer letters in his manuscripts, probably for reference purposes with the intention of removing them before final publication. However, he died before completing his work, and they were inadvertently included in his "Catalogus Britannicus." Baily removed them all in his revised edition of Flamsteed's catalogue (*BF*, p. 397).

About eighty years before Baily published his version of Flamsteed's catalogue, John Bevis had embarked on the publication of his own stellar atlas and catalogue, *Uranometria Britannica*. In compiling this work, he made extensive use not only of Flamsteed's authorized catalogue of 1725 but also of the pirated edition of 1712. Like Baily after him, he noted the extraneous letters the Astronomer Royal had added in many of the constellations. Bevis realized that they had nothing to do with Bayer's lettering system, so he removed most of them, as he did with i Aql. Occasionally, however, he added letters — usually lower-case gothic characters — to distinguish the brighter stars that Bayer had left unlettered. For example, noting that Flamsteed had erroneously designated his 50 And as Upsilon, Bevis removed the letter Upsilon and labeled the star a; he felt a star of its brightness — 4th magnitude — merited a letter. None of Bevis's letters, though, have survived in the literature, primarily because only a few incomplete copies of his atlas and catalogue were ever printed and his work remained largely unknown.

Although Bevis and Baily sought to expunge Flamsteed's so-called extraneous or superfluous letters, it is possible they were mistaken in their belief that Flamsteed assigned them haphazardly and intended eventually to remove them. Like i, many of them designate Bayer's *informes;* they appear extensively in Flamsteed's manuscripts; and they are included in both Halley's edition of his catalogue and the authorized edition of 1725. Based on these facts, one could argue that Flamsteed intended to retain them permanently. In the case of 12 Aql, Gould restored its i since he felt a star of its brightness merited a letter (*Uran. Arg.,* p. 95).

j

This letter is not used in Aquila.

k, 9 Aql in Scutum as Eta Sct, HR 7149, 4.83V, HEV 1, HF 11*, B 16*, BF 2552, H 11 in Scutum, G 33 as Eta Sct.

Bayer did not letter this star. Although he included this star in his atlas, he considered it an *informis* and left it unlettered. Hevelius observed it and included it in his catalogue within his newly devised constellation Scutum. The letter was added by Flamsteed but removed by Baily, who considered it superfluous (see i). Although Flamsteed did not recognize Scutum as a separate constellation — he included its stars in Aquila, as Bayer and Ptolemy had done — he did mention that it had been devised by Hevelius. Flamsteed erroneously designated two stars as k Aql — this star and his 37. Gould relettered this star Eta Sct because of its brightness.

k, 37, HR 7430, 5.12V, *SA,* HF 36, BV 36, B 125 as K, BF 2640*, BAC 6703*, H 54*, G 59.

Flamsteed, not Bayer, lettered this star. While the Halley edition of Flamsteed's catalogue leaves this star unlettered, it and 9 Aql are labeled k in the authorized edition of 1725. Although Baily and Heis removed 9's k, surprisingly, they retained it for 37. This was the only time one of Flamsteed's extraneous letters ever found its way into their catalogues. Bode labeled this star capital K, but the more scrupulous Bevis and Argelander (*UN,* p. 76) left it unlettered. Gould also left it unlettered. He considered it almost a 6th-magnitude star, too faint to merit a letter.

l, 71, HR 7884, 4.32V, *HRP,* B 280 as I, G 146*.

Bayer did not letter this star. Bode designated it I, possibly as a misprint or typographical error for l (ell). Gould agreed that the star's brightness merited a letter, but he felt it should follow Bayer's letters in alphabetical order, so he relettered it l (ell). He skipped k to avoid confusion with Flamsteed's k, above.

m, 1 Aql in Scutum, HR 6973, 3.85V, HEV 2, B 1*, BF 2509, G 14 as Alpha Sct.
n, 3 Aql in Scutum, HR 7032, 4.90V, HEV 3, B 3*, BF 2522, G 21 as Epsilon Sct.
o, 2 Aql in Scutum, HR 7020, 4.72V, HEV 4, B 2*, BF 2518, G 19 as Delta Sct.

Bayer did not letter these three stars. Although he included them in his atlas, he considered them *informes* and left them unlettered. The letters were added by Flamsteed but removed by Baily, who considered them superfluous (see i). The stars were later relettered Alpha, Epsilon, and Delta Sct by Gould because of their brightness. See k, 9 Aql, above.

A*RA*, A*RA*

Altar

Ara (Figure 6) is a Ptolemaic constellation that Bayer lettered from Alpha to Theta. Since it is below the horizon in Europe, Bayer did not observe it. Consequently, his chart does not conform to the constellation's actual appearance. Lacaille redesigned and relettered it from Alpha to Sigma.

In the beginning, according to Greek mythology, there is Chaos — shapelessness, darkness, and confusion. From Chaos emerges Gaea (Ge) or Mother Earth, who gives birth to Uranus, the personification of Heaven. From the union of Gaea and Uranus emerge the twelve mighty Titans including Cronus (Saturn among the Romans), Iapetus and his sons Prometheus and Atlas, and Oceanus. Fearing their awesome power, Uranus imprisons the Titans inside Mother Earth, but Gaea helps her son Cronus escape. Cronus succeeds in maiming and overpowering Uranus, freeing the imprisoned Titans, and becoming ruler of the Universe. He marries his sister — the Titan Rhea — who bears him the deities Hestia, Demeter, Hera, Hades, Poseidon, and Zeus. Remembering the curse uttered by his father that he, too, would be overthrown by his children, Cronus swallows his first five offspring as they are born. Rhea succeeds in saving her sixth child, Zeus, and secrets him in Crete. Zeus eventually grows to manhood, overthrows Cronus, and forces him to disgorge the deities he had swallowed. Led by Zeus, the gods take an oath on the Altar (Ara) to form an alliance against the Titans, now led by Atlas. After a ten-year struggle, the gods, symbolizing intelligence, emerge victorious against the Titans, the epitome of brute force. They consign the Titans to Tartarus, the lowest part of the Underworld, and in a thanksgiving gesture they place the Altar in the heavens. The victorious gods divide the world among themselves. Zeus (Jupiter among the Romans) becomes king of the gods and the sky god. His sister and wife, Hera (Juno), becomes queen of the gods and patron of marriage. Their sister Hestia (Vesta) becomes the goddess of the hearth and domestic life. Another sister Demeter (Ceres) becomes goddess of grain and agriculture. Their brother Poseidon (Neptune) becomes god of the sea. Zeus's seven children are also deified — Hephaestus (Vulcan), the god of fire; Ares (Mars), the god of war; Aphrodite (Venus), the goddess of love and beauty; Athena (Minerva), the goddess of wisdom; Hermes (Mercury), the god of speed and messenger of the gods; Apollo, god of the sun; and Artemis (Diana), the twin sister of Apollo, goddess of hunting and the moon. These deities evolve as the twelve major Olympian gods of Greek and Roman mythology.

The Altar in the heavens was supposed to remind humans that before they initiate important undertakings they should offer sacrifices to the gods. Some scholars have claimed that the constellation can be traced back to the Middle East, but as far as is presently known, no Mesopotamian or Near Eastern legend or myth was associated directly with Ara.

An explanation for Ara's particular position in the heavens is suggested by a recent researcher who notes that although Ara's stars "are of mediocre brightness, ... it marks the point in the horizon from which the summer Milky Way streams upwards like a glowing, writhing column of smoke."

See Reinhold, *Essentials of Greek and Roman Classics,* pp. 329-35; Ridpath, *Star Tales,* pp. 28-29; Condos, *Star Myths,* pp. 37-38; Staal, *New Patterns in the Sky,* pp. 229-30; Rogers, "Origins of the Ancient Constellations," p. 83. For the significance of the number twelve in myths and legends, see Apus.

Table 11. The lettered stars of Ara

	Magnitudes			Catalogue Numbers				
Letter	Bayer	Lacaille	Visual	Flamsteed	Lacaille	Other	HR	HD
Alpha		3	2.95		1436		6510	158427
Beta		3	2.85		1423		6461	157244
Gamma		3	3.34		1422		6462	157246
Delta		4	3.62		1433		6500	158094
Epsilon^1+		4	4.06		1402		6295	152980
Epsilon^2		6	5.29		1405		6314	153580
Zeta		4	3.13		1399		6285	152786
Eta		5	3.76		1386		6229	151249
Theta		4	3.66		1480		6743	165024
Iota		6	5.25		1424		6451	157042
Kappa^1+		6	5.23		1426		6468	157457
Kappa^2		6	5.92		1430		6478	157662
Lambda		6	4.77		1448		6569	160032
Mu		6	5.15		1453		6585	160691
Nu^1+		6	5.92		1463		6622	161783
Nu^2		6	6.09		1464		6632	161917
Pi		6	5.25		1444		6549	159492
Rho+		6	6.33		1397		6274	152478
Sigma		6	4.59		1443		6537	159217

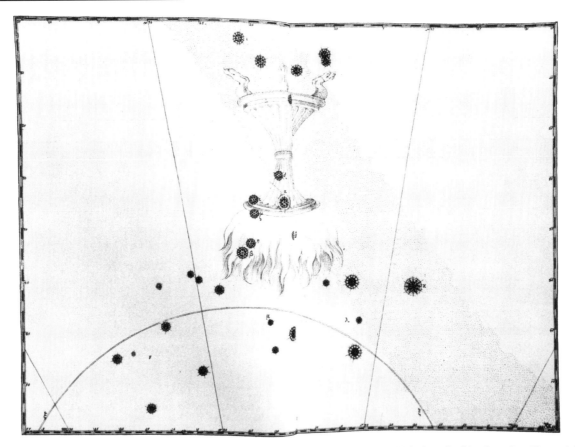

Figure 6. Bayer's Ara. The line Xi (ξ) is the celestial Antarctic Circle. The letter Iota (ι) is the tail of the Scorpion; Kappa (κ) is the foot of the Centaur; Lambda (λ) is the Southern Triangle; Mu (μ) is Grus; and Nu (ν) is Pavo. For the significance of the Antarctic and Arctic Circles, see Figure 36, Draco.

The Lost, Missing, or Troublesome Stars of Ara

Epsilon[1], HR 6295, 4.06V, *SA, HRP, BS,* CA 1402 as Epsilon, BR 5900*, L 7050*, G 25*.
Epsilon[2], HR 6314, 5.29V, *SA, HRP, BS,* CA 1405 as Epsilon, BR 5921*, L 7073*, G 31*.

Lacaille designated these two neighboring stars Epsilon.

Kappa[1], HR 6468, 5.23V, *SA* as Kappa, *HRP* as Kappa, *BS* as Kappa, CA 1426 as Kappa, BR 6060*, L 7253 as Kappa, BAC 5859*, G 53 as Kappa.
Kappa[2], HR 6478, 5.92V, CA 1430 as Kappa, BR 6067*, L 7262, BAC 5865*, G 54.

Lacaille designated these two nearly adjacent stars Kappa. Gould removed Kappa from CA 1430 because of its faintness, noting, though, that the light from both stars appeared as one image to his naked eye.

Nu[1], HR 6622, 5.92V, CA 1463 as Nu, BR 6204*, L 7426, BAC 6009*, G 77.
Nu[2], HR 6632, 6.09V, CA 1464 as Nu, BR 6208*, L 7428, BAC 6014*, G 78, GC 24208 as Upsilon[2].

Lacaille designated these two neighboring stars Nu. Gould dropped their letters because of their faintness. In *SA,* they are labeled as Upsilon, an obvious error for Nu since the Greek letters are similar in appearance, ν (Nu) and υ (Upsilon). The same error appears in Becvar's atlases and in Boss's *General Catalogue,* where it may have originated. Boss's GC 24187 is Nu[1] but is unlabeled in his catalogue, where the HD number is given incorrectly as HD 161763; it should be HD 161783. Both stars are mislabeled Upsilon in Tirion's *SA* but are left unlabeled in his *Uranometria 2000.0* and in the second edition of *SA*. See Hoffleit, "Discordances," p. 44 and notes in *BS*.

Xi

Omicron

Lacaille did not use these two letters in Ara.

Rho, HR 6274, 6.33V, CA 1397*, BR 5882*, L 7024, BAC 5664*, B 12 as Rho-1, G 21.

Lacaille designated this star Rho, but Bode called it Rho-1 and its easterly neighbor Rho-2 (HR 6289, 5.55V, CA 1401, BR 5897, L 7045, B 15, G 24), which Lacaille observed but left unlettered. Gould dropped both their letters because of their faintness.

Tau

This letter is not used in Ara.

Upsilon

See Nu.

ARIES, ARI

Ram

Aries (figures 7, 8, and 9) is a relative newcomer to the skies. Its position in the heavens was originally held by the Sumerian constellation LU.HUN.GA (Akkadian Agru), the Hired Farm Worker. He was usually depicted on Mesopotamian documents, seals, and monuments holding a prod to guide the Bull of Heaven (Taurus) in plowing (Triangulum) the fields (Great Square of Pegasus). Many years afterward, *Mul-Apin,* a late Babylonian text of 687 B.C., identified LU.HUN.GA, the Hired Man, with the god Dumuzi or Tammuz. Among his many attributes were those associated with agriculture, and especially fertilization. He was known, for instance, as Dumzi-Amaushumgalanna, "the power in the date palm to produce new fruit." He was also associated with the rising of sap in trees and with new life in plants and vegetables, attributes which were indeed appropriate for a god identified with a constellation that marked the vernal equinox. LU.HUN.GA became the first constellation in the zodiac about thirty-seven hundred years ago when, because of precession, the vernal equinox shifted away from the Bull, where it had been since about 4200 B.C., and toward the Hired Man's position in the sky.

The arrival of the vernal equinox was a significant event for the people of the ancient Near East since it was one of the key factors in determining the new year. The first day of the year was fixed when the new moon nearest the equinox was first sighted. This marked the beginning of the first month, Nisanu (March-April). In Babylon, the New Year or Akitu Festival lasted eleven days and was considered the highlight of both the religious and civil calendar. On the fifth day, the masmasu-priest, together with a slaughterer, sacrificed a ram in what was called the kuppuru-ritual. The animal was beheaded, its body rubbed against the walls of the temple of Nabu and then thrown into the Euphrates. This was done to purify the temple and, symbolically, Babylon itself, with the dead ram supposedly absorbing all the sins and impurities of the past year (compare the Passover ritual described in Exodus 13:12).

The sacrifice of a young ram or male sheep — later called the paschal lamb — in the spring or birthing season, was an old established custom, not only in Babylon but also among the Semitic tribes in the area from Arabia to Canaan around the Fertile Crescent. This custom, with its attendant sacred rites and ceremonies, gave such prominence to the ram that it eventually replaced the Hired Farm Hand in the heavens. When exactly this occurred is unknown, for as late as the fifth century B.C., Mesopotamian texts still refer to the first sign of the zodiac as LU.HUN.GA.

Since there is no specific mention of the Ram as a constellation in early Mesopotamian documents, scholars have speculated over precisely how this constellation came into being. In addition to the theory posited above, historians have proposed that it arose as the result of some confusion in interpreting cuneiform symbols. The constellation the Hired Farm Worker or Hireling is written syllabically in Sumerian cuneiform as LU.HUN.GA or in its shortened form as Lu, which is a homophone for another, different cuneiform symbol meaning sheep (*immeru*) in Akkadian. Possible confusion between these two different symbols — both sounding the same but with different meanings — may have been the source for associating the Ram with the Hired Farm Worker.

Greek mythology contains several legends and variations of legends associated with Aries. One story relates that King Athamas of Boeotia marries a minor goddess, Nephele (Cloud), who returns to heaven, leaving behind her husband; their son, Phrixus; and their daughter, Helle. Athamas then marries Ino, who, for one reason or another, seeks the death of her two step-children. According to one account, she plots to create a famine in order to convince her husband that only the sacrifice of her step-son will save the kingdom. In another account, Ino attempts to seduce Phrixus, is rebuffed, and seeks revenge. In any case, just before Athamas is about to sacrifice his son, their mother, Nephele, appears with a golden, winged ram to fly her children to safety. Unfortunately, Helle falls off the ram and into the sea, which is henceforth named in her honor, the Hellespont (Dardanelles). Phrixus lands safely in Colchis on the eastern shore of the Black Sea and sacrifices the ram to the gods, who set its image in the heavens. He presents the hide with its gold fleece to the king of Colchis, with whom it remains until it is seized by Jason and his band of Argonauts.

Aries is a Ptolemaic constellation that Bayer included in his *Uranometria,* where its stars are lettered from Alpha to Tau.

See van der Waerden, "History of the Zodiac," pp. 219-20; van der Waerden, "Babylonian Astronomy. II. The Thirty-Six Stars," pp. 14-15; O'Neil, *Time and the Calendar,* pp. 60-61; Hartner, "The Earliest History of the Constellations in the Near East," p. 10; Pritchard, *Anc. Near East. Texts,* pp. 331-34; Langdon, *Babylonian Menologies,* pp. 67-68; de Vaux, *The Early History of Israel,* pp. 366-69; Frankfort, *Kingship and the Gods,* chap. 22; Roux, *Ancient Iraq,* pp. 331-35; Hooke, *Myth, Ritual, and Kingship,* pp. 38-40; Gaster, *Thespis,* chap. 3; Hunger and Pingree, *Mul.Apin,* pp. 30, 40, 140; Krupp, *Beyond the Blue Horizon,* p. 132; Jacobsen, *Toward the Image of Tammuz,* chap. 6; Gleadow, *The Origin of the Zodiac,* pp. 173-74; Condos, *Star Myths,* pp. 43-47; Gantz, *Early Greek Myths,* pp. 176-180; Ridpath, *Star Tales,* pp. 29-31; Staal, *New Patterns in the Sky,* pp. 36-41. For Jason and the Argonauts, see Carina; see also Libra, Orion, Pegasus, Taurus, and Triangulum.

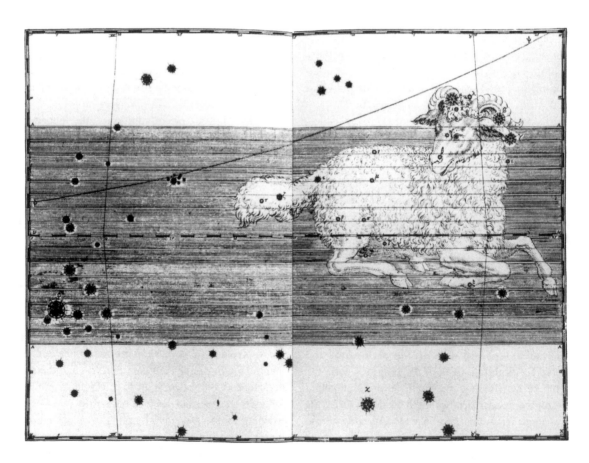

Figure 7. Bayer's Aries. The line Psi (ψ) is the Tropic of Cancer; the line Omega (ω) is the ecliptic; and the line A, extending 8° on both sides of the ecliptic, is the zodiac, the apparent path of the sun, moon, and planets around the celestial globe. The letter Upsilon (υ) is the Pleiades; Phi (φ) is the head of Taurus; and Chi (χ) is Cetus.

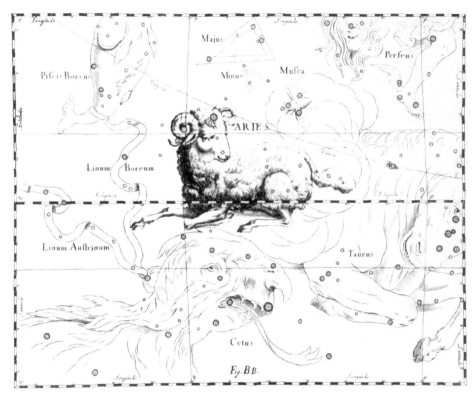

Figure 8. Hevelius's Aries. Above the Ram is the asterism Musca, the Northern Fly. Between the Fly and the Ram is Triangulum Minus, the Little Triangle, one of the twelve constellations devised by Hevelius (see Canes Venatici). The figures in Hevelius's atlas are drawn backward, as they would appear on a celestial globe.

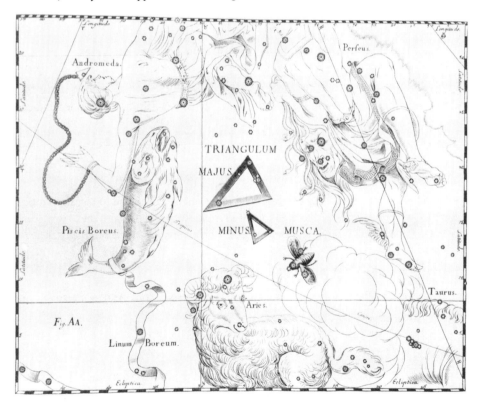

Figure 9. Hevelius's Triangulum Majus, Triangulum Minus, and Musca. The group of six stars on the Tropic of Cancer and to the right of the Ram's tail is the Pleiades.

Lost Stars

Table 12. The lettered stars of Aries

	MAGNITUDES			CATALOGUE NUMBERS				
Letter	Bayer	Lacaille	Visual	Flamsteed	Lacaille	Other	HR	HD
Alpha	3		2.00	13			617	12929
Beta	3		2.64	6			553	11636
Gamma^{1,2}+	3		3.88c	5			545/6	11503/2
Delta	4		4.35	57			951	19787
Epsilon	5		4.63c	48			887/8	18519/20
Zeta	5		4.89	58			972	20150
Eta	6		5.27	17			646	13555
Theta^1+	6		5.62	22			669	14191
Theta^2			6.84	23				14305
Iota	6		5.10	8			563	11909
Kappa	6		5.03	12			613	12869
Lambda	6		4.79	9			569	11973
Mu	6		5.69	34			793	16811
Nu	6		5.30	32			773	16432
Xi+	6		5.47	24			702	14951
Omicron	6		5.77	37			809	17036
Pi	6		5.22	42			836	17543
Rho^1+			6.9	44				18091
Rho^2			5.91	45			867	18191
Rho^3	6		5.63	46			869	18256
Sigma	6		5.49	43			847	17769
Tau^1+	6		5.28	61			1005	20756
Tau^2			5.09	63			1015	20893
c+			3.63	41		B 119	838	17573

The Lost, Missing, or Troublesome Stars of Aries

Gamma^1,Gamma^2, 5, HR 545/6, 4.83V, 4.75V, comb. mag. 3.88V, *SA, HRP* as Gamma, *BS,* ADS 1507.

Bayer described Gamma as a single 3rd-magnitude star. Later astronomers added indices to identify the components of this double star.

Theta^1, 22, HR 669, 5.62V, *SA* as Theta, *HRP* as Theta, *BS* as Theta, HF 21 as Theta, BV 21 as Theta, BF 287 as Theta, H 27 as Theta.
Theta^2, 23, HD 14305, 6.9V, BV 23, BF 290.

Bayer's *Uranometria* describes only one 6th-magnitude star as Theta, and the pirated edition of Flamsteed's catalogue labels 22 as Theta. Flamsteed's catalogue of 1725, however, designates these two nearly adjacent stars, 22 and 23, as Theta-1 and Theta-2. Bevis and Baily rejected 23, believing that the brighter star, 22, was synonymous with Bayer's Theta.

Xi, 24, HR 702, 5.47V, *SA, HRP, BS,* HF 23*, BV 14*, BF 300*.

This star is the same as Psi Cet. For Bayer's duplicate stars, see Alpha And; see also Psi Cet.

Rho^1, 44, HD 18091, 6.9V, BF 363*, BAC 898*, LL 5372*.
Rho^2, 45, HR 867, 5.91V, *BS,* BF 364*, BAC 901*, LL 5401*, the variable RZ.
Rho^3, 46, HR 869, 5.63V, *SA* as Rho, *HRP* as Rho, *BS,* HF 45 as Rho, BF 366*, BAC 903*, LL 5412*, H 54 as Rho.

Bayer described only one 6th-magnitude star as Rho, and the pirated edition of Flamsteed's catalogue labels 46 as Rho. Flamsteed's catalogue of 1725, though, designates these three neighboring stars, 44, 45, and 46, as Rho-1, Rho-2, and Rho-3. Argelander, seeing only 46 with his naked eye, assumed it alone was Bayer's Rho (*UN,* p. 51). Heis dropped 44 as being obviously too faint for the naked eye; he himself could not see it. He noted that the combined light of 45 and 46 was synonymous with Bayer's Rho. Actually, he did not mention 45, but he equated BAC 903 and LL 5401 (misprinted in his catalogue as 4501) with 46, Rho. BAC 903 is synonymous with 46, Rho³, and LL 5401 is synonymous with 45, Rho².

Tau¹, 61, HR 1005, 5.28V, *SA* as Tau, *HRP* as Tau, *BS,* HF 60 as Tau, BV 62, BF 421*, H 74 as Tau.
Tau², 63, HR 1015, 5.09V, *BS,* HF 62 as Tau, BV 63 as Tau, BF 425*, H 76.

Bayer's *Uranometria* describes only one 6th-magnitude star as Tau, but Flamsteed's 1725 catalogue designates these two neighboring stars Tau-1 and Tau-2. Bevis, however, felt that 63, the brighter star, was Bayer's Tau. Argelander, believing 61 brighter than 63, labeled it Tau (*UN*, p. 52). Heis agreed with Argelander since he too judged 61 brighter than 63. Oddly enough, the pirated edition of Flamsteed's catalogue labels three stars as Tau: 61, 63, and 65 (HR 1027, 6.08V, HF 64 as Tau), but specifically notes that 61 is Bayer's Tau.

c, 41, HR 838, 3.63V, *SA,* B 119*.

Bayer did not letter this star. Although he included this star in his atlas, he considered it an *informis* and left it unlettered (see i Aql). The letter was added by Bode, who considered it part of the asterism Musca or Northern Fly, just above the Ram's back. In fact, Bode called the constellation Aries et Musca and labeled Musca's four stars as follows:

        a, 35, HR 801, 4.66V, B 106.
        b, 39, HR 824, 4.51V, B 112.
        c, 41, HR 838, 3.63V, B 119.
        d, 33, HR 782, 5.30V, B 101.

The asterism first appeared as Apes — a typographical error for Apis — the Bee — on Petrus Kaerius's globe of 1613, which Warner (*Sky Explored,* p. 204) suggested was based on the work of Petrus Plancius, a prominent and influential Calvinist minister in Amsterdam who meant to depaganize the heavens by creating constellations with Biblical significance. Plancius may have intended the Bee to represent the bees that clustered in the carcass of the lion that Sampson slew (Judges 14:8). Thinking along similar lines, Kepler's son-in-law, Jacob Bartsch, who helped Julius Schiller prepare his *Coelum Stellatum Christianum,* depicted this small group of stars as Vespa, the Wasp, on his planisphere of 1624. He suggested (*Usus Astronomicus,* p. 57) that the constellation symbolized for him either the bees of Sampson or the wasps or flies of Beelzebub (Luke 11:15). Bartsch noted that *ba'al zebub* literally means *deus muscarum* in Hebrew, lord or god of the flies. Hevelius and Flamsteed depicted it as Musca on their atlases but did not include it as a specific constellation in their catalogues. See Columba and Monoceros.

The attempt to depaganize the heavens was not limited just to the newly devised constellations. In his *Usus Astronomicus* (pp. 55-57), Bartsch sought to interpret many of the classic Ptolemaic constellations from a Biblical point of view. He declared that as far as he was concerned, Lyra represented David's Harp, Sagitta was Jonathan's Javelin, Hercules was Sampson, and so on.

Schiller's *Coelum Stellatum Christianum* of 1627 was an attempt to depaganize the heavens completely. Unlike Plancius, who proposed Biblical references only for newly devised constellations, Schiller, with the help of Bartsch and Bayer, revised all the constellations — including the forty-eight classical constellations of Ptolemy. He eliminated their former mythological names, redrew them completely, and rechristened them in honor of the apostles, the saints, and various theological symbols drawn from the Old and New Testaments. Although beautifully drawn and more accurate even than Bayer's *Uranometria,* the atlas, together with its accompanying star lists, never became popular. In the first place, its attempt to redraw, let alone rename, all the constellations as well as all the twelve signs of the ecliptic was too radical to gain widespread acceptance. Moreover, unlike Bayer's *Uranometria,* it was drawn to represent the way the sky would appear on a stellar globe, that is, backward to the earth-bound observer. Consequently, it was not a very practical tool for the observational astronomer.

Auriga, Aur

Charioteer

Auriga (Figure 10) is a Ptolemaic constellation that Bayer lettered from Alpha to Psi.

Auriga, Latin for charioteer, is usually shown on stellar atlases as holding a whip and reins in one hand and a goat with two kids in the other. Several legends in Greek mythology are connected to this constellation. In one legend, the Charioteer is depicted as Erichthonius, son of Hephaestus, the god of fire and an early king of Athens. Ericthonius is credited with hitching four horses to a chariot, thus inventing the four-horse chariot, or *quadriga* in Latin. Zeus rewards him by placing his image among the stars. Another legend portrays him as Myrtilus (Myrtilos, Myrtilis), son of the god Hermes and charioteer of King Oenomaus of Elis, a city in southwestern Greece. Myrtilus is the odd man out in a love triangle involving the king's daughter, Hippodamia, and her lover, Pelops. The princess inveigles Myrtilus to arrange an accident to kill her father, who stands in the way of her marriage to Pelops. After Oenomaus's death, Pelops kills Myrtilus, who is placed among the stars by his divine father. Just before his death, Myrtilus curses Pelops and all his descendants. Pelops becomes king of Elis and forges the city-state into a mighty military power, eventually giving his name to the southern part of Greece — Peloponnesus. But some would say that Myrtilus's curse haunted the peninsula, leading eventually to internecine fighting among the city-states that culminated in the disastrous Peloponnesian War of the fifth century B.C., which, in turn, led to Greece's conquest by a foreigner, Philip of Macedon, in 338 B.C.

Another legend dealing with this constellation relates to the goat and kids in the Charioteer's hand. According to this myth, when Rhea saves her baby Zeus from being devoured by his father, Cronos, she brings her son to Crete. Far away from the wrath of his father, Zeus is suckled by Amalthea, a goat that had just given birth to two kids. When Zeus later wages war against his father and the other Titans, an oracle tells him that he must wear a goat's skin if he wishes to be victorious. (In an alternate version of the legend, the Oracle advises Zeus to carry a shield made of goat-skin.) Heeding this advice, Zeus defeats his enemies and forevermore a goat-skin shield, the aegis (αγις), derived from the Greek word for goat (αιξ), is one of his symbols. As a sign of gratitude, Zeus places the image of a goat and her two kids in the heavens.

Scholars believe that these two basically unrelated accounts—one concerning a charioteer and another concerning Zeus and goats — indicate that the constellation Auriga was probably an amalgam of two separate constellations that were somehow mixed together. Others have suggested that the constellation can be traced to the Middle East. A Babylonian text from the seventh century B.C. mentions GIS.GIGIR (Akkadian Narkabtu), the Chariot, a constellation located not far from Auriga in the vicinity of Perseus and Taurus. This may have been the source for Auriga, but as far as is presently known there are no Mesopotamian myths or legends related to either Auriga or the Chariot. In Auriga's place in the heavens, the Babylonians envisioned a constellation they called the Crook.

See Condos, *Star Myths,* pp. 49-54; Ridpath, *Star Tales,* pp. 31-33; Staal, *New Patterns in the Sky,* pp. 78-80; Reiner and Pingree, *Baby. Plan. Omens,* Part 2, 11-12; Hunger and Pingree, *Mul.Apin,* p. 137. For Zeus's war with the Titans, see Ara; see also Ursa Major for another chariot or wagon in the sky.

Lettered Stars - Auriga

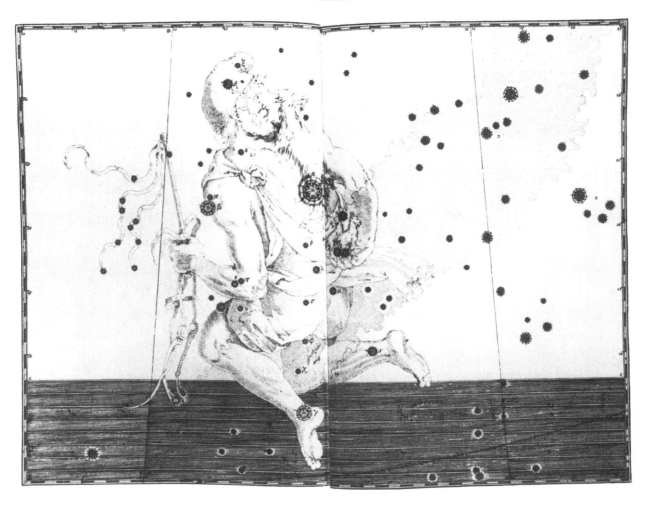

Figure 10. Bayer's Auriga. The ten stars in the whip should have been lettered Psi (ψ), but the letter was inadvertently omitted. The line A running through the zodiac is the Tropic of Cancer; B is the ecliptic; and C is the northern limit of the zodiac, 8° north of the ecliptic. The letter D is Perseus and E is the star Algol, Beta Per, in the head of Medusa, one of the Gorgons that Perseus slew.

Table 13. The lettered stars of Auriga

	MAGNITUDES			CATALOGUE NUMBERS				
Letter	Bayer	Lacaille	Visual	Flamsteed	Lacaille	Other	HR	HD
Alpha	1		0.08	13			1708	34029
Beta	2		1.90	34			2088	40183
Gamma+	2		1.65	23			1791	35497
Delta	4		3.72	33			2077	40035
Epsilon	4		2.99	7			1605	31964
Zeta	4		3.75	8			1612	32068
Eta	4		3.17	10			1641	32630
Theta	4		2.62	37			2095	40312
Iota	4		2.69	3			1577	31398
Kappa	4		4.35	44			2219	43039
Lambda	5		4.71	15			1729	34411
Mu	5		4.86	11			1689	33641
Nu	5		3.97	32			2012	39003
Xi	6		4.99	30			2029	39283
Omicron	6		5.47	27			1971	38104
Pi	6		4.26	35			2091	40239
Rho	6		5.23	20			1749	34759
Sigma	6		4.99	21			1773	35186
Tau	6		4.52	29			1995	38656
Upsilon	6		4.74	31			2011	38944
Phi	6		5.07	24			1805	35620
Chi	6		4.76	25			1843	36371
Psi^1+	6		4.91	46			2289	44537
Psi^2	6		4.79	50			2427	47174
Psi^3	6		5.20	52			2420	47100
Psi^4	6		5.02	55			2459	47914
Psi^5	6		5.25	56			2483	48682
Psi^6	6		5.22	57			2487	48781
Psi^7	6		5.02	58			2516	49520
Psi^8	6		6.30	60			2541	50037
Psi^8			6.48	61			2547	50204
Psi^9	6		5.87				2568	50658
Psi^{10}	6		4.90	16 Lyn			2585	50973
Omega+			4.94	4			1592	31647

The Lost, Missing, or Troublesome Stars of Auriga

Gamma, 23, HR 1791, 1.65V, *SA, HRP, BS.*

Like Ptolemy, Bayer noted that this star was synonymous with Beta Tau. For Bayer's duplicate stars, see Alpha And.

Psi^1, 46, HR 2289, 4.91V, *SA, HRP, BS,* H 108*.
Psi^2, 50, HR 2427, 4.79V, *SA, HRP, BS,* H 117*.
Psi^3, 52, HR 2420, 5.20V, *SA, HRP, BS,* H 116*.
Psi^4, 55, HR 2459, 5.02V, *SA, HRP, BS,* H 121*.
Psi^5, 56, HR 2483, 5.25V, *SA, HRP, BS,* H 123*.
Psi^6, 57, HR 2487, 5.22V, *SA, HRP, BS,* H 124*.
Psi^7, 58, HR 2516, 5.02V, *SA, HRP, BS,* H 126*.
Psi^8, 60, HR 2541, 6.30V, H 129*.
Psi^8, 61, HR 2547, 6.48V, *SA, HRP, BS,* H 129*.
Psi^9, HR 2568, 5.87V, *SA, HRP, BS,* H 132*.

Psi[10], 16 Lyn in Lynx, HR 2585, 4.90V, *HRP,* H 133*.

In his star list, Bayer described Psi as a group of ten 6th-magnitude stars, but he inadvertently omitted the letter Psi from his star chart. Flamsteed, working only with Bayer's charts, happened to include all but one of the Psi stars in his own catalogue but without designating any of them Psi. As far as he knew, Psi did not exist in Auriga (*BF,* p. 399). Bevis was among the first to assign indices to Psi. He labeled them as follows:

Psi-1, 46, BV 52.
Psi-2, 47, HR 2338, 5.90V, BV 53.
Psi-3, 50, BV 55.
Psi-4, 52, BV 56.
Psi-5, 51, HR 2419, 5.69V, BV 57.
Psi-6, 55, BV 59.
Psi-7, 56, BV 63.
Psi-8, 58, BV 64.
Psi-9, 59, HR 2539, 6.12V, BV 65.
Psi-10, 60, BV 66.

In Bevis's catalogue, BV 65 is labeled as Flamsteed's 58, a typographical error for 59; and on Tabula XII of his atlas, the positions of Flamsteed's 55 and 56 are incorrectly plotted. Argelander disagreed somewhat with Bevis's selections for Psi. His selections are those listed in the heading, above, and most authorities have accepted them. Argelander's Psi's, by and large, conform more closely to Bayer's chart than Bevis's. Since 60 and 61 are almost adjacent, Argelander felt that Bayer must have seen their combined light as one star, so he designated both of them Psi[8] (*UN,* pp. 18-19). See also the notes to HR 2541 and HR 2547 in *HRP* and *BS.*

Omega, 4, HR 1592, 4.94V, *SA, BS,* BV 4 as a, B 25*, BF 630.
f, 1 Aur in Perseus, HR 1533, 4.88V, BV 1, B 12*, BF 605.
g, 2, HR 1551, 4.78V, BV 2, B 15*, BF 610.

Bayer did not letter these three stars. Although he included them in his atlas, he considered them *informes* between Auriga and Perseus and left them unlettered. Flamsteed assigned the three letters but Baily removed them as superfluous. Bevis also removed Flamsteed's letters, but he relabeled Omega as a (see i Aql). Flamsteed's inclusion of Omega in this constellation poses an interesting and rather perplexing problem. Baily explained Flamsteed's confusion over Bayer's multiple letterings, like Psi, above, by noting that Flamsteed probably worked with one of the eight editions of Bayer's *Uranometria* that included only his charts, not his star lists. If this supposition is correct — and most of the evidence seems to support it — why did Flamsteed assign the letter Omega to his 4 Aur? The last Greek letter on Bayer's chart for Auriga is Chi while the last Greek letter on his star list is Psi. If Baily's supposition is correct, Flamsteed should have assigned Psi to 4 Aur, not Omega. Is it possible that Flamsteed just happened to pick Omega by chance instead of Psi? See *BF,* pp. 399-400; Warner, *Sky Explored,* p. 19.

Flamsteed assigned the letters f and g since Bayer's last capital Roman letter in Auriga was E, Medusa's head in Perseus. For Bayer's use of Roman capital letters, see P Cyg.

Bootes, Boo

Herdsman, Bear Keeper, Bear Driver

Bootes (Figure 11) is a Ptolemaic constellation that Bayer lettered from Alpha to Omega and from A to k.

Like other classical constellations, the legends and myths dealing with Bootes are many and varied. According to one legend, Zeus, the archetypical philandering husband, betrays his wife, Hera, with Callisto, the beautiful daughter of King Lycaon of Arcadia. A son, Arcas, is born of this dalliance, and King Lycaon takes the boy to raise. Hera, angered by the affair, turns Callisto into a bear. One day, Zeus visits Arcadia and invites himself to dinner at the king's residence. Uncertain whether the visitor is Zeus or an imposter, Lycaon decides to test his guest. He kills his grandson, Arcas, cuts him up, and serves him as the main course of the meal, thinking that if his guest is indeed Zeus, he will recognize the meat as that of his son. As might be expected from a god, Zeus recognizes Arcas and restores him to life, but not before he strikes Lycaon dead in retaliation for what he did to his son. Zeus then turns the care of his son over to Maia — a Pleiad, one of the seven daughters of the Titan Atlas — who raises him to manhood. One day, while hunting in the forest, Arcas comes upon a bear who happens to be Callisto. Unaware that the bear is his mother, he chases her into the temple of Zeus, where both of them are killed for violating the sanctity of the holy site. Zeus takes pity on them and places the bear and Arcas among the stars. Henceforth, Arcas came to be known as the Bear Keeper or Protector, Arctophylax, of the Great Bear — Ursa Major.

In an entirely different legend, Dionysus — the Roman Bacchus, god of wine — on one of his jaunts on Earth, is so impressed with the piety and goodness of Icarius of Attica that he rewards him by teaching him how to cultivate grapes and make wine. Icarius decides to spread his good fortune among his neighbors. He fills his ox-drawn wagon with wine skins and proceeds to distribute the wine. Unfortunately, his neighbors imbibe too much and fall into a drunken stupor. Thinking that he poisoned them, friends and relatives kill Icarius. When Icarius fails to return home, his daughter, Erigone, together with his faithful dog, Maera, searches high and low until they eventually find his body buried under a tree. Unable to restrain their grief, Erigone and Maera kill themselves. Zeus, taking pity on the family, places all three in the heavens — Icarius as Bootes, Erigone as Virgo, and Maera as Canis Minor. In the seventeenth century, the Polish astronomer Hevelius, aware perhaps of the legend of Icarius and his faithful hound, devised the constellation Canes Venatici, the Hunting Dogs, and placed their leash in the hand of Bootes.

A third legend associated with this constellation recounts the story of Philomelus, the son of Demeter — the Roman Ceres, goddess of grain and agriculture. According to this myth, Philomelus purchases two oxen and invents the plow. So impressed is his goddess mother with his inventiveness that she calls him Bootes and places him among the stars. The name Bootes probably comes from the Greek *boe* (βοη, cry out or shout) or possibly from *bous* (βους, cow or ox) and refers to the shouting of a plowman to his team of oxen or to Arcas, or Arctophylax, who is yelling and shouting to control the Bear.

In his chart of Bootes, Bayer depicted the Herdsman holding a farmer's sickle in one hand and a shepherd's crook in the other. These bucolic implements symbolize the agricultural and pastoral nature of the constellation. Arcadia, the epitome of rural happiness and contentment, was named in honor of its most famous son, Arcas, before his resurrection in the heavens as Bootes.

Some scholars have suggested that Bootes can be traced back to the ancient Middle East, possibly because the Mesopotamians saw in its place in the night sky the constellation SU.PA or SUDUN (Akkadian Niru) — the Yoke — a distant connection perhaps to the legend of Philomelus with his oxen and plow. Otherwise there is no Mesopotamian myth or legend that relates to this constellation.

See Condos, *Star Myths*, pp. 55-60; Ridpath, *Star*

Tales pp. 34-35; Staal, *New Patterns in the Sky,* pp. 152-56; Hunger and Pingree, *Mul.Apin,* p. 137; Reiner and Pingree, *Baby. Plan. Omens,* Part 2, p. 15. See also Canes Venatici and Ursa Major.

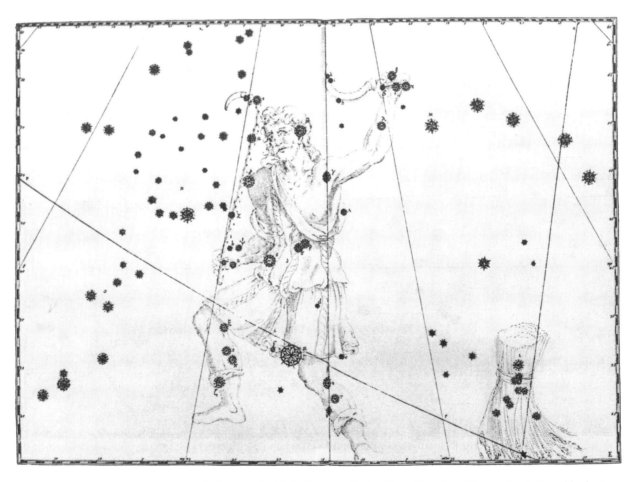

Figure 11. Bayer's Bootes. The sheaf of wheat, marked L, is the constellation Coma Berenices. Bayer refers to it on this chart as Azimeth, a bundle of grain *(spicarum manipulus),* or Coma Berenices. The line P is the Tropic of Cancer and Q is the equinoctial colure. The letter M is the tail of the Big Bear; N is the Northern Crown; and O is the head of the Serpent.

Table 14. The lettered stars of Bootes

		MAGNITUDES			CATALOGUE NUMBERS			
Letter	**Bayer**	**Lacaille**	**Visual**	**Flamsteed**	**Lacaille**	**Other**	**HR**	**HD**
Alpha	1		-0.04	16			5340	124897
Beta	3		3.50	42			5602	133208
Gamma	3		3.03	27			5435	127762
Delta	3		3.47	49			5681	135722
Epsilon	3		2.37c	36			5505/6	129988/9
Zeta	3		3.78c	30			5477/8	129246/7
Eta	3		2.68	8			5235	121370
Theta	4		4.05	23			5404	126660
Iota	4		4.75	21			5350	125161
Kappa^{1,2}+	4		4.44c	17			5328/9	124674/5
Lambda	4		4.18	19			5351	125162
Mu^{1,2}+	4		4.33c	51			5733/4	137391/2
Nu^1+	4		5.02	52			5763	138481
Nu^2			5.02	53			5774	138629
Xi	4		4.55	37			5544	131156
Omicron	4		4.60	35			5502	129972
Pi^{1,2}+	4		4.53c	29			5475/6	129174/5
Rho	4		3.58	25			5429	127665
Sigma	4		4.46	28			5447	128167
Tau	4		4.50	4			5185	120136
Upsilon	4		4.07	5			5200	120477
Phi	5		5.24	54			5823	139641
Chi	5		5.26	48			5676	135502
Psi	5		4.54	43			5616	133582
Omega	5		4.81	41			5600	133124
A+	5		4.81				5361	125351
b	6		5.67	46			5638	134320
c	6		4.93	45			5634	134083
d	6		4.83	12			5304	123999
e+	6		4.91	6			5201	120539
f	6		5.39	22			5405	126661
g	6		5.59	24			5420	127243
h+	6		5.74	38			5533	130945
i+	6		4.76	44			5618	133640
k+	6		5.57	47			5627	133962

The Lost, Missing, or Troublesome Stars of Bootes

Kappa^1, Kappa^2, 17, HR 5328/9, 6.69V, 4.54V, comb. mag. 4.44V, *SA, HRP* as Kappa, *BS*, ADS 9173.

Bayer described Kappa as a single 4th-magnitude star. Later astronomers added the indices to identify the components of this double star.

Mu^1, Mu^2, 51, HR 5733/4, 4.31V, 6.50V, comb. mag. 4.33V, *SA, HRP* as Mu, *BS*, ADS 9626.

Bayer described Mu as a single 4th-magnitude star. Later astronomers added the indices to identify the components of this double star.

Nu^1, 52, HR 5763, 5.02V, *SA, HRP* as Nu, *BS*, HF 53 as Nu, BV 48 as Nu, BF 2116*, H 139 as Nu.

Nu², 53, HR 5774, 5.02V, *SA, HRP* as Nu, *BS,* HF 54 as Nu, BV 49 as Nu, BF 2122*, H 139 as Nu.

Bayer described only one 4th-magnitude star as Nu, but Flamsteed designated these two adjacent stars, his 52 and 53, as Nu-1 and Nu-2. Bevis believed Bayer must have meant the combined light of both stars, so he designated them both simply as Nu, as did Argelander (*UN,* p. 30). Like Ptolemy, Bayer noted that Nu was synonymous with Psi Her. For Bayer's duplicate stars, see Alpha And.

Pi¹, Pi², 29, HR 5475/6, 4.94V, 5.81V, comb. mag. 4.53V, *SA, HRP* as Pi, *BS,* ADS 9338.

Bayer described Pi as a single 4th-magnitude star. Later astronomers added the indices to identify the components of this double star.

A, HR 5361, 4.81V, *SA, HRP,* BV 15*, BF 1959*, BAC 4747*.

Although Flamsteed observed this star, it, along with 457 others, was inadvertently omitted from his printed catalogue and hence has no Flamsteed number. Bevis was among the first to identify this star with Bayer's A, and Baily included it in his revised edition of Flamsteed's catalogue (*BF,* p. 392). See also Baily, "A Catalogue of 564 Stars."

b, 46, HR 5638, 5.67V, *SA, HRP.*

c, 45, HR 5634, 4.93V, *SA, HRP.*

d, 12, HR 5304, 4.83V, *SA, HRP.*

e, 6, HR 5201, 4.91V, *SA, HRP,* BF 1900*, H 10*.

Bayer described e as a 6th-magnitude star. Flamsteed designated his 10 (HR 5255, 5.76V, HF 11 as e, BF 1920) as e, but Baily switched e to 6 since he felt it conformed better to Bayer's chart and Bayer's magnitude. Flamsteed had estimated 6's magnitude as 5½ but 10's magnitude as 7, one gradation dimmer than Bayer's estimate of e; whereas his 6 was only one-half gradation brighter than Bayer's estimate of e. Both Argelander (*UN,* p. 27) and Heis agreed with Baily that Bayer's e was synonymous with 6.

f, 22, HR 5405, 5.39V, *SA, HRP.*

g, 24, HR 5420, 5.59V, *SA, HRP.*

h¹, 33, HR 5468, 5.39V, HF 34, B 250*, BF 2004, H 74.
h², 38, HR 5533, 5.74V, *SA* as h, *HRP* as h, HF 39 as h, BV 14 as h, B 301*, BF 2032 as h, H 96 as h.

Bayer described h as a single 6th-magnitude star. In the pirated edition of Flamsteed's catalogue, 38 is designated as h. For some inexplicable reason, however, Flamsteed's "Catalogus Britannicus" of 1725 assigns h-1 to 33 and h-2 to 38, although the two stars are 2½° apart. The latter is Bayer's h; the former is one of his *informes.* Bayer included it in his atlas but left it unlettered since he considered it outside the figure of the Herdsman. Bevis and Baily considered h-1 one of Flamsteed's superfluous letters and dropped it. See i Aql.

i, 44, HR 5618, 4.76V, *SA, HRP,* HF 45*, BV 20*, B 358*, BF 2062*.

Flamsteed's manuscript catalogue, as well as the pirated edition of his catalogue, equates 44 with Bayer's i. However, the letter is omitted from Flamsteed's catalogue of 1725. Bevis correctly identified 44 with i.

j

Bayer did not use this letter.

k, 47, HR 5627, 5.57V, *SA, HRP,* HF 47*, BV 18*, B 370*, BF 2067*.

In the pirated edition of Flamsteed's catalogue, Bayer's k is equated with 47 and both its equatorial and ecliptical coordinates are listed, but in Flamsteed's "Catalogus Britannicus" of 1725, the letter k is omitted and only 47's right ascension is noted. Bevis correctly identified 47 as Bayer's k.

CAELUM, CAE

Sculptor's or Engraver's Burin or Chisel

Lacaille devised and lettered Caelum (figures 2c and 2d) from Alpha to Zeta. In the chart accompanying his preliminary catalogue, which was written in French, Lacaille called this constellation les Burins and depicted two crossed burins tied together with a ribbon. He explained that the constellation represented the Engraver's Burin (Le Burin du Graveur) and that the figure on the chart depicted a burin and a graver (un burin & une echope). In his catalogue of 1763, which is written in Latin, the same chart with the two burins is shown but with the inscription, in singular, Caelum Scalptorium, Engraving or Engraver's Burin. Several years later, Bode sought to reconcile the picture with the inscription by calling the constellation Caela Scalptoris, the Engraver's Burins, but the name did not survive in the literature.

In devising constellations to describe the stars in the southern skies, Lacaille intended to honor various aspects of the arts and sciences of the Enlightenment that was spreading throughout Western Europe in the eighteenth century. Caelum was meant to be a celestial monument to the artistry and skill of engravers.

See Lacaille, "Table des Ascensions," pp. 588-89; Bode, *Allgemeine Beschreibung der Gestirne,* p. 78.

Table 15. The lettered stars of Caelum

	MAGNITUDES			CATALOGUE NUMBERS				
Letter	Bayer	Lacaille	Visual	Flamsteed	Lacaille	Other	HR	HD
Alpha		5	4.45		360		1502	29875
Beta		5	5.05		361		1503	29992
Gamma^1+		5	4.55		385		1652	32831
Gamma^2		6	6.34		386		1653	32846
Delta		5	5.07		348		1443	28873
Zeta		6	6.37		370		1539	30608
Nu+			6.07			B 18	1557	30985

The Lost, Missing, or Troublesom Stars of Caelum

Gamma[1], HR 1652, 4.55V, *SA* as Gamma, *HRP* as Gamma, *BS,* CA 385 as Gamma, BR 858*, L 1712*, BAC 1573*, G 28 as Gamma.

Gamma[2], HR 1653, 6.43V, *BS,* CA 386 as Gamma, BR 860*, L 1713*, BAC 1574*.

Lacaille split this close pair with his telescope and called each component Gamma, but Gould, after noting that both stars appeared as one image to the naked eye, designated only the brighter one Gamma.

Epsilon

Lacaille did not use this letter in Caelum.

Nu, HR 1557, 6.07V, *SA,* CA 373, BR 799, L 1626, BAC 1506, B 18*, G 18.

Although Lacaille observed and catalogued this star, he left it unlettered. The letter was added by Bode, who extended Lacaille's letters to Rho, but with the exception of Nu, the other letters are no longer in use. Gould dropped Nu because of its dimness.

Camelopardalis, Cam

Giraffe

Camelopardalis (Figure 12) is a modern constellation devised after the appearance of Ptolemy's catalogue with its classic forty-eight constellations. It first appeared on Petrus Kaerius's globe of 1613. Warner believed that, based on its style and appearance, the globe was probably based on the work of Petrus Plancius. Bartsch also depicted Camelopardalis on his planisphere of 1624 and noted that it was the giraffe, "an animal as tall as a camel, the color of a panther, with bovine feet, that was recently devised from unformed *(informes)* stars about the pole between Cassiopeia's footstool and Auriga." He asserted that for him it represented the camel that brought Rebecca (Camelus Rebeccae) to Isaac (Genesis 25).

Plancius, whom Warner credited with first forming this constellation, was a Calvinist minister who certainly approved of placing Biblical symbols in the heavens. He himself had devised, among others, Noah's Dove (Columba); Sampson's Bees (see c Ari); and Monoceros, which he mistook for the Old Testament *re'em,* a giant prehistoric ox. Camelopardalis is a giraffe, not Rebecca's camel, and it is unlikely that Plancius, a learned clergyman trained in the classics, would confuse the two beasts. Instead of commemorating the camel as Bartsch suggested, Plancius more likely was commemorating the new creatures that Europeans were discovering during the Age of Exploration. He himself was deeply involved, both personally and financially, in the Dutch search for the northeast passage and in the first Dutch voyages to the East Indies.

Plancius may also have been indirectly responsible for forming the twelve new southern constellations that Pieter Dircksen Keyzer (Petrus Theodorus) and his colleague Frederick de Houtman devised. Keyzer had attended the school that Plancius conducted for Dutch seamen in astronomy, geography, and navigation, and a preliminary copy of Keyzer's catalogue of southern stars was probably forwarded to Plancius. Among the twelve new constellations in Keyzer's catalogue were the Flying Fish (Piscis Volans), Toucan (Tucana), Peacock (Pavo), Bird of Paradise (Apus Indica), Heron (Grus), and Dolphin (Dorado). These, together with the Giraffe (Camelopardalis), were among the new, exotic creatures that Plancius and his contemporaries were encountering in the sixteenth century.

Camelopardalis is derived in both Hebrew (נמרי גמל, *gamal nimeri*) and Greek from words meaning spotted camel. It is not mentioned in either the Old or New Testament. It first appears in Biblical literature in the third century B.C. in the Septuagint, where the editors, in their supposed haste to translate the Bible from Hebrew to Greek in seventy days for Ptolemy II of Egypt, erroneously translated the Hebrew *zemer* (chamois, deer, זמר) in Deuteronomy 14:5 as giraffe (καμηλοπαρδαλιν). The error was later copied into the Latin Vulgate of Saint Jerome as camelopardalum (the accusative case of *camelopardalus,* a second declension noun; *camelopardalis* is third declension). As a result, some Hebrew dictionaries still list giraffe as a synonym for *zemer* although there is no linguistic basis for this assertion.

Camelopardalis was included in the works of Hevelius and Flamsteed. The latter also included in his catalogue of 1725 the other "modern" constellations of Canes Venatici, Lacerta, Leo Minor, Lynx, Monoceros, Sextans, and Vulpecula although they were omitted from his first catalogue, which Halley had printed in 1712 against Flamsteed's wishes (see Introduction). Flamsteed or his editors — he died in 1719 — probably included these new constellations in the later catalogue in order to distinguish it as much as possible from Halley's edition. The authorized version contains about 250 additional stars.

Baily lettered Camelopardalis in the *British Association Catalogue (BAC)* from Alpha to Gamma. All of Baily's lettered stars are listed in the *BS,* and there are no missing or problem lettered stars in the constellation.

See Warner, *Sky Explored,* p. 204; Bartsch, *Usus*

Astronomicus, p. 52; Masselman, *Cradle of Colonialism,* pp. 83-86, 90-91, 102, and *passim*; Knobel, "Frederick de Houtman's Catalogue;" Wagman, "Who Numbered Flamsteed's Stars?" For Bartsch's role in depaganizing the heavens, see c Ari.

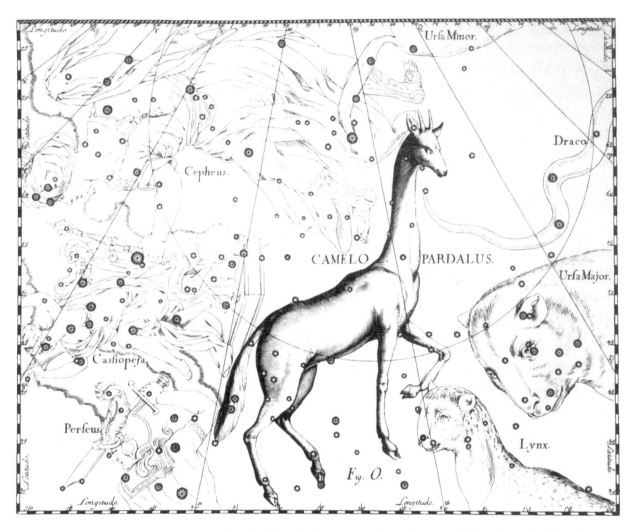

Figure 12. Hevelius's Camelopardalis. Below the Giraffe is Lynx, one of the twelve constellations devised by Hevelius.

Table 16. The lettered stars of Camelopardalis

	MAGNITUDES			CATALOGUE NUMBERS				
Letter	Bayer	Lacaille	Visual	Flamsteed	Lacaille	Other	HR	HD
Alpha			4.29	9		BAC 1474	1542	30614
Beta			4.03	10		BAC 1536	1603	31910
Gamma			4.63			BAC 1137	1148	23401

CANCER, CNC

Crab

Cancer (Figure 13) is a Ptolemaic constellation that Bayer lettered from Alpha to Omega and from A to d. Cancer dates back at least to the seventh century B.C. At that time it was called in Sumerian NAGAR or AL.LUL, in Akkadian Alluttu, which translates as crab. It may have referred to the entire constellation as we know it today or just to Praesepe, the Beehive Cluster, in the center of the constellation.

Although there are at present no known Mesopotamia legends concerning the celestial Crab, one did arise in ancient Greece that involves the legendary hero Hercules. Like so many other myths, those associated with Hercules are many and varied. According to one account, Zeus, always on the lookout for a pretty woman, seduces Alcmene, granddaughter of Perseus, by assuming the shape of her husband, Amphitryon of Thebes. When Alcmene gives birth to a son, her husband apparently adopts the baby and names it Alcides, after his own father. In the meantime, Hera, Zeus's wife, develops an abiding hatred for the baby, a living reminder of her husband's infidelity. When Alcides is barely a year old, Hera sends two deadly snakes to destroy the child, but little Alcides, displaying the strength that he would later become famous for, kills both serpents with his bare hands.

In adulthood, Alcides defeats the Minyans, the traditional enemies of the people of Thebes, and is rewarded with the hand of Princess Megara. He loves Megara deeply, and she bears him three sons. But Hera's hatred for Alcides has not waned, and she casts a spell on him that turns him into a raging mad man. Totally unaware of his actions, Alcides slays his beloved family. When his sanity returns, he is shocked by what he has done. Grief stricken and seeking some form of penance, he seeks help from the Oracle of Delphi. Probably under the influence of Hera, the Oracle tells him that to atone for his unspeakable crime, he must submit to his cousin — some accounts say his half-brother — Eurystheus, King of Mycenae. At this point in his life, Alcides adopts the name Hercules — bound to Hera. Eurystheus, at the urging of Hera, imposes twelve seemingly impossible labors on Hercules. The second of these tasks requires Hercules to destroy the monstrous water snake, Hydra. Hydra is a seemingly indestructible creature that lives in the swamps of Lerna. She is a fierce fighter with nine heads, one of which is immortal. Hercules' original plan of action is to chop off Hydra's heads, but as soon as he cuts one off, two more grew back in its place. While Hercules is furiously battling Hydra, Hera encourages a giant crab that dwells in the swamp to join in the fray and attack Hercules. But the only offensive the crab can mount is to bite the foot of Hercules, who, in a fit of anger, steps on the crab and crushes it. He then employs a new mode of attack against Hydra. As soon as he chops off one of the Water Snake's heads, his nephew and faithful companion, Iolaus, applies a burning torch to the stump, cauterizing and sealing the wound, thus preventing any regrowth. Hercules finally succeeds in cutting off the ninth and immortal head and buries it. Although frustrated and angered by the victory of Hercules, Hera rewards the crab's selfless sacrifice on her behalf and places the Crab in the heavens among the stars.

Yet another Greek legend is associated with the constellation Cancer. In this story of the battle between the gods and the Titans for the control of the universe, Hephaestus, Dionysus, and the Satyrs ride to the battle on asses. When from a great distance the Titans hear the braying of the asses, they think they are about to be attacked by fearsome monsters and flee the battlefield. To honor the asses, the gods set them in the sky. They are depicted as two stars — the Northern Ass (Gamma Cnc) and the Southern Ass (Delta Cnc) — on either side of the star cluster in the center of Cancer, Praesepe, which can be translated from the Latin as either Stall or Hive. On some star charts Gamma and Delta are shown as two asses eating from the Manger or Stall.

Some scholars have suggested that the Crab's position in the sky has special astronomical significance. About four thousand years ago, when the constellation was probably first devised, Cancer was the first constellation that the sun passed through after the summer solstice, June 21-22, the sun's highest or most northern point in the sky. For several days after a solstice, the sun just seems to hesitate or stand still before beginning its journey toward the south. Actually the sun is still moving, very, very slowly, but to the ancients, who were without any of the instruments available to modern astronomers, its slight movement was undetectable. They felt that the sun's hesitation or apparent uncertainty resembled the movement of a crab. Therefore, the Mesopotamians gave the name Cancer to the stars that occupy the area of the sky immediately following the summer solstice.

See O'Neil, *Time and the Calendar,* p. 56; van der Waerden, "Babylonian Astronomy. II. The Thirty-Six Stars," p. 13; van der Waerden, "History of the Zodiac," pp. 219-20; Langdon, *Babylonian Menologies,* p. 5; Reiner and Pingree, *Baby. Plan. Omens,* Part 2, pp. 10, 13; Hunger and Pingree, *Mul.Apin,* 20 and *passim;* Edith Hamilton, *Mythology,* pp. 224-31; Condos, *Star Myths,* pp. 61-64; Ridpath, *Star Tales,* pp. 37-38; Staal, *New Patterns in the Sky,* pp. 145-48. For the battle between the gods and the Titans, see Ara; for the twelve labors of Hercules, see Vulpecula.

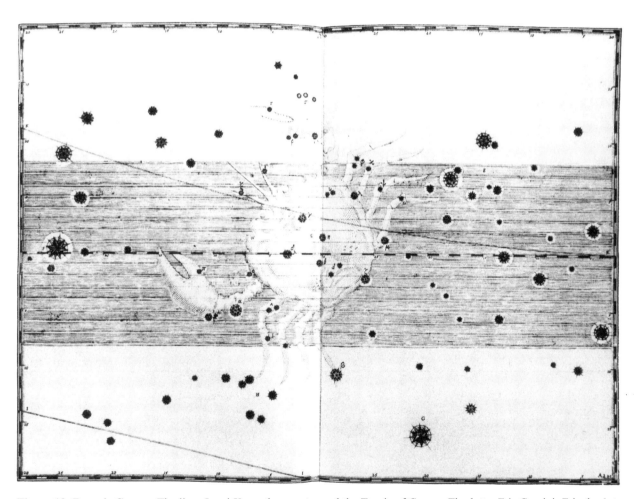

Figure 13. Bayer's Cancer. The lines I and K are the equator and the Tropic of Cancer. The letter E is Gemini; F is the 1st-magnitude star Regulus, Alpha Leo; G is the 1st-magnitude star Procyon, Alpha CMi; and H is the head of Hydra. The very faint object to the left of the letter Epsilon (ε) is Praesepe, the Beehive Cluster. The shaded area is the zodiac.

Table 17. The lettered stars of Cancer

	MAGNITUDES			CATALOGUE NUMBERS				
Letter	Bayer	Lacaille	Visual	Flamsteed	Lacaille	Other	HR	HD
Alpha[1]+			5.41	60			3550	76351
Alpha[2]	3		4.25	65			3572	76756
Beta	3		3.52	17			3249	69267
Gamma	4		4.66	43			3449	74198
Delta	4		3.94	47			3461	74442
Epsilon+	4		6.30	41			3429	73731
Zeta[1,2]+	4		4.67c	16			3208/9/10	68257/5/6
Eta	5		5.33	33			3366	72292
Theta	5		5.35	31			3357	72094
Iota[1]+	5		4.09c	48			3474/5	74738/9
Iota[2]			5.39	57			3532	75959
Kappa	5		5.24	76			3623	78316
Lambda	5		5.98	19			3268	70011
Mu[1]+			5.99	9			3169	66875
Mu[2]	5		5.30	10			3176	67228
Nu	6		5.45	69			3595	77350
Xi	6		5.14	77			3627	78515
Omicron[1]+	6		5.20	62			3561	76543
Omicron[2]			5.67	63			3565	76582
Pi[1]+			6.51	81			3650	79096
Pi[2]	6		5.34	82			3669	79554
Rho[1]+	6		5.95	55			3522	75732
Rho[2]	6		5.22	58			3540	76219
Sigma[1]+	6		5.66	51			3519	75698
Sigma[2]	6		5.45	59			3555	76398
Sigma[3]	6		5.20	64			3575	76813
Tau	6		5.43	72			3621	78235
Upsilon[1]+	6		5.75	30			3355	72041
Upsilon[2]	6		6.36	32			3369	72324
Phi[1]+	6		5.57	22			3304	71093
Phi[2]	6		5.56c	23			3310/1	71150/1
Chi	6		5.14	18			3262	69897
Psi+	6		5.73	14			3191	67767
Omega[1]+	6		5.83	2			3124	65714
Omega[2]			6.31	4			3132	65856
A[1]+	6		5.62	45			3450	74228
A[2]	6		5.87	50			3481	74873
b	6		5.66	49			3465	74521
c+	6		5.88	36			3406	73143
d[1]+	6		5.84	20			3284	70569
d[2]	6		6.14	25			3299	71030

The Lost, Missing, or Troublesome Stars of Cancer

Alpha¹, 60, HR 3550, 5.41V, HF 56, BV 74, BF 1259, H 66*.
Alpha², 65, HR 3572, 4.25V, *SA* as Alpha, *HRP* as Alpha, *BS* as Alpha, HF 61 as Alpha, BV 78 as Alpha, BF 1268 as Alpha, H 74*.

Bayer's *Uranometria* describes Alpha as a single 3rd-magnitude star, and Halley's edition of Flamsteed's catalogue equates 65 with Alpha. In Flamsteed's 1725 catalogue, however, these two neighboring stars, 60 and 65, are designated Alpha-1 and Alpha-2. Bevis felt only the brighter star, 65, was Bayer's Alpha and Baily agreed. Baily reasoned that the disparity between Flamsteed's magnitude for this star, 6, and Bayer's magnitude, 3, was so great that Alpha-1 could not possibly be synonymous with Alpha. Argelander, unable to see 60, agreed (*UN,* p. 64), but Heis observed both stars with his naked eye and restored Alpha to 60.

It should be noted that Flamsteed's catalogue of 1725 describes 60's magnitude as 4½, but when Flamsteed observed the star he noted its magnitude as 6, the figure Baily accepted since it more nearly matches the star's actual magnitude.

Epsilon, 41, HR 3429, 6.30V, *SA, HRP, BS,* BV 46*, B 103*, BF 1217*, BAC 2922*, H 47*.

Since Bayer described Epsilon as a 4th-magnitude "patch of nebulosity in the chest" of the Crab, he meant the entire Beehive Cluster, M 41, Praesepe, rather than a particular star. Flamsteed, using a telescope, resolved the stars in the cluster and equated Epsilon with his 41, which he described as a 7th-magnitude star in the midst of the nebulosity. Since then, astronomers have wavered between Bayer's and Flamsteed's positions. Bevis and Bode agreed with Flamsteed. Baily felt that Bayer intended the letter Epsilon to apply to the whole cluster, but he went along with Flamsteed's decision to equate Epsilon with a single star, his 41. Argelander, on the other hand, felt that Epsilon applied to the entire cluster, and he therefore left out its Flamsteed number in his catalogue (*UN,* p. 63). Heis compromised: he equated Epsilon with 41 but described it, nonetheless, as *cumulus,* a mass of stars.

Zeta¹, 16, HR 3208/9, 5.44V, 6.20V, *SA, HRP* as Zeta, *BS.*
Zeta², 16, HR 3210, 6.01V, *SA, HRP* as Zeta, *BS.*

Bayer described Zeta as a single 4th-magnitude star. Later astronomers added the indices to identify the components of this triple star, ADS 6650. The combined magnitude of all three is 4.67V.

Iota¹, 48, HR 3474/5, 6.57V, 4.02V, comb. mag. 4.09V, *SA, HRP* as Iota, *BS* as Iota, HF 45 as Iota, BV 38 as Iota, BF 1230 as Iota, H 53 as Iota.
Iota², 57, HR 3532, 5.39V, *SA,* BV 47 as Sigma-2, BF 1252 as Sigma², H 57.

Bayer's *Uranometria* describes Iota as a single 5th-magnitude star, and Halley's edition of Flamsteed's catalogue specifically designates 48 as Bayer's Iota. But for some inexplicable reason, Flamsteed's "Catalogus Britannicus" of 1725 labels his 57 as Iota-2 although it is 3° from his 48, Iota-1, which is the star synonymous with Bayer's Iota. Bevis and Baily, noting the apparent error, dropped Iota from 57 and relabeled it Sigma² (see Sigma). Although Tirion labeled 57 as Iota² in his *SA,* he removed its letter in *Uranometria 2000.0* and in the second edition of *SA.*

Mu¹, 9, HR 3168, 5.99V, *BS,* HF 8, BV 7, BF 1131*, H 12*.
Mu², 10, HR 3176, 5.03V, *SA* as Mu, *HRP* as Mu, *BS,* HF 9 as Mu, BV 11 as Mu, BF 1133*, H 13*.

Bayer described Mu as a single 5th-magnitude star. Halley's pirated edition of Flamsteed's catalogue equates 10 with Mu. Flamsteed's catalogue of 1725, however, labels these two neighboring stars, 9 and 10, as Mu-1 and Mu-2. Bevis designated only the brighter star, 10, as Mu, and Argelander agreed. Unable to see 9, Argelander felt that 10 must have been the star Bayer described as Mu (*UN,* p. 63). Heis, though, observed both stars and restored 9's letter.

Omicron¹, 62, HR 3561, 5.20V, *SA, HRP* as Omicron, *BS,* HF 58*, BV 71 as Omicron, BF 1261*, H 69 as Omicron.
Omicron², 63, HR 3565, 5.67V, *SA, BS,* HF 59*, BV 72, BF 1262*, H 71.

Bayer described Omicron as a single 6th-magnitude star, but Flamsteed designated these two neighboring stars, 62 and 63, as Omicron-1 and Omicron-2. Bevis and Argelander, though, felt that 62, the brighter star, was Bayer's Omicron (*UN,* p.64). But Baily agreed with Flamsteed, as have most modern astronomers.

Pi¹, 81, HR 3650, 6.51V, *BS,* BF 71 as Pi, BV 83, B 212*, BF 1300*, H 89.
Pi², 82, HR 3669, 5.34V, *SA* as Pi, *HRP* as Pi, *BS,* HF 72, BV 85, B 216*, BF 1305*, H 91 as Pi.

Bayer described Pi as a single 6th-magnitude star, and Flamsteed designated his 81 as Pi. Bode thought either 81 or 82 could be Pi, so he labeled them Pi-1 and Pi-2. Baily believed that 82 was Bayer's Pi, but since there appeared to be some doubt among astronomers, he designated 81 and 82 as Bode had done. Argelander, unable to see 81, felt that Bayer must have meant 82 (*UN,* p. 64). Although Heis observed both stars, he agreed with Argelander that only 82 was synonymous with Bayer's Pi. Bevis's atlas (Tabula XXV) shows 81 and 82 with the letter Pi placed between them, but his catalogue erroneously designates 83 (HD 80218, 6.60V, BV 86) as Pi.

Rho¹, 55, HR 3522, 5.95V, *SA, HRP, BS,* BF 1250 as Rho², H 61*.
Rho², 58, HR 3540, 5.22V, *SA, HRP, BS,* BF 1255 as Rho³, H 64*.

Bayer noted that Rho was a couple of 6th-magnitude stars in the lower part of the Crab's right claw. Since there are two pairs in the area, Flamsteed, not certain which one Bayer meant, designated his stars as follows:

53, Rho-1, HR 3521, 6.23V, BF 1247 as Rho¹.
55, Rho-2, HR 3522, 5.95V, BF 1250 as Rho².
56, Rho-3, nonexistent, BF, Table B, p. 645.
58, Rho-4, HR 3540, 5.22V, BF 1255 as Rho³.
67, Rho-5, HR 3589, 6.07V, BF 1272.
70, Rho-6, HR 3601, 6.38V, BF 1279.

The two pairs that Flamsteed observed are: (1) 55 and 58, with 53 adjacent to 55, and (2) 67 and 70. Star 56 is nonexistent, its coordinates arising from a 1^m clock error in determining its time of transit. Baily felt that 53, together with the adjacent 55, formed one half of Bayer's pair and 58 the other half, and that they should be designated Rho¹, Rho², and Rho³. Argelander, unable to see 53 with his naked eye, dropped its letter and relettered the remaining two as in the heading above (*UN,* p. 64).

Sigma¹, 51, HR 3519, 5.66V, *SA, HRP, BS,* HF 48*, BF 1245, H 60*.
Sigma², 59, HR 3555, 5.45V, *SA, HRP, BS,* HF 55*, BF 1256, H 67*.
Sigma³, 64, HR 3575, 5.20V, *SA, HRP, BS,* HF 60*, BF 1265, H 75*.
Sigma⁴, 66, HR 3587, 5.82V, HF 62*, BF 1269, H 77.

Bayer noted that Sigma was a group of three 6th-magnitude stars. Noticing four stars in the area, Flamsteed designated his 51, 59, 64, and 66 as Sigma-1, Sigma-2, Sigma-3, and Sigma-4. Bevis and Baily, however, believed another group of stars, about 2° south, was synonymous with Bayer's three Sigma's and should be designated as follows:

Sigma¹, 46, HR 3464, 6.13V, BV 32, BF 1225.
Sigma², 57, HR 3532, 5.39V, BV 47, BF 1252.
Sigma³, 61, HR 3563, 6.29V, BV 54, BF 1260.

Argelander, however, felt that Flamsteed's group without 66, which he could not see, was equivalent to Bayer's three Sigma's (*UN,* p. 64). Although Heis saw all four of Flamsteed's stars, he agreed with Argelander's designations. A comparison between Bayer's chart and modern atlases indicates that Argelander was probably correct. On Bayer's chart, the distance between Tau and the most easterly of the Sigma group is about 3°. On modern charts, the distance between Tau and 64, the most easterly of the Argelander group, is also about 3°, but the distance between Tau and 61, the most easterly of Baily's group, is only about 2°.

Upsilon¹, 30, HR 3355, 5.75V, *SA, HRP, BS,* BF 1187 as Upsilon³, H 35*.
Upsilon², 32, HR 3369, 6.36V, *SA, HRP, BS,* BF 1191, H 39*.

Bayer noted that Upsilon was a couple of 6th-magnitude stars. Not sure which pair of stars Bayer meant, Flamsteed selected two pairs and designated all four stars as Upsilon:

24, Upsilon-1, HR 3312/3, 7.02V, 7.81V, comb. mag. 6.58V, BF 1175 as Upsilon¹.
28, Upsilon-2, HR 3329, 6.10V, BF 1181 as Upsilon².

and

30, Upsilon-3, HR 3355, 5.75V, BF 1187 as Upsilon³.
32, Upsilon-4, HR 3369, 6.36V, BF 1191.

Baily felt that 24 was the westerly star of Bayer's pair and that either 28 or 30 was the easterly; consequently, he designated these three as Upsilon¹, Upsilon², and Upsilon³. Argelander, on the other hand, asserted that 30 and 32 were the stars Bayer meant, probably because he could see neither 24 nor 28 with his naked eye (*UN,* p. 63).

Phi¹, 22, HR 3304, 5.57V, *SA, HRP, BS.*
Phi², 23, HR 3310/1, 6.32V, 6.30V, *SA, HRP, BS.*

Bayer described Phi as a couple of 6th-magnitude stars. Flamsteed designated three stars, his 22, 23, and 26, as Phi-1, Phi-2, and Phi-3, but Baily pointed out that 26 does not exist. See 26 Cnc in Part Two.

Psi¹, 13, HD 67690, 6.41V, HF 11 as Psi, BV 9, BF 1140*.
Psi², 14, HR 3191, 5.73V, *SA* as Psi, *HRP* as Psi, *BS,* HF 12 as Psi, BV 10 as Psi, BF 1141*, H 15 as Psi.

Bayer described only one 6th-magnitude star as Psi. But in Halley's edition of Flamsteed's catalogue, two neighboring stars, 13 and 14, are designated as Psi. Flamsteed's 1725 catalogue not only designates 13 and 14 as Psi-1 and Psi-2 but also designates 15 (HR 3215, 5.64V, BF 1147) as Psi-3, although it is over 4° away from the other two. It was erroneously included with this group because it is Psi Gem, and it was somehow confused with Psi Cnc (see Psi Gem). Bevis corrected this by removing the Psi's from 13 and 15, the latter because it is Psi Gem and the former because of its dimness. Baily, uncertain which was Bayer's Psi, designated 13 and 14 as Psi¹ and Psi², but Argelander felt that 14 alone was Psi since he could not see 13 with his naked eye (*UN,* p. 63).

Omega¹, 2, HR 3124, 5.83V, *SA* as Omega, *HRP* as Omega, *BS,* HF 2 as Omega, BV 1 as Omega, BF 1113*, H 4 as Omega.
Omega², 4, HR 3132, 6.31V, *BS,* HF 4 as Omega, BV 2, BF 1116*, H 7.

Bayer described Omega as a single 6th-magnitude star, but Flamsteed designated these two neighboring stars, 2 and 4, as Omega-1 and Omega-2. Bevis labeled only the brighter star, 2, as Omega. Inasmuch as he could not see 4, Argelander agreed (*UN,* p. 63). Heis observed both stars but kept Bevis and Argelander's designations.

A¹, 45, HR 3450, 5.62V, *SA, HRP.*
A², 50, HR 3481, 5.87V, *SA, HRP.*

Bayer described A as two 6th-magnitude stars, and Flamsteed designated his 45 and 50 as A-1 and A-2.

b, 49, HR 3465, 5.66V, *SA, HRP.*

c¹, 36, HR 3406, 5.88V, *SA* as c, *HRP* as c, HF 36 as c, BV 59 as c, BF 1202*, H 44 as c.
c², 37, HR 3412, 6.53V, HF 37 as c, BV 60, BF 1208*.

Bayer described only one 6th-magnitude star as c, but Flamsteed designated these two neighboring stars, 36 and 37, as c-1 and c-2. Bevis labeled only the brighter star, 36, as c. Inasmuch as 37 was invisible to his naked eye, Argelander agreed (*UN,* p. 63). Flamsteed also designated his 42 (HD 73785, 6.85V, HF 40 as c, BV 45, BF 1218) as c although it is over 10° from the others. This was an obvious error that Bevis corrected.

d¹, 20, HR 3284, 5.84V, *SA, HRP,* H 27*.
d², 25, HR 3299, 6.14V, *SA, HRP,* H 29*.

Bayer noted that d was a *binae* or pair of 6th-magnitude stars, and Flamsteed designated his 20 and 25 as d-1 and d-2. Owing to a typographical error, Argelander's catalogue describes these stars as Delta¹ and Delta² (*UN,* p. 63), but they are correctly labeled in his atlas (Chart IX).

o, 38, HD 73575, 6.66V, HF 38*, BV 41, RAS 1048*, BF 1211.

Flamsteed lettered this star, but Bevis and Baily removed it as superfluous. See i Aql.

CANES VENATICI, CVn

Hunting Dogs

Although Hevelius is usually credited with devising Canes Venatici (figures 14 and 15), the constellation had first appeared 150 years earlier in the works of Peter Apian. In his planisphere of 1533, Apian depicted Bootes with two hounds leashed to his right hand, while in his planisphere of 1536 there were three hounds leashed to his left hand. Hevelius was apparently unaware of Apian's charts since he claimed credit for devising the Hunting Dogs. All in all, Hevelius created twelve new constellations. Like Keyzer and Houtman, he considered the number twelve to hold special significance. The twelve new constellations he devised are:

1 and 2. Canes Venatici. He considered each dog a separate constellation.
3. Lacerta.
4. Leo Minor.
5. Lynx.
6. Scutum.
7. Sextans.
8 and 9. Vulpecula cum Ansere. He considered the Little Fox and the Goose separate constellations.
10. Cerberus. He placed this monstrous three-headed hound, the mythological guardian of the Underworld, in Hercules' left hand.
11. Mons Maenalus. He placed Mt. Maenalus at the feet of Bootes since the latter was said to live on this mountain in Arcadia.
12. Triangulum Minus. Since this constellation was formed from three stars right below Triangulum, he claimed he could not think of a more appropriate name than "Little or Lesser Triangle."

The last three constellations are no longer cited in astronomical works probably because Hevelius did not devise individual star charts for them: Cerberus was included in the chart with Hercules; Mons Maenalus, with Bootes; and Triangulum Minus, with Triangulum.

Hevelius noted that he positioned the two Hunting Dogs under the Great Bear's tail and leashed to Bootes's upraised left hand since Bootes has been called the Clamorer, the Shouter, and the Hunter. Consequently, he depicted him shouting encouragement to his Hounds as they pursue Ursa Major. It should be noted, though, that Hevelius's view of Bootes contrasted sharply with the traditonal view that held that Bootes is Arctophylax, Protector of the Bear.

Hevelius named the Hunting Dogs Asterion and Chara. He chose Asterion, he asserted, because of the numerous small stars, *astra,* in that sector of the sky that dot the dog's body and because the name would not be unfamiliar to classical scholars since it had once belonged to a mythical king of Crete. The second dog he called Chara because she is female and her speed endears (*grata & chara*) her to her master. Flamsteed included this constellation in his atlas and catalogue, and Baily lettered two of its brighter stars Alpha and Beta.

See Warner, *Sky Explored,* pp. 9-10, 116; Kunitzsch, "Peter Apian," pp. 117-24; Allen, *Star Names,* pp. 114-15. For Hevelius's description of these twelve constellations, see his *Prodromus Astronomica,* pp. 114-17; for the significance of the number twelve, see Apus; see also Bootes and Vulpecula.

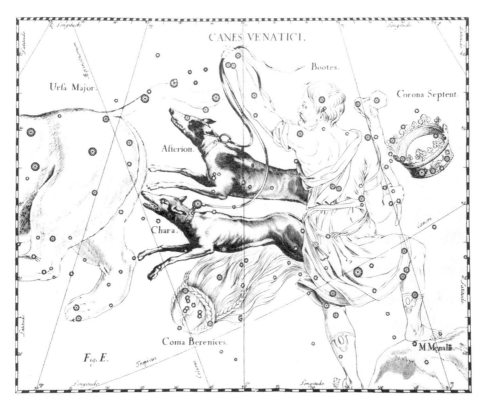

Figure 14. Hevelius's Canes Venatici. Hevelius's chart depicts Coma Berenices as a head of hair and shows Bootes standing on Mount Maenalus, one of the twelve constellations he devised. All the constellations in Hevelius's atlas are drawn backwards, as they would appear on a celestial globe.

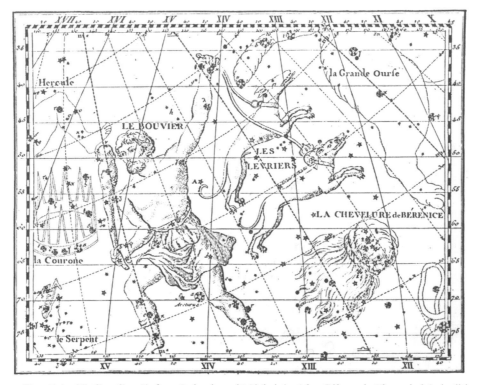

Figure 15. Canes Venatici with Cor Caroli, from Lalande and Méchain's *Atlas Céleste de Flamstéed,* 3rd edition. Their chart depicts the Hunting Dogs as Greyhounds (Les Levriers) and shows on Chara's collar a crowned heart, the asterism Cor Caroli, the Heart of Charles I.

Table 18. The lettered stars of Canes Venatici

	MAGNITUDES			CATALOGUE NUMBERS				
Letter	Bayer	Lacaille	Visual	Flamsteed	Lacaille	Other	HR	HD
Alpha[1]+			5.60			BAC 4345	4914	112412
Alpha[2]			2.90	12		BAC 4346	4915	112413
Beta			4.26	8		BAC 4235	4785	109358

The Lost, Missing, or Troublesome Stars of Canis Venatici

Alpha[1], HR 4914, 5.60V, *SA, BS,* BAC 4345.

Alpha[2], 12, HR 4915, 2.90V, *SA, BS,* BAC 4346 as Alpha, Cor Caroli Regis Martyris (The Heart of Charles, the Martyred King).

Bayer included this star in his chart of Bootes, but he considered it an *informis* between the Herdsman and Ursa Major and left it unlettered. Baily, noting it was a double, ADS 8706, designated only its brighter component as Alpha. Later astronomers added the indices.

Devised by Charles Scarborough, the court physician, Cor Caroli was originally a one-star constellation that first appeared in place of Canes Venatici on the planisphere of Francis Lamb in 1673. It was created in honor of Charles I, who had been executed by Parliament for treason during the Puritan Revolution. Later astronomers, like John Bevis (Tabula V), included it as a small asterism within the Hunting Dogs, usually next to the collar of the more southerly of the two Dogs, Chara (Warner, *Sky Explored,* p. 150).

CANIS MAJOR, CMA

Greater, Larger, or Big Dog

Canis Major (Figure 16) is a Ptolemaic constellation that Bayer lettered from Alpha to Omicron. Other astronomers added additional letters, especially in the southern sector of the constellation, which is barely visible from Central Europe.

This group of stars was known in Mesopotamia as the Bow (Sumerian BAN or GIS.BAN), and Sirius, its brightest star or lucida, was known as the Arrow (Sumerian KAK.SI.SA). The constellation may also have included parts of Canis Minor, especially its lucida, Procyon. According to Akkadian mythology, Tiamat, the goddess of the ever-changing, constantly moving ocean, which was the symbol of primeval chaos and evil, seeks to destroy the other gods. When they learn that she is preparing to marshal her forces against them, the gods all cringe in terror and despair until one of their number, Marduk, son of the fresh-water god Enki (Ea), agrees to serve as their champion. In a fierce battle, with the future of civilization hanging in the balance, Marduk succeeds in destroying Tiamat with his mighty bow and arrow:

> *He released the arrow, it tore her belly,*
> *It cut through her insides, splitting the heart.*
> (The Creation Epic, *IV:101-2*)

This is undoubtedly no ordinary bow that Marduk uses to kill the fearsome monster, but the new, powerful composite bow of wood strengthened with bone and sinew that was introduced into Mesopotamia as early as the third millennium B.C. In commemoration of Marduk's victory and of the innovative technology that produced his weapons, the Mesopotamians placed Marduk's bow and arrow in the heavens. It was appropriate indeed that the brightest star in the firmament, the Bow-Star (Sirius), was associated with Marduk, who, as a result of his victory over Tiamat, rose to become the principal deity of Babylon. Although originally the Semitic version of Ninurta, the Sumerian war god, Marduk was worshipped as the chief and most prominent god in the Mesopotamian pantheon. He was also associated with the most brilliant planet that dominated the night sky, Jupiter (Akkadian Nebiru). It was appropriate, too, that the weapon used to destroy Tiamat, the *marru*, became the symbol of Marduk and the etymological root of his name. *Marru* is Akkadian for arrowhead, spear point, or, possibly, spade and it is depicted on numerous Mesopotamian monuments and boundary stones. See Orion for a claim that the heavenly bow belongs to the Mighty Hunter, not to Marduk.

Although some scholars have claimed that Canis Major originated in Mesopotamia, no Middle Eastern astral legends or myths have been uncovered that refer to a dog in this area of the heavens. The Babylonians had a constellation the Dog, but it was located south of Hercules, far from the Big Dog. As far as is definitely known, the first reference to Canis Major appears in Greek mythology. Zeus, in the guise of a bull, abducts the beautiful Europa from her home in Sidon along the Levantine coast and brings her to Crete, where, according to legend, she gives her name to the whole continent. Before leaving Europa, Zeus, concerned for her safety, presents her with a dog of wondrous speed, Laelaps, and a spear that never misses its mark. These presents eventually pass to Cephalus, grandson of the wind god, Aeolus. At the request of the citizens of Thebes, Cephalus set Laelaps in pursuit of a voracious fox that is victimizing the populace. Like Laelaps, the fox is also magical — he cannot be caught. The dog chases the fox, but to no avail. Zeus intervenes and turns the fox to stone and promotes Laelaps to the heavens as Canis Major.

Another Greek legend asserts that the Great Dog belongs to the Mighty Hunter, Orion. The Dog faithfully accompanies Orion on all his hunting expeditions. When Orion dies, the gods place not only the Mighty Hunter in the sky but all that was associated with him — his dogs, Canis Major and Canis Minor; and his prey, Lepus, the Hare.

See van der Waerden, "Babylonian Astronomy. II. The Thirty-Six Stars." pp. 13, 16; Jacobsen, *Toward the Image of Tammuz,* pp. 35-36, 166-67, 339; Pritchard, *Anc. Near East. Texts,* pp. 67, 69, 72; McNeill, *The Rise of the West,* pp. 118-19; Pritchard, *Anc. Near East Picts.,* plates 453-54, 519-20; van Buren, *Symbols of the Gods,* pp. 14-15; Hunger and Pingree, *Mul.Apin,* pp. 11, 32, 127, 138; Reiner and Pingree, *Baby. Plan. Omens,* Part 2, chap. 2; Kramer, *Sumerian Mythology,* pp. 76-78; Condos, *Star Myths,* pp. 65-67; Hamilton, *Mythology,* pp. 100-5; Ridpath, *Star Tales,* pp. 40-41; Staal, *New Patterns in the Sky,* pp. 85-89.

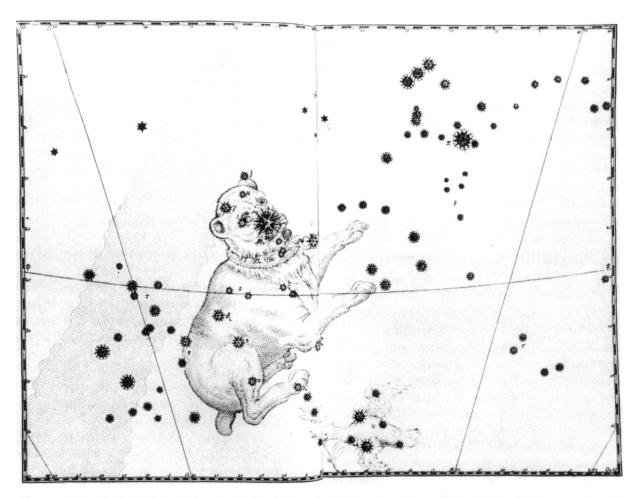

Figure 16. Bayer's Canis Major. Below the Big Dog is Columba, the Dove, which is marked as Upsilon (υ). Bayer noted in his *Uranometria* that Columba was a relatively new constellation. The line Phi (φ) is the Tropic of Capricorn. The letter Pi (π) is the 1st-magnitude star Rigel, Beta Ori, in the Giant's foot; Rho (ρ) is Lepus; Sigma (σ) is Eridanus; and Tau (τ) is Argo Navis.

Table 19. The lettered stars of Canis Major

	MAGNITUDES			CATALOGUE NUMBERS				
Letter	Bayer	Lacaille	Visual	Flamsteed	Lacaille	Other	HR	HD
Alpha	1		-1.46	9			2491	48915
Beta	2		1.98	2			2294	44743
Gamma	3		4.12	23			2657	53244
Delta	3		1.84	25			2693	54605
Epsilon	3		1.50	21			2618	52089
Zeta	3		3.02	1			2282	44402
Eta	3		2.45	31			2827	58350
Theta	4		4.07	14			2574	50778
Iota	4		4.37	20			2596	51309
Kappa+	4		3.96	13			2538	50013
Lambda+	4		4.48				2361	45813
Mu	5		5.00	18			2593	51250
Nu^1+	5		5.70	6			2423	47138
Nu^2	5		3.95	7			2429	47205
Nu^3	5		4.43	8			2443	47442
Xi^1+	5		4.33	4			2387	46328
Xi^2	5		4.54	5			2414	46933
$Omicron^1$+	5		3.87	16			2580	50877
$Omicron^2$	5		3.02	24			2653	53138
Pi+			4.68	19			2590	51199
Rho+			6.08	12			2509	49333
Sigma+			3.47	22		B 130	2646	52877
Tau+			4.40	30		B 169	2782	57061
Omega+			3.85	28		B 162	2749	56139

The Lost, Missing, or Troublesome Stars of Canis Major

$Kappa^1$, 10, HR 2492, 5.20V, BV 22, BF 964, H 30.
$Kappa^2$, 13, HR 2538, 3.96V, *SA* as Kappa, *HRP* as Kappa, *BS* as Kappa, HF 14 as Kappa, BV 30 as Kappa, BF 969 as Kappa, H 39 as Kappa.

Bayer described Kappa as a single 4th-magnitude star. In Halley's edition of Flamsteed's catalogue, 13 is equated with Kappa but in Flamsteed's catalogue of 1725, 10 and 13 are designated Kappa-1 and Kappa-2, although they are about 2° apart. Bevis labeled only the brighter star, 13, as Kappa. Baily agreed and dropped 10's letter.

Lambda, HR 2361, 4.48V, *SA, HRP, BS,* BV 12*.

Flamsteed mistakenly identified his 3 CMa as Bayer's Lambda. The star he observed was actually Delta Col (HR 2296, 3.85V, BV 11, BF 902, BAC 2066); he did not observe Lambda. Baily corrected the error in his revised edition of Flamsteed's catalogue. Bevis had earlier noted Flamsteed's error and was among the first to identify this star with Bayer's Lambda.

Nu^1, 6, HR 2423, 5.70V, *SA, HRP, BS.*
Nu^2, 7, HR 2429, 3.95V, *SA, HRP, BS.*
Nu^3, 8, HR 2443, 4.43V, *SA, HRP, BS.*

Bayer described Nu as a group of three 5th-magnitude stars, and Flamsteed designated his 6, 7, and 8 as Nu-1, Nu-2, and Nu-3.

Xi¹, 4, HR 2387, 4.33V, *SA, HRP, BS.*
Xi¹, 5, HR 2414, 4.54V, *SA, HRP, BS.*

Bayer described Xi as a couple of 5th-magnitude stars, and Flamsteed designated his 4 and 5 as Xi-1 and Xi-2.

Omicron¹, 16, HR 2580, 3.87V, *SA. HRP, BS.*
Omicron², 24, HR 2653, 3.02V, *SA, HRP, BS.*

Bayer described Omicron as a couple of 5th-magnitude stars, and Flamsteed designated his 16 and 24 as Omicron-1 and Omicron-2.

Pi¹, 15, HR 2571, 4.83V, BF 974, H 41, G 87.
Pi², 17, HR 2588, 5.74V, BF 977, G 94.
Pi³, 19, HR 2590, 4.68V, *SA* as Pi, *HRP* as Pi, *BS* as Pi, BF 980, H 44, G 95 as Pi.

Bayer did not use the letter Pi in Canis Major. It was first used by Flamsteed, who designated this group of three closely placed stars as Pi-1, Pi-2, and Pi-3. Baily considered them superfluous letters and removed them (see i Aql). Since they were not Bayer letters, both Argelander (*UN,* p. 95) and Heis also rejected them, but Gould retained 19 as Pi because of its brightness.

Rho, 12, HR 2509, 6.08V, BV 20, BF 965, G 76.

Bayer did not use Rho in Canis Major. It was added by Flamsteed but removed by Bevis and Baily as superfluous. See i Aql.

Bode added additional Greek and Roman letters in Canis Major, but only Sigma, Tau, and Omega have become established in the literature, primarily because Gould included them in his catalogue. Gould felt stars of their brightness merited letters.

Sigma, 22, HR 2646, 3.47V, *SA, HRP, BS,* B 130*, G 114*.

Tau¹, 29, HR 2781, 4.98V, B 168*, G 150, the variable UW.
Tau², 30, HR 2782, 4.40V, *SA* as Tau, *HRP* as Tau, *BS* as Tau, B 169*, G 151 as Tau.

Bode described these two neighboring stars as Tau-1 and Tau-2. Gould retained the letter Tau for 30, but dropped 29's letter because of its dimness; he considered its magnitude 5.6. Oddly enough, although Gould was especially concerned with magnitude and variability, he failed to notice that 29 is a variable, probably an eclipsing binary, with a range of 4.84V to 5.33V in about 4½ days.

Omega, 28, HR 2749, 3.85V, *SA, HRP, BS,* B 162*, G 140*.

Lacaille lettered additional stars in Canis Major, using lower-case Roman letters from a to k. None of these has survived in the literature primarily because Baily, who edited Lacaille's extended catalogue of 9,766 stars and the *BAC*, removed all of Lacaille's Roman letters except those in the three components of Argo Navis.

a, 24, HR 2653, 3.02V, CA 591*, BR 1445*, L 2588 as Omicron², BAC 2318 as Omicron², G 115 as Omicron².

Lacaille relettered Flamsteed's Omicron-2 as a, but Baily restored it. See Omicron.

b, 22, HR 2646, 3.47V, CA 590*, BR 1437*, L 2581 as Sigma, BAC 2309, G 114 as Sigma.

Baily relettered this star Sigma, as Bode had suggested (see Sigma). This is the only instance in which Baily acknowledged a Bode letter.

c, 16, HR 2580, 3.87V, CA 580*, BR 1393*, L 2506 as Omicron¹, BAC 2267 as Omicron¹, G 92 as Omicron¹.

Lacaille relettered Flamsteed's Omicron-1 as c, but Baily restored it. See Omicron.

d, 30, HR 2782, 4.40V, CA 626*, L 2721, BAC 2418, G 151 as Tau.

Gould relettered this star Tau, as Bode had suggested. See Tau.

e, 27, HR 2745, 4.66V, CA 612*, L 2674, BAC 2388, G 139, the variable EW.

e, 28, HR 2749, 3.85V, CA 614*, BR 1514 as e^2, L 2681, BAC 2391, G 140 as Omega.

Gould relettered this star Omega as Bode had suggested (see Omega). Brisbane observed CA 614 and lettered it e^2; he did not observe its neighbor to the north, CA 612. Lacaille did not use indices to distinguish closely placed stars of the same letter.

f, in Columba, HR 2424, 5.59V, CA 546*, BR 1291 as f^1, L 2359, BAC 2172, G 108 in Columba.

Gould included this star unlettered in Columba.

f, in Columba, HR 2446, 5.27V, CA 552*, BR 1303 as f^2, L 2376, BAC 2180, G 112 in Columba.

Gould included this star unlettered in Columba.

g, in Puppis, HR 2549, 4.99V, CA 573*, BR 1378*, L 2486, BAC 2252, G 37 in Puppis.

Gould included this star unlettered in Puppis.

h, HR 2545, 5.70V, CA 572*, BR 1375 as h^2, L 2479, BAC 2251, G 83.

Brisbane's h^1 (BR 1374) is made up of components B and C (comb. mag. 8.2V) of this multiple-star system, Herschel V 108. Lacaille observed only component A, CA 572.

i

Lacaille did not use this letter in Canis Major.

j

Lacaille did not use this letter.

k, in Puppis, HR 2873, 5.77V, CA 650*, BR 1624*, L 2823 erroneously as k^1 Pup, BAC 2478, G 96 in Puppis.

Baily included this star in Puppis erroneously as k^1 Pup; Gould left it unlettered. See k Pup.

k, in Puppis, HR 2881, 4.65V, CA 654*, BR 1634*, L 2834 erroneously as k^2 Pup, BAC 2484, G 100 in Puppis.

Baily included this star in Puppis erroneously as k^2 Pup; Gould left it unlettered since he considered it dimmer than 5th magnitude. See k Pup.

Canis Minor, CMi

Lesser, Smaller, or Little Dog

Canis Minor (Figure 17) is a Ptolemaic constellation that Bayer lettered from Alpha to Eta.

In classical astral mythology, Canis Minor was sometimes confused with Canis Major. The Roman poet Hyginus, for instance, declared: "The stories we have recounted...about Canis Major apply to him [Canis Minor] as well *(Poetic Astronomy,* 2.36)." According to legend, Canis Minor is one of Orion's hunting dogs. Upon Orion's death, the gods place both of his dogs — Canis Major and Canis Minor — together with their master among the stars.

In another legend Canis Minor is Maera, the faithful dog of Icarius, the first human the gods teach to cultivate grapes and produce wine. When Icarius good-naturedly distributes wine to his neighbors, they imbibe so much that they become dead drunk. Their relatives, thinking that Icarius poisoned them, kill Icarius and bury him under a tree. Maera, who accompanied his master, returns home, and, tugging on Icarius's daughter's dress with his teeth, leads her to her father's grave. Unable to contain her grief, she hangs herself under the very tree where Icarius was buried. Maera, too, ends his life by throwing himself into a well. Zeus, taking pity on the ill-fated family, sets Icarius and his daughter among the stars in the heavens. Nor does Zeus forget Maera. As a reward for his faithful service, the loyal hound is placed in the sky.

See Condos, *Sky Myths,* pp. 69-70; Ridpath, *Star Tales,* pp. 42-43. See Bootes for the legend of Icarius.

Table 20. The lettered stars of Canis Minor

	Magnitudes			Catalogue Numbers				
Letter	Bayer	Lacaille	Visual	Flamsteed	Lacaille	Other	HR	HD
Alpha	1		0.38	10			2943	61421
Beta	3		2.90	3			2845	58715
Gamma	5		4.32	4			2854	58972
Delta^1+	5		5.25	7			2880	59881
Delta^2	5		5.59	8			2887	60111
Delta^3			5.81	9			2901	60357
Epsilon	6		4.99	2			2828	58367
Zeta	6		5.14	13			3059	63975
Eta	6		5.25	5			2851	58923

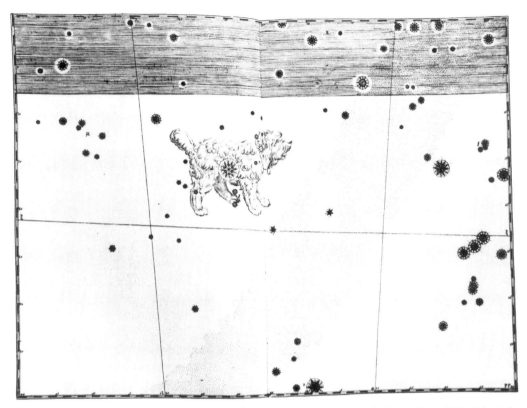

Figure 17. Bayer's Canis Minor. The horizontal line Xi (ξ) is the celestial equator. The letter Theta (θ) is Orion's head; Kappa (κ) is Gemini; Lambda (λ), at the top left of the chart, is Cancer; Mu (μ) is Hydra's head; and Nu (ν) is the 1st-magnitude star Sirius, Alpha CMa. The letter Iota (ι) is Orion's belt, the three nearly adjacent stars near the right edge of the chart; the letter is missing from some copies of the *Uranometria*. The shaded area across the top is the zodiac.

The Lost, Missing, or Troublesome Stars of Canis Minor

Delta^1, 7, HR 2880, 5.25V, *SA, HRP, BS,* BV 7*, BF 1056*, H 13*, G 21*.
Delta^2, 8, HR 2887, 5.59V, *SA, HRP, BS,* BV 8*, BF 1059*, H 14*, G 22*.
Delta^3, 9, HR 2901, 5.81V, *SA, HRP, BS,* BV 9, BF 1060*, H 16*, G 23.

Bayer described Delta as a couple of 5th-magnitude stars in the Lesser Dog's rear left paw. When Flamsteed examined the area, he found not two but three stars — one to the south and two nearly adjacent stars to the north. He designated the southerly one, his 7, as Delta-1. Since he was not sure which of the two northerly ones Bayer meant, or whether Bayer meant the combined light of both, he designated his 8 and 9 as Delta-2 and Delta-3. Bevis, though, felt that Bayer's Delta's referred to the two brighter stars, 7 and 8. Argelander agreed since he was unable to see Delta-3 with his naked eye (*UN*, p. 62). Heis, however, observed the star and restored its letter, but Gould removed it because of its dimness.

Omicron, 6, HR 2864, 4.54V, BF 1050.
Pi, 11, HR 3008, 5.30V, BF 1081.

Bayer did not letter Omicron and Pi. Although he included these stars in his atlas, he considered them *informes* between Canis Minor and Gemini and did not letter them. Flamsteed assigned them letters, but Baily removed their letters as superfluous. See i Aql.

CAPRICORNUS, CAP

Goat, Sea Goat

A Ptolemaic constellation that Bayer lettered from Alpha to Omega and from A to c, Capricornus (Figure 18) or the Goat-Fish (Sumerian SUHUR.MAS) evolved from the larger constellation Ibex. Ibex originally marked the winter solstice, the beginning of the rainy season on the Arabian Peninsula. As noted earlier (see Aquarius), the Ibex was the emblem of the Sumerian fresh-water god Enki. Consequently, when Ibex was later split into two smaller constellations, Aquarius and Capricornus, both retained their association with Enki. Aquarius was Enki himself, carrying a vase filled with the water of life. Capricornus was depicted with a fish tail, a reminder of how the great Ibex was often seen with its lower body submerged in the water and its horns protruding like reeds above the surface of the marshes. Representing a smaller member of the goat family, Capricornus became a symbol of Enki and its likeness was carved on numerous monuments and boundary stones that date back at least to the latter half of the second millennium B.C. The constellation's first appearance in documents as a separate constellation, however, occurred several hundred years later in the eighth century.

The Greeks believed that Capricornus was Pan, half-man and half-goat. According to myth, during Zeus's battle with the Titans Pan blows his conch horn, which frightens the Titans and throws them into a ***pan***ic, thus saving the day for Zeus and the gods. On another occasion, Pan saves Zeus in his battle with the monster Typhon, a huge Titan with one hundred dragonheads. Grateful to Pan for his faithful service to the gods, Zeus rewards him with a place in the heavens.

See Hartner, "The Earliest History of the Constellations in the Near East," p. 14; Reiner and Pingree, *Baby. Plan. Omens,* Part 2, p. 14; van der Waerden, "Babylonian Astronomy. II. The Thirty-Six Stars," pp. 14, 23; van der Waerden, "History of the Zodiac," pp. 219-20; O'Neil, *Time and the Calendars,* pp. 51-52, 59; Pritchard, *Anc. Near East Picts.,* plate 520; Ridpath, *Star Myths,* pp. 43-44; Gantz, *Early Greek Myth,* pp. 48-49.

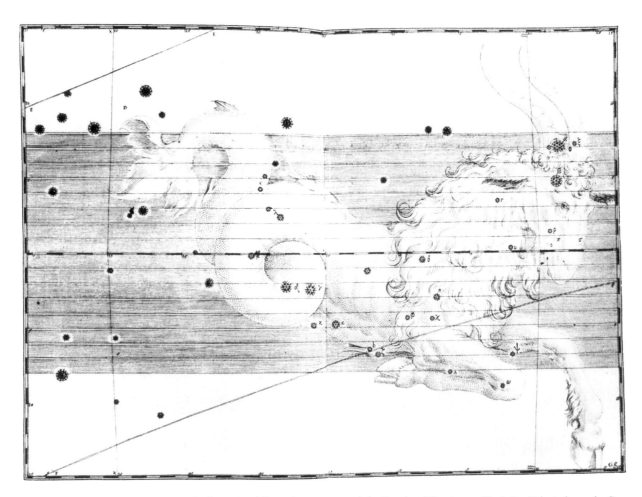

Figure 18. Bayer's Capricornus. The lines E and F are the equator and the Tropic of Capricorn. The letter D just above the Sea Goat's tail is the constellation Aquarius. The shaded area is the zodiac.

Table 21. The lettered stars of Capricornus

		MAGNITUDES			CATALOGUE NUMBERS			
Letter	Bayer	Lacaille	Visual	Flamsteed	Lacaille	Other	HR	HD
Alpha[1]+			4.24	5			7747	192876
Alpha[2]	3		3.57	6			7754	192947
Beta[1,2]+	3		3.18c	9			7775/6	193452/95
Gamma	3		3.68	40			8278	206088
Delta	3		2.87	49			8322	207098
Epsilon	4		4.68	39			8260	205637
Zeta	5		3.74	34			8204	204075
Eta	5		4.84	22			8060	200499
Theta	5		4.07	23			8075	200761
Iota	5		4.28	32			8167	203387
Kappa	5		4.73	43			8288	206453
Lambda	5		5.58	48			8319	207052
Mu	5		5.08	51			8351	207958
Nu	6		4.76	8			7773	193432
Xi[1]+			6.34	1			7712	191753
Xi[2]	6		5.85	2			7715	191862
Omicron	6		5.58c	12			7829/30	195093/4
Pi	6		5.25	10			7814	194636
Rho	6		4.78	11			7822	194943
Sigma	6		5.28	7			7761	193150
Tau+	6		5.22	14			7889	196662
Upsilon	6		5.10	15			7900	196777
Phi	6		5.24	28			8127	203320
Chi+	6		6.02	25			8087	201184
Psi	6		4.14	16			7936	197692
Omega	6		4.11	18			7980	198542
A	6		4.50	24			8080	200914
b	6		4.51	36			8213	204381
c[1]+	6		5.09	46			8311	206834
c[2]	6		6.00	47			8318	207005

The Lost, Missing, or Troublesome Stars of Capricornus

Alpha[1], 5, HR 7747, 4.24V, *SA, HRP, BS,* ADS 13632.
Alpha[2], 6, HR 7754, 3.57V, *SA, HRP, BS,* ADS 13645.

Bayer described Alpha as a 3rd-magnitude *duplex* or double star. Flamsteed also noted it was a double and designated its components Alpha-1 and Alpha-2. Actually, this is a multiple-star system with at least nine components.

Beta[1], 9, HR 7776, 3.08V, *SA* as Beta, *HRP* as Beta, *BS,* H 7 as Beta, G 14 as Beta.
Beta[2], HR 7775, 6.10V, *BS,* G 13.

Bayer described Beta as a single 3rd-magnitude star. Later astronomers added the indices to identify the components of this double star, ADS 13717. For HR 7775's designation as Beta[2], see Hoffleit, "Discordances," p. 47.

Xi[1], 1, HR 7712, 6.34V, *SA, BS,* BV 3 as a, BF 2736*.
Xi[2], 2, HR 7715, 5.85V, *SA, HRP* as Xi, *BS,* BV 4 as Xi, BF 2738*, H 1 as Xi.

Bayer noted that Xi was a single 6th-magnitude star, and Flamsteed designated his 2 as Xi. Baily, however, felt that Xi might be either 1 or 2, so he designated them Xi^1 and Xi^2. Bevis labeled only 2, the brighter star, as Xi. Argelander also designated only 2 as Xi since he could not see 1 with his naked eye (*UN,* p. 110).

Tau^1, 13, HD 196348, 6.76V, BV 17, BF 2798*.
Tau^2, 14, HR 7889, 5.22V, *SA* as Tau, *HRP* as Tau, *BS,* BV 20 as Tau, BF 2804*, H 17 as Tau.

Bayer noted that Tau was a single 6th-magnitude star, and Flamsteed designated his 14 as Tau, as did Bevis. Baily, however, felt that Tau might be either 14 or its close but faint neighbor 13, so he designated them Tau^1 and Tau^2. Argelander designated only 14, the brighter one, as Tau since he could not see 13 (*UN,* p. 111).

Chi^1, 25, HR 8087, 6.02V, *SA* as Chi, *HRP* as Chi, *BS* as Chi, BV 28 as Chi, BF 2881 as Chi. H 38 as Chi, G 76 as Chi.
Chi^2, 26, HD 201301, 6.8V, BV 30, BF 2882, G 77.
Chi^3, 27, HR 8091, 6.25V, BV 29, BF 2883, H 39, G 78.

Although Bayer described Chi as a single 6th-magnitude star, Flamsteed designated these three neighboring stars, his 25, 26, and 27, as Chi-1, Chi-2, and Chi-3. Bevis and Baily, however, felt that only the brightest of the group, 25, was synonymous with Bayer's Chi. Argelander agreed since he could see only this star (*UN,* p. 111).

A, 24, HR 8080, 4.50V, *SA, HRP.*

b, 36, HR 8213, 4.51V, *SA, HRP.*

c^1, 46, HR 8311, 5.09V, *SA, HRP* as c, BV 54 as c-2, BF 2962*, H 57*, G 118 as c.
c^2, 47, HR 8318, 6.00V, *SA,* BV 55 as c-3, BF 2967*, H 58*, G 120.

Bayer noted that c was a group of three 6th-magnitude stars and placed the letter c between the upper two. Working only with Bayer's chart, Flamsteed designated his 46 and 47 as c-1 and c-2. Bevis and Baily later searched for the third star but admitted they could find none that would conform to Bayer's chart. Bevis designated 46 and 47 as c-2 and c-3, since the missing c, which would have been c-1, preceded the others on Bayer's chart. Although Argelander observed both 46 and 47, he felt that only the brighter one, 46, should be designated c (*UN,* p. 111). Heis disagreed with him, concurring instead with Flamsteed's designation of 46 and 47 as c^1 and c^2.

CARINA, CAR

Ship's Keel

The constellation Argo Navis, the Ship Argo, commemorates the first seagoing vessel of ancient Greece. The vessel is built for Jason at the outset of his quest for the Golden Fleece. Jason is the son of the rightful king of Boeotia, Aeson, who has been deposed by his brother, Pelias. When Jason grows to manhood, he confronts his uncle, who agrees to abdicate in favor of Aeson if Jason can prove himself a hero and bring him the Golden Fleece. The fleece is the pure golden wool of an extraordinary winged ram created by the god Hermes. Jason contracts with Argus to build him a ship that will endure what seemed then a voyage to the ends of the earth since the fleece was located in far away Colchis, on the eastern shore of the Black Sea. The ship is constructed with the help of Athena, who directs that the prow contain wood from a sacred tree that will advise the adventurers. When completed the ship is christened the Argo, either in honor of its builder (Argus or Argos) or from the Greek word for swift (argos, αργος). Jason gathers together a band of fifty of the greatest heroes in Greece including Hercules, Castor and Pollux, Theseus, and Orpheus. After a series of twelve hair-raising adventures, Jason and the Argonauts finally succeed in seizing the Golden Fleece and eventually dedicating it to the gods.

Although it has been suggested that the constellation had its origin in the ancient Near East, there are no astral legends or myths associated with Argo Navis. In its place in the sky, the Mesopotamians had several constellations: the Bow, the Arrow, Eridu (NUN.KI, Star of Eridu), Ninmah, and the Harrow.

Ptolemy described Argo as a single constellation, which Bayer lettered from Alpha to Omega and from A to s. Since it is mostly below the horizon in Europe, Bayer's chart (Figure 19) does not conform to the constellation's actual appearance in the heavens. As a consequence, Lacaille (Figure 2) decided to redesign and reletter it and, because of its size, break it up into four separate constellations: Carina (Keel), Puppis (Stern), Vela (Sails), and Pyxis (Compass). He used, though, only one set of Greek letters (Alpha to Omega) for the bright stars in Carina, Puppis, and Vela and continued to designate these Greek-lettered stars as part of Argo, thus preserving the unity of the original constellation. But he treated the dimmer Roman-lettered stars differently. First of all, he gave each of Argo's three new components (Carina, Puppis, and Vela) its own set of Roman letters. Secondly, he noted specifically which component of Argo each star belonged to, something he did not do with the Greek-lettered stars. His catalogue reads, for example: CA 734, D, Argus in Carina; CA 735, K, Argus in Puppi; CA 807, f, Argus in Velis. Although Pyxis was originally part of Argo Navis, Lacaille treated it differently from the other three components (see Pyxis). It was Gould, a century later, who finally dropped the designation Argo and referred to the Greek-lettered stars by the individual constellations in which they were located. In what may be described as a fitting epitaph for Argo Navis, Gould wrote:

> *The total abandonment of the venerable constellation* Argo *has caused me much regret; yet it will not cease to exist, being represented by its legitimate descendants. To retain it in name, while abandoning it in fact, can be productive of no good, and needlessly adds one to the list of constellations, at the same time that it complicates the system of nomenclature (Gould,* Uran. Arg., *p.62).*

When Lacaille first divided the Ship in his preliminary catalogue of 1756, he called the main section of the Ship le Corps, the Hull. In his catalogue of 1763, he translated this into Latin as Carina, the Keel.

See Staal, *New Patterns in the Sky,* pp. 101-9; Ridpath, *Star Tales,* pp. 139-40; Condos, *Star Myths,* pp. 39-42; Hunger and Pingree, *Mul.Apin,* p. 138; Reiner and Pingree, *Baby. Plan. Omens,* Part 2, pp. 11, 14; Lacaille, "Table des Ascensions," p. 590. For the story of the Golden Fleece, see Aries.

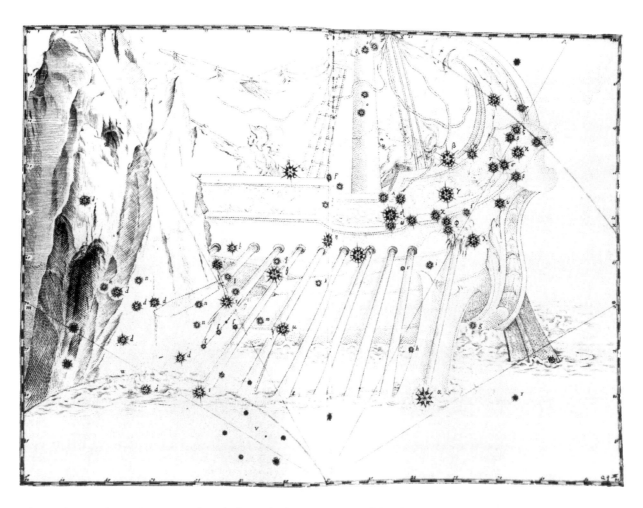

Figure 19. Bayer's Navis. Bayer's chart depicts only the stern or rear of the Ship. Although Lacaille's planisphere (Figure 2) shows a single constellation, a ship like Bayer's with its bow broken off, Lacaille's catalogue divides Navis into three parts: Carina (Keel), Puppis (Stern), and Vela (Sails). The lines W and X are the Antarctic Circle and the Tropic of Capricorn. The letter V is the constellation Volans, the Flying Fish. Bayer was uncertain about the location of the two stars marked T; he noted that some authorities placed them in the beam of the Ship.

Table 22. The lettered stars of Carina

		MAGNITUDES			CATALOGUE NUMBERS			
Letter	Bayer	Lacaille	Visual	Flamsteed	Lacaille	Other	HR	HD
Alpha		1	-0.72		519		2326	45348
Beta		1	1.68		848		3685	80007
Epsilon		2	1.86		761		3307	71129
Eta+		2	6.21		968		4210	93308
Theta		3	2.76		964		4199	93030
Iota		2	2.25		849		3699	80404
Upsilon		3	2.97c		894		3890/1	85123/4
Chi		4	3.47		721		3117	65575
Omega		4	3.32		920		4037	89080
a		5	3.44		835		3659	79351
b¹+		5	4.92		821		3582	77002
b²		5	5.16		823		3598	77370
c		6	3.84		818		3571	76728
d		5	4.33		791		3457	74375
e¹+		6	5.26		779		3415	73390
e²		6	4.86		778		3414	73389
f		6	4.49		803		3498	75311
g		6	4.34		844		3696	80230
h		5	4.08		873		3825	83183
i		5	3.97		838		3663	79447
k		6	4.81		853		3728	81101
l		5	3.69		890		3884	84810
m		6	4.52		882		3856	83944
n+		5	6.05		865		3761	82068
o+		6	6.20		905		3957	87238
p		4	3.32		943		4140	91465
q		5	3.40		922		4050	89388
r		6	4.45		947		4159	91942
s		6	3.82		935		4114	90853
t¹+		6	5.08		950		4164	92063
t²		6	4.66		952		4177	92397
u		5	3.78		979		4257	94510
w+			4.57			G 224	4200	93070
x		6	3.91		997		4337	96918
y		6	4.60		1001		4352	97534
z¹+		6	4.61		993		4325	96566
z²		6	5.13		998		4338	96919
A+			4.40			G 18	2554	50337
B		6	4.76		745		3220	68456
C		6	5.16		751		3260	69863
D¹+		6	4.82		732		3159	66591
D²		6	6.30		734		3186	67536
D³		6	6.28		747		3217	68423
E		6	4.71		833		3642	78764
F+		6	6.14		852		3695	80194
G		6	4.48		834		3643	78791
H		6	5.47		880		3821	83095
I		5	4.00		936		4102	90589
K		6	4.74		944		4138	91375
L		6	4.99		929		4089	90264
M		6	5.16		918		4025	88981
N+		6	4.39		?		2435	47306
O+		6	5.61		574		2526	49877
P+		6	5.96		608		2698	54732
Q		6	4.94		666		2934	61248

The Lost, Missing, or Troublesome Stars of Carina

Listed below are all of Lacaille's Greek-lettered stars of Argo Navis from Alpha to Omega with a note indicating whether they are in Carina, Puppis, or Vela. With the exception of Omicron, these stars are all properly listed in *SA, HRP,* and *BS.*

Alpha, in Carina, HR 2326, -0.72V, CA 519*, BR 1241*, L 2291*, G 7*, Canopus.

Beta, in Carina, HR 3685, 1.68V, CA 848*, BR 2425*, L 3791*, G 123*.

Gamma¹, in Vela, HR 3206, 4.27V, *SA, BS,* BR 1916*, BAC 2754, G 8.
Gamma², in Vela, HR 3207, 1.78V, *SA, HRP* as Gamma, *BS,* CA 737 as Gamma, BR 1917*, L 3185 as Gamma, BAC 2755 as Gamma, G 9 as Gamma.

Lacaille described Gamma as a single 2nd-magnitude star. Later astronomers added the indices to distinguish the components of this multiple-star system. HR 3206 is component B; HR 3207 is A. Brisbane (BR 1918) designated component C, magnitude 8.2V, as Gamma³. The other elements in this system are components D and E, magnitudes 9.1V and 12.5V. The combined magnitude of the entire system is 1.85V.

Delta, in Vela, HR 3485, 1.96V, CA 796*, BR 2194*, L 3532*, G 65*.

Epsilon, in Carina, HR 3307, 1.86V, CA 761*, BR 2012*, L 3327*, G 89*.

Zeta, in Puppis, HR 3165, 2.25V, CA 729*, BR 1876*, L 3136*, G 248*.

Eta, in Carina, HR 4210, 6.21V, CA 968*, BR 3198*, L 4457*, G 231*.

This is the famous variable. When Lacaille observed it, it had reached magnitude 2. It attained maximum brilliance in 1847 when it was observed at about -0.8 and was the second brightest star in the night sky after Sirius. Within twenty years, it gradually faded from sight. See *Burnham's Celestial Handbook,* I, 466-71.

Theta, in Carina, HR 4199, 2.76V, CA 964*, BR 3184*, L 4447*, G 223*.

Iota, in Carina, HR 3699, 2.25V, CA 849*, BR 2429*, L 3792*, G 127*.

Kappa, in Vela, HR 3734, 2.50V, CA 851*, BR 2459*, L 3816*, G 129*.

Lambda, in Vela, HR 3634, 2.21V, CA 830*, BR 2346*, L 3699*, G 100*.

Mu, in Vela, HR 4216, 2.69V, CA 970*, BR 3206*, L 4461*, G 229*.

Nu, in Puppis, HR 2451, 3.17V, CA 557*, BR 1310*, L 2386*, G 20*.

Xi, 7, in Puppis, HR 3045, 3.34V, CA 691*, BR 1763*, L 2994*, G 191*.

Omicron, in Vela, HR 3447, 3.62V, CA 786*, BR 2148*, L 3482*, G 56*.

HRP erroneously calls this star Sigma, while *BS* designates it Omicron. *HRP* and *BS* erroneously designate HR 3034, 4.50V, in Puppis, as Omicron; it should be o (oh). In effect, *BS* designates two Omicrons: one in Vela, the other in Puppis. The former is Lacaille's Omicron; the latter is his o (oh) Pup. *SA* correctly distinguishes between the two. The error is a result of the similarity in appearance of the two letters, Omicron (o) and oh (o).

Pi, in Puppis, HR 2773, 2.70V, CA 627*, BR 1536*, L 2720*, G 82*.

Rho, 15, in Puppis, HR 3185, 2.81V, CA 731*, BR 1892*, L 3153*, G 253*.

Sigma, in Puppis, HR 2878, 3.25V, CA 655*, BR 1631*, L 2837*, G 99*.

Tau, in Puppis, HR 2553, 2.93V, CA 579*, BR 1383*, L 2505*, G 39*.

Lost Stars

Upsilon, in Carina, HR 3890/1, 2.96V, 6.03V, comb. mag. 2.97V, CA 894*, BR 2682*, L 4051*, G 160*.

Phi, in Vela, HR 3940, 3.54V, CA 901*, BR 2752*, L 4093*, G 171*.

Chi, in Carina, HR 3117, 3.47V, CA 721*, BR 1835*, L 3102*, G 65*.

Psi, in Vela, HR 3786, 3.60V, CA 864*, BR 2519*, L 3885*, G 140*.

Omega, in Carina, HR 4037, 3.32V, CA 920*, BR 2924*, L 4243*, G 185*.

While using one set of Greek letters for the brightest stars in the old constellation Argo Navis, Lacaille assigned separate sets of Roman letters for each of its new components — Carina, Puppis, and Vela. In Carina, he used a to z and A to S.

a, HR 3659, 3.44V, *SA, HRP, BS,* CA 835*, BR 2388*, L 3738*, G 117*.

b^1, HR 3582, 4.92V, *SA, HRP,* CA 821 as b, BR 2293*, L 3639*, G 109*.
b^2, HR 3598, 5.16V, *SA, HRP,* CA 823 as b, BR 2311*, L 3661*, G 110*.

Lacaille designated these two neighboring stars b.

c, HR 3571, 3.84V, *SA, HRP,* CA 818*, BR 2281*, L 3626*, G 108*.

d, HR 3457, 4.33V, *SA, HRP,* CA 791*, BR 2163*, L 3504*, G 99*.

e^1, HR 3415, 5.62V, *SA, HRP,* CA 779 as e, BR 2113 as e^2, L 3452 as e^2, BAC 2920*, G 95*.
e^2, HR 3414, 4.86V, *SA, HRP,* CA 778 as e, BR 2112 as e^1, L 3451 as e^1, BAC 2921*, G 96*.

Lacaille designated these two nearly adjacent stars e. These two stars are moving in opposite directions. When Baily prepared the *BAC*, he found that CA 778 had passed CA 779 in right ascension and consequently he switched their indices.

f, HR 3498, 4.49V, *SA, HRP,* CA 803*, BR 2217*, L 3554*, G 103*.

g, HR 3696, 4.34V, *SA, HRP,* CA 844*, BR 2424*, L 3782*, G 125*.

h, HR 3825, 4.08V, *SA, HRP,* CA 873*, BR 2565*, L 3949*, G 147*.

i, HR 3663, 3.97V, *SA, HRP,* CA 838*, BR 2394*, L 3753*, G 119*.

j

Lacaille did not use this letter.

k, HR 3728, 4.81V, *SA, HRP,* CA 853*, BR 2461*, L 3823*, G 132*.

l, HR 3884, 3.69V, *SA, HRP,* CA 890*, BR 2664*, L 4033*, G 157*.

m, HR 3856, 4.52V, *SA, HRP,* CA 882*, BR 2607*, L 3987*, G 150*.

n, HR 3761, 6.05V, *SA,* CA 865*, BR 2513*, L 3890*, BAC 3249*, G 137.

Gould dropped this star's letter because of its dimness.

o, HR 3957, 6.20V, CA 905*, BR 2806, L 4138, R 185*, BAC 3441, G 169.

Brisbane dropped this star's letter noting that the area contained a "cluster of small stars." Baily dropped its letter because of its dimness. It is located on the edge of open cluster NGC 3114.

p, HR 4140, 3.32V, *SA, HRP,* CA 943*, BR 3072*, L 4348*, G 203*.

q, HR 4050, 3.40V, *SA, HRP,* CA 922*, BR 2935*, L 4249*, G 187*.

r, HR 4159, 4.45V, *SA, HRP,* CA 947*, BR 3099*, L 4373*, G 208*.

s, HR 4114, 3.82V, *SA, HRP,* CA 935*, BR 3031*, L 4314*, G 196*.

t¹, HR 4164, 5.08V, *SA, HRP,* CA 950 as t, BR 3112*, L 4380*, G 210*.
t², HR 4177, 4.66V, *SA, HRP,* CA 952 as t, BR 3127*, L 4396*, G 213*.

    Lacaille designated these two neighboring stars t.

u, HR 4257, 3.78V, *SA, HRP,* CA 979*, BR 3274*, L 4515*, G 246*.

v

    Lacaille did not use this letter.

w, HR 4200, 4.57V, *SA, HRP,* CA 963, BR 3185, L 4446, G 224*.

    Lacaille did not letter this star. Although he observed and catalogued it, he left it unlettered since he considered it only 6th magnitude. Gould designated it w because he felt a star of its brightness merited a letter.

x, HR 4337, 3.91V, *SA, HRP,* CA 997*, BR 3416*, L 4627*, G 260*.

y, HR 4352, 4.60V, *SA, HRP,* CA 1001*, BR 3462*, L 4652*, G 263*.

z¹, HR 4325, 4.61V, *SA, HRP* as z, CA 993 as z, BR 3402*, L 4611*, BAC 3805*, G 257 as z.
z², HR 4338, 5.13V, *SA,* CA 998 as z, BR 3419*, L 4629*, BAC 3820*, G 261, the variable V 371 with a range of 5.12V to 5.19V.

    Lacaille designated these two neighboring stars z. Since Gould considered CA 998 a 6th-magnitude star, he dropped its letter.

A, HR 2554, 4.40V, *SA, HRP,* CA 581 as B Pup, BR 1388 as B Pup, L 2511, BAC 2259 erroneously as B Car, G 18*.

    Lacaille did not use the letter A in Carina. Lacaille considered this star to be in Puppis and lettered it B Pup. In Lacaille's expanded catalogue, Baily included it in Carina and left it unlettered; but in the *BAC*, he erroneously labeled it B Car. Gould, however, relettered it A Car because he felt a star of its brightness merited a letter.

B, HR 3220, 4.76V, *SA, HRP,* CA 745*, BR 1934*, L 3222*, G 82*.

C, HR 3260, 5.16V, *SA, HRP,* CA 751*, BR 1971*, L 3275*, G 84*.

D¹, HR 3159, 4.82V, *SA, HRP* as D, CA 732 as D, BR 1883*, L 3154, BAC 2713*, G 77 as D.
D², HR 3186, 6.30V, *SA,* CA 734 as D, BR 1906*, L 3178, BAC 2738*, G 79.
D³, HR 3217, 6.28V, CA 747 as D, BR 1935 erroneously as D, L 3224, BAC 2768*, G 81.

    Lacaille designated these three neighboring stars D. Gould, however, dropped the letters from CA 734 and 747 because of their dimness.

E, HR 3642, 4.71V, *SA, HRP,* CA 833*, BR 2369*, L 3730*, G 115*.

F, in Chamaeleon, HR 3695, 6.14V, CA 852*, L 3817, BAC 3191*, G 12 in Chamaeleon.

    Although Lacaille placed this star in Carina, Gould included it in Chamaeleon and left it unlettered because of its dimness. With one or two very rare exceptions, Gould established the modern boundaries of all the southern constellations.

G, HR 3643, 4.48V, *SA, HRP,* CA 834*, BR 2374*, L 3736*, G 116*.

H, HR 3821, 5.47V, *SA, HRP,* CA 880*, BR 2573*, L 3968*, G 146*.

I, HR 4102, 4.00V, *SA, HRP,* CA 936*, BR 3025*, L 4319*, G 193*.

J

    Lacaille did not use this letter.

K, HR 4138, 4.74V, *SA, HRP,* CA 944*, BR 3074*, L 4357*, G 202*.

L, HR 4089, 4.99V, *SA, HRP,* CA 929*, BR 2999*, L 4296*, G 191*.

M, HR 4025, 5.16V, *SA, HRP,* CA 918*, BR 2918*, L 4233*, G 184*.

N, HR 2435, 4.39V, *SA, HRP,* CA 556, BR 1302, L 2383*, BAC 2176, G 11*.

Lacaille designated CA 555 as N, but neither Brisbane nor Baily could locate it. Thinking Lacaille had made a clock error that resulted in erroneous coordinates, they eliminated CA 555 from their catalogues. Both Baily and Gould felt that the star he had observed and meant to label N was CA 556, which he had left unlabeled in his catalogue.

O, HR 2526, 5.61V, *SA,* CA 574*, BR 1376 erroneously as o (lower-case oh), L 2490, BAC 2250*, G 17.

Gould dropped this star's letter since he considered it less than 6th magnitude.

P, HR 2698, 5.96V, CA 608*, BR 1488*, L 2651, BAC 2353*, G 28.

Gould dropped this star's letter since he considered it less than 6th magnitude. The P in *SA* is P Vel (HR 4110, 4.66V), not P Car. See P Vel.

Q, HR 2934, 4.94V, *SA, HRP,* CA 666*, BR 1674*, L 2902, BAC 2524*, G 50*.

R, HR 2862, 5.10V, CA 652*, BR 1619*, L 2829, BAC 2476*, G 44.
S, HR 2652, 5.14V, CA 594*, BR 1451*, L 2601, BAC 2321*, G 23.

Gould dropped the letters from these two stars, R and S, to avoid confusing them with variables (see R Cen). Since they are fainter than 5th magnitude, he left them unlettered.

Cassiopeia, Cas

Cassiopeia, Queen, Lady in the Chair

Cassiopeia (Figure 20) is a Ptolemaic constellation that Bayer lettered from Alpha to Omega and A.

Although some scholars believe the origin of Cassiopeia can be traced back to the Levant, no Middle Eastern astral myths or legends have been discovered thus far that relate to this constellation. In Mesopotamia, Cassiopeia's place in the heavens was occupied by two constellations, the Horse and the Panther.

According to Greek mythology, Cassiopeia, the wife of King Cepheus of Ethiopia, boasts that she is more beautiful than the Nereids, the sea nymphs. The Nereids appeal to the sea god Poseidon (Roman Neptune) to punish Cassiopeia for her arrogance and vanity. Poseidon is more than willing to comply with their request since he himself is married to a Nereid, Amphitrite, queen of the seas. As punishment, Poseidon sends huge waves and the sea monster Cetus, to ravage the coast. On the advice of an oracle, King Cepheus chains his daughter Andromeda to a rock along the coast to appease the sea monster and to bring peace to his land. Before Cetus can devour Andromeda, she is rescued by the hero Perseus. Although Cassiopeia, along with Cepheus, Andromeda, Perseus, and Cetus, is placed among the stars, the Nereids eventually get their revenge. The Queen, though depicted as sitting on her throne, is in the ignominious position of being upside down half the time as the constellation makes its daily rotation around the north celestial pole.

See Hunger and Pingree, *Mul.Apin,* p.138; Condos, *Star Myths,* pp. 75-77; Ridpath, *Star Tales,* pp. 45-47; Staal, *New Patterns in the Sky,* pp. 14-18. See Andromeda, Cepheus, Cetus, and Perseus.

Lost Stars

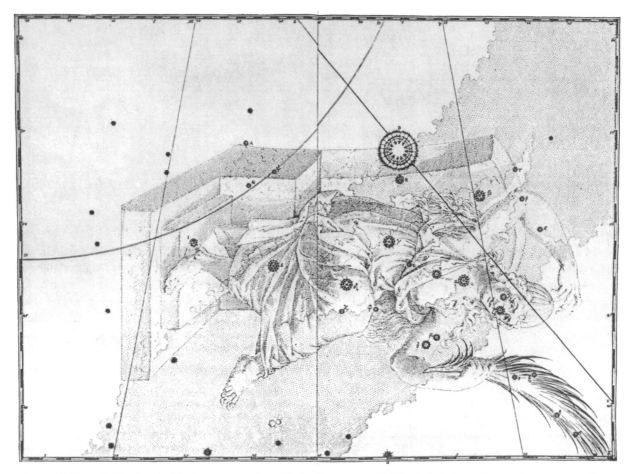

Figure 20. Bayer's Cassiopeia. The large star marked B is the Supernova of 1572, Tycho's Star. The two fuzzy, nearly adjacent stars above Cassiopeia's right toe are the Double Cluster, Chi (χ) and h Per. The lines D and E are the Arctic Circle and the equinoctial colure. The letter C near the bottom center of the chart represents Perseus's sword. On some copies of the *Uranometria* the letter is missing.

Table 23. The lettered stars of Cassiopeia

	MAGNITUDES			CATALOGUE NUMBERS				
Letter	Bayer	Lacaille	Visual	Flamsteed	Lacaille	Other	HR	HD
Alpha	3		2.23	18			168	3712
Beta	3		2.27	11			21	432
Gamma	3		2.47	27			264	5394
Delta	3		2.68	37			403	8538
Epsilon	3		3.38	45			542	11415
Zeta	4		3.66	17			153	3360
Eta	4		3.44	24			219	4614
Theta	4		4.33	33			343	6961
Iota+	4		4.52				707	15089
Kappa	4		4.16	15			130	2905
Lambda	5		4.73	14			123	2772
Mu	5		5.17	30			321	6582
Nu	6		4.89	25			223	4636
Xi	6		4.80	19			179	3901
Omicron	6		4.54	22			193	4180
Pi	6		4.94	20			184	4058
Rho	6		4.54	7			9045	224014
Sigma	6		4.88	8			9071	224572
Tau	6		4.87	5			9008	223165
Upsilon¹+	6		4.83	26			253	5234
Upsilon²			4.63	28			265	5395
Phi+	6		4.98	34			382	7927
Chi	6		4.71	39			442	9408
Psi	6		4.74	36			399	8491
Omega+	6		4.99	46			548	11529
A+	6		4.54	48			575	12111
B+							92	

The Lost, Missing, or Troublesome Stars of Cassiopeia

Iota, HR 707, 4.52V, *SA, HRP, BS,* HF 54*, BV 65 and Tabula X*, BF 292, B 157*, BAC 744, H 113*, W 51*.

Although Flamsteed observed Iota, it was among the 458 stars that were inexplicably omitted from his printed catalogue (*BF*, p. 392). Halley was among the first to equate this star with Bayer's Iota in his unauthorized edition of Flamsteed's catalogue. Bevis, Bode, Argelander (*UN,* p. 4), and Heis all properly identified this star as Iota. Surprisingly, although Baily included it in his revised edition of Flamsteed's catalogue and in the *BAC*, he did not equate it with Iota.

Upsilon¹, 26, HR 253, 4.83V, *SA, HRP, BS,* BV 25, BF 85*.
Upsilon², 28, HR 265, 4.63V, *SA, HRP, BS,* BV 26 as Upsilon, BF 89*.

Bayer described Upsilon as a single 6th-magnitude star, but Flamsteed designated these two nearly adjacent stars, his 26 and 28, as Upsilon-1 and Upsilon-2. Except for Bevis, who labeled only the brighter star, 28, as Upsilon, most authorities have accepted Flamsteed's designations.

Phi, 34, HR 382, 4.98V, *SA, HRP, BS,* HF 34*, BV 31*, B 118*, BF 152*.

In Halley's pirated edition of Flamsteed's catalogue, published in 1712, Phi is properly equated with 34, but in Flamsteed's catalogue of 1725, his "Catalogus Britannicus," 34 is left unlettered. Bevis caught the omission and corrected it.

Omega, 46, HR 548, 4.99V, *SA, HRP, BS,* BV 55, H 99*.
A, 48, HR 575, 4.54V, *SA, HRP, BS,* BV 64, H 103*.

Confusion persists concerning Omega and A. In Halley's 1712 edition of Flamsteed's catalogue, Omega and A are designated as follows:

Omega, 43 (HR 478, 5.59V, HF 41 as Omega, BV 49 as Omega, B 134 as Omega, BF 190 as Omega).
A, 38 (HR 427, 5.81V, HF 37 as A, BV 52 as A, B 126 as A, BF 167 as A).

In the authorized edition of Flamsteed's catalogue, the "Catalogus Britannicus," there is no reference whatsoever either to Omega or A. In the years that followed, Bevis, Bode, and Baily accepted Halley's designations that 43 and 38 were synonymous with Bayer's Omega and A, but Argelander thought them in error. Since 38 and 43 are relatively faint stars, he switched Omega and A to 46 and 48 (*UN,* p. 3). Although most authorities have accepted this switch, it presents some interesting problems. Bayer noted that Omega and A are 6th-magnitude stars, while 46 and 48 are considerably brighter. On the other hand, Bevis and Bode's proposed stars, 38 and 43, are closer to Bayer's magnitude. In addition, Bayer's chart shows Omega and A forming an equilateral triangle with Psi; so, too, do 38 and 43, but not 46 and 48. Moreover, Bayer's chart shows Omega about 1½° east of Psi; so, too, is 43, but not 46, which is about 3° away. Who, then, has correctly identified Bayer's stars, Bevis and Bode or Argelander? See Hoffleit, "Discordances," p. 47. Bevis's catalogue inadvertently omits any reference to Omega and A, but they are included and labeled properly on his chart (Tabula X).

B, HR 92, *SA, HRP, BS.*

This is the Supernova of November 1572, Tycho's Star. At its brightest it reached an estimated magnitude of -4, but faded beyond visibility within sixteen months. Nothing remains of the Nova but some faint wisps of nebulosity (see *Burnham's Celestial Handbook,* I, 503-19). For Bayer's use of Roman capitals other than A, see P Cyg and H Gem.

c, 43, HR 478, 5.59V, BV 49, BF 190.
d, 4, HR 8904, 4.98V, BV 9, BF 3215.
d, 46, HR 548, 4.99V, BV 55, BF 212.
e, 1, HR 8797, 4.85V, BV 1, BF 3171.
e, 48, HR 575, 4.54V, BV 64, BF 223.
f, 50, HR 580, 3.98V, BV 68, BF 228.

Bayer did not assign letters to these six stars. Flamsteed assigned them letters, but Bevis and Baily removed them as superfluous (see i Aql). Bevis and Bode thought 43 was Omega, while Argelander relettered 46 and 48 as Omega and A (see Omega and A). Bayer included 50 in his atlas, but he considered it an *informis,* an unformed star, between Cassiopeia and Cepheus and left it unlettered.

CENTAURUS, CEN

Centaur

Centaurus (Figure 21) is a Ptolemaic constellation that Bayer lettered from Alpha to Omega and from A to q. Since it is mostly below the horizon in Central Europe, Bayer's chart does not conform to the constellation's actual appearance in the heavens. As a consequence, Lacaille redesigned and relettered it from Alpha to Omega, from a to z, and from A to Z. Bayer included Crux as an asterism in Centaurus, but Lacaille separated it and assigned it its own set of letters.

Although some scholars claim that Centaurus dates back to the Middle East, no astral legends or myths have thus far been found that refer to this constellation. In the Centaur's section of the heavens, the Mesopotamians recognized the constellation EN.TE.NA.BAR.HUM (Akkadian Habasiranu), which has not yet been translated.

According to Greek mythology, centaurs are creatures half-man and half-horse, human to the waist and equine below. They are the offspring of Ixion, King of Thessaly, and a cloud that he thinks is actually the goddess Hera. Hera's husband, Zeus, is so furious that Ixion would even dare to think of having an affair with his wife that he confines him to the lowest regions of hell, bound by serpents to a continuously revolving wheel. The association of Ixion to centaurs probably occurred because the Thessalians were reputed to be the first mortals to tame horses.

In another legend, the centaur in the sky is Chiron, son of Cronos — King of the Titans — and the sea nymph Philyra. When Cronos's wife, Rhea, catches him with Philyra, Cronos turns himself into a horse and gallops away, leaving behind a hybrid child, Chiron. When he attains adulthood, Chiron becomes famous for his righteousness, piety, and scholarship. He is the teacher of many famous heroes in Greek mythology — Achilles, Jason, and Asclepius, the god of medicine. Chirion is killed accidentally by one of Hercules' poisoned arrows and placed in the heavens where he is depicted as placing an offering, Lupus, the Wolf, on Ara, the Altar — a final tribute to his righteousness and piety.

See Hunger and Pingree, *Mul.Apin,* p. 138; Reiner and Pingree, *Baby. Plan. Omens,* Part 2, pp. 11-12; Condon, *Star Myths,* pp. 79-82; Ridpath, *Star Tales,* pp. 47-48; Stall, *New Patterns in the Sky,* pp. 170-71; Schwab, *Gods & Heroes,* p. 210.

Table 24. The lettered stars of Centaurus

	Magnitudes			Catalogue Numbers				
Letter	Bayer	Lacaille	Visual	Flamsteed	Lacaille	Other	HR	HD
Alpha^1+		1	-0.01		1227		5459	128620
Alpha^2		4	1.33		1226		5460	128621
Beta		1	0.61		1185		5267	122451
Gamma		3	2.17		1098		4819	110304
Delta		3	2.60		1064		4621	105435
Epsilon		3	2.30		1155		5132	118716
Zeta		3	2.55		1177		5231	121263
Eta		3	2.31		1219		5440	127972
Theta		3	2.06	5	1192		5288	123139
Iota		3	2.75		1143		5028	115892

(table continued on next page)

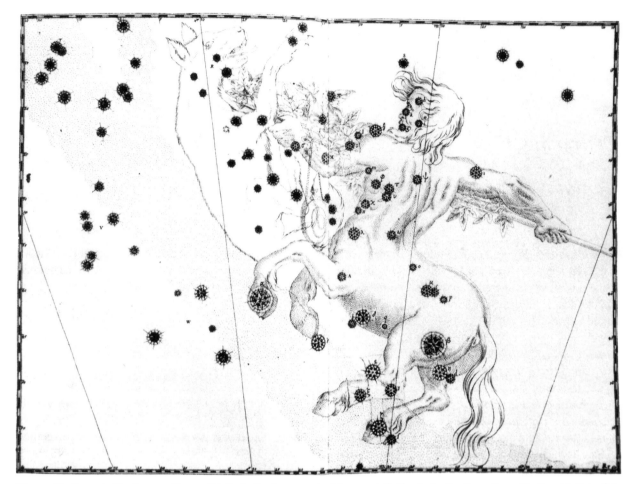

Figure 21. Bayer's Centaurus. Drawn faintly in the lower part of the constellation is the Southern Cross, Epsilon (ε), Zeta (ζ), Nu (ν), and Xi (ξ). The letter R is Lupus; S is the tail of Hydra; T is the tail of Scorpius; V is Ara; and W is Triangulum Australe.

Table 24. The lettered stars of Centaurus *(continued)*

	MAGNITUDES			CATALOGUE NUMBERS				
Letter	Bayer	Lacaille	Visual	Flamsteed	Lacaille	Other	HR	HD
Kappa		3	3.13		1255		5576	132200
Lambda		4	3.13		1025		4467	100841
Mu		4	3.04		1166		5193	120324
Nu		4	3.41		1165		5190	120307
Xi^1+		6	4.85		1121		4933	113314
Xi^2		5	4.27		1126		4942	113791
Omicron^1+			5.13			G 37	4441	100261
Omicron^2			5.15			G 38	4442	100262
Pi		4	3.89		1004		4390	98718
Rho		4	3.96		1068		4638	105937
Sigma		5	3.91		1086		4743	108483
Tau		5	3.86		1093		4802	109787
Upsilon^1+		5	3.87		1183		5249	121790
Upsilon^2		6	4.34		1184		5260	122223
Phi		5	3.83		1182		5248	121743
Chi		5	4.36		1191		5285	122980
Psi		5	4.05		1205		5367	125473
Omega+		Neb			1148			
a		5	4.42		1207		5378	125823

(table continued on next page)

Table 24. The lettered stars of Centaurus *(continued)*

	MAGNITUDES			CATALOGUE NUMBERS				
Letter	Bayer	Lacaille	Visual	Flamsteed	Lacaille	Other	HR	HD
b		5	4.00		1232		5471	129116
c^1+		5	4.05		1234		5485	129456
c^2		6	4.92		1237		5489	129685
d		5	3.88		1150		5089	117440
e		6	4.33		1111		4888	111915
f		6	4.71		1123		4940	113703
g+		5	4.19	2	1167		5192	120323
h+		5	4.73	4	1173		5221	120955
i+		5	4.23	1	1162		5168	119756
j+			4.32			G 69	4537	102776
k+		5	4.32c	3	1171		5210/1	120709/10
l		6	4.64		1095		4817	110073
m+			4.53			G 214	5041	116243
n		5	4.27		1112		4889	111968
o+		6	5.55		1085		4735	108323
p		6	4.91		1109		4874	111597
q+		6	7.1		1122			113398
r		6	5.10		1139		5006	115310
s+		6	6.45		1151		5098	117718
t+		6	5.83		1156		5128	118646
u		6	5.44		1087		4748	108541
v+			4.33			G 336	5358	125288
w+			4.66			G 136	4831	110458
x^1+		6	5.32		1080		4712	107832
x^2		6	5.73		1081		4724	108114
y		6	5.54		1174		5222	120987
z		6	5.15		1164		5174	119921
A		6	4.62		1022		4460	100673
B+		6	4.46		1044		4546	102964
C^1+		6	5.71		1023		4463	100733
C^2			5.25			G 45	4466	100825
C^3		6	5.44		1027		4476	101067
D		6	5.31		1069		4652	106321
E		6	5.34		1063		4620	105416
F		6	5.00		1074		4682	107079
G		6	4.82		1083		4732	108257
H		6	5.16		1117		4913	112409
I+		6	5.73		1108		4872	111588
J+			4.53			G 208	5035	116087
K		6	5.06		1149		5071	117150
L+		6	6.06		1129		4965	114365
M		6	4.65		1161		5172	119834
N		6	5.25		1169		5207	120642
O^1+		6	4.65		1106		4848	110956
O^2		5	3.95c		1114		4898/9	112092/91
P+		6	6.14		1178		5234	121336
Q		6	5.01		1158		5141	118991

The Lost, Missing, or Troublesome Stars of Centaurus

Alpha¹, HR 5459, -0.01V, *SA* as Alpha, *HRP, BS,* CA 1227 as Alpha, BR 4991 as Alpha², L 6017 as Alpha², G 363 as Alpha, Rigel Kent.

Alpha², HR 5460, 1.33V, *HRP, BS,* CA 1226 as Alpha, BR 4990 as Alpha¹, L 6014 as Alpha¹, G 364 as Alpha.

The third brightest star in the heavens, Alpha Cen was discovered to be a multiple-star system in 1689 when Father Jean Richaud split it into two components. Lacaille designated each of its components Alpha. The change in indices is a result of the relatively short orbital period of A and B, about eighty years, during which time they switch positions. Actually, this is a triple-star system. Component A is HR 5459 and component B is HR 5460. Discovered in 1911 by R. T. A. Innes, component C, Proxima Centauri, about 11th magnitude, is almost 2° southwest from Alpha, and, after the sun, is the nearest star to Earth, at 4.22 light-years away. See *Burnham's Celestial Handbook,* I, 549-52; *Sky Cat.,* II, 160.

Xi¹, HR 4933, 4.85V, *SA, HRP, BS,* CA 1121 as Xi, BR 4299*, L 5370*, G 165*.
Xi², HR 4942, 4.27V, *SA, HRP, BS,* CA 1126 as Xi, BR 4321*, L 5396*, G 173*.

Lacaille designated these two neighboring stars Xi.

Omicron¹, HR 4441, 5.13V, *SA, HRP, BS,* CA 1017, BR 3631, L 4774, G 37*.
Omicron², HR 4442, 5.15V, *SA, HRP, BS,* CA 1018, BR 3633, L 4775, G 38*.

Lacaille did not letter this star, and he omitted the letter Omicron from this constellation. Although he observed and catalogued these two nearly adjacent stars, he left them unlettered; he considered them 6th magnitude. Gould designated them Omicron since he felt their combined light was brighter than 5th magnitude. He stated erroneously that Lacaille's Omicron was in Hydra (*Uran. Arg.,* p. 83). The star Gould thought was Omicron Cen in Hydra is actually Lacaille's o (oh) Cen, HR 4735, 5.55V. See o.

Upsilon¹, HR 5249, 3.87V, *SA, HRP, BS,* CA 1183 as Upsilon, BR 4707*, L 5770*, G 297*.
Upsilon², HR 5260, 4.34V, *SA, HRP, BS,* CA 1184 as Upsilon, BR 4729*, L 5782*, G 303*.

Lacaille designated these two neighboring stars Upsilon.

Omega, *SA,* CA 1148*, L 5533, BAC 4485*, G 219*.

This is the globular cluster NGC 5139 with a total magnitude of 3.65V. See Xi Tuc for Lacaille's nebulous objects.

a, HR 5378, 4.42V, *SA, HRP,* CA 1207*, BR 4883*, L 5911*, G 342*.

b, HR 5471, 4.00V, *SA, HRP,* CA 1232*, BR 5011*, L 6048, G 368*.

c¹, HR 5485, 4.05V, *SA, HRP,* CA 1234 as c, BR 5029*, L 6063, G 371*.
c², HR 5489, 4.92V, *SA, HRP,* CA 1237 as c, BR 5039*, L 6071, G 372*.

Lacaille designated these two nearly adjacent stars c.

d, HR 5089, 3.88V, *SA, HRP,* CA 1150*, BR 4496*, L 5569, G 227*.

e, HR 4888, 4.33V, *SA, HRP,* CA 1111*, BR 4225*, L 5308, G 149*.

f, HR 4940, 4.71V, *SA, HRP,* CA 1123*, BR 4316*, L 5390, G 171*.

g, 2, HR 5192, 4.19V, *SA, HRP,* CA 1167*, BR 4647*, L 5688*, BAC 4603, G 274*.
h, 4, HR 5221, 4.73V, *SA, HRP,* CA 1173*, BR 4669*, L 5725, BAC 4629*, G 284*.
i, 1, HR 5168, 4.23V, *SA, HRP,* CA 1162*, BR 4619*, L 5668, BAC 4579*, G 265*.

These three stars, along with k and Theta, are among the few for which Lacaille retained their Bayer letters, probably because, located in the northern extremity of the Centaur (his head and shoulders), they are, by and large, properly positioned on Bayer's chart. They are, moreover, the only stars in Centaurus that Flamsteed observed and numbered.

j, HR 4537, 4.32V, *SA, HRP, BS,* CA 1041, BR 3787, L 4903, G 69*.

Although Lacaille observed and catalogued this star, he did not letter it since he considered it only 6th magnitude. Gould designated it j because he felt a star of its brightness merited a letter.

k, 3, HR 5210/1, 4.56V, 6.06V, comb. mag. 4.32V, *SA, HRP,* CA 1171*, BR 4662*, L 5708*, BAC 4623*, G 280/1*.

See g, h, i.

l, HR 4817, 4.64V, *SA, HRP,* CA 1095*, BR 4135*, L 5231, G 132*.

m, HR 5041, 4.53V, *SA, HRP,* CA 1145, BR 4428, L 5500, G 214*.

Although Lacaille observed and catalogued this star, he left it unlettered since he considered it only 6th magnitude. Gould, mistakenly thinking that Lacaille had omitted the letter m from Centaurus (*Uran. Arg.,* p. 83), designated this star m because of its brightness. Lacaille had assigned m to HR 4979 (4.85V, CA 1134 as m, BR 4361 as m, L 5429, G 185), but because of Gould's oversight, it has lost its letter designation; Gould observed it but left it unlettered.

n, HR 4889, 4.27V, *SA, HRP,* CA 1112*, BR 4232, L 5312, G 150*.

o, in Hydra, HR 4735, 5.55V, CA 1085*, BR 4056*, L 5154, BAC 4192, G 150 in Hydra.

Baily included this star in Hydra and left it unlettered.

p, HR 4874, 4.91V, *SA, HRP,* CA 1109*, BR 4212*, L 5296, G 143*.

q, HD 113398, 7.1V, CA 1122*, BR 4302*, L 5376, BAC 4369.

Gould not only dropped this star's letter but also excluded the star from his *Uran. Arg.* because it was less than 7th magnitude.

r, HR 5006, 5.10V, *SA, HRP,* CA 1139*, BR 4386*, L 5466, G 195*.

s, in Hydra, HR 5098, 6.45V, CA 1151*, BR 4519*, L 5578, BAC 4517, G 349 in Hydra.
t, in Hydra, HR 5128, 5.83V, CA 1156*, BR 4571*, L 5623, BAC 4548, G 354 in Hydra.

Baily included these two stars in Hydra and left them unlettered.

u, HR 4748, 5.44V, *SA, HRP,* CA 1087*, BR 4068*, L 5164, G 122*.

v, HR 5358, 4.33V, *SA, HRP,* CA 1200 as V, BR 4847 as V, L 5879, G 336*.

Lacaille did not assign this star lower-case v. He observed it and designated it V, but Gould relettered it v to avoid confusion with variables, which start with the letter R.

w, HR 4831, 4.66V, *SA, HRP,* CA 1100, BR 4161, L 5250, G 136*.

Lacaille did not letter this star w. Although he observed and catalogued it, he left it unlettered since he considered it 6th magnitude. Gould mistakenly thought that Lacaille had omitted the letter w from Centaurus (*Uran. Arg.,* p. 83) and, consequently, he designated this star w because of its brightness. Lacaille had assigned w to HR 4973 (5.25V, CA 1133 as w, BR 4353 as w, L 5422, G 182), but because of Gould's oversight, it lost its letter designation; Gould observed it but left it unlettered.

x^1, HR 4712, 5.32V, *HRP, BS,* CA 1080 as x, BR 4035*, L 5129, BAC 4174 erroneously as Kappa^1, G 113*.
x^2, HR 4724, 5.73V, *HRP, BS,* CA 1081 as x, BR 4046*, L 5142, BAC 4183 erroneously as Kappa^2, G 118*.

Lacaille designated these two neighboring stars x. Some confusion arose over these stars because of a typographical error in the *BAC* that referred to them as Kappa. Kappa and x are similar in appearance, κ, x. See Hoffleit, "Discordances," p. 48.

y, HR 5222, 5.54V, *SA, HRP,* CA 1174*, BR 4671*, L 5726, G 286*.

z, HR 5174, 5.15V, *SA, HRP,* CA 1164*, BR 4627*, L 5676, G 267*.

A, HR 4460, 4.62V, *SA, HRP,* CA 1022*, BR 3657*, L 4794, G 42*.

B, HR 4546, 4.46V, *SA, HRP,* CA 1044*, BR 3796*, L 4941*, BAC 4007, G 71*.

Probably because of an oversight, Baily did not remove this star's Roman letter from Lacaille's expanded catalogue. As he stated in the preface to the *BAC* (pp. 62-63), he intended to drop all of Lacaille's Roman letters except those in the three newly created subdivisions of Argo Navis.

C¹, HR 4463, 5.71V, *SA, HRP,* CA 1023 as C, BR 3660*, L 4796, BAC 3936, G 44*.
C², HR 4466, 5.25V, *SA, HRP,* BR 3663, BAC 3938, G 45*.
C³, HR 4476, 5.44V, *SA, HRP,* CA 1027 as C, BR 3684 as C², L 4815, BAC 3951, G 53*.

Lacaille designated two neighboring stars, CA 1023 and CA 1027, as C. Gould, noting there was a third star — HR 4466 — between them, labeled all three stars C¹, C², and C³.

D, HR 4652, 5.31V, *SA, HRP,* CA 1069*, BR 3967*, L 5069, G 103*.

E, HR 4620, 5.34V, *SA, HRP,* CA 1063*, BR 3932*, L 5031, G 93*.

F, HR 4682, 5.00V, *SA, HRP,* CA 1074*, BR 3995*, L 5092*, G 108*.

G, HR 4732, 4.82V, *SA, HRP,* CA 1083*, BR 4052*, L 5150, G 119*.

H, HR 4913, 5.16V, *SA, HRP,* CA 1117*, BR 4254*, L 5331, G 156*.

I, HR 4872, 5.73V, CA 1108*, BR 4210*, L 5294, BAC 4307, G 142.

Gould dropped this star's letter because he considered it less than 6th magnitude, hence not worthy of being lettered.

J, HR 5035, 4.53V, *SA, HRP,* CA 1144 as Z, BR 4421 as Z², L 5492, G 208*.

Lacaille did not letter this star J. He observed it and designated it Z, but Gould relabeled it J to avoid confusion with variables.

J is a double star, Finsen 208. Brisbane's Z¹ (BR 4420) is HR 5034 (6.18V, G 207), component B of this multiple star system; HR 5035 is component A. Gould noted that the combined light of both components appeared as one object to his naked eye.

K, HR 5071, 5.06V, *SA, HRP,* CA 1149*, BR 4476*, L 5552, G 224*.

L, HR 4965, 6.06V, CA 1129*, BR 4346*, L 5413, BAC 4400, G 179.

Gould dropped this star's letter because of its dimness.

M, HR 5172, 4.65V, *SA, HRP,* CA 1161*, BR 4618*, L 5664, G 266*.

N, HR 5207, 5.25V, *SA, HRP,* CA 1169*, BR 4656*, L 5700, G 277*.

O¹, in Crux, HR 4848, 4.65V, CA 1106 as O, BR 4182*, L 5273, BAC 4284, G 45 in Crux.
O², in Crux, HR 4898/9, 4.03V, 5.17V, comb. mag. 3.95V, CA 1114 as O, BR 4237*, L 5317, BAC 4325, G 52 as Mu Cru in Crux.

Lacaille assigned O to this pair of neighboring stars in Centaurus. Gould, however, included the pair in Crux, dropped the letter from CA 1106, and relettered CA 1114 as Mu Cru because of its brightness. He did not reletter CA 1106 because he considered it less than 5th magnitude.

P, HR 5234, 6.14V, CA 1178*, BR 4685*, L 5741, BAC 4641, G 290.

Gould dropped this star's letter because of its dimness.

Q, HR 5141, 5.01V, *SA, HRP,* CA 1158*, BR 4582*, L 5632, G 250*.

R, HR 5297, 4.75V, CA 1195*, BR 4779*, L 5827, BAC 4695, G 321.
S¹, HR 4944, 5.99V, CA 1124 as S, BR 4319*, L 5392, BAC 4380, G 174.
S², HR 4975, 4.60V, CA 1132 as S, BR 4354*, L 5418, BAC 4412, G 183.
S³, HR 4989, 4.92V, CA 1137 as S, BR 4370*, L 5437, BAC 4422, G 191.
T, HR 5140, 5.38V, CA 1157*, BR 4580*, L 5627, BAC 4557, G 251.

Lacaille designated these five stars as noted above, but Gould dropped their letters to avoid confusion with variables, which Argelander had just begun to label with letters beginning with R. Since Gould considered these stars all less than 5th magnitude, he did not reletter them.

U

Lacaille did not use this letter.

V, HR 5358, 4.33V, CA 1200.

Gould relettered this star v to avoid confusion with variables. See v.

W

Lacaille did not use this letter in Centaurus.

X, HR 5316, 5.07V, CA 1198*, BR 4810*, L 5850, BAC 4709, G 328.
Y, HR 5371, 4.92V, CA 1204*, BR 4864 as Y¹, L 5893, BAC 4749, G 340.

Lacaille designated these two stars X and Y, but Gould dropped their letters to avoid confusion with variables (see R, S, T). HR 5371 is a quadruple-star system, Dunlop 159. Brisbane's Y², BR 4865, is component B, magnitude 7.1V.

Z, HR 5035, 4.53V, CA 1144.

Gould relettered this star J to avoid confusion with variables. See J.

Cepheus, Cep

Cepheus, King

Cepheus (Figure 22) is a Ptolemaic constellation that Bayer lettered from Alpha to Rho.

Although some scholars have suggested that Cepheus may have originated in the Middle East, there are no astral legends or myths to support this supposition. In its place in the Mesopotamian night sky was the constellation the Panther.

The ancient Greeks believed that Zeus, always on the lookout for a pretty girl, becomes infatuated with Io, the beautiful daughter of the king of Argos. When Zeus's wife Hera approaches the amorous couple, Zeus quickly changes Io into a heifer, hoping to hide his relationship with her. Hera suspects correctly what her philandering husband has been up to. She sends a gadfly to torment Io the heifer across half the known world until Zeus eventually succeeds in restoring Io to her human form. Along the way, Io gives her name to the Ionian Sea, one of the areas she crosses. The offspring that results from her affair with Zeus becomes the ancestor of a line of kings that eventually produce Cepheus II, King of Ethiopia or, according to some accounts, King of Phoenicia, or possibly, King of Media (Persia). Cepheus marries Cassiopeia, whose bragging about her own beauty so angers the sea nymphs and Poseidon that the sea god dispatches the monster Cetus to ravish the country. Hoping to stop Cetus's continuing attacks on his land, Cepheus consults the Oracle of Ammon, who advises the king that the only way to stop the attacks is to offer his daughter as a sacrifice to Cetus. The king reluctantly has his daughter, Andromeda, chained to a rock along the coast, but just as Cetus is about to devour her, Perseus slays the monster and as a reward for her rescue, he asks for Andromeda in marriage. King Cepheus readily consents and arranges for the ceremony. At the marriage feast, the king's brother, Phineus, arrives and claims that Andromeda was promised to him. A battle ensues between Perseus and Phineus and his followers in which Perseus emerges victorious. Eventually, the goddess Athena places Cepheus, together with Andromeda, Cassiopeia, Perseus, and Cetus, among the stars in the heavens.

See Hunger and Pingree, *Mul.Apin,* p. 138; Hamilton, *Mythology,* pp. 95-99, 468; Ridpath, *Star Tales,* pp. 49-50; Condos, *Star Myths,* pp. 83-84.

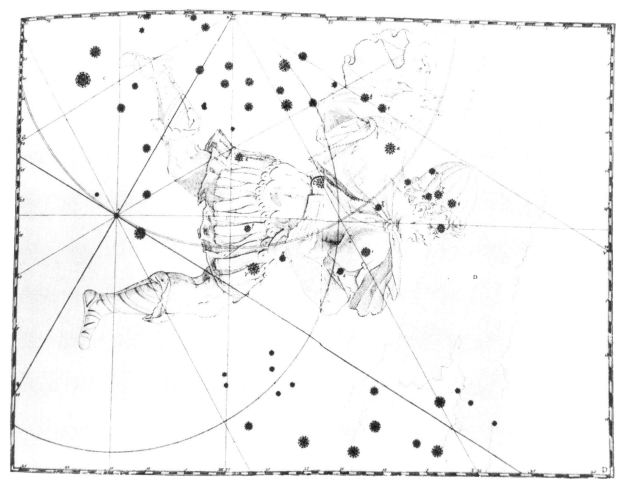

Figure 22. Bayer's Cepheus. The points A and B are the celestial north pole and the ecliptic north pole. The letter C is Ursa Minor and D is the Milky Way.

Lost Stars

Table 25. The lettered stars of Cepheus

	MAGNITUDES			CATALOGUE NUMBERS				
Letter	Bayer	Lacaille	Visual	Flamsteed	Lacaille	Other	HR	HD
Alpha	3		2.44	5			8162	203280
Beta	3		3.23	8			8238	205021
Gamma	3		3.21	35			8974	222404
Delta	4		3.75	27			8571	213306
Epsilon	4		4.19	23			8494	211336
Zeta	4		3.35	21			8465	210745
Eta	4		3.43	3			7957	198149
Theta	4		4.22	2			7850	195725
Iota	4		3.52	32			8694	216228
Kappa	4		4.39	1			7750	192907
Lambda	5		5.04	22			8469	210839
Mu+	5		4.08				8316	206936
Nu+	5		4.29	10			8334	207260
Xi^1+			6.2					209791
Xi^2	5		4.29	17			8417	209790
Omicron	5		4.75	34			8872	219916
Pi	5		4.41	33			8819	218658
Rho^1+			5.83	28			8578	213403
Rho^2	5		5.52	29			8591	213798

The Lost, Missing, or Troublesome Stars of Cepheus

Mu, HR 8316, 4.08V, *SA, HRP, BS,* H 43*, W 12*, Herschel's Garnet Star, a semi-regular variable with a range of 3.43V to 5.1V.

Some confusion surrounds the identity of this star. Bayer described it as 5th magnitude, and Flamsteed designated his 13 (HR 8371, 5.80V, BF 2997 as Mu, H 55) as Mu, and Baily agreed. But Argelander, unable to see 13 with his naked eye, designated 14 (HR 8406, 5.56V, H 61) as Mu (*UN,* p. 12). Heis saw both stars but felt Mu was synonymous with the variable HR 8316 because, of the three stars, it conformed best to Bayer's chart and its magnitude was closest to Bayer's. See Hoffleit, "Discordances," p. 48; also, Bevis's comments about Mu, Tabula IV.

Nu, 10, HR 8334, 4.29V, *SA, HRP, BS,* BV 16*, BF 2984*, H 46*.

Bayer described Nu as a 5th-magnitude star, and Flamsteed designated his 15 (HD 209744, 6.70V, BF 3025) as Nu. However, because Bevis felt that this star was much too dim for naked-eye observation and did not conform well to Bayer's chart, he switched Nu to Flamsteed's 10, and Baily agreed. See Bevis's comments about Nu, Tabula IV.

Xi^1, Xi^2, 17, HD 209791, HR 8417, 6.2V, 4.29V, *SA, HRP* as Xi, *BS* as Xi, H 63 as Xi, ADS 15600.

Bayer described Xi as a single 5th-magnitude star. Later astronomers added the indices to identify the components of this multiple-star system.

Rho^1, 28, HR 8578, 5.83V, *BS,* H 86 as Rho.
Rho^2, 29, HR 8591, 5.52V, *SA* as Rho, *HRP* as Rho, *BS,* BV 43 as Rho, H 87 as Rho.

Bayer noted that Rho was a single 5th-magnitude star, and Flamsteed designated his 29 as Rho. Argelander, though, thought that Bayer must have seen the combined light of 28 and 29 as Rho since the two stars are nearly adjacent (*UN,* p. 12).

Cetus, Cet

Sea Monster, Whale

Cetus (Figure 23) is a Ptolemaic constellation that Bayer lettered from Alpha to Psi. The Oxford Greek-English lexicon defines Cetus (κητος) as "any sea monster or huge fish." Some have suggested a Middle Eastern source for Cetus, but no Mesopotamian or Levantine astral myth or legend can be traced directly to this constellation.

At the very beginning of time, according to Greek mythology, the Earth (Gaea) and the Sea (Pontus) produce a son and daughter, Phorcys and Ceto. They, in turn, give birth to numerous non-human monsters. The Gorgons — three horrible sisters with snakes for hair, scales for skin, wings, claws, and huge sharp teeth — can turn a man to stone with just one glance. The three gray-haired Graiae, sisters to the Gorgons, have but one tooth and one eye between them. Phorcys and Ceto's other offspring include Echidna, half-woman and half-snake, and Ladon, a serpent. This generation produces even more hideous monsters: Orthos, a two-headed hound; Hydra, a water snake with nine heads; Cerberus, a three-headed hound, the guardian of the Underworld; Chimaera, a fire-breathing, three-headed monster with the forepart of a lion, the body of a goat, and the tail of a snake; Sphinx, a winged creature with the forepart of a woman and the hindpart of a lion; and the Nemean Lion, a monstrous beast that cannot be harmed by human weapons. Cetus, like these creatures, may very well be an offspring or descendant of Phorcys and Ceto.

The sea monster Cetus is sent by the sea god Poseidon to ravage the coastline of King Cepheus's kingdom as punishment for Queen Cassiopeia's boasts about her beauty. To appease the gods, Cepheus must offer his daughter, Andromeda, as a sacrifice to Cetus. Perseus, however, arrives in the nick of time, slays the monster, and marries Andromeda. Athena places all five characters in the heavens as an eternal reminder of one of the great legends in Greek mythology.

See Condos, *Star Myths,* pp. 85-86, 233; Schwab, *Gods & Heroes,* pp. 64-65, 172; Gayley, *Classic Myths,* pp. 55-57, 486; Gantz, *Early Greek Myth,* pp. 19-25, Table 2; Hamilton, *Mythology,* p. 48; Ridpath, *Star Tales,* pp. 50-51; Staal, *New Patterns in the Sky,* pp. 33-35. See also Andromeda.

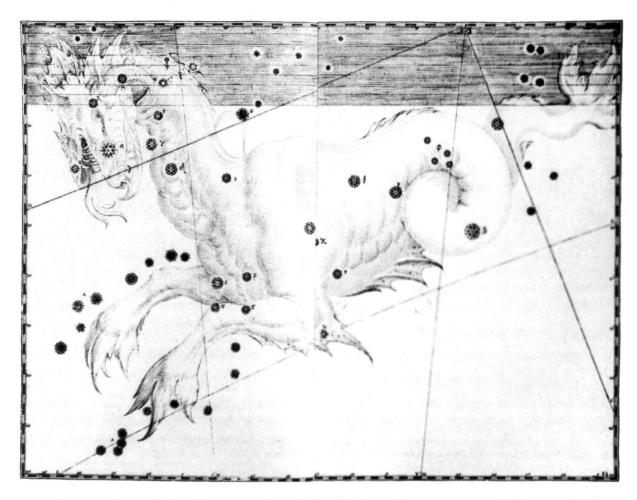

Figure 23. Bayer's Cetus. The letter Omega (ω) is the knot in the ribbon that binds together the two fish in Pisces, which Bayer called *Nodus coelestis,* the Knot of Heaven. The letter A is Eridanus. The letter B is the First Point of Aries, ♈, and it is so marked on the chart where the vernal equinox, C, meets the celestial equator, D. The line E, parallel to the equator, is the Tropic of Capricorn. The shaded area across the top is the zodiac.

Table 26. The lettered stars of Cetus

	MAGNITUDES			CATALOGUE NUMBERS				
Letter	Bayer	Lacaille	Visual	Flamsteed	Lacaille	Other	HR	HD
Alpha	2		2.53	92			911	18884
Beta	2		2.04	16			188	4128
Gamma	3		3.47	86			804	16970
Delta	3		4.07	82			779	16582
Epsilon	3		4.84	83			781	16620
Zeta	3		3.73	55			539	11353
Eta	3		3.45	31			334	6805
Theta	3		3.60	45			402	8512
Iota	3		3.56	8			74	1522
Kappa^1+	4		4.83	96			996	20630
Kappa^2			5.69	97			1007	20791
Lambda	4		4.70	91			896	18604
Mu	4		4.27	87			813	17094
Nu	4		4.86	78			754	16161
Xi^1+	4		4.37	65			649	13611
Xi^2	4		4.28	73			718	15318
Omicron	4		3.04	68			681	14386
Pi	4		4.25	89			811	17081
Rho	4		4.89	72			708	15130
Sigma	4		4.75	76			740	15798
Tau	4		3.50	52			509	10700
Upsilon+	4		4.00	59			585	12274
Phi^1+	5		4.76	17			194	4188
Phi^2	5		5.19	19			235	4813
Phi^3	5		5.31	22			267	5437
Phi^4	5		5.61	23			279	5722
Chi	5		4.67	53			531	11171
Psi+	6		5.47	24Ari			702	14951
f+			4.89	6			33	693

The Lost, Missing, or Troublesome Stars of Cetus

Kappa^1, 96, HR 996, 4.83V, *SA* as Kappa, *HRP* as Kappa, *BS*, HF 77 as Kappa, BF 420*, H 161 as Kappa, G 319 as Kappa.
Kappa^2, 97, HR 1007, 5.69V, *BS*, HF 78, BF 424*, H 162, G 321.

Bayer's *Uranometria* describes Kappa as a single 4th-magnitude star, and in Halley's edition of Flamsteed's catalogue, 96 is designated Kappa. In Flamsteed's "Catalogus Britannicus" of 1725, however, both 96 and 97 are labeled Kappa-1 and Kappa-2. Argelander, though, agreed with Halley that only the brighter of the two, 96, was equivalent to Bayer's Kappa (*UN*, p. 88).

Bevis (Tabula XXIII) asserted that Kappa was synonymous with Bayer's g Tau and Baily agreed, but Argelander and most others felt that g is HR 958, 5.56V. A comparison between Bayer's *Uranometria* and modern atlases indicates that Bevis was probably correct. Bayer's charts, for example, show both stars, g Tau and Kappa Cet, in the same position: long. 13° in Taurus, lat. 14½° south of the ecliptic. Also, on modern charts HR 958 is 4° southwest of Omicron Tau while 96, Kappa, is 6° southwest, which corresponds almost exactly with Bayer's position for g Tau. Finally, in Schiller's *Coelum Stellatum Christianum,* Bayer had an opportunity to correct and revise his *Uranometria.* The new work removes the star from Saint Andrew (Taurus) and designates it as 26 Saints Joachim and Anne (Cetus). It is, moreover, specifically described as Bayer's 23 (Kappa) Cet, with the same coordinates adjusted for precession from 1600, the epoch of *Uranometria,* to 1625, the epoch of *Coelum Stellatum Christianum.* For Bayer's duplicate stars, see Alpha And; see also g Tau.

Julius Schiller sought to "depaganize" the heavens in his star atlas *Coelum Stellatum Christianum* by removing all the pagan constellations and replacing them with Christian saints and Biblical references. For example, he replaced Cetus with the constellation Saints Joachim and Ann; and Taurus with Saint Andrew. See c Ari.

Xi¹, 65, HR 649, 4.37V, *SA, HRP, BS,* BF 273*.
Xi², 73, HR 718, 4.28V, *SA, HRP, BS,* BF 309*.

Bayer described Xi as a pair of 4th-magnitude stars, and Flamsteed designated his 65 and 73 as Xi-1 and Xi-2. Baily suggested that Xi¹ might be either 65 or its neighbor 64 (HR 635, 5.63V), but most authorities have rejected 64 because of its dimness.

Upsilon¹, 56, HR 565, 4.85V, BF 237, H 81.
Upsilon², 59, HR 585, 4.00V, *SA* as Upsilon, *HRP* as Upsilon, *BS* as Upsilon, BF 246 as Upsilon, H 83 as Upsilon.

Although Bayer described only one 4th-magnitude star as Upsilon, Flamsteed designated these two stars, his 56 and 59, about 1° apart, as Upsilon-1 and Upsilon-2. Baily and Argelander, though, felt that only the brighter star, 59, was synonymous with Bayer's Upsilon (*UN,* p. 86).

Phi¹, 17, HR 194, 4.76V, *SA, HRP, BS,* BV 13*, BF 63*, H 29 and Corrigenda, which mistakes Psi¹ for Phi¹.
Phi², 19, HR 235, 5.19V, *SA, HRP, BS,* BF 83*, H 37*.
Phi³, 22, HR 267, 5.31V, *SA, HRP, BS,* BF 94*, H 41*.
Phi⁴, 23, HR 279, 5.61V, *SA, HRP, BS,* BF 97*, H 42*.

Bayer described Phi as a group of four 5th-magnitude stars, and Flamsteed designated his 17, 19, 22, and 23 as Phi-1, Phi-2, Phi-3 and Phi-4. Although Baily noted that these four stars did not conform to Bayer's chart, he, Heis, and most astronomers accepted Flamsteed's designations. But Bevis did not. He designated only 17 as Phi-1; the others, he stated, do not "agree with any stars in the modern catalogues (Tabula XXXIV)." Argelander went one step farther and refused to designate any star as Phi, either in his atlas or catalogue (*UN,* p. 85).

Psi

This star is the same as Xi, 24 Ari, HR 702 (5.47V, BF 300 as Xi Ari). It is one of Bayer's duplicate stars. In Schiller's *Coelum Stellatum Christianum,* Bayer removed Psi from Saints Joachim and Anne (Cetus) and designated it 28 Saint Peter (Aries). Bevis (Tabula XXII) also noted the duplication. For Bayer's duplicate stars, see Alpha And.

f, 6, HR 33, 4.89V, *SA,* BV 4, BF 3304.
g, 2, HR 9098, 4.55V, BV 2, BF 3290.
h, 7, HR 48, 4.44V, BV 3, BF 3308.

Bayer did not letter these stars. Although he included these three stars in his atlas, he considered them *informes,* unformed stars between Cetus and Aquarius and left them unlettered. Flamsteed designated them f, g, and h, but Bevis and Baily removed the letters as superfluous (see i Aql). Flamsteed started with f since Bayer's last letter in Cetus was E, the Tropic of Capricorn. For Bayer's use of upper-case Roman letters, see P Cyg. The letter f has somehow managed to survive in the literature, primarily in the atlases of Becvar and Tirion. The latter inluded it in his *Uranometria 2000.0,* but not in his second edition of *SA.*

CHAMAELEON, CHA

Chameleon

Keyzer and Houtman devised the constellation Chamaeleon (Figure 24). Houtman called it Het Chameljeon, the Chameleon, in his catalogue of 1603. As with all of Keyzer and Houtman's new constellations, Bayer depicted it unlettered on his chart of the southern skies. Lacaille assigned it letters from Alpha to Pi.

When Keyzer and Houtman devised their twelve new constellations in the southern skies, many of the figures they depicted represented new forms of life that Europeans were encountering for the first time as they traveled throughout the world on their voyages of exploration, conquest, and colonization.

Dorado, the Coryphaena
Piscis Volans, the Flying Fish
Apus, the Bird of Paradise
Pavo, the Peacock
Tucana, the Toucan
Indus, the Indian
Chamaeleon, the Chameleon

The chameleon, a member of the reptile family, is mostly found in Madagascar and tropical Africa, although one species has migrated to southern Spain. Houtman was familiar with these creatures since he had been to Madagascar and published a small handbook of the Malayan and Madagascar languages.

See Houtman, *Spraeckende Woordboeck Inde Maleysche ende Madagaskarche Talen;* Staal, *New Patterns in the Sky,* pp. 244-45; Knobel, "On Frederick de Houtman's Catalogue," p. 420. See also Camelopardalis.

Table 27. The lettered stars of Chamaeleon

	Magnitudes			Catalogue Numbers				
Letter	Bayer	Lacaille	Visual	Flamsteed	Lacaille	Other	HR	HD
Alpha		5	4.07		770		3318	71243
Beta		5	4.26		1071		4674	106911
Gamma		5	4.11		958		4174	92305
Delta^1+		6	5.47		976		4231	93779
Delta^2		5	4.45		978		4234	93845
Epsilon		5	4.91		1052		4583	104174
Zeta		6	5.11		892		3860	83979
Eta		5	5.47		817		3502	75416
Theta		5	4.35		774		3340	71707
Iota		6	5.36		881		3795	82554
Kappa		6	5.04		1059		4605	104902
Lambda+		6	5.18		1061		4617	105340
Mu^1+		6	5.52		917		3983	87971
Mu^2		6	6.60		921		3997	88351
Nu		6	5.45		899		3902	85396
Pi^1+		6	5.65		1028		4479	101132
Pi^2		6	6.46		1033			101805

Lost Stars

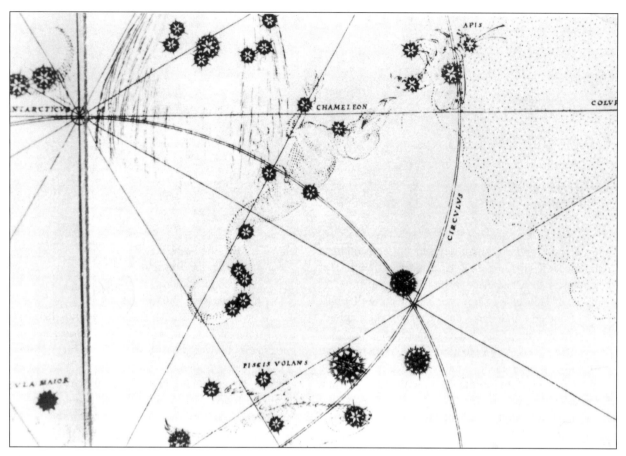

Figure 24. Bayer's Chamaeleon. This is an enlarged view of Chamaeleon from Bayer's Tabula XLIX, his planisphere of the southern heavens. It appears as though a smiling Chamaeleon is about to feast on the hapless Apis (Musca).

The Lost, Missing, or Troublesome Stars of Chamaeleon

Delta¹, HR 4231, 5.47V, *SA, HRP, BS,* CA 976 as Delta, BR 3243, L 4509*, G 25*.
Delta², HR 4234, 4.45V, *SA, HRP, BS,* CA 978 as Delta, BR 3247*, L 4513*, G 26*.

Lacaille designated these two adjacent stars Delta. Brisbane inadvertently omitted the letter designation, Delta¹, for BR 3243.

Lambda, in Musca, HR 4617, 5.18V, CA 1061*, BR 3927*, L 5028*, BAC 4082*, G 29 in Musca.

Gould included this star in Musca and left it unlettered because of its dimness; he considered its magnitude 5.8.

Mu¹, HR 3983, 5.52V, *SA* as Mu, *HRP* as Mu, *BS,* CA 917 as Mu, BR 2880*, L 4232, BAC 3480*, G 19 as Mu.
Mu², HR 3997, 6.60V, *BS,* CA 921 as Mu, BR 2901*, L 4246, BAC 3493*, G 20.

Lacaille designated these two neighboring stars Mu. Gould, however, dropped the letter from CA 921 since he considered it a 7th-magnitude star.

Xi
Omicron

Lacaille did not use these letters in Chamaeleon.

Pi¹, HR 4479, 5.65V, *SA* as Pi, *HRP* as Pi, *BS* as Pi, CA 1028 as Pi, BR 3691*, L 4831*, BAC 3957*, G 32 as Pi.
Pi², in Musca, HD 101805, 6.46V, CA 1033 as Pi, BR 3733*, L 4866*, BAC 3972*, G 13 in Musca.

Lacaille designated these two neighboring stars Pi. Gould included Pi² in Musca and, since it was less than 6th magnitude, left it unlettered.

Circinus, Cir

Pair of Compass Dividers

Lacaille devised and lettered Circinus (figures 2a and 2b) from Alpha to Theta. All his lettered stars are listed in the *BS,* and there are no missing or problem lettered stars in the constellation. In the chart accompanying the preliminary edition of his catalogue, Lacaille drew a pair of compass dividers and labeled it in French le Compas. It represented the compass of the geometer (Le Compas du Géomètre) to distinguish it from the other heavenly compass he devised, Pyxis, the Mariner's Compass. In the second edition of his catalogue, he translated le Compas into Latin as Circinus.

Like the other constellations devised by Lacaille, Circinus represented the arts and sciences that eventually produced the European Enlightenment of the eighteenth century. Circinus symbolized the new developments in mathematics that had been introduced by René Descartes (1596-1650) and his contemporary Pierre de Fermat (1601-1665). Working independently of each other, they discovered that algebraic expressions and formulas could be expressed in geometric terms by plotting coordinates as points on a graph. In essence, their work led to the development of analytic geometry.

See Lacaille, "Table des Ascensions," p. 589; Wolf, *History of Science,* I, 196-202.

Table 28. The lettered stars of Circinus

	MAGNITUDES			CATALOGUE NUMBERS				
Letter	Bayer	Lacaille	Visual	Flamsteed	Lacaille	Other	HR	HD
Alpha		4	3.19		1225		5463	128898
Beta		5	4.07		1271		5670	135379
Gamma		6	4.51		1280		5704	136415
Delta		6	5.09		1270		5664	135240
Epsilon		6	4.86		1269		5666	135291
Zeta		6	6.09		1245		5539	131058
Eta		6	5.17		1257		5593	132905
Theta		6	5.11		1249		5551	131492

COLUMBA, COL

Dove, Noah's Dove

Columba (Figure 25) is a modern constellation that was devised by Petrus Plancius. It first appeared on his map of 1592 where it is labeled Columba Nohae, Noah's Dove. As part of his campaign to depaganize the heavens, the Reverend Plancius also rechristened Argo Navis as Arca Noachi, Noah's Ark. Bayer depicted Columba as an asterism on his chart of Canis Major (Figure 16) but otherwise left it unlettered. In the same year, Houtman listed it in his catalogue as (Noah's) Dove with the Olive Branch (De Duyve met den Olijftack), a reference to the Dove returning to Noah with proof that the flood waters were receding. Bartsch referred to it in his *Usus Astronomicus* of 1624 as Columba Nohae, Noah's Dove. Halley included it in both his catalogue and chart of the southern skies. Although Hevelius and Flamsteed depicted Columba on their charts, they excluded it from their catalogues. Lacaille included it on his planisphere of the southern skies and lettered its stars in his catalogue from Alpha to Sigma.

See Warner, *Sky Explored,* pp. 202, 204; Bartsch, *Usus Astronomicus,* p. 63; Baily, "Catalogues of Ptolemy et al.," p. 171.

Table 29. The lettered stars of Columba

	MAGNITUDES			CATALOGUE NUMBERS				
Letter	**Bayer**	**Lacaille**	**Visual**	**Flamsteed**	**Lacaille**	**Other**	**HR**	**HD**
Alpha		2	2.64		434		1956	37795
Beta		3	3.12		452		2040	39425
Gamma		4	4.36		465		2106	40494
Delta		4	3.85	3 CMa	510		2296	44762
Epsilon		4	3.87		419		1862	36597
Eta		5	3.96		471		2120	40808
Theta		5	5.02		485		2177	42167
Kappa		5	4.37		497		2256	43785
Lambda		5	4.87		453		2056	39764
Mu		6	5.17		444		1996	38666
Nu^1+		6	6.16		426		1926	37430
Nu^2		6	5.31		427		1935	37495
Xi+			4.97			G 61	2087	40176
Omicron		6	4.83		403		1743	34642
Pi^1+		6	6.12		486		2171	42078
Pi^2		6	5.50		488		2181	42303
Rho+		6	6.35		480		2157	41700
Sigma		6	5.50		462		2092	40248

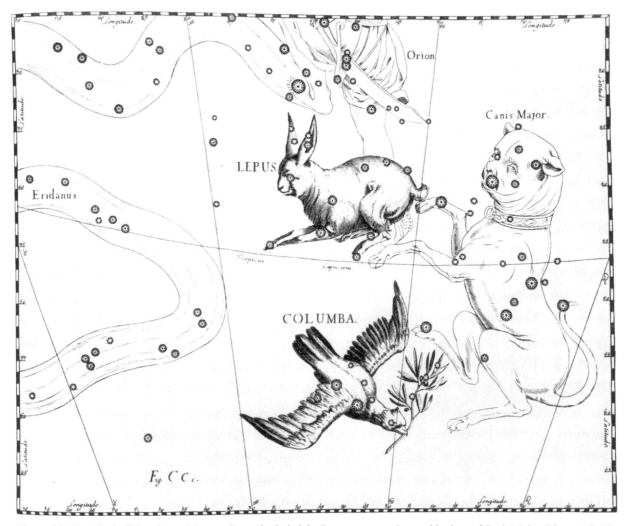

Figure 25. Hevelius's Columba and Lepus. Bayer included the Dove as an asterism on his chart of Canis Major (Figure 16). The line under the Hare's feet is the Tropic of Capricorn.

The Lost, Missing, or Troublesome Stars of Columba

Zeta
Iota

    Lacaille did not use these two letters in Columba.

Nu¹, HR 1926, 6.16V, *SA, HRP, BS,* CA 426 as Nu, BR 996*, L 1911, G 32*.
Nu², HR 1935, 5.31V, *SA, HRP, BS,* CA 427 as Nu, BR 998*, L 1915, G 34*.

    Lacaille designated these two neighboring stars Nu.

Xi, HR 2087, 4.97V, *SA, HRP, BS,* CA 461, BR 1089, L 2069, G 61*.

    Although Lacaille observed and catalogued this star, he left it unlettered since he considered it 6th magnitude. Gould designated it Xi because he felt a star of its brightness merited a letter.

Pi¹, HR 2171, 6.12V, *SA, HRP, BS,* CA 486 as Pi, BR 1144*, L 2154*, G 78*.
Pi², HR 2181, 5.50V, *SA, HRP, BS,* CA 488 as Pi, BR 1153*, L 2164*, G 80*.

    Lacaille designated these two neighboring stars Pi.

Rho, in Puppis, HR 2157, 6.35V, CA 480*, BR 1131*, L 2137, BAC 1964, G 1 in Puppis.

    Baily included this star in Puppis and left it unlettered.

COMA BERENICES, COM

Berenice's Hair

While Ptolemy mentioned Coma Berenices as an asterism near Leo, Tycho was the first to consider it a separate, independent constellation. Bayer did not consider it so, nor did he letter its stars. He depicted it twice — as a sheaf of wheat, "Azimeth," on his chart of Bootes (Figure 11) and as a head of hair on his chart of Ursa Major (Figure 76). On the other hand, Hevelius (Figure 26) and Flamsteed (Figure 86) regarded it as a separate constellation and included it in both their atlases and catalogues. Baily assigned Greek letters from Alpha to Gamma to Coma's brighter stars.

Bayer's depiction of Coma Berenices as *azimeth*, Arabic for sheath of wheat, is somewhat of a puzzle. Perhaps its name can be explained by the asterism's proximity to the constellation Virgo, who is usually depicted holding a spike of wheat.

Coma Berenices is one of the few figures in the sky that can be traced to a real historic personage. Berenice II was the daughter of King Magas of Cyrene (Libya) and the designated heir to the throne. At her father's deathbed, the fourteen-year-old Berenice was promised in marriage to her cousin, Ptolemy III of Egypt. However, the queen mother, Apama II, had different plans. Hoping to be queen herself, Apama offered the throne to Demetrius the Fair, who, like Berenice and Ptolemy III, was a grandson of Ptolemy I. Shortly thereafter, the youthful Berenice became involved in a plot that resulted in the assassination of Demetrius. As the undisputed queen of Cyrene, she married her betrothed, Ptolemy III, King of Egypt (246-221 B.C.), thus uniting the two kingdoms. As a result of the union, Ptolemy III was given the name Euergetes I, "Benefactor [of Egypt]." Ptolemy was an ambitious ruler anxious to extend the boundaries of his realm. Such an opportunity for expansion arose when his sister's husband, Antiochus of Syria, died. On the pretext of protecting his sister — also named Berenice — and her interests, Ptolemy led an army accompanied by African elephants into Syria in 246 B.C. Upon her husband's departure, Berenice II vowed she would offer her hair to the gods if her husband returned victoriously. With the help of his army, his elephants, and his gods, Ptolemy successfully conquered the entire Phoenician coast. When he returned home safely, Berenice cut her hair and placed the locks in the temple of Arsinoe-Aphrodite. Soon after, the hair mysteriously vanished. The astronomer-mathematician Conon of Samos, who was in Alexandria at the time and who was hoping to win favor at court, announced that the gods had placed Berenice's hair as a cluster of stars in the heavens.

See Toomer, *Ptolemy's Almagest*, p. 368; Allen, *Star Names*, pp. 170-71, 467; Cook *et al.*, *Cambridge Ancient History*, VII, 302, 712-18; Mozel, "The Real Berenice's Hair"; Brahe, *Opera Omnia*, II, 273; III, 358; Condos, *Star Myths*, pp. 126-28; Ridpath, *Star Tales*, pp. 53-54.

Table 30. The lettered stars of Coma Berenices

	Magnitudes			Catalogue Numbers				
Letter	Bayer	Lacaille	Visual	Flamsteed	Lacaille	Other	HR	HD
Alpha			4.32c	42		BAC 4406	4968/9	114378/9
Beta			4.26	43		BAC 4421	4983	114710
Gamma			4.36	15		BAC 4195	4737	108381

Figure 26. Hevelius's Coma Berenices. In addition to Berenices's Hair, this chart also depicts Bootes with his Hunting Dogs standing on Mons (Mount) Maenalus. The two Hunting Dogs and Mons Maenalus are three of the twelve new constellations devised by Hevelius.

The Lost, Missing, or Troublesome Stars of Coma Berenices

a, 16, HR 4738, 5.00V, BF 1723.
b, 14, HR 4733, 4.95V, BF 1721.
c, 15, HR 4737, 4.36V, BF 1722.
d, 17, HR 4752, 5.29V, BF 1728.
e, 12, HR 4707, 4.81V, BF 1713.
f, 13, HR 4717, 5.18V, BF 1717.
g, 21, HR 4766, 5.46V, BF 1735.
h, 7, HR 4667, 4.95V, BF 1695.
k, 23, HR 4789, 4.81V, BF 1744.

Flamsteed lettered these nine stars in Coma Berenices, but Baily removed their letters as superfluous. See i Aql and *BF*, p. 398.

Corona Australis, CrA

Southern Crown

Corona Australis (Figure 27) is a Ptolemaic constellation that Bayer lettered from Alpha to Nu. Since it is mostly below the horizon in Europe, Bayer's chart does not conform to the constellation's actual appearance. Consequently, Lacaille redesigned and relettered it from Alpha to Lambda.

This constellation has been known as the Southern Crown since earliest times, but astronomers have used a variety of Latin synonyms for "southern." Bayer, for example, called it Corona Meridionalis, Austrina, or Notia. Lacaille labeled it Corona Australis, as have most other authorities including Eugène Delporte in his *Atlas Céleste* of 1930 for the International Astronomical Union.

There are no astral myths or legends associated with the Southern Crown. Some writers have suggested that the crown belongs to Sagittarius since it is right next to the Archer in the sky, and since centaurs like Sagittarius are said to have worn wreaths. Others have proposed that the constellation represents the crown worn by Bacchus as he leads his mother out of the Underworld.

See Ridpath, *Star Tales,* p. 55; Allen, *Star Names,* pp. 172-74. See also Corona Borealis.

Table 31. The lettered stars of Corona Australis

	Magnitudes			Catalogue Numbers				
Letter	**Bayer**	**Lacaille**	**Visual**	**Flamsteed**	**Lacaille**	**Other**	**HR**	**HD**
Alpha		5	4.11		1577		7254	178253
Beta		5	4.11		1579		7259	178345
Gamma		5	4.21c		1573		7226/7	177474/5
Delta		5	4.59		1574		7242	177873
Epsilon		6	4.87		1563		7152	175813
Zeta		6	4.75		1566		7188	176638
Eta^1+		6	5.49		1546		7062	173715
Eta^2		6	5.61		1549		7068	173861
Theta		5	4.64		1527		6951	170845
Kappa^{1,2}+		6	5.46c		1528		6952/3	170868/67
Lambda		6	5.13		1538		7021	172777
Mu+			5.24			B 15	7050	173540

Lettered Stars - Corona Australis

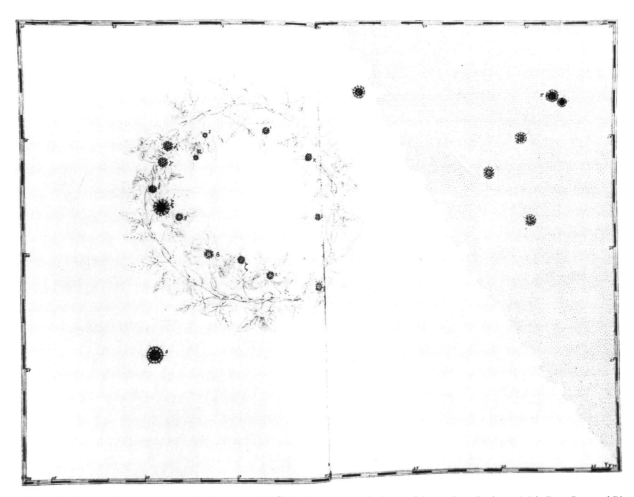

Figure 27. Bayer's Corona Australis. The letter Xi (ξ) is Alpha Sgr in the knee of the Archer; Omicron (o) is Beta Sgr; and Pi (π) is the pair of stars dubbed the Cat's Eyes in the tail of Scorpius.

The Lost, Missing, or Troublesome Stars of Corona Australis

Eta¹, HR 7062, 5.49V, *SA, HRP, BS,* CA 1546 as Eta, BR 6491*, L 7852, BAC 6381*, G 25*.
Eta², HR 7068, 5.61V, *SA, HRP, BS,* CA 1549 as Eta, BR 6493*, L 7859, BAC 6385*, G 26*.

Lacaille designated these two neighboring stars Eta.

Iota

Lacaille did not use this letter in Corona Australis.

Kappa¹, HR 6952, 6.32V, *SA, HRP* as Kappa, *BS,* BR 6428*, G 16 as Kappa.
Kappa², HR 6953, 5.65V, *SA, HRP* as Kappa, *BS,* CA 1528 as Kappa, BR 6429*, L 7758, BAC 6298 as Kappa, G 17 as Kappa.

Lacaille described Kappa as a single 6th-magnitude star. Later astronomers added indices to distinguish the components of this ouble-star system, Dunlop 222.

Mu, HR 7050, 5.24V, *SA, HRP, BS,* L 7846, B 15*, G 24*.

Lacaille did not letter this star. Although he observed it, he left it unlettered; he considered it 6th magnitude. Mu was first assigned to this star by Bode, who employed additional letters in the constellation, none of which has survived in the literature except Mu. Mu owes its survival primarily to Gould, who retained Bode's designation since he felt a star of its brightness merited a letter.

Corona Borealis, CrB

Northern Crown

Corona Borealis (Figure 28), the Northern Crown, a Ptolemaic constellation, is one of the few constellations that bears some resemblance to its namesake. It is made up of a group of stars forming a circlet or wreath. Bayer lettered its stars from Alpha to Upsilon.

There are several Greek astral legends associated with the Northern Crown. One legend relates that King Minos of Crete blames the Athenians for the untimely death of his young son and heir, Androgeus. Minos avenges his son's death by invading Greece and capturing Athens. He demands that the Athenians send him seven young men and seven maidens every nine years as a sacrificial offering to the Minotaur (Bull of Minos). The Minotaur, a half-human and half-bull creature, is the offspring of Minos's wife, Pasiphae, and a bull. The creature is consigned to live in the Labyrinth, an extraordinary, complex maze designed by the architect Daedalus to entrap any who enter. When the time comes to send the fourteen youngsters to Crete, Theseus, the son and heir of King Aegeus of Athens, volunteers to go. Upon Theseus's arrival in Crete, King Minos's daughter, Ariadne, falls madly in love with him. She proposes to save his life if he will promise to marry her. At this point, the legends vary. Some say that the god Dionysus (Roman Liber or Bacchus), himself in love with Ariadne, presents her with a golden crown fashioned by the god Hephaestus (Vulcan) and encrusted with precious gems. Ariadne, in turn, gives it to Theseus, who, after killing the Minotaur, manages to find his way out of the Labyrinth by the brilliance of the crown that sparkled and glowed in the dark. After Ariadne's death, Dionysus places the crown in the heavens. In another variation of the legend, Ariadne saves Theseus by giving him a ball of thread that he attaches to the door of the Labyrinth. After killing the Minotaur with his bare hands, Theseus rewinds the thread to find his way out of the Labyrinth. Upon his escape, he quickly gathers his thirteen Athenian comrades and sails away, abandoning Ariadne and his promise to marry her.

Theseus not only forgets his promise to Ariadne, he also forgets his promise to his father, King Aegeus. He had told Aegeus that if he should succeed in slaying the Minotaur, his ship upon its return would have a white sail instead of the black sail that was used to carry the sacrificial victims to Crete. Theseus forgets to change sails. When Aegeus sees the black sail approaching, he assumes his son is dead and throws himself into the sea, which henceforth bears his name — Aegean.

In the meantime, Ariadne, grief stricken at the loss of her beloved, is consoled by the god Dionysus, who expresses his deep and abiding love for her. Ariadne agrees to marry Dionysus, and on their wedding day he presents her with a crown designed by the gods. Still another legend relates that at the wedding festival attended by all the gods, Aphrodite (Venus) presents the bejeweled crown to the bride as a wedding gift. It is later placed in the heavens as Ariadne's Crown.

Like many legends, the tale of the Minotaur has some basis in history. Some time around 1900 B.C., a people from Asia Minor conquered Crete and developed a powerful trading, military, and cultural empire centered around their capital city, Knossos. Under their legendary King Minos, they dominated the islands in the eastern Mediterranean and southern Greece until the sixteenth century B.C. It was, in all likelihood, the lingering memory of Minoan hegemony over the Greek mainland that formed the basis for the legend of the Minotaur and Minoan dominance of Athens.

All of the legends cited thus far are associated, in one way or another, with Crete and the Cretan princess Ariadne. But there is one other astral legend related by the Latin poet Hyginus that is entirely different. Cadmus, King of Thebes, weds Harmonia, daughter of Venus and Mars. They have several children including Semele, who grows up to become a most beautiful young woman. She attracts the attention of Jupiter (Zeus), who courts her. Hoping to put an end to this affair between Semele and her unfaithful husband, Juno (Hera), appears in disguise to Semele and tricks her into believing that the man she is involved with is not who he claims to be — king of the sky and the gods. At the next meeting of the lovers, Semele forces Jupiter to take an oath to prove that he is indeed the King of Heaven. Bound by his oath, Jupiter appears before Semele in all his terrible glory, with clapping thunder and bolts of fiery lighting. Semele is instantly struck dead. But before her death, she bears Jupiter a son, Bacchus (Dionysus). Bacchus becomes the god of wine and teaches mankind how to cultivate the grape and make wine. Remembering what had happened to his mother, Bacchus receives Jupiter's permission to lead Semele out of the Underworld. Upon his approach to the entrance of the Underworld, he removes from his head an elegant crown that his great-grandmother Venus, had given him. He thinks that such a beautiful piece of jewelry should not be polluted by contact with the spirits of the dead. After safely guiding Semele out the Underworld, Bacchus places the crown in the heavens as Corona Borealis. Since this legend is so radically different from those relating to Ariadne's crown, one scholar has postulated that perhaps Hyginus was mistaken in his *Poetic Astronomy* (2.5) in designating Bacchus's crown as Corona Borealis. He believes that Hyginus may have been confused and that he should have placed it in Corona Australis.

See Hamilton, *Mythology,* pp. 211-15; Condos, *Star Myths,* pp. 87-91; Gayley, *Classic Myths,* pp. 71-73, 89; Schwab, *Gods & Heroes,* pp. 204-6; Ridpath, *Star Tales,* pp. 55-56.

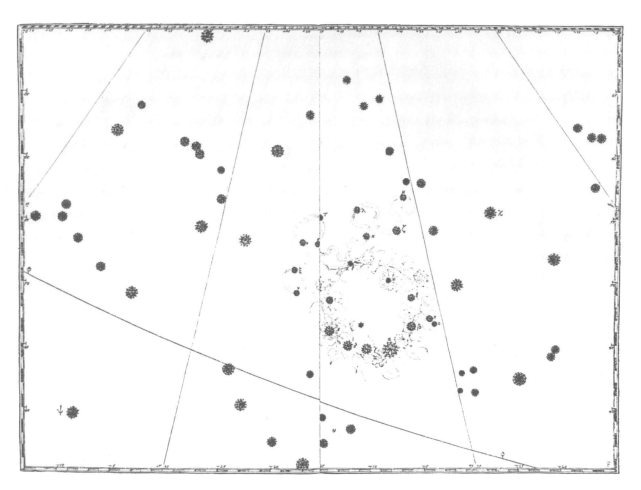

Figure 28. Bayer's Corona Borealis. The three Greek letters Chi (χ), Psi (ψ), and Omega (ω) are not meant to identify stars in the Northern Crown. They are, respectively, the head of Bootes, the head of Hercules, and the head of Serpens. The line Phi (φ) is the Tropic of Cancer.

Table 32. The lettered stars of Corona Borealis

	MAGNITUDES			CATALOGUE NUMBERS				
Letter	Bayer	Lacaille	Visual	Flamsteed	Lacaille	Other	HR	HD
Alpha	2		2.23	5			5793	139006
Beta	4		3.68	3			5747	137909
Gamma	4		3.84	8			5849	140436
Delta	4		4.63	10			5889	141714
Epsilon	4		4.15	13			5947	143107
Zeta[1,2]+	4		4.69c	7			5833/4	139891/2
Eta	5		4.98c	2			5727/8	137107/8
Theta	5		4.14	4			5778	138749
Iota	5		4.99	14			5971	143807
Kappa	5		4.82	11			5901	142091
Lambda	5		5.45	12			5936	142908
Mu	5		5.11	6			5800	139153
Nu[1]+	5		5.20	20			6107	147749
Nu[2]			5.39	21			6108	147767
Xi	5		4.85	19			6103	147677
Omicron	6		5.51	1			5709	136512
Pi	6		5.56	9			5855	140716
Rho	6		5.41	15			5968	143761
Sigma	6		5.22c	17			6063/4	146361/2
Tau	6		4.76	16			6018	145328
Upsilon	6		5.78	18			6074	146738

The Lost, Missing, or Troublesome Stars of Corona Borealis

Zeta[1], Zeta[2], 7, HR 5833/4, 6.00V, 5.07V, *SA, HRP* as Zeta, *BS,* ADS 9737.

Bayer described Zeta as a single 4th-magnitude star. Later astronomers added the indices to identify the components of this double star.

Nu[1], 20, HR 6107, 5.20V, *SA, HRP* as Nu, *BS,* HF 21 as Nu, BF 2252*, H 31 as Nu.
Nu[2], 21, HR 6108, 5.39V, *SA, HRP* as Nu, *BS,* BF 2253*, H 31 as Nu.

Bayer's *Uranometria* describes Nu as a single 5th-magnitude star, and in Halley's pirated edition of Flamsteed's catalogue, 20 is equated with Nu. In Flamsteed's "Catalogus Britannicus" of 1725, however, 20 and its close neighbor 21 are designated Nu-1 and Nu-2. Argelander, on the other hand, felt that the combined light of both stars was equivalent to Bayer's Nu and that the use of indices was unnecessary (*UN*, p. 31).

CORVUS, CRV

Raven, Crow

Corvus (Figure 29) is a Ptolemaic constellation that Bayer lettered from Alpha to Eta. All its lettered stars are listed in the *BS,* and there are no missing or problem lettered stars in the constellation.

The ancient Mesopotamians recognized Corvus as early as the latter half of the second millennium B.C. and possibly earlier. It was called in Sumerian UGA (.MUSEN) or Aribu in Akkadian, which translates as the Raven. It probably owes its place in the heavens to the story of the flood as recounted in Tablet XI of the Epic of Gilgamesh. After six days of horrendous storms, the rains finally stop, and on the seventh day, Utnapishtim, the Mesopotamian Noah, releases a raven from his ark to search for dry land. As Utnapishtim relates:

> *Then I sent forth and set free a raven.*
> *The raven went forth and seeing that the waters had diminished,*
> *He eats, circles, caws, and turns not round.*

The raven thus becomes the first creature to find dry land. A similar account regarding the raven is mentioned in Genesis 8:6-7.

The depiction of the raven as a messenger of hope can be traced to the belief that it was the symbol of the Sumerian god Ninshubur, the messenger and vizier of the goddess Inanna (Akkadian Ishtar), the queen or lady of heaven. According to the Epic of Gilgamesh, after the gods decide to destroy man with the deluge, they recoil in horror at the enormity of the destruction they wrought. As "the gods cowered like dogs," Ishtar alone spoke out, regretting what had been done to humanity:

> *How could I bespeak evil [of man] in the Assembly of the Gods,*
> *Ordering battle for the destruction of my people,*
> *When it is I myself who give birth to my people!*

It was indeed appropriate that Inanna's messenger, the Raven, should bring renewed hope to the human race and be rewarded with a place in the firmament.

See van der Waerden, "Babylonian Astronomy. II. The Thirty-Six Stars," pp. 11, 13-14; Reiner and Pingree, *Baby. Plan. Omens,* Part 2, p. 15; Langdon, *Semitic Mythology,* p. 177; Jacobsen, *Toward the Image of Tammuz,* p. 323; Pritchard, *Anc. Near East. Texts,* pp. 52-57, 94-95.

Lost Stars

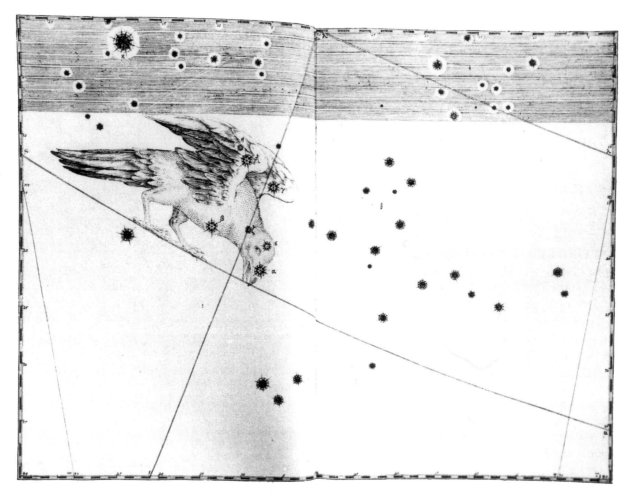

Figure 29. Bayer's Corvus. The letter Theta (θ) identifies the stars in the constellation Crater. Iota (ι) is Leo; and Kappa (κ) is the 1st-magnitude star Spica, Alpha Vir. The lines Lambda (λ), Mu (μ), and Nu (ν) are, respectively, the celestial equator, the Tropic of Capricorn, and the equinoctial colure. At the top center of the chart, where the equinoctial colure meets the equator at the ecliptic, is the First Point of Libra, ♎, the autumnal equinox. The shaded area across the top is the zodiac.

Table 33. The lettered stars of Corvus

	Magnitudes			Catalogue Numbers				
Letter	Bayer	Lacaille	Visual	Flamsteed	Lacaille	Other	HR	HD
Alpha	3		4.02	1			4623	105452
Beta	3		2.65	9			4786	109379
Gamma	3		2.59	4			4662	106625
Delta	3		2.95	7			4757	108767
Epsilon	4		3.00	2			4630	105707
Zeta	5		5.21	5			4696	107348
Eta	5		4.31	8			4775	109085

CRATER, CRT

Cup, Bowl

Bayer lettered the stars of the Ptolemaic constellation Crater (Figure 30) from Alpha to Lambda. Flamsteed included in this constellation all those stars in Hydra south of Crater, thus creating, for all intents and purposes, a new constellation, which he called Hydra and Crater.

In Greek mythology, the three constellations Crater, Corvus, and Hydra are linked together. The sun god Apollo, preparing to make an offering to Zeus, asks the crow, his messenger, to fetch some water. Taking a cup in his beak, the crow flies off in search of fresh water. On his way, the bird spots a fig tree and decides to wait until its fruit ripens. Finally, after several days he remembers his mission and decides to return to Apollo, but in order to account for his delay, he seizes a water snake and tells the sun god that the snake prevented him from filling the cup. Apollo sees through this lie, and as punishment, ordains that henceforth all crows will thirst during the period when figs are ripening. Apollo places the Crow (Corvus), the Cup (Crater), and the Water Snake (Hydra) in the sky as a warning to those who would dare disobey the gods.

Poetic Astronomy (2.40) by the Greek poet Hyginus relates another legend associated with Crater. The inhabitants of the city of Eleusa on the European side of the Hellespont are suffering from a plague. Demiphon, the ruler, consults an oracle who advises that the city's troubles will end if a girl of noble birth is sacrificed each year to the gods. Demiphon decides to use a lottery to select the girls but purposely omits the names of his own daughters. One nobleman, Mastusius, protests this unfair lottery. In a fit of anger, Demiphon thereupon chooses Mastusius's daughter to be sacrificed. Vowing vengeance, Mastusius invites the king and his daughters to his home for a grand feast. While the king is purposely delayed, Mastusius kills the king's daughters and mixes their blood with wine in a wine jar. When Demiphon learns what has occurred, he orders Mastusius thrown into the sea. Hyginus concluded the legend by noting that astronomers placed the image of the wine jar or mixing bowl, Crater (κρατηρ), in the heavens to warn humans against committing evil deeds. He also noted that the port near the city of Eleusa was called Crater.

See Condos, *Star Myths,* pp. 119-23, 243; Staal, *New Patterns in the Sky,* pp. 160-62; Ridpath, *Star Tales,* pp. 57-58. Condos cites (p.243) a classical author who noted that on the Chersonese there was actually a port named Crateres Achaion, Cups or Craters of the Greeks.

Lost Stars

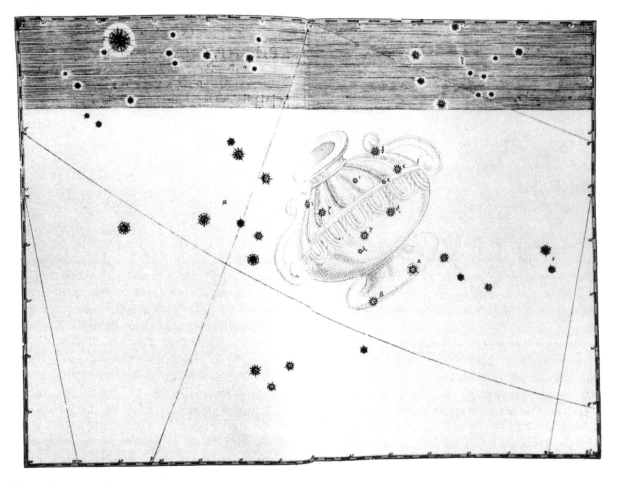

Figure 30. Bayer's Crater. The lines Pi (π), Rho (ρ), and Sigma (σ) are respectively, the celestial equator, the equinoctial colure, and the Tropic of Capricorn. At the top center of the chart is the First Point of Libra, ♎, where the equinoctial colure meets the equator in the constellation Virgo. When the sun reaches this point in September on its yearly trip around the celestial globe, it is the time of the autumnal equinox. The letter Mu (μ) is Corvus; Nu (ν) is Hydra; Xi (ξ) is Leo; and Omicron (ο) is Spica, Alpha Vir. The shaded area across the top is the zodiac.

Table 34. The lettered stars of Crater

	Magnitudes			Catalogue Numbers				
Letter	Bayer	Lacaille	Visual	Flamsteed	Lacaille	Other	HR	HD
Alpha	4		4.08	7			4287	95272
Beta	4		4.48	11			4343	97277
Gamma	4		4.08	15			4405	99211
Delta	4		3.56	12			4382	98430
Epsilon	4		4.83	14			4402	99167
Zeta	4		4.73	27			4514	102070
Eta	4		5.18	30			4567	103632
Theta	4		4.70	21			4468	100889
Iota+	5		5.48	24			4488	101198
Kappa	6		5.94	16			4416	99564
Lambda	6		5.09	13			4395	98991
Psi+			6.13			B 22	4347	97411

The Lost, Missing, or Troublesome Stars of Crater

Iota, 8 Crt in Hydra, HR 4302, 6.23V, B 245 as i Hya, BF 1570.

Bayer did not letter this star. Flamsteed designated this star Iota, but Bode relettered it i Hya, and Baily removed the i altogether. This star should not be confused either with Iota, 35 Hya (HR 3845, 3.91V) or with Iota, 24 Crt (HR 4488, 5.48V), which is Bayer's Iota Crt. For Flamsteed's extraneous letters, see i Aql.

Psi, HR 4347, 6.13V, *SA, BS,* B 22*, H 11, G 16.

Bayer did not letter this star. The letter was added by Bode, who employed additional letters in Crater, none of which has survived in the literature except Psi. Psi probably owes its survival to its inclusion in the atlases of Becvar and Tirion.

CRUX, CRU

Cross, Southern Cross

Ptolemy included the stars that later became Crux (Figure 31) as part of the hind legs of the Centaur. Iberian seamen in the fifteenth or sixteenth century devised Crux as they explored new lands in the Southern Hemisphere. Vespucci noted Crux's stars in his *Mundus Novus* of 1504. Pedro de Medina referred to it as The Cross in his treatise on navigation of 1554. It appeared on Jodocus Hondius Sr.'s globe of 1600, which drew heavily on the work of Keyzer and Houtman. In his catalogue, Houtman called the constellation De Cruzero, an indication of its Iberian origin. Bayer (Figure 21) noted that Crux was a modern asterism within Centaurus and lettered its stars Epsilon, Nu, Xi, and Zeta Cen, which are respectively Lacaille's Gamma, Delta, Beta, and Alpha Cru. Lacaille (Figure 2a) mapped Crux as a separate constellation with its own set of letters from Alpha to Lambda.

Petrus Plancius, in his efforts to depaganize the heavens, included a cross on some of his early maps and globes in the area just below Eridanus. In his globe of 1589, he labeled it Crux in Latin and Σταυρος (Stauros, Cross) in Greek, which Warner misread as Σιανρος (Sianros). Bartsch noted in his *Usus Astronomicus* that this cross, located south of the River Eridanus, was included within the bounds of the newly formed constellation Phoenix and should not be confused with Crux, the Spanish Cross, "Hispan. Cruzero."

See Warner, *Sky Explored,* pp. 173, 201, 255; Bartsch, *Usus Astronomicus,* p. 66.

Table 35. The lettered stars of Crux

	Magnitudes			Catalogue Numbers				
Letter	Bayer	Lacaille	Visual	Flamsteed	Lacaille	Other	HR	HD
Alpha[1,2]+		1	0.79c		1082		4730/1	108248/9
Beta		2	1.25		1107		4853	111123
Gamma		2	1.60c		1088		4763/4	108903/25
Delta		3	2.80		1070		4656	106490
Epsilon		4	3.59		1076		4700	107446
Zeta		6	4.04		1073		4679	106983
Eta		5	4.15		1060		4616	105211
Theta[1]+		6	4.33		1054		4599	104671
Theta[2]		6	4.72		1057		4603	104841
Iota		6	4.69		1103		4842	110829
Kappa+		Neb			1110			
Lambda+		6	4.62		1113		4897	112078
Mu[1,2]+			3.95c			G 53	4898/9	112092/91

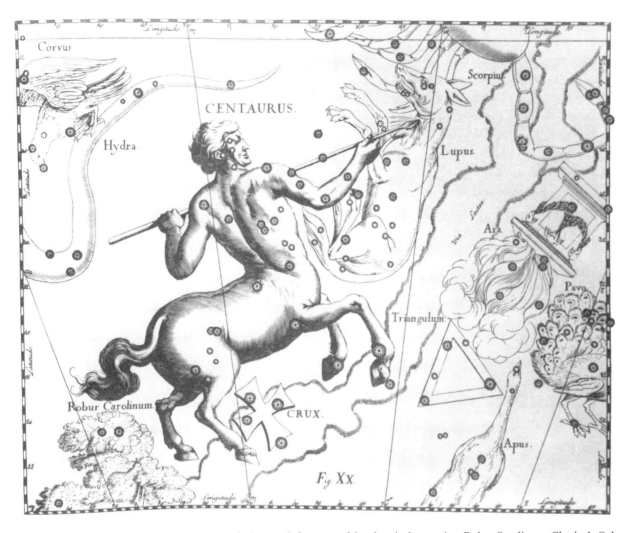

Figure 31. Hevelius's Crux and Centaur. In the bottom left corner of the chart is the asterism Robur Carolinum, Charles's Oak, devised by Halley in 1678 to commemorate the tree where Charles II supposedly hid from the Roundheads after his defeat at Worcester in 1651 (see Figure 65). Lacaille did not include the asterism either in his atlas or catalogue but instead placed most of its stars in Carina.

The Lost, Missing, or Troublesome Stars of Crux

Alpha^1, Alpha^2, HR 4730/1, 1.58V, 2.09V, comb. mag. 0.79V, *SA, HRP, BS,* CA 1082 as Alpha, BR 4050/1*, L 5148 as Alpha, BAC 4187 as Alpha, G 26/7 as Alpha, Acrux.

Lacaille described Alpha as a single 1st-magnitude star. Later astronomers added the indices to identify components A and B. The star HR 4729, 4.86V, is component C of this triple-star system. It was observed by Lacaille, Brisbane, and Gould but left unlettered (L 5147, BR 4049, G 25). Lacaille was able to observe C since it was considerably distant, about 1½′, from the other two, which were close together and whose combined light appeared to him as one object.

Theta^1, HR 4599, 4.33V, *SA, HRP, BS,* CA 1054 as Theta, BR 3892*, L 4990*, G 6*.
Theta^2, HR 4603, 4.72V, *SA, HRP, BS,* CA 1057 as Theta, BR 3901*, L 4999*, G 7*.

Lacaille designated these two nearly adjacent stars Theta.

Kappa, HR 4890, 5.90V, *HRP, BS,* G 50*.

In his catalogue of 1763, Lacaille identified Kappa as a nebulous object (CA 1110 as Kappa, L 5306). Brisbane described it as "a cluster of twelve or fourteen small stars in the form of a rhomboid, very close together" (BR 4227 as Kappa). In his observations of the area, Lacaille also identified several individual stars within and around the nebulosity. Like Lacaille, Gould referred to the entire cluster as Kappa (*Uran. Arg.,* p. 269), but in addition, he designated one of its stars as Kappa, G 50, as did the editors of *HRP* and *BS*. CA 1110 is the open cluster NGC 4755, the Jewel Box, which is identified in *SA* as Kappa. See Xi Tuc.

Lambda, HR 4897, 4.62V, *HRP, BS,* CA 1113*, BR 4236*, L 5136, BAC 4324*, G 51*.

SA does not identify this star with any letter. Tirion does, however, properly label this star in his *Uranometria 2000.0,* and in the 2nd edition of *SA*.

Mu^1, Mu^2, HR 4898/9, 4.03V, 5.17V, comb. mag. 3.95V, *SA, HRP* only HR 4898 as Mu, *BS,* CA 1114 as O Cen, BR 4237 as O^2 Cen, BR 4238, L 5317, G 52 as Mu, G 53.

Lacaille did not letter this star in Crux. This star was originally Lacaille's O^2 Cen. Gould included it in Crux and relettered it Mu Cru since he felt a star of its brightness merited a letter. Although he designated only G 52 as Mu, he noted that G 52 and G 53 appeared as one image to his naked eye. Lacaille considered it a single 5th-magnitude star. Later astronomers added the indices to identify the components of this double star, Dunlop 126. See O Cen.

Cygnus, Cyg

Swan

Bayer lettered the Ptolemaic constellation Cygnus (Figure 32) from Alpha to Omega and from A to g.

According to Greek mythology, Nemesis, daughter of Night (Nyx) and goddess of divine retribution, catches the eye of the ever-philandering Zeus. To escape his attention, she changes into a swan. Zeus does the same and follows her in hot pursuit. He eventually overtakes and ravishes her. He flies back to heaven as a swan and places the image of a swan among the stars. Nemesis later produces an egg from which hatches the famous Helen of Troy. A variation of this legend says that Zeus turns himself into a swan and flies onto Nemesis lap. When she falls asleep, Zeus rapes her and flies back to heaven. In still another variation Zeus, in the shape of a swan, seduces Leda, Queen of Sparta, who produces an egg from which hatch Helen and the twins Castor and Pollux.

Cygnus is one of the few constellations that bears a striking resemblance to its namesake. It appears in the heavens as a swan in flight, with outstretched wings and a long, thin neck. During the time of the winter solstice, around December 22, the Christmas season, the constellation begins to set in the western sky and takes on the appearance of a cross. Some have called it the Northern Cross, and it actually bears more of a resemblance to a cross than the constellation Crux, the Southern Cross. Bayer remarked that since antiquity it has been called the Cross.

See Condos, *Star Myths,* pp. 93-95; Ridpath, *Star Tales,* pp. 59-61; Staal, *New Patterns in the Sky,* pp. 175-77; Allen, *Star Names,* pp. 192-95. See P, below, for the Northern Cross as a symbol of Christianity and the Crucifixion.

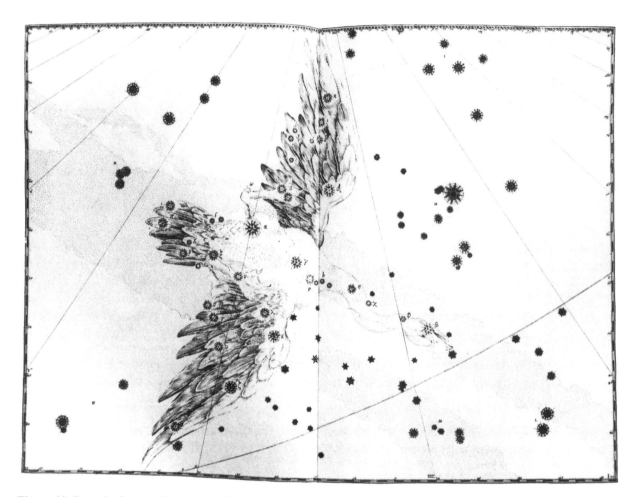

Figure 32. Bayer's Cygnus. Bayer noted that the stars Alpha (α), Beta (β), Gamma (γ), Delta (δ), and Epsilon (ε) form the (Northern) Cross. He went on to state that if P, the Nova of 1600, were to be added to the group, it could represent the crucified Jesus hanging from the Cross. The letter H is the constellation Lyra; I is the head of Draco; K is Cepheus; L is the tail of Aquila; M is Sagitta; N is the leg of Pegasus; and P is the Nova of 1600. The line O is the Tropic of Cancer.

Table 36. The lettered stars of Cygnus

	Magnitudes			Catalogue Numbers				
Letter	Bayer	Lacaille	Visual	Flamsteed	Lacaille	Other	HR	HD
Alpha	2		1.25	50			7924	197345
Beta[1,2]+	3		3.10c	6			7417/8	183912/14
Gamma	3		2.20	37			7796	194093
Delta	3		2.87	18			7528	186882
Epsilon	3		2.46	53			7949	197989
Zeta	3		3.20	64			8115	202109
Eta	4		3.89	21			7615	188947
Theta	4		4.48	13			7469	185395
Iota[1]+			5.75	7			7408	183534
Iota[2]	4		3.79	10			7420	184006
Kappa	4		3.77	1			7328	181276
Lambda	4		4.53	54			7963	198183
Mu[1,2]+	4		4.50c	78			8309/10	206826/7
Nu	4		3.94	58			8028	199629

(table continued on next page)

Table 36. The lettered stars of Cygnus *(continued)*

	MAGNITUDES			CATALOGUE NUMBERS				
Letter	Bayer	Lacaille	Visual	Flamsteed	Lacaille	Other	HR	HD
Xi	4		3.72	62			8079	200905
Omicron¹+			4.83	30			7730	192514
Omicron¹	4		3.79	31			7735	192577
Omicron²	4		3.98	32			7751	192909
Pi¹+	4		4.67	80			8301	206672
Pi²	4		4.23	81			8335	207330
Rho	4		4.02	73			8252	205435
Sigma	4		4.23	67			8143	202850
Tau	4		3.72	65			8130	202444
Upsilon	4		4.43	66			8146	202904
Phi	5		4.69	12			7478	185734
Chi+	5		4.23				7564	187796
Psi	5		4.92	24			7619	189037
Omega¹+	5		4.95	45			7844	195556
Omega²			5.44	46			7851	195774
A	5		5.00	68			8154	203064
b¹+	6		5.36	27			7689	191026
b²	6		4.93	28			7708	191610
b³	6		4.97	29			7736	192640
c+	6		5.56c	16			7503/4	186408/27
d+	6		5.03	20			7576	188056
e+	6		5.05	26			7660	190147
f¹+	6		4.74	59			8047	200120
f²	6		4.55	63			8089	201251
g	6		5.24	71			8228	204771
P+	3		4.81	34			7763	193237

The Lost, Missing, or Troublesome Stars of Cygnus

Beta¹, Beta², 6, HR 7417/8, 3.08V, 5.11V, comb. mag. 3.10V, *SA* as Beta, *HRP* as Beta, *BS,* ADS 12540, Albireo.

Bayer described Beta as a single 3rd-magnitude star. Later astronomers added the indices to identify the components of this double star.

Iota¹, 7, HR 7408, 5.75V, *BS,* BV 24, BF 2647*, H 8.
Iota², 10, HR 7420, 3.79V, *SA* as Iota, *HRP* as Iota, *BS,* BV 23 as Iota, BF 2657*, H 11 as Iota.

Bayer described only one 4th-magnitude star as Iota, and Flamsteed designated his 10 as Iota. Baily, however, felt that 10 and its smaller neighbor, 7, were equivalent to Iota, so he designated them Iota¹ and Iota². Bevis and Argelander, on the other hand, agreed with Flamsteed that only 10, the brighter star, was synonymous with Iota (*UN,* p.38).

Mu¹, Mu², 78, HR 8309/10, 4.73V, 6.08V, *SA, HRP* as Mu, *BS,* HF 104 as Mu, BV 64 as Mu, B 383 as Mu, BF 2973 as Mu, ADS 15270.

Bayer described Mu as a single 4th-magnitude star. Later astronomers added the indices to identify the components of this double star.

Although 78 is labeled Mu in Halley's edition of Flamsteed's catalogue and although it is identified as Mu in the body of Flamsteed's *Historia Coelestis,* it is inadvertently omitted from his "Catalogus Britannicus." Bevis (Tabula IX) and Bode caught the omission and equated 78 with Bayer's Mu. Bode also noted that 78 was a double star.

Omicron¹, 30, 31, HR 7730, HR 7735, 4.83V, 3.79V, *SA* equates 31 with Omicron¹, *HRP, BS* equates 31 with Omicron, HF 47 and 48 as Omicron, BAC 6962 and 6965 equate 30 and 31 with Omicron¹ and Omicron², H 70 equates 31 with Omicron¹.

Omicron², 32, HR 7751, 3.98V, *SA, HRP, BS* as Omicron, HF 49 as Omicron, BAC 6983, H 75*.

Considerable confusion surrounds these three stars—30, 31, and 32. Bayer described Omicron as a couple of 4th-magnitude stars. Working with a telescope, Flamsteed discovered that the westerly of the two has a small, close companion preceding it. In Halley's edition of Flamsteed's catalogue, these two stars, 30 and 31, are called *duplex* (double), are bracketed together, and are designated Omicron. The other half of Bayer's pair, 32, about ½° to the east, is also labeled Omicron. When Flamsteed's authorized catalogue was printed in 1725, these designations changed: 30 is labeled Omicron-1; and its larger companion, 31, is labeled Omicron-2. The last star, 32, is left without any letter although it is the eastern half of the pair that Bayer had originally designated as Omicron. In other words, Flamsteed's 1725 catalogue drops Omicron from one of the two stars that Bayer had designated and instead applies the letter to a much smaller star that is not on Bayer's atlas. Baily agreed with this arrangement, but Argelander suggested returning to Bayer's original proposal. He designated 31 as Omicron¹ and 32 as Omicron²; moreover, unable to separate with his naked eye 31 from its close companion, 30, he omitted 30 altogether (*UN*, p. 40). Heis, though, felt that Bayer had seen the combined light of both stars, and so he equated Omicron¹ with 30 and 31. Actually, Heis did not mention 30, but he equated BAC 6962 and 6965 with 31, and these two stars are synonymous with 30 and 31. See Hoffleit, "Discordances," p. 49; *BS*, p. 435.

Pi¹, 80, HR 8301, 4.67V, *SA, HRP, BS.*
Pi², 81, HR 8335, 4.23V, *SA, HRP, BS.*

Bayer noted that Pi was a *binae* or pair of 4th-magnitude stars, and Flamsteed designated his 80 and 81 as Pi-1 and Pi-2.

Chi, HR 7564, 4.23V, *SA, HRP, BS,* BV 11*, H 35*, W 18*, a long-term variable star with a magnitude range of 3.3 to 14.2 over a period of 408 days.

Bayer described Chi as a 5th-magnitude star, and Flamsteed designated his 17 (HR 7534, 4.99V, BF 2687 as Chi, BAC 6784 as Chi) as Chi. Baily noted that he had some question as to whether 17 was synonymous with Chi, but he made no change. Bevis (Tabula IX) decided to reassign Chi to the variable located 1° to the southwest, which more nearly conforms to the position of Chi on Bayer's chart. Argelander (*UN,* p. 39) and most modern authorities agreed with Bevis.

Omega¹, 45, HR 7844, 4.95V, *SA, HRP, BS,* BV 53 as Omega, H 98*.
Omega², 46, HR 7851, 5.44V, *SA, HRP, BS,* BV 57 as Omega, H 100*.

Bayer noted that Omega was a 5th-magnitude *duplex* or double star surrounded by nebulosity. When Flamsteed observed the area, he saw three stars, which he designated as follows:

Omega-1, 43 (HR 7828, 5.69V, BF 2792 as Omega¹).
Omega-2, 45 (HR 7844, 4.95V, BF 2799 as Omega²).
Omega-3, 46 (HR 7851, 5.44V, BF 2802 as Omega³).

Although 43 is almost 1° away from the others and probably not the star observed by Bayer, Baily went along with Flamsteed's designations because, as he noted, there was some doubt about which of the three was equivalent to Omega. Argelander had no such doubts. Unable to see 43, he dropped it and designated 45 and 46 as Omega¹ and Omega² (*UN*, p. 41). Like Argelander, Bevis dropped 43's (BV 52) letter. On his chart (Tabula IX), he showed 45 and 46 as Omega-1 and Omega-2, but in his catalogue, he listed them simply as Omega.

As noted above, Bayer described Omega as nebulous. Although there are no nebulous objects near 45 or 46, the area is one of the richest in the Milky Way with many closely packed stars surrounding 45 and 46. For Bayer's nebulous objects, see z Her.

A, 68, HR 8154, 5.00V, *SA, HRP.*

b¹, 27, HR 7689, 5.36V, *SA, HRP.*
b², 28, HR 7708, 4.93V, *SA, HRP.*
b³, 29, HR 7736, 4.97V, *SA, HRP.*

Bayer noted that b was a group of three 6th-magnitude stars, and Flamsteed designated his 27, 28, and 29 as b-1, b-2, and b-3.

c, 16, HR 7503/4, 5.96V, 6.20V, comb. mag. 5.56V, *SA, HRP,* HF 20*, BV 31*, B 72*, BF 2682*.

In Halley's edition of Flamsteed's catalogue, 16 is properly identified as Bayer's c, but in Flamsteed's 1725 catalogue, it is erroneously labeled c-1. Bevis and Bode corrected it. Bode also noted that 16 was a double star. For Flamsteed's c-2, see e.

d, 20, HR 7576, 5.03V, *SA, HRP,* HF 26*, BV 46*, B 98*, BF 2711*.

In Halley's 1712 edition of Flamsteed's catalogue, 20 is properly equated with Bayer's d, but in Flamsteed's authorized catalogue of 1725, the letter is omitted. Bevis and Bode discovered the omission and corrected it.

e, 26, HR 7660, 5.05V, *SA, HRP,* HF 35*, BV 43*, B 131*, BF 2734*.

In Halley's edition of Flamsteed's catalogue, 26 is properly identified as Bayer's e, but in Flamsteed's 1725 catalogue, it is erroneously labeled c-2. Bevis and Bode corrected it. See c.

f¹, 59, HR 8047, 4.74V, *SA, HRP.*
f², 63, HR 8089, 4.55V, *SA, HRP.*

Bayer noted that f was a couple of 6th-magnitude stars, and Flamsteed designated his 59 and 63 as f-1 and f-2.

g, 71, HR 8228, 5.24V, *SA, HRP.*

h, 39, HR 7806, 4.43V, BF 2781.
i, 41, HR 7834, 4.01V, BF 2789.
k, 52, HR 7942, 4.22V, BF 2834.
l, 47, HR 7866, 4.61V, BF 2803.
m, 35, HR 7770, 5.17V, BF 2772.

Bayer did not letter these stars h, i, k, l, and m. Although he included these stars in his atlas, he left them unlettered since he considered them *informes* outside the border of Cygnus. The letters were added by Flamsteed but removed by Baily as superfluous. See i Aql.

P, 34, HR 7763, 4.81V, *SA, HRP, BS.*

This is the Nova of August 1600, which at its brightest reached 3rd magnitude, where it remained for the next six years. Flamsteed noted that by 1690 it had dimmed to 6th magnitude. See *Burnham's Celestial Handbook,* II, 772-73.

Bayer's *Uranometria* notes that the stars Alpha, Beta, Gamma, Delta, and Epsilon constitute a cross. The first three letters form the verticle beam while Delta and Epsilon form the cross beam. If P is added, it forms the body of Christ hanging from the Cross.

After using Greek and lower-case Roman letters to designate the fixed stars, Bayer switched to capital Roman letters to identify miscellaneous objects on his sky charts. In Cygnus, for example, since the last star was designated g, he identified the neighboring constellation Lyra as H, Draco as I, Cepheus as K, Aquila as L, Sagitta as M, Perseus as N, the Tropic of Cancer as O, and the Nova of 1600 as P. Bayer, however, was not always consistent. Sometimes, when he did not exhaust all the Greek letters on his stars, he would continue using them for miscellaneous objects. In Corona Borealis, for instance, his last star was Upsilon, but then he designated the Tropic of Cancer as Phi, Bootes as Chi, Hercules as Psi, and Serpens as Omega. See also B Cas and H Gem.

Delphinus, Del

Dolphin

Bayer lettered the Ptolemaic constellation Delphinus (Figure 33) from Alpha to Kappa.

Delphinus represents the mammalian Dolphin, and several legends describe how this constellation found its place among the stars. These accounts reflect the belief — still held by many — that dolphins are friendly and helpful to humans. Poseidon, god of the sea, seeks to wed Amphitrite, one of the sea nymphs, or Nereids. She rebuffs his advances and hides with her sister Nereids. Undaunted by her resistance, Poseidon sends a dolphin to search for her. The dolphin finds Amphitrite and convinces her to marry Poseidon, who subsequently rewards the dolphin with a place in the heavens.

A second legend recounts the adventures of the seventh-century B.C. poet and musician Arion, who becomes so famous that he amasses a considerable fortune. As Arion sails from Sicily to Greece on one of his travels, the ship's sailors — some say his own slaves — decide to seize his money and throw him overboard. As a final request, Arion asks that he be allowed to play his cithara or lyre. As he plays and sings, he attracts a number of dolphins. When he sees them cavorting around the ship, Arion leaps overboard, and one of the mammals carries him safely to shore. A variation of the legend claims that Apollo, god of music, hears Arion's music and is so enchanted with it that he orders the dolphins to carry Arion to safety. As a reward, Apollo sets the image of the dolphin among the stars.

See Condos, *Star Myths,* pp. 97-100; Ridpath, *Star Tales,* pp. 61-62; Staal, *New Patterns in the Sky,* pp. 189-90.

Table 37. The lettered stars of Delphinus

	MAGNITUDES			CATALOGUE NUMBERS				
Letter	Bayer	Lacaille	Visual	Flamsteed	Lacaille	Other	HR	HD
Alpha	3		3.77	9			7906	196867
Beta	3		3.63	6			7882	196524
Gamma[1,2+]	3		4.12c	12			7947/8	197963/4
Delta	3		4.43	11			7928	197461
Epsilon	3		4.03	2			7852	195810
Zeta	5		4.68	4			7871	196180
Eta	6		5.38	3			7858	195943
Theta	6		5.72	8			7892	196725
Iota	6		5.43	5			7883	196544
Kappa	6		5.05	7			7896	196755

Lettered Stars - Delphinus

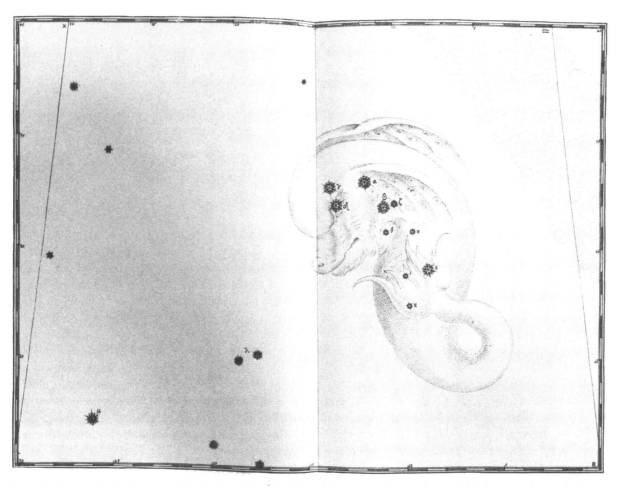

Figure 33. Bayer's Delphinus. The letters Lambda (λ) and Mu (μ) represent the nearby constellations Equuleus and Pegasus.

The Lost, Missing, or Troublesome Stars of Delphinus

Gamma¹, Gamma², 12, HR 7947/8, 5.14V, 4.27V, comb. mag. 4.12V, *SA, HRP* as Gamma, *BS,* ADS 14279.

   Bayer described Gamma as a single 3rd-magnitude star. Later astronomers added the indices to identify the components of this double star.

DORADO, DOR

Swordfish, Goldfish

Keyzer and Houtman devised the constellation Dorado (figures 34 and 35). Houtman's catalogue calls it Den Dorado. Bayer depicted it on his chart of the southern sky, and Lacaille lettered it from Alpha to Pi.

Dorado is often translated as the Goldfish, but on Bayer's chart it is depicted as a relatively large fish, which Allen suggested is the tropical Coryphaena or dolphin that should not be confused with Delphinus, the mammalian dolphin. Coryphaena can grow up to five feet long and has iridescent coloring, giving the impression that it can change its colors. The constellation's alternate name Xiphias, or the Swordfish, probably appeared for the first time in Bartsch's *Usus Astronomicus* of 1624, where it is referred to as "Dorado, the Spanish Fish, the Goldfish (Aurata), and at other times as Xiphias or Swordfish (Gladius)." Three years later, Bartsch's father-in-law, Kepler, referred to the constellation as "Dorado, Xiphias" in his Rudolphine Tables. For a possible explanation of the difference in names, see Apus.

The south eliptic pole is located within Dorado.

Before Dorado was devised, Petrus Plancius, in his maps of 1592 and 1594, placed the figure Polophylax, or Guardian of the Pole, in the area later occupied by Dorado. Plancius devised the bearded, elderly figure of Polophylax to be the southern counterpart to Draco, the guardian of the north ecliptic pole. A few years later Keyzer and Houtman replaced Polophylax with their Dorado, which may have been transformed into Xiphias because a swordfish was felt to be a more suitable guard than a goldfish. Besides Draco and Polophylax, at least one other guardian was assigned to protect the creatures that abide in the heavens. Since classical times, Bootes has often been referred to as Arctophylax, or Guardian of the Great Bear (Ursa Major).

See Allen, *Star Names,* pp. 201-2; Bartsch, *Usus Astronomicus,* p. 66; Kepler, *Gesammalte Werke,* X, 140; Warner, *Sky Explored,* pp. 202-3; Staal, *New Patterns in the Sky,* pp. 244-45. For another explanation of the different names associated with Dorado, see Apus.

Table 38. The lettered stars of Dorado

	Magnitudes			Catalogue Numbers				
Letter	Bayer	Lacaille	Visual	Flamsteed	Lacaille	Other	HR	HD
Alpha		3	3.27		356		1465	29305
Beta		4	3.76		436		1922	37350
Gamma		4	4.25		327		1338	27290
Delta		5	4.35		455		2015	39014
Epsilon		5	5.11		468		2064	39844
Zeta		5	4.72		392		1674	33262
Eta^1+		6	5.71		494		2194	42525
Eta^2		6	5.01		504		2245	43455
Theta		6	4.83		409		1744	34649
Kappa		6	5.27		371		1530	30478
Lambda		6	5.14		420		1836	36189
Mu+		6	8.92		396			33599
Nu		6	5.06		502		2221	43107

(table continued on next page)

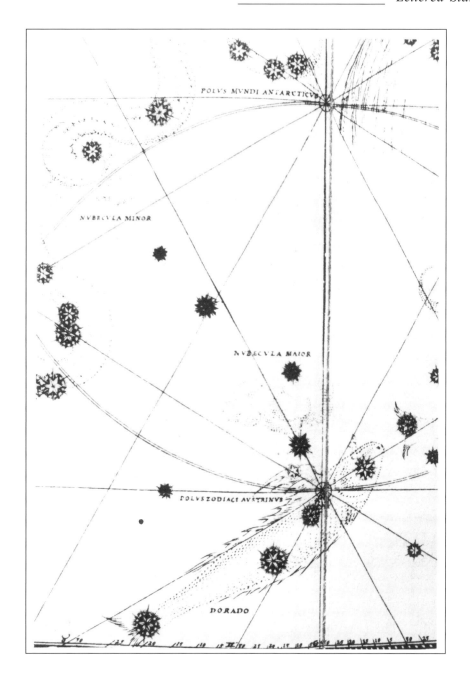

Figure 34. Bayer's Dorado. This is an enlarged view of Dorado from Bayer's Tabula XLIX, his planisphere of the southern heavens. Above the fish is Nubecula Maior and Nubecula Minor, the Large and Small Magellenic Clouds. The chart shows the ecliptic south pole in Dorado and the celestial south pole directly above it next to the tail of Apus. See Draco, Figure 36.

Table 38. The lettered stars of Dorado *(continued)*

	MAGNITUDES			CATALOGUE NUMBERS				
Letter	Bayer	Lacaille	Visual	Flamsteed	Lacaille	Other	HR	HD
Pi^1+		6	5.56		540		2352	45669
Pi^2		6	5.38		547		2377	46116
G+			5.34				1917	37297

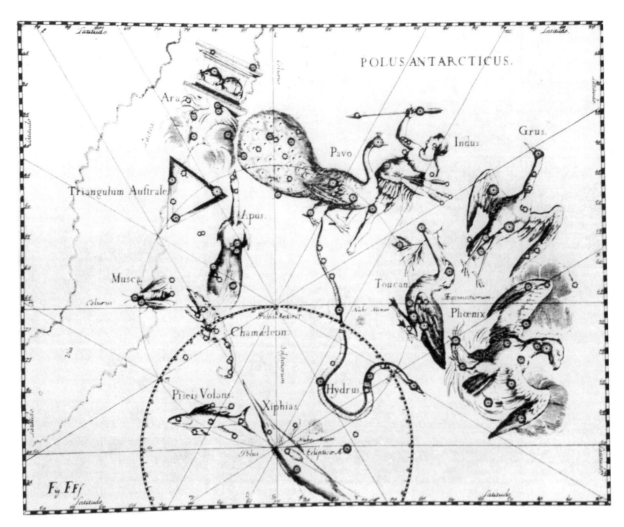

Figure 35. Hevelius's Dorado. Hevelius's planisphere of the southern heavens transforms Dorado, the peaceful Goldfish, into Xiphias, the deadly Swordfish. Both the ecliptic south pole, by Xiphias, and the celestial south pole, directly above, are depicted on this chart. See Draco, Figure 36. All the constellations in Hevelius's atlas are drawn backward, as they would appear on a celestial globe.

The Lost, Missing, or Troublesome Stars of Dorado

Eta¹, HR 2194, 5.71V, *SA, HRP, SA,* CA 494 as Eta, BR 1174*, L 2203, BAC 2003*, G 38*.
Eta², HR 2245, 5.01V, *SA, HRP, BS,* CA 504 as Eta, BR 1195*, L 2230, BAC 2031*, G 40*.

Lacaille designated these two neighboring stars Eta. He had originally designated a third star as Eta (CA 513, BR 1223 as Eta³, L 2275, BAC 2065 as Eta³), but Baily noted in the *BAC* that it was probably nonexistent. Neither Gould, Maclear, Rumker, Stone, nor Taylor could locate it.

Iota

Lacaille did not use this letter in Dorado.

Mu, HD 33599, 8.92V, CA 396*, BR 891*, L 1766*, BAC 1612*, GC 6303, SAO 249196.

This is the dimmest star in Lacaille's catalogue, and Gould therefore dropped its letter and excluded it from his catalogue of the southern skies. For an interesting comment on the identity and magnitude of this star, see Gould, *Uran. Arg.,* pp. 262-63.

Xi
Omicron

Lacaille did not use these two letters in Dorado.

Pi¹, HR 2352, 5.56V, *SA, HRP, BS,* CA 540 as Pi, BR 1259*, L 2340*, G 41*.
Pi², HR 2377, 5.38V, *SA, HRP, BS,* CA 547 as Pi, BR 1275*, L 2368*, G 42*.

Lacaille designated these two neighboring stars Pi.

G, HR 1917, 5.34V, *SA,* CA 437, BR 1002, L 1949, G 28.

Although Lacaille observed and catalogued this star, he left it unlettered since he considered it 6th magnitude. In *Sky Cat. 2000.0* (II, 207), this star, a spectroscopic binary, is referred to as 28 Dor. Since it is Gould's 28 Dor in his *Uran. Arg.,* this is probably the source of the G and the 28.

Draco, Dra

Dragon

Bayer lettered the Ptolemaic constellation Draco (Figure 36) from Alpha to Omega and from A to i. His chart — Tabula III — contains a couple of typographical errors: (1) Iota and i are interchanged and (2) Omicron is omitted.

Several legends pertain to Draco. At the marriage ceremony of Zeus and Hera, Gaea, goddess of the Earth, presents the bride with a tree that bears golden apples. Hera asks Gaea to plant the tree in a garden by the Atlas Mountains and sets the Hesperides, the daughters of Atlas, to guard it. Instead of guarding the golden apples, however, they steal some. Hera decides a more vigilant caretaker is needed. She enlists the service of the dragon Ladon, an offspring of Phorcys and Ceto, whose union produced a whole brood of monsters — Ladon, the Gorgons, the Graiae, Echidna, and possibly Cetus. Ladon is killed as he attempts to protect the apples from Hercules — or from the Titan Atlas, depending on the story — who is assigned to steal the apples as his eleventh labor. Mindful of Ladon's efforts on her behalf, Hera rewards the dragon with a place in the heavens, where it remains today, guarding the north ecliptic pole.

In another legend, Draco symbolizes the serpent thrown at Athena as she and the other gods battle the Titans for control of the universe at the beginning of time. Athena catches the serpent and flings it far into the heavens.

Many cultures have considered the snake a guardian and protector and have extended to it medicinal and healing powers. The people of ancient Mesopotamia associated the snake with long life and rejuvenation. The ancient Greeks and Romans believed that the snake was sacred to Asclepius, the god of medicine.

See Condos, *Star Myths,* pp.101-3; Ridpath, *Star Tales,* pp. 64-65; Hamilton, *Mythology,* p. 233; Schwab, *Gods & Heroes,* pp. 171-73; Gantz, *Early Greek Myth,* pp. 19-25 and Table 2. For Draco's role as guardian of the north ecliptic pole, see Dorado; for the monstrous offspring of Phorcys and Ceto, see Cetus; for the serpent's association with rejuvenation, see Ophiuchus.

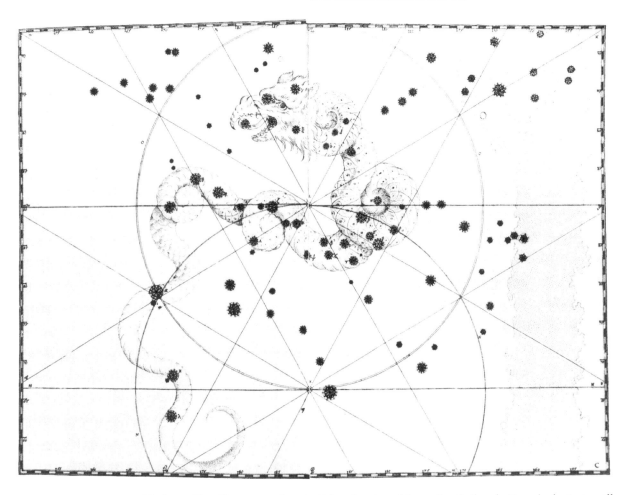

Figure 36. Bayer's Draco. The letters K and L represent the celestial north pole and the north ecliptic pole. Bayer's charts, as well as those of Hevelius, are all oriented toward the north ecliptic pole. The letters M and N are the equinoctial colure and the Arctic Circle. The letter O marks the northern ecliptic circle. It is distinguished on all of Bayer's charts by a double-lined circle. (For the southern ecliptic circle, which is also double-lined, see Bayer's Tabula XLIX, Figure 3.) The 2nd-magnitude star slightly to the right of the celestial pole is Alpha UMi, the Pole Star, or Stella Maris, the Star of the Sea.

The Arctic Circle circumscribes the celestial north pole, which is 23½° distant from the ecliptic north pole. The difference between the two poles represents the 23½° angle of deviation of Earth's axis and the celestial sphere from the plane of Earth's annual orbit about the sun. That is, as seen from Earth, the sun's apparent path across the sky, the ecliptic, cuts a plane through the celestial equator at a 23½° angle. Ninety degrees north of the ecliptic is the north ecliptic pole—located in Draco at 18^h right ascension, +66½° declination. The angle of deviation represents the 23½° tilt of Earth's axis from the vertical.

Lost Stars

Table 39. The lettered stars of Draco

	MAGNITUDES			CATALOGUE NUMBERS				
Letter	Bayer	Lacaille	Visual	Flamsteed	Lacaille	Other	HR	HD
Alpha	2		3.65	11			5291	123299
Beta	3		2.79	23			6536	159181
Gamma	3		2.23	33			6705	164058
Delta	3		3.07	57			7310	180711
Epsilon	3		3.83	63			7582	188119
Zeta	3		3.17	22			6396	155763
Eta	3		2.74	14			6132	148387
Theta	3		4.01	13			5986	144284
Iota	3		3.29	12			5744	137759
Kappa	3		3.87	5			4787	109387
Lambda	3		3.84	1			4434	100029
Mu	4		4.92c	21			6369/70	154905/6
Nu^1+	4		4.88	24			6554	159541
Nu^2			4.87	25			6555	159560
Xi	4		3.75	32			6688	163588
Omicron	4		4.66	47			7125	175306
Pi	4		4.59	58			7371	182564
Rho	4		4.51	67			7685	190940
Sigma	4		4.68	61			7462	185144
Tau	4		4.45	60			7352	181984
Upsilon	4		4.82	52			7180	176524
Phi	4		4.22	43			6920	170000
Chi	4		3.57	44			6927	170153
Psi^1+	4		4.58c	31			6636/7	162003/4
Psi^2			5.48	34			6725	164613
Omega	4		4.80	28			6596	160922
A	4		5.00	15			6161	149212
b	5		4.98	39			6923	170073
c	5		5.04	46			7049	173524
d	5		4.77	45			6978	171635
e	5		5.27	64			7676	190544
f	5		5.05	27			6566	159966
g	5		4.83	18			6223	151101
h^1+	5		4.89	19			6315	153597
h^2			6.42	20			6319	153697
i	5		4.65	10			5226	121130

The Lost, Missing, or Troublesome Stars of Draco

Nu^1, Nu^2, 24/5, HR 6554/5, 4.88V, 4.87V, comb. mag. 4.21V, *SA, HRP* as Nu, *BS,* BV 69/70 as Nu, BF 2416/7*, H 110 as Nu, ADS 10628.

Nu appeared to Bayer as a single 4th-magnitude star, but Flamsteed, using a telescope, separated the components of this binary system and designated them Nu-1 and Nu-2. Bevis and Argelander, however, equated the combined light of both stars with Nu (*UN,* p. 9).

Psi^1, 31, HR 6636/7, 4.58V, 5.79V, *SA* as Psi, *HRP* as Psi, *BS,* BV 34 as Psi, BF 2458*, H 121 as Psi.
Psi^2, 34, HR 6725, 5.48V, *BS,* BV 32, BF 2485*, H 129.

Bayer described Psi as a single 4th-magnitude star, but Flamsteed designated these two neighboring stars, his 31 and 34, as Psi-1 and Psi-2. Bevis and Argelander, however, felt that only 31, the brighter of the two, was synonymous with Psi (*UN,* p. 91).

A, 15, HR 6161, 5.00V, *SA, HRP.*

b, 39, HR 6923, 4.98V, *SA, HRP.*

c, 46, HR 7049, 5.04V, *SA, HRP.*

d, 45, HR 6978, 4.77V, *SA, HRP.*

e, 64, HR 7676, 5.27V, *SA, HRP.*

f, 27, HR 6566, 5.05V, *SA, HRP.*

g, 18, HR 6223, 4.83V, *SA, HRP.*

h^1, 19, HR 6315, 4.89V, *SA* as h, *HRP* as h, HF 13 as h, B 121 as h, BF 2357*, H 90 as h.
h^2, 20, HR 6319, 6.42V, HF 14 as h, B 122, BF 2358*.

Bayer considered h a single star of the 5th magnitude. In Halley's 1712 pirated edition of Flamsteed's catalogue, 19 and 20 are referred to as *duplex,* are bracketed together, and are each labeled h. In Flamsteed's authorized catalogue, his "Catalogus Britannicus," they are still bracketed together, but they are no longer referred to as *duplex* and only 20, the dimmer of the two, is labeled h. Bode felt that 19, a much brighter star, was equivalent to Bayer's h. Baily agreed, but because there was some doubt, he reverted to the 1712 catalogue designation by labeling 19 and 20 as h^1 and h^2. Argelander, unable to see 20 with his naked eye since the two stars are virtually adjacent, believed only 19 was synonymous with h (*UN,* p. 8). In all likelihood, Bayer saw the combined light of both stars as one image.

i, 10, HR 5226, 4.65V, *SA, HRP.*

Equuleus, Equ

Little Horse

Equuleus (Figure 37) is a Ptolemaic constellation that Bayer lettered from Alpha to Delta. He described the constellation as Equus Minor, Equus Prior, or simply, Equuleus.

The origin of Equuleus has puzzled scholars for centuries. It is not mentioned in any classical Greek or Roman myths. Nor does any classical poet or playwright mention it. Allen has suggested Equuleus may have been devised by Hipparchus of Nicaea, a second-century B.C. astronomer who prepared the first stellar catalogue. The first mention of the constellation was in Ptolemy's catalogue, where it is referred to as the bust or figurehead of a horse (ιππος, hippos). Since its appearance, it has been associated with various figures in mythology. Most recently, one scholar has suggested that it represents Thetis, the daughter of the centaur Chiron, who is seduced by Aeolus, the grandson of Deucalion, the Greek Noah. She tries to hide from her pious father by asking the gods to change her into a mare, Hippe. The goddess Artemis, the protector of maidens, takes pity on her and places her in the heavens.

See Allen, *Star Names,* pp. 212-14; Toomer, *Ptolemy's Almagest,* p. 358; Ridpath, *Star Tales,* pp. 65-66. See also Pegasus for additional information about Thetis-Hippe.

Table 40. The lettered stars of Equuleus

	MAGNITUDES			CATALOGUE NUMBERS				
Letter	Bayer	Lacaille	Visual	Flamsteed	Lacaille	Other	HR	HD
Alpha	4		3.92	8			8131	202447
Beta	4		5.16	10			8178	203562
Gamma	4		4.69	5			8097	201601
Delta	4		4.49	7			8123	202275
Epsilon+			5.23	1		B 1	8034	199766
Lambda+			6.64c	2		B 2		200256

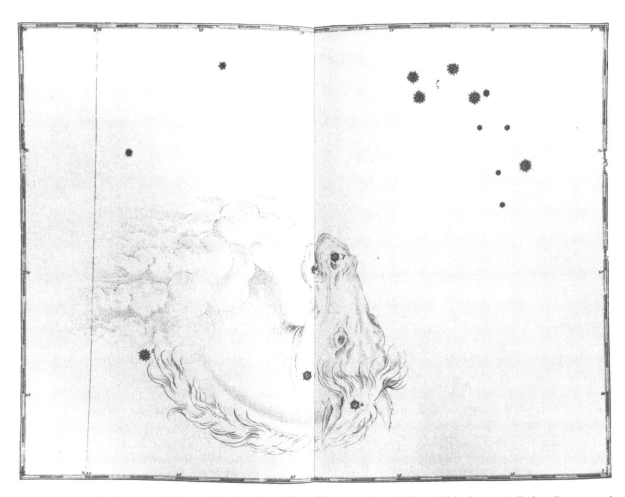

Figure 37. Bayer's Equuleus. The letters Epsilon (ε) and Zeta (ζ) represent stars in the neighboring constellations Pegasus and Delphinus.

The Lost, Missing, or Troublesome Stars of Equuleus

Epsilon, 1, HR 8034, 5.23V, *SA, BS,* B 1*, BAC 7276, H 2, G 4.
Lambda, 2, HD 200256, comb. mag. 6.64V, *SA,* B 2*, BAC 7302, H 3, GC 29361, ADS 14556.

Bayer did not letter these two stars in Equuleus. The letters were first introduced by Bode, who added letters beyond Bayer's Delta, none of which has survived in the literature except these two. Both Argelander (*UN,* p. 80) and Heis claimed they observed 2 and listed its coordinates correctly in their catalogues, but the magnitude of the star is beyond the visibility of the naked eye. They confused it with its brighter neighbor, about 1/2° away, HR 8038, 5.99V, BAC 7285. Adding to the confusion, Heis's catalogue erroneously equates BAC 7255 (HR 8010, 6.05V) with 1 and BAC 7276 with 2. Gould noted the error in his *Uran. Arg.* (p. 34, nos. 627/8; p. 38; p. 232, G 5; p. 338). Lambda is the double star Struve 2742; its two components, A and B, are each magnitude 7.4V.

ERIDANUS, ERI

River Eridanus

Eridanus (Figure 38) is a Ptolemaic constellation that Bayer lettered from Alpha to Omega and from A to d. Lacaille designated additional stars to z, especially in the southern sector of the constellation, which is below the horizon in Central Europe.

Scholars have been puzzled for ages as to which of the great rivers of the ancient world should be associated with Eridanus. Since it is a southern constellation, most have felt that Eridanus refers to the Nile, although Ptolemy — himself an Egyptian — called it simply Potamos (ποταμος), or River. Others have suggested the Po, the Rhine, the Rhone, or even the Ebro in Spain as possible candidates. Nor has the derivation of the name Eridanus provided any clue to the identity of the River since it does not in any way resemble the name of any of the great rivers of the world. The first use of the name occurs in *Phenomena* (270 B.C.), a poem by the Greek poet Aratus.

> *The grim monster [Cetus] lies by the River of Stars:*
> *for here flows the remnant of the river ERIDANUS*
> *under the feet of the gods, a river of tears*
> *that reaches at last the left foot of Orion* (338-41).

Some scholars have suggested that its name may have been derived from the city of Eridu (modern Abu Shahrein, near the Shatt-al-Arab, Iraq), which emerged as an urban center along the banks of the Euphrates some five thousand years ago. They point out that Eridu was the home of Enki, the Sumerian god of fresh water; that he was the chief god of the city; and that Eridu was the cult center for his worship. It was, moreover, the cultural center of ancient Sumer and the cradle of Mesopotamian civilization. But as far as is presently known, there is no Mesopotamian equivalent to the constellation Eridanus nor is it mentioned in any astronomical texts — at least no references to it have been uncovered to date.

Although the ancient Greeks and Romans were unsure where the River was, they did associate it with the death of Phaethon, the human son of Helios-Apollo, the sun god — and the mortal Clymene. Phaethon tricks his father into allowing him to drive his sun-chariot across the heavens to light up the day. Unable to master the great steeds that pull the sun, Phaethon loses control and the chariot careens crazily through the sky. Zeus puts an end to the mad ride by striking Phaethon with one of his thunderbolts and the hapless Phaeton falls burning and smoldering into the River Eridanus, thus explaining Aratus's "river of tears".

See Allen, *Star Names,* pp. 215-17; Ridpath, *Star Tales,* pp. 66-68; Toomer, *Ptolemy's Almagest,* p. 384; Langdon, *Semitic Mythology,* p. 310; Lombardo, *Sky Signs,* p. 15; Jacobsen, *Toward the Image of Tammuz,* pp. 21-22; Kramer, *Sumerian Mythology,* pp. 62, 65; Hamilton, *Mythology,* pp. 180-84.

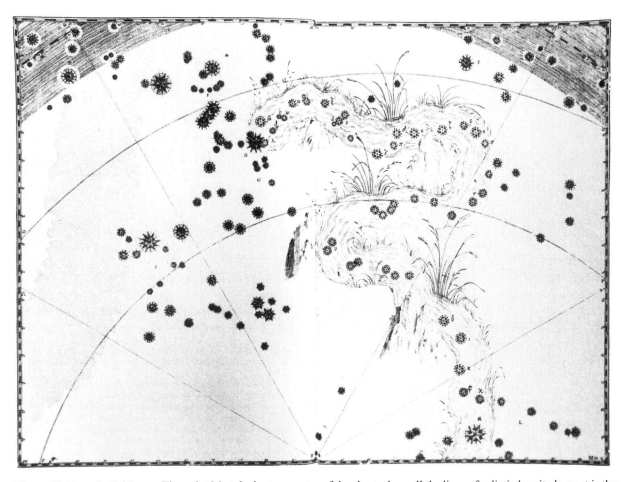

Figure 38. Bayer's Eridanus. The point M at the bottom center of the chart where all the lines of ecliptic longitude meet is the south ecliptic pole. The line N is the ecliptic; O is the celestial equator; and P is the Tropic of Capricorn. The letter E is the star Alpha Cet in the snout of the Whale; F is Orion; G is Lepus; H is the feet of Gemini; I is Canis Major; K is Dorado; and L is Phoenix. The shaded area at the top corners is the zodiac.

Table 41. The lettered stars of Eridanus

	Magnitudes			Catalogue Numbers				
Letter	Bayer	Lacaille	Visual	Flamsteed	Lacaille	Other	HR	HD
Alpha	1		0.46				472	10144
Beta	3		2.79	67			1666	33111
Gamma	3		2.95	34			1231	25025
Delta	3		3.54	23			1136	23249
Epsilon	3		3.73	18			1084	22049
Zeta	3		4.80	13			984	20320
Eta	3		3.89	3			874	18322
Theta[1,2]+	3		2.90c				897/8	18622/3
Iota	3		4.11				794	16815
Kappa	3		4.25				721	15371
Lambda	4		4.27	69			1679	33328
Mu	4		4.02	57			1520	30211
Nu	4		3.93	48			1463	29248

(table continued on next page)

Table 41. The lettered stars of Eridanus (continued)

	MAGNITUDES				CATALOGUE NUMBERS			
Letter	Bayer	Lacaille	Visual	Flamsteed	Lacaille	Other	HR	HD
Xi	4		5.17	42			1383	27861
Omicron¹+	4		4.04	38			1298	26574
Omicron²	4		4.43	40			1325	26965
Pi	4		4.42	26			1162	23614
Rho¹+			5.75	8			907	18784
Rho²	4		5.32	9			917	18953
Rho³			5.26	10			925	19107
Sigma+	4							
Tau¹+	4		4.47	1			818	17206
Tau²	4		4.75	2			850	17824
Tau³	4		4.09	11			919	18978
Tau⁴	4		3.69	16			1003	20720
Tau⁵	4		4.27	19			1088	22203
Tau⁶	4		4.23	27			1173	23754
Tau⁷	4		5.24	28			1181	23878
Tau⁸	4		4.65	33			1213	24587
Tau⁹	4		4.66	36			1240	25267
Upsilon¹+	4		4.51	50			1453	29085
Upsilon²	4		3.82	52			1464	29291
Upsilon³	4		3.96	43			1393	28028
Upsilon⁴	4		3.56	41			1347	27376
Phi	4		3.56				674	14228
Chi	4		3.70				566	11937
Psi	5		4.81	65			1617	32249
Omega	5		4.39	61			1560	31109
A	5		4.87	39			1318	26846
b	6		5.51	62			1582	31512
c	6		5.23	51			1474	29391
d+	6		5.85				1363	27563
e		4	4.27		253		1008	20794
f+		4	4.35c		290		1189/90	24071/2
g+		4	4.17		291		1195	24160
h+		5	4.59		282		1143	23319
i+		5	5.11		295		1214	24626
k+		5	4.66	36	305		1240	25267
l+			3.87	53		G 254	1481	29503
m+		6	5.24	28	286		1181	23878
n+		5	4.23	27	284		1173	23754
o¹+		6	6.87		326			27518
o²		6	6.01		328		1374	27710
p		6	5.07c		104		486/7	10360/1
q¹+		6	5.52		111		506	10647
q²		6	5.04		115		520	10939
r+		6	5.82		235		929	19141
s		5	4.75		184		789	16754
t+		6	6.01		186		805	16975
u+		6	5.93		243		968	20121
v+			4.73	17		G 97	1070	21790
w+			4.68c	32		G 148	1211/2	24554/5
x+		6			264			
y		5	4.58		273		1106	22663
z+		6	6.12		265		1076	21899

The Lost, Missing, or Troublesome Stars of Eridanus

Theta¹, Theta², HR 897/8, 3.42V, 4.42V, comb. mag. 2.90V, *SA, HRP* as Theta, *BS,* HA 22 as Theta, G 48/9 as Theta.

Bayer designated only one 3rd-magnitude star as Theta. Later astronomers added the indices to identify the components of this double star.

Until the Age of Exploration, this star, Theta, was considered the end of the stream. Ptolemy referred to it as "the last star of the river, the bright one," and in the Middle Ages, Arab astronomers referred to it as *akhir al-nahr,* "the River's End," from which was derived the name Achernar. During the fifteenth and sixteenth centuries as European seamen sailed into the tropics and viewed southern skies for the first time, astronomers extended the River's course some 20° farther south than Theta, from about -40° to about -60°. There they discovered a 1st-magnitude star, the ninth brightest in the heavens, Alpha Eri (HR 472, 0.46V), which became the new *lucida,* the brightest star of the constellation. Houtman, the first to describe it in his 1603 catalogue of southern skies, called it "Acarnar [*sic*]," the star at "the end of the Nile." Henceforth, Theta Eri lost its prominence. As Halley noted in his catalogue of the southern skies of 1673, it used to be "the end of the river in the Old Catalogue [of Ptolemy]." Theta not only lost its prominence, it also lost its original name. Forever after, it was called Acamar, a somewhat jumbled, corrupted form of Achernar.

Ptolemy's reference to Theta as "the bright one" has long puzzled astronomers. He categorized it as a 1st-magnitude star, but at least since the time of Tycho and Bayer, it has been regarded as 3rd magnitude. Actually, it is a double star, Piazzi 2, with a combined magnitude of 2.90V. See Toomer, *Ptolemy's Almagest,* p. 386; *Burnham's Celestial Handbook,* II, 889; Allen, *Star Names,* p. 219; Kunitzsch and Smart, *Short Guide to Modern Star Names,* p. 36; Knobel, "On Frederick de Houtman's Catalogue," p. 422; Baily, "Catalogues of Ptolemy *et al.,*" p. 170.

Omicron¹, 38, HR 1298, 4.04V, *SA, HRP, BS,* HF 37 as Omicron, BF 519 as Omicron, BAC 1290*, H 93*.
Omicron², 40, HR 1325, 4.43V, *SA, HRP, BS,* HF 39 as Omicron, BF 531 as d, BAC 1309*, H 97*.

Considerable confusion surrounds these two stars and d. Bayer described Omicron as a couple of 4th-magnitude stars and he depicted them on his chart, together with d, on a line running from southeast to northwest — the two Omicrons to the northwest and d to the southeast. In Halley's 1712 edition of Flamsteed's catalogue, 38 and 40 are properly equated with Bayer's two Omicrons. In Flamsteed's 1725 catalogue, however, only the northwestern star, 38, is labeled Omicron; the other half of the Omicron pair, 40, is mistakenly labeled d. In other words, Flamsteed's 1725 catalogue designates 38 as Omicron when it should be Omicron-1, designates 40 as d when it should be Omicron-2, and leaves d (HR 1363, 5.85V) without any number since Flamsteed never observed this star. Baily allowed the confusion between the Omicrons and d to pass unnoticed into his revised edition of Flamsteed's catalogue although he later corrected it in the *BAC*. See d.

Rho¹, 8, HR 907, 5.75V, *SA, HRP, BS,* BF 388*, G 52*.
Rho², 9, HR 917, 5.32V, *SA, HRP, BS,* BF 393*, H 21 as Rho, G 55*.
Rho³, 10, HR 925, 5.26V, *SA, HRP, BS,* BF 397*, H 23, G 57*.

Bayer described Rho as a single 4th-magnitude star, but Flamsteed designated these three neighboring stars, his 8, 9, and 10, as Rho-1, Rho-2, and Rho-3. Although Baily felt only 10 was synonymous with Rho, he kept Flamsteed's designations. Argelander, on the other hand, who could not see 8, equated only 9 with Rho (*UN,* p. 88), but Gould left all three as Rho.

Sigma

This star is Ptolemy's 17 Eri, which Bayer lettered Sigma. It is apparently one of Bayer's lost stars. Flamsteed could not locate it, nor could any other astronomer. Toomer (*Ptolemy's Almagest,* p. 385) and Hoffleit (*BS,* p. 378) tentatively have suggested it might be HR 859, 6.32V, but Bayer, like Ptolemy, described Sigma as a 4th-magnitude star and HR 859 is much fainter. It does, however, correspond to the Sigma on Bayer's chart and it is visible, though barely so, to the naked eye. Although Argelander could not see HR 859, Heis observed it and included it in his catalogue but without suggesting it might be Sigma (H 10). For Bayer's apparent ability to see the unseeable, see Chi Per.

The Sigma equated with Lacaille's i (HR 1214, 5.11V) in *SA* is not Bayer's Sigma. It is found in Becvar's atlases and catalogue but not in any of the catalogues of the leading astronomers of the southern skies: Brisbane, Gould, Lacaille, Rumker, Maclear, Piazzi, Taylor, or Stone. Tirion removed it from his *Uranometria 2000.0* and the second edition of *SA*.

Tau¹, 1, HR 818, 4.47V, *SA, HRP, BS,* BV 9*, BF 352*.
Tau², 2, HR 850, 4.75V, *SA, HRP, BS,* BV 10*, BF 362*.
Tau³, 11, HR 919, 4.09V, *SA, HRP, BS,* BV 14*, BF 396*.
Tau⁴, 16, HR 1003, 3.69V, *SA, HRP, BS,* BV 19*, BF 426*.
Tau⁵, 19, HR 1088, 4.27V, *SA, HRP, BS,* BV 30*, BF 443*.
Tau⁶, 27, HR 1173, 4.23V, *SA, HRP, BS,* BV 34*, BF 478*.
Tau⁷, 28, HR 1181, 5.24V, *SA, HRP, BS,* BV 33*, BF 481*.
Tau⁸, 33, HR 1213, 4.65V, *SA, HRP, BS,* BV 37*, BF 491*.
Tau⁹, 36, HR 1240, 4.66V, *SA, HRP, BS,* BV 42*, BF 501*.

Bayer described Tau as a group of nine 4th-magnitude stars and placed the letter Tau between the upper two stars on his chart. Working just with Bayer's chart, Flamsteed designated only these two stars, his 1 and 2, as Tau-1 and Tau-2. Bevis added the other seven Tau's as in the heading above. Baily and most modern authorities have accepted these designations.

Upsilon¹, 50, HR 1453, 4.51V, *SA, HRP, BS,* CA 349 as Upsilon, BR 732 as Upsilon², L 1513 as Upsilon⁶, BAC 1422 as Upsilon⁶, H 114*, G 243*.

Upsilon², 52, HR 1464, 3.82V, *SA, HRP, BS,* CA 352 as Upsilon, BR 740 as Upsilon³, L 1529 as Upsilon⁷, BAC 1433 as Upsilon⁷, H 117*, G 251*.

Upsilon³, 43, HR 1393, 3.96V, *SA, HRP* erroneously as d, *BS,* CA 335 erroneously as d, BR 699 erroneously as d, L 1441 as Upsilon⁵, BAC 1372 as Upsilon⁵, H 104*, G 219 erroneously as d.

Upsilon⁴, 41, HR 1347, 3.56V, *SA, BS,* CA 325 erroneously as Xi, BR 681 erroneously as Xi, L 1411*, BAC 1333*, H 99*, G 204 as X.

Upsilon generates considerable confusion. Ptolemy listed in his catalogue seven stars of the 4th magnitude near the very end of the River. Bayer did exactly the same and placed the letter Upsilon next to the upper two stars in the group. Thinking these were the only two stars Bayer meant as Upsilon, Flamsteed designated his 50 and 52 as Upsilon-1 and Upsilon-2. Lacaille went one step farther. He assigned Upsilon to four stars, the two Flamsteed had identified and two neighboring smaller stars:

HD 28720, 7.2V, CA 343 as Upsilon, BR 720 as Upsilon¹.
50, HR 1453, 4.51V, CA 349 as Upsilon, BR 732 as Upsilon².
52, HR 1464, 3.82V, CA 352 as Upsilon, BR 740 as Upsilon³.
HR 1476, 6.30V, CA 353 as Upsilon, BR 742 as Upsilon⁴.

Baily, for his part, decided that he could identify all seven of Bayer's Upsilon's:

Upsilon¹, HR 1143, 4.59V, L 1198 as Upsilon¹, BAC 1159 as Upsilon¹, Lacaille's h (CA 282).
Upsilon², HR 1195, 4.17V, L 1248 as Upsilon², BAC 1201 as Upsilon², Lacaille's g (CA 291).
Upsilon³, HR 1214, 5.11V, L 1275 as Upsilon³, BAC 1220 as Upsilon³, Lacaille's i (CA 295).
Upsilon⁴, 41, HR 1347, 3.56V, BF 546 as Upsilon⁴, L 1411 as Upsilon⁴, BAC 1333 as Upsilon⁴, Lacaille's Xi (CA 325).
Upsilon⁵, 43, HR 1393, 3.96V, BF 567 as Upsilon⁵, L 1441 as Upsilon⁵, BAC 1372 as Upsilon⁵, Lacaille's d, (CA 335).
Upsilon⁶, 50, HR 1453, 4.51V, BF 590 as Upsilon⁶, L 1513 as Upsilon⁶, BAC 1422 as Upsilon⁶, Lacaille's Upsilon[-2], CA 349.
Upsilon⁷, 52, HR 1464, 3.82V, BF 597 as Upsilon⁷, L 1529 as Upsilon⁷, BAC 1433 as Upsilon⁷, Lacaille's Upsilon[-3], CA 352.

Baily's Upsilon's conform somewhat to Bayer's chart, but he numbered his indices in order of right ascension, shocking many astronomers. Baily's stars read from south to north, but most other astronomers felt the numbers should more logically read from north to south, following the natural flow of the River as in Ptolemy's catalogue. Argelander proposed to set matters right by renumbering Baily's indices. But because of his limited view of the southern skies from Bonn, he kept only the four northerly stars, those that bore Flamsteed's numbers 41, 43, 50, and 52. He assigned Upsilon¹ and Upsilon² to 50 and 52, as Flamsteed had done; and Upsilon³ and Upsilon⁴ to 43 and 41 (*UN*, p. 90). See Pi Ori, where Argelander also renumbered Baily's indices because he felt Baily's adherence to a rigid policy of numbering indices by right ascension undermined logic as well as astronomical custom and tradition.

In editing Ptolemy's catalogue, Professor Toomer suggested that the stars Bayer labeled as Upsilon were the ones Baily proposed, except that Lacaille's f (HR 1189/90, 5.42V, 4.86V, CA 290) should replace his i, CA 295 (*Ptolemy's Almagest,* pp. 385-86). Baily at one time did propose that f was one of the Upsilon's, but he later changed his mind (see f). Toomer believed that these stars were synonymous with Ptolemy's seven stars near the end of the River.

Eighty years before Baily sought to identify Bayer's Upsilon's, Bevis had selected the very same stars he did, but his indices were somewhat different. Bevis's Upsilon's were:

Upsilon-1, 50, BV 57.
Upsilon-2, 52, BV 58.
Upsilon-3, 43, BV 48.
Upsilon-4, 41, BV 44.
Upsilon-5, BV 29, Lacaille's i (CA 295).
Upsilon-6, BV 25, Lacaille's g (CA 291).
Upsilon-7, BV 17, Lacaille's h (CA 282).

Although Bevis agreed with Baily on which stars were Bayer's Upsilon's, Bevis's indices tended to follow the flow of the River. In fact, the first four are the same stars that Argelander chose.

Although Baily felt that Lacaille's g, h, and i or maybe f were the three remaining Upsilon's, for all intents and purposes, Bayer's seven have been permanently reduced to four; the other three have forever lost their Upsilon identity. As a matter of fact, Gould decided to retain only 50 and 52 as Upsilon^1 and Upsilon^2 since he felt the others, especially 41 and 43, were too far away to be considered part of the same group. See *Uran. Arg.,* pp. 84-87; for the confusion between Upsilon^3 and d, see d; for the confusion between X, Xi, and 41, see X; see also f, g, h, and i.

A, 39, HR 1318, 4.87V, *SA, HRP.*

b, 62, HR 1582, 5.51V, *SA, HRP.*

c, 51, HR 1474, 5.23V, *SA, HRP.*

d, HR 1363, 5.85V, *SA,* H 101*, G 210.

Some confusion surrounds d. Flamsteed's 1725 catalogue, his "Catalogus Britannicus," erroneously assigns d to his 40, which is Omicron^2 (see Omicron). Lacaille also erred with d. Although Bayer had assigned d to a star near Omicron, Lacaille mistakenly assigned d to another star (CA 335) almost 30° away, Upsilon^3, 43. And Gould, unaware of the mistake, agreed with Lacaille (*Uran. Arg.,* p. 86 and G 219 as d). Argelander was among the first to assign d to a star whose position conformed to Bayer's chart, HR 1363 (*UN,* p. 90). Gould later caught his own error, noting there were two d's in Eridanus, but he stubbornly insisted on keeping d synonymous with 43 as Lacaille had proposed. "No confusion," he asserted, "is...likely to arise from the employment of the letter [d] for a star situated in a distant part of the sky, although within the limits of the same constellation" (*Uran. Arg.,* p. 274). But the two d's in Eridanus have indeed caused considerable confusion. See Upsilon^3 and X.

Although Bayer's letters stopped with d, Lacaille added additional letters from d — which he duplicated — to z. Baily originally planned to eliminate all of Lacaille's lower-case Roman letters except those in the three subdivisions of Argo Navis, but he retained many of them in Eridanus.

e, HR 1008, 4.27V, *SA, HRP,* CA 253*, BR 530, L 1060*, BAC 1044, G 82*.

f, HR 1189/90, 5.42V, 4.86V, comb. mag. 4.35V, *SA, HRP,* CA 290*, BR 610*, L 1244*, BAC 1199, G 135/6*.

This is possibly one of Bayer's three lost Upsilon's. Baily originally thought it was synonymous with Upsilon^3, but he later changed his mind and equated Upsilon^3 with i. Toomer has recently suggested that Baily's first proposal was probably correct and that f, not i, should be one of the stars in the River. See *BF,* p. 400; Toomer, *Ptolemy's Almagest,* p. 386.

g, HR 1195, 4.17V, *SA, HRP,* BV 25 as Upsilon-6, CA 291*, BR 612*, L 1248 as Upsilon^2, BAC 1201 as Upsilon^2, G 138*.

This is possibly one of Bayer's three lost Upsilon's. Bevis thought it was synonymous with Upsilon-6 while Baily equated it with Upsilon^2.

h, HR 1143, 4.59V, *SA, HRP,* BV 17 as Upsilon-7, CA 282*, BR 591*, L 1198 as Upsilon^1, BAC 1159 as Upsilon^1, G 124*.

This is possibly one of Bayer's three lost Upsilon's. Bevis thought it was synonymous with Upsilon-7, while Baily equated it with Upsilon^1.

Flamsteed's "Catalogus Britannicus" of 1725 designates his 67 as h. This is either a copying or typographical error for Beta, which Bevis and Bode corrected (BV 83, B 398, BF 667). In Halley's 1712 edition of Flamsteed's catalogue, 67 is properly equated with Bayer's Beta (HF 66).

i, HR 1214, 5.11V, *SA, HRP,* BV 29 as Upsilon-5, CA 295*, BR 620*, L 1275 as Upsilon^3, BAC 1220 as Upsilon^3, G 151*.

This is possibly one of Bayer's three lost Upsilon's. Bevis thought it was synonymous with Upsilon-5, and Baily equated it with Upsilon^3.

j

Lacaille did not use this letter.

k, 36, HR 1240, 4.66V, BV 42 as Tau-9, CA 305*, L 1312 as Tau^9, G 161 as Tau^9.

Baily, believing 36 was synonymous with Tau^9, dropped Lacaille's k and relettered it Tau^9. Bevis had earlier equated 36 with Tau-9.

l, 53, HR 1481, 3.87V, *SA, HRP,* G 254*.

Lacaille did not letter this star nor use l in Eridanus. Gould assigned l to Flamsteed's 53 since he felt a star of its brightness merited a letter.

Although Lacaille did not use the letter l to identify any star in Eridanus, the letter somehow found its way into his star chart for this constellation through a series of copying and typographical errors. The first blunder occurred when Lacaille's editor erroneously equated his CA 294 in his catalogue with Iota, a mistake for Flamsteed's 33, Tau^8, probably because of their similarity in appearance (ι, τ). This, in turn, led to a double blunder. For when preparing his star chart, the editor misread the erroneous Iota in his catalogue for its look-alike, the letter l (ι, l). See Hoffleit, "Discordances," p. 51 and Gould, *Uran. Arg.,* pp. 86-87.

It is interesting to note that like several other astronomers — such as Hevelius, Flamsteed, and Baily — Lacaille died before he had a chance to complete his magnum opus, leaving its final form to friends and editors.

m, 28, HR 1181, 5.24V, BV 33 as Tau-7, CA 286*, BR 603*, L 1226 as Tau^7, G 130 as Tau^7.

Baily, believing 28 was synonymous with Tau^7, dropped Lacaille's m and relettered it Tau^7. Bevis had earlier come to the same conclusion.

n, 27, HR 1173, 4.23V, BV 34 as Tau-6, CA 284*, BR 597*, L 1220 as Tau^6, G 128 as Tau^6.

Baily, believing 27 was synonymous with Tau^6, dropped Lacaille's n and relettered it Tau^6. Bevis had earlier come to the same conclusion.

o^1, HD 27518, 6.87V, CA 326 as o, BR 684*, L 1415, BAC 1340.
o^2, HR 1374, 6.01V, CA 328 as o, BR 690*, L 1422, BAC 1355, G 214.

Lacaille assigned o to these two neighboring stars. Baily dropped their letters because of their dimness, and Gould excluded CA 326 from his *Uran. Arg.* since he considered it less than 7th magnitude.

p, HR 486/7, 5.86V, 5.82V, comb. mag. 5.07V, *SA, HRP, BS,* CA 104*, BR 243*, L 495*, BAC 521, G 3/4*.

q^1, HR 506, 5.52V, *SA, HRP,* CA 111 as q, BR 249 mistakenly as q, L 506*, G 5*.
q^2, HR 520, 5.04V, *SA, HRP,* CA 115 as q, BR 254 mistakenly as q, L 523*, G 6*.

Lacaille designated these two neighboring stars q. Brisbane's catalogue usually includes superscripts, but they are absent from these two stars.

r, HR 929, 5.82V, CA 235*, BR 466*, L 794, BAC 961, G 58.

    Baily dropped this star's letter because of its dimness.

s, HR 789, 4.75V, *SA, HRP,* CA 184*, BR 383*, L 827*, BAC 828, G 18*.

t, in Fornax, HR 805, 6.01V, CA 186*, BR 387*, L 841, BAC 840, G 40 in Fornax.

    Baily included this star in Fornax and left it unlettered because of its dimness.

u, HR 968, 5.93V, CA 243*, BR 501*, L 1016, BAC 1004, G 71.

    Baily dropped this star's letter because of its dimness.

v, 17, HR 1070, 4.73V, *SA, HRP,* G 97*.

    Lacaille did not letter this star. It was added by Gould, who felt a star of 17's brightness merited a letter.

w, 32, HR 1211/2, 6.14V, 4.79V, comb. mag. 4.68V, *SA, HRP,* G 148*.

    Lacaille did not letter this star in Eridanus. It was added by Gould, who felt a star of 32's brightness merited a letter.

x, CA 264*, L 1116, BAC 1088*.

    This is one of Lacaille's nonexistent stars; there is no star in the sky at the location given in his catalogue. He observed it on November 7, 1751, but because of an error in recording its position, he generated an incorrect set of coordinates. See *Uran. Arg.,* p. 86.

y, HR 1106, 4.58V, *SA, HRP,* CA 273*, BR 578*, L 1161*, BAC 1125, G 110*.

z, HR 1076, 6.12V, CA 265*, BR 561*, L 1125, BAC 1093, G 100.

    Baily dropped this star's letter because of its dimness.

X, 41, HR 1347, 3.56V, BV 44 as Upsilon-4, CA 325 erroneously as Xi, BR 681 erroneously as Xi, L 1411 as Upsilon^4, B 286*, H 99 as Upsilon^4, G 204*.

    Lacaille did not letter this star X. He observed it, equated it with Flamsteed's 41, and lettered it Xi. This was a mistake. Lacaille was using Flamsteed's 1725 catalogue, where *42* is equated with Bayer's Xi. Lacaille's eye caught the wrong line, an easy thing to do since the stars in this edition of Flamsteed's catalogue are not numbered. Bode, who added Roman capitals in Eridanus from A to Z, relettered 41 as X, but Bevis and Baily equated 41 with Bayer's Upsilon^4 and Argelander agreed. Gould, however, felt it was too far away from Upsilon^1 and Upsilon^2, so he retained Bode's X. For the same reason, he also relettered Argelander's Upsilon^3, 43, as Lacaille's d. But in doing this, the usually careful Gould erred twice. In the first place, he repeated Lacaille's error of equating d with 43, forgetting that Bayer had already assigned d to a star near Omicron (see Omicron). Secondly, he used the letter X for 41 although it is not variable and although he agreed to reserve the letters R to Z exclusively for variables. Oddly enough, 41 was later found to be a spectroscopic binary, and there is some variability in its light. In any case, though, it should not be confused with the variable X Eri, at approximately 2^h30^m, -41½°. Modern authorities have rejected Gould's X and d for 41 and 43 and adopted Argelander's designations of Upsilon^4 and Upsilon^3. See Upsilon and d; *Uran. Arg.,* p. 86; *HRP,* p. 203.

Fornax, For

Chemical Furnace

Lacaille devised and lettered the constellation Fornax (figures 2c and 2d) from Alpha to Omega. In his preliminary catalogue, he called it le Fourneau chymique (Chemical Furnace) with its alembic and receiver. On the chart accompanying his catalogue, he called it simply le Fourneau. In his catalogue of 1763, it is labeled in Latin Fornax Chimiae (Chemical Furnace).

In placing Fornax in the heavens, Lacaille was celebrating the birth of the science of modern chemistry. In the past, chemistry was primarily associated with alchemy, the search for the philosopher's stone that would turn base metals into gold. By the seventeenth century, this was changing. Robert Boyle (1627-91), the father of modern chemistry, rejected the Aristotelian concept that all things were made up of four basic elements — fire, air, earth, and water. He asserted a more complex theory about the multiplicity of matter that provided the groundwork for the later development of chemical elements. His contemporary Johann Glauber (1604-68), often considered the co-father of modern chemistry, experimented with the various reagents of inorganic chemistry such as hydrochloric acid, nitric acid, muriatic acid, and *aqua regia,* a mixture of nitric and hydrochloric acid. He produced these acids by mixing various products like alum, salt, vitriol, and saltpetre and then heating them in a brick furnace containing an iron still which emptied into a large receiver. This was a somewhat primitive form of the alembic and receiver that Lacaille pictured on his chart of the southern skies.

See Wolf, *History of Science,* I, 329-32, 336-41.

Table 42. The lettered stars of Fornax

		Magnitudes			Catalogue Numbers			
Letter	Bayer	Lacaille	Visual	Flamsteed	Lacaille	Other	HR	HD
Alpha+		3	3.87	12 Eri	239		963	20010
Beta		5	4.46		204		841	17652
Gamma^1+		5	6.14		205		844	17713
Gamma^2		6	5.39		206		845	17729
Delta		5	5.00		279		1134	23227
Epsilon		6	5.89		232		914	18907
Zeta		6	5.71		224		901	18692
Eta^1+		6	6.51		200		835	17528
Eta^2		6	5.92		208		848	17793
Eta^3		6	5.47		210		851	17829
Theta+		6	6.22		191		817	17168
Iota^1+		6	5.75		178		767	16307
Iota^2		6	5.83		182		777	16538
Kappa		6	5.20		160		695	14802
Lambda^1+		6	5.90		176		744	15975
Lambda^2		6	5.79		180		772	16417
Mu		6	5.28		151		652	13709
Nu		6	4.69		139		612	12767
Omicron+		6	6.19		237		943	19545
Pi		6	5.35		134		594	12438
Rho		6	5.54		288		1184	23940
Sigma		6	5.90		285		1171	23738
Tau		6	6.01		274		1114	22789
Upsilon^1+		6	6.7		203			17627
Upsilon^2			8.5					
Phi+			5.14			G 28	724	15427
Chi^1+		6	6.39		258		1042	21423
Chi^2		6	5.71		262		1054	21574
Chi^3		6	6.50		263		1058	21635
Psi		6	5.92		215		863	18149
Omega		6	4.90		177		749	16046

The Lost, Missing, or Troublesome Stars of Fornax

Alpha, 12 Eri in Fornax, HR 963, 3.87V, *SA, HRP, BS,* CA 239*, BR 493*, L 1000 in Eridanus, BAC 997 erroneously as Alpha Eri, G 72*.

This is Flamsteed's 12 Eridani. Lacaille diverted part of the River to the Furnace and designated this star, the *lucida* of the constellation, as Alpha For.

Gamma^1, HR 844, 6.14V, *SA, HRP, BS,* CA 205 as Gamma, BR 417 as Gamma, L 890*, BAC 880 as Gamma, G 50*.
Gamma^2, HR 845, 5.39V, *SA, HRP, BS,* CA 206 as Gamma, L 892*, BAC 883, G 51*.

Lacaille designated these two stars Gamma although they are almost 3½° apart.

Eta^1, HR 835, 6.51V, *SA, HRP, BS,* CA 200 as Eta, BR 409*, L 879, G 46*.
Eta^2, HR 848, 5.92V, *SA, HRP, BS,* CA 208 as Eta, BR 421*, L 897, G 52*.
Eta^3, HR 851, 5.47V, *SA, HRP, BS,* CA 210 as Eta, BR 423*, L 899, G 53*.

Lacaille designated these three neighboring stars Eta.

Theta, HR 817, 6.22V, CA 191*, BR 399*, L 855*, BAC 855, G 44.

Gould dropped this star's letter because of its dimness.

Iota^1, HR 767, 5.75V, *SA, HRP, BS,* CA 178 as Iota, BR 372*, L 798, G 35*.
Iota^2, HR 777, 5.83V, *SA, HRP, BS,* CA 182 as Iota, BR 376*, L 811, G 37*.

Lacaille designated these two neighboring stars Iota.

Lambda^1, HR 744, 5.90V, *SA, HRP, BS,* CA 176 as Lambda, BR 366*, L 781, G 33*.
Lambda^2, HR 772, 5.79V, *SA, HRP, BS,* CA 180 as Lambda, BR 374*, L 805, G 36*.

Lacaille designated these two neighboring stars Lambda.

Xi

Lacaille did not use this letter in Fornax.

Omicron, HR 943, 6.19V, CA 237*, BR 479*, L 984, BAC 978, G 69.

Baily dropped this star's letter because of its dimness.

Upsilon^1, Upsilon^2, HD 17267, GC 3381, 6.7V, 8.5V, *SA*, CA 203 as Upsilon, BR 413 as Upsilon, L 887, BAC 878 as Upsilon, G 48.

Lacaille described Upsilon as a single 6th-magnitude star. Later astronomers added the indices to identify the components of this double star, h 3532. Gould removed its letter because of its dimness. Upsilon^2, SAO 193927, is not listed in *The Henry Draper Catalogue*; it is listed in Boss's *General Catalogue*. GC 3380 is Upsilon^1.

Phi, HR 724, 5.14V, *SA, HRP, BS,* CA 167, BR 354, L 749, G 28*.

Although Lacaille observed and catalogued this star, he left it unlettered since he considered it 6th magnitude. Gould designated it Phi because he felt a star of its brightness merited a letter.

Chi^1, HR 1042, 6.39V, *SA, HRP, BS,* CA 258 as Chi, BR 549*, L 1101, BAC 1074*, G 89*.
Chi^2, HR 1054, 5.71V, *SA, HRP, BS,* CA 262 as Chi, BR 555*, L 1108, BAC 1082*, G 91*.
Chi^3, HR 1058, 6.50V, *SA, HRP, BS,* CA 263 as Chi, BR 556*, L 1111, BAC 1085 as Chi^2, G 92*.

Lacaille designated these three neighboring stars Chi. Although Baily described all three stars as 6th magnitude, he dropped only CA 262's letter in the *BAC* and relettered CA 263 as Chi^2.

Gemini, Gem

Twins

A Ptolemaic constellation that Bayer labeled from Alpha to Omega and from A to g, Gemini (Figure 39) was devised in the ancient Near East at least four thousand years ago. Alpha and Beta Gem, Castor and Pollux, were called in Sumerian MAS.TAB.BA.GAL.GAL, the Great Twins, and they appeared in Mesopotamian astronomical documents of the latter half of the second millennium B.C. The earliest reference to the twins was found in a collection of clay tablets discovered in 1930 in the village of Ras Shamra, along the Syrian coast, about 125 miles north of Beirut. Thirty-five hundred years ago, Ras Shamra was the flourishing Canaanite city of Ugarit, whose inhabitants worshipped El as the chief god of their pantheon. According to Canaanite mythology, El fathers two sons, the Heavenly Twins, Shachar and Shalem, Semitic names meaning Dawn and Sunset, respectively. The twins became associated with the pruning of the grape vine, which took place in the month of Sivan (May-June) when the sun appeared to move through the constellation Gemini. This period culminated in the Canaanite Festival of the Firstfruits — a prototype of the Jewish Pentecost or Shavuot (Leviticus 23:15-21; Deuteronomy 16:9-12) — when the Twins were singled out for special adoration. The poem "Dawn and Sunset," an account of their miraculous birth, was recited in their honor. According to this poem, they are not actually twins but siblings, conceived in the wombs of two mortal women, the wives of one mortal man who was honored with this gift El had bestowed on his family. When the children are born, he reports to El:

> *My two wives, O El, have given birth,*
> *(And) oh, what they have born!*
> *My two children are Dawn and Sunset!*

Other Near Eastern sources also cite these two deities. Shachar, for example, is mentioned in Isaiah (14:12):

> *How you are fallen from heaven,*
> *O Day Star, son of Dawn (Shachar,* שחר*)!*

Early Babylonian and Assyrian texts also refer to Shachar as an astral deity, and his name appears as well in Phoenician and Carthaginian documents. His brother Shalem was known both to the Phoenicians and Canaanites, and some scholars have suspected that Jerusalem (City of Shalem), founded by Canaanites sometime before 1400 B.C., may have been named in his honor.

As noted earlier, the Twins were especially honored in the month of Sivan, the time for pruning the grape vine and for the Canaanite Festival of the Firstfruits, which later coincided with the Jewish Shavuot or Pentecost. Shavuot is a period of special rejoicing for the Israelites since it celebrates an anticipated bountiful harvest and since it marks the traditional time of the Lord's revelation of the Commandments on Mt. Sinai. It is also on the Pentecost that the Holy Ghost descends on Christ's apostles causing them to speak in tongues to the various foreigners then visiting Jerusalem for the festival. Touched by the spirit of God, they enthusiastically seek to convert strangers to the teachings of Jesus. Some skeptics claim that "they...were filled with new wine" (Acts 2:1-15). In sum, it is perhaps more than coincidental that these various occasions for celebrating and rejoicing and their association with wine all occur within Sivan, the month that was originally marked in the Canaanite calendar as the time for pruning the grape vine, when the Great Twins were the reigning constellation in the heavens.

In addition to MAS.TAB.BA.GAL.GAL, the Great Twins, Mesopotamian documents also refer to MAS.TAB.BA.TUR.TUR, the Small Twins. As previously noted, while the Great Twins were Alpha and Beta Gem, the Small Twins were either Iota and Nu Gem or possibly Zeta and Lambda Gem.

When the Greeks adopted Gemini as one of their constellations, their myth portrayed the birth of Dawn and Sunset, but with a slight but significant variation. Not surprisingly, instead of two women bearing separate children from one man, the Greek legend provides for two men and one woman. Zeus, according to the legend, seduces Leda, Queen of Sparta, who also sleeps that night with her husband King Tyndareus. The union with Zeus leads to the birth of Polydeuces (Roman Pollux) while that with Tyndareus produces Castor.

See O'Neil, *Time and the Calendars,* p. 56; van der Waerden, "Babylonian Astronomy. II. The Thirty-Six Stars," pp. 13-14, 21; van der Waerden, "History of the Zodiac," pp. 219-20; Langdon, *Babylonian Menologies,* p. 4; Gaster, *Thespis,* pp. 64-65, 225-56; Kramer, *Mythologies of the Ancient World,* pp. 188-89; Reimer and Pingree, *Baby. Plan. Omens,* Part 2, p. 13; Ridpath, *Star Tales,* pp. 68-70.

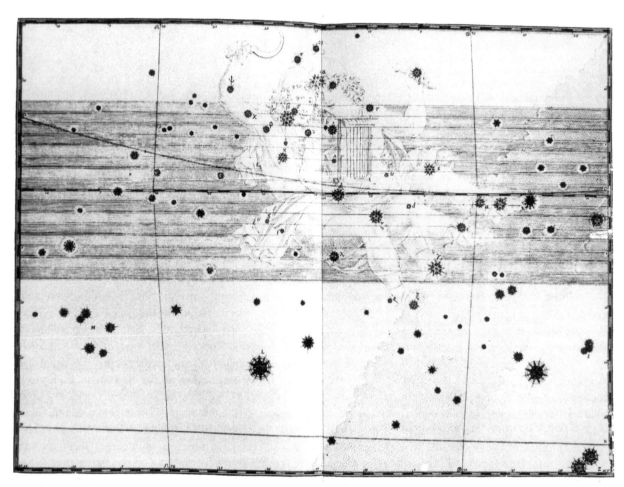

Figure 39. Bayer's Gemini. The lines O and N are the Tropic of Cancer and the celestial equator. The letter I is the head of Orion; K, to the left of the Twins, is Cancer; L is the 1st-magnitude star Procyon, Alpha CMi; and M is the head of Hydra. Although included on Bayer's chart, the letters I through O are neither cited nor explained on his star list for Gemini. The shaded area is the zodiac.

Table 43. The lettered stars of Gemini

	MAGNITUDES			CATALOGUE NUMBERS				
Letter	Bayer	Lacaille	Visual	Flamsteed	Lacaille	Other	HR	HD
Alpha+	2		1.57c	66			2890/1	60178/9
Beta+	2		1.14	78			2990	62509
Gamma	2		1.93	24			2421	47105
Delta	3		3.53	55			2777	56986
Epsilon	3		2.98	27			2473	48329
Zeta	3		3.79	43			2650	52973
Eta	3		3.28	7			2216	42995
Theta	4		3.60	34			2540	50019
Iota	4		3.79	60			2821	58207
Kappa	4		3.57	77			2985	62345
Lambda	4		3.58	54			2763	56357
Mu	4		2.88	13			2286	44478
Nu	4		4.15	18			2343	45542
Xi^1+			4.49	30			2478	48433
Xi^2	4		3.36	31			2484	48737
Omicron	5		4.90	71			2930	61110
Pi	5		5.14	80			3013	62898
Rho	5		4.18	62			2852	58946
Sigma	5		4.28	75			2973	62044
Tau	5		4.41	46			2697	54719
Upsilon	5		4.06	69			2905	60522
Phi	5		4.97	83			3067	64145
Chi+	5		4.94			6 Cnc	3149	66216
Psi+	5		5.64			15 Cnc	3215	68351
Omega+	6		5.18	42			2630	52497
A	6		5.03	57			2808	57727
b^1+	6		5.05	64			2857	59037
b^2			5.01	65			2861	59148
c	6		5.31	76			2983	62285
d	6		5.27	36			2529	49908
e	6		4.65	38			2564	50635
f	6		5.05	74			2938	61338
g	6		4.88	81			3003	62721
H+	3		4.16	1			2134	41116

The Lost, Missing, or Troublesome Stars of Gemini

Alpha, 66, HR 2890/1, 1.58V, 1.59V, comb. mag. 1.57V, *SA, HRP, BS,* ADS 6175, Castor.
Beta, 78, HR 2990, 1.14V, *SA, HRP, BS,* Pollux.

Over the years, astronomers have wondered why Bayer assigned Alpha to Castor when Pollux is brighter. Some have even suggested that Castor was once brighter than Pollux but has gradually grown dimmer. The explanation is much simpler and goes to the very heart of Bayer's lettering system. As Argelander noted, Bayer did not arrange his stars in absolute order of magnitude from the brightest to the faintest. He arranged them by class; that is, he grouped all stars of the same magnitude together, not bothering to distinguish degree of brightness within each magnitude class. He then assigned them letters according to right ascension or, more likely, their position within the constellation. In Draco, for example, the first star of the 4th-magnitude class is Mu (4.92V). Bayer assigned this letter to the first 4th-magnitude star in the Dragon's head, at the very tip of his tongue, the beginning of the constellation. In assigning letters in (Greek) alphabetical order, he proceeded down the Dragon's body to the stars he classified of the 4th-magnitude class: Nu (4.12V) in the Dragon's jaw; Xi (3.75V) in his cheeks; Omicron (4.66V) in his first coil; Pi (4.59V) in his second coil, and so forth.

In the case of Castor and Pollux, Ptolemy, Tycho Brahe, and even Hevelius classified them both as 2nd-magnitude stars. Bayer, as Argelander described him, was "catalogo Tychonis seductus" and, consequently, he too classified them as 2nd magnitude. Since Castor precedes Pollux and since it is the first 2nd-magnitude star in the constellation, Bayer designated it Alpha. See Argelander, *Fide Uran.*, pp. 14-23; Swerdlow, "A Star Catalogue Used by Johannes Bayer," pp. 191-92; Werner and Schmeidler, *Synopsis*, pp. 16-17.

Xi^1, 30, HR 2478, 4.49V, HF 33, BV 32, BF 946, H 19.
Xi^2, 31, HR 2484, 3.36V, *SA* as Xi, *HRP* as Xi, H 22 as Xi, HF 34 as XI, BV 34 as Xi, BF 947 as Xi, H 22 as Xi.

Bayer described Xi as a single 4th-magnitude star. Although Halley's edition of Flamsteed's catalogue designates only 31 as Xi, Flamsteed's 1725 catalogue labels these two nearly adjacent stars, his 30 and 31, as Xi-1 and Xi-2. Bevis, though, agreed with Halley's edition that only the brighter star, 31, was equivalent to Bayer's Xi. Baily felt the same. He too dropped the letter Xi from 30.

Chi

See 6 Cnc in Part Two.

Psi, 15 Cnc in Cancer, HR 3215, 5.64V, *HRP, BS,* BF 1147, H 106*.

Flamsteed included Psi in Cancer, as did Baily. Argelander (*UN,* p. 62) and Heis, however, faithful to Bayer, returned Psi to Gemini. When the International Astronomical Union fixed permanent boundaries for the eighty-eight recognized constellations in 1930, it sided with Flamsteed by switching Psi to Cancer, thus marking one of the very rare occasions when it moved a star with a Bayer letter from one of his constellations to another. See also Psi Cnc.

Omega^1, 42, HR 2630, 5.18V, *SA* as Omega, *HRP* as Omega, *BS* as Omega, BV 43 as Omega, BF 983 as Omega, H 38*.
Omega^2, 44, HR 2659, 5.93V, HF 46, BV 46, BF 998, H 45*.

Bayer described Omega as a single 6th-magnitude star, and Halley's 1712 edition of Flamsteed's catalogue designates 42 as Omega. Flamsteed's 1725 "Catalogus Britannicus," however, labels these two stars, 42 and 44, nearly 2° apart, as Omega-1 and Omega-2. Bevis, though, labeled only the brighter star, 42, as Omega, and Baily agreed. He dropped the letter from 44 since he felt 42 conformed better to Bayer's chart. Since Argelander could not see 44 with his naked eyed, he omitted it from his catalogue (*UN,* p. 61). Heis, on the other hand, observed both stars and restored 44's letter.

A, 57, HR 2808, 5.03V, *SA, HRP.*

b^1, 64, HR 2857, 5.05V, *SA,* HF 65 as b, BV 64 as b, BF 1044*, H 70*.
b^2, 65, HR 2861, 5.01V, *SA, HRP* as b, HF 66, BV 65, BF 1046*, H 72*.

Bayer described b as a single 6th-magnitude star, and Halley's edition of Flamsteed's catalogue designates 64 as b. Flamsteed's authorized catalogue, however, designates these two nearly adjacent stars, 64 and 65, as b-1 and b-2. Bevis labeled only 64 as b. Argelander, too, thought Bayer's b was 64 (*UN, p. 61)*, but Heis disagreed. He labeled 64 and 65 as b^1 and b^2.

c, 76, HR 2983, 5.31V, *SA, HRP.*

d, 36, HR 2529, 5.27V, *SA, HRP.*

e, 38, HR 2564, 4.65V, *SA, HRP.*

f, 74, HR 2938, 5.05V, *SA, HRP.*

g, 81, HR 3003, 4.88V, *SA, HRP.*

H, 1, HR 2134, 4.16V, BF 837, H 1.

Ptolemy placed this star, although relatively bright, outside the constellation Gemini. He considered it amorphous, an unformed star. Tycho included it as the last star in the Twins and mistakenly referred to it as Propus. Propus is not this star but 7, Eta Gem (HR 2216, 3.28V), about 2½° to the southeast (Toomer, *Ptolemy's Almagest*, p. 365). Bayer agreed with Ptolemy that HR 2134 was an *informis,* but he mistakenly labeled it a 3rd-magnitude star and copied Tycho's error of designating it Propus. Since he considered it an *informis,* he labeled it H instead of using a Greek or lower-case Roman letter. It is the only *informis* that Bayer lettered, probably because of its brightness and because he thought it was a named star. Flamsteed also referred to it as H in his catalogue, but Baily removed the letter since he felt the star was outside the constellation and therefore did not merit a letter.

k, 68, HR 2886, 5.25V, BF 1055.
l, 85, HR 3086, 5.35V, BF 1106.

Bayer did not letter these two stars. Although he included them in his atlas, he considered them *informes* outside the constellation Gemini and left them unlettered. Flamsteed added the letters, but they were removed by Baily as superfluous. See i Aql.

m, 48, HR 2706, 5.85V, BV 50, BF 1005.
n, 52, HR 2725, 5.82V, BV 53, BF 1013.
o, 45, HR 2684, 5.44V, BV 51, BF 998.
p, 63, HR 2846, 5.22V, BV 68, BF 1042.
q, 56, HR 2795, 5.10V, BV 63, BF 1030.
r, 61, HR 2837, 5.93V, BV 67, BF 1039.

Bayer did not letter these six stars. The letters were added by Flamsteed but removed by Bevis and Baily as superfluous. See i Aql.

s, 62, HR 2852, 4.18V, HF 63 as Rho, BV 61 as Rho, BF 1041 as Rho.

This star's letter should be Rho. It is either a copying or typographical error in Flamsteed's 1725 catalogue. Bevis and Baily caught the error and corrected it. It is properly identified as Rho in Halley's pirated edition of Flamsteed's catalogue.

GRUS, GRU

Crane

Keyzer and Houtman devised the constellation Grus (Figure 40). In his catalogue of 1603, Houtman referred to it as Den Reygher (The Heron). In Petrus Plancius's globe of 1598 it is labeled Krane Grus, the Dutch and Latin words, respectively, for Crane. On his chart of the southern skies, Bayer also designated this constellation Grus. Lacaille lettered the stars in Grus from Alpha to Phi.

See Warner, *Sky Explored,* p. 204; for an explanation of the differences in nomenclature of the constellations devised by Keyzer and Houtman, see Apus.

Table 44. The lettered stars of Grus

	MAGNITUDES			CATALOGUE NUMBERS				
Letter	Bayer	Lacaille	Visual	Flamsteed	Lacaille	Other	HR	HD
Alpha		2	1.74		1781		8425	209952
Beta		3	2.10		1823		8636	214952
Gamma		3	3.01		1762		8353	207971
Delta^1+		4	3.97		1804		8556	213009
Delta^2		5	4.11		1805		8560	213080
Epsilon		4	3.49		1835		8675	215789
Zeta		5	4.12		1847		8747	217364
Eta		5	4.85		1827		8655	215369
Theta		5	4.28		1857		8787	218227
Iota		5	3.90		1865		8820	218670
Kappa+			5.37			G 83	8774	217902
Lambda		5	4.46		1780		8411	209688
Mu^1+		5	4.79		1787		8486	211088
Mu^2		6	5.10		1790		8488	211202
Nu+			5.47			G 41	8552	212953
Xi+			5.29			G 2	8229	204783
Omicron+			5.52			G 107	8907	220729
Pi^1+			6.62			BR 7158	8521	212087
Pi^2		6	5.62		1797		8524	212132
Rho		6	4.85		1826		8644	215104
Sigma^1+		6	6.28		1814		8600	214085
Sigma^2		6	5.86		1815		8602	214150
Tau^1+		6	6.04		1840		8700	216435
Tau^2		6	6.5c		1841			216655/6
Tau^3		6	5.70		1843		8722	216823
Upsilon		6	5.61		1859		8790	218242
Phi		6	5.53		1878		8859	219693

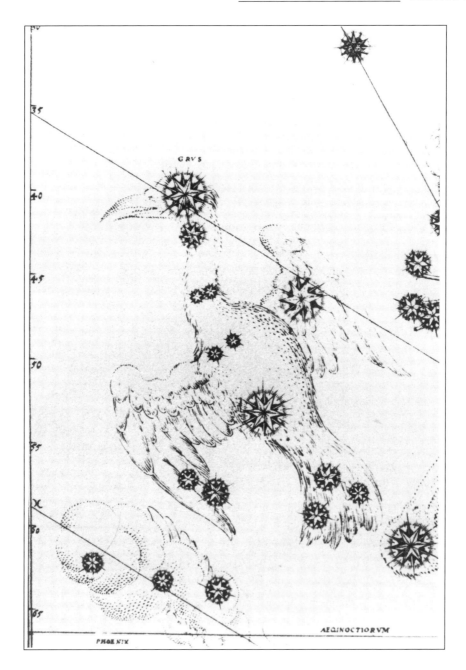

Figure 40. Bayer's Grus. This is an enlarged view of Grus from Bayer's Tabula XLIX, his planisphere of the southern heavens.

Lost Stars

The Lost, Missing, or Troublesome Stars of Grus

Delta¹, HR 8556, 3.97V, *SA, HRP, BS,* CA 1804 as Delta, BR 7172*, L 9138*, G 42*.
Delta², HR 8560, 4.11V, *SA, HRP, BS,* CA 1805 as Delta, BR 7173*, L 9140*, G 43*.

Lacaille designated these two nearly adjacent stars Delta.

Kappa, HR 8774, 5.37V, *SA, HRP, BS,* L 9353, G 83*.

Although Lacaille observed and catalogued this star, he left it unlettered since he considered it 6th magnitude. Gould designated it Kappa since he felt a star of its brightness merited a letter.

Mu¹, HR 8486, 4.79V, *SA, HRP, BS,* CA 1787 as Mu, BR 7146*, L 9069*, G 35*.
Mu², HR 8488, 5.10V, *SA, HRP, BS,* CA 1790 as Mu, BR 7148*, L 9075*, G 36*.

Lacaille designated these two neighboring stars Mu.

Nu, HR 8552, 5.47V, *SA, HRP, BS,* CA 1803, BR 7171, L 9136, G 41*.

Although Lacaille observed and catalogued this star, he left it unlettered; he considered it a 6th-magnitude *informis* between the Crane and the Southern Fish. Gould included it in Grus and designated it Nu since he felt a star of its brightness merited a letter.

Xi, HR 8229, 5.29V, *SA, HRP, BS,* CA 1738, BR 7036, L 8833, G 2*.

Although Lacaille observed and catalogued this star, he included it in Microscopium and left it unlettered since he considered it 6th magnitude. Gould switched it to Grus and designated it Xi since he felt a star of its brightness merited a letter.

Omicron, HR 8907, 5.52V, *SA, HRP, BS,* CA 1889, BR 7287, L 9470, G 107*.

Although Lacaille observed and catalogued this star, he left it unlettered; he considered it 6th magnitude. Gould designated it Omicron since he felt a star of its brightness merited a letter.

Pi¹, HR 8521, 6.62V, *SA, HRP, BS,* BR 7158*, L 9107, G 38*.
Pi², HR 8524, 5.62V, *SA, HRP, BS,* CA 1797 as Pi, BR 7159*, L 9108, BAC 7794 as Pi, G 39*.

Lacaille observed both of these stars, but in his catalogue of 1763, he included only the easterly of the two, CA 1797, and designated it Pi. Brisbane was the first to describe these two nearly adjacent stars as Pi¹ and Pi². Gould agreed, noting that their combined light appeared as one image to his naked eye.

Sigma¹, HR 8600, 6.28V, *SA, BS,* CA 1814 as Sigma, BR 7183*, L 9181, BAC 7869*, G 50.
Sigma², HR 8602, 5.86V, *SA, BS,* CA 1815 as Sigma, BR 7184*, L 9183, BAC 7873*, G 51.

Lacaille described these two nearly adjacent stars as Sigma. Gould dropped their letters because he considered them fainter than 6th magnitude.

Tau¹, HR 8700, 6.04V, *SA, HRP, BS,* CA 1840 as Tau, BR 7219*, L 9289, BAC 7969*, G 70*.
Tau², HD 216655/6, 7.03V, 6.67V, comb. mag. 6.5V, *SA,* CA 1841 as Tau, BR 7220*, BR 7221, L 9295, BAC 7979*, G 71*, G 72.
Tau³, HR 8722, 5.70V, *SA, HRP, BS,* CA 1843 as Tau, L 9305, G 75*.

Lacaille designated these three neighboring stars Tau. Gould retained their letters although he considered all three fainter than 6th magnitude. Boss equated his GC 31953 with HD 216655 and his GC 31952 with HD 216656, but he assigned Tau² only to the latter star. HD 216655/6 is a quadruple star system:

> HD 216655, GC 31953, Components AB, 7.6V, 7.9V, Innes 22.
> HD 216656, GC 31952 as Tau², Components CD, 7.4V, 7.6V, Bos 2506.

The multiplicity of this star system has caused some confusion in stellar identification. Brisbane, for instance, observed that his BR 7220 was a double star, which he equated with CA 1841 as Tau². He did not notice any duplicity in BR 7221 nor did he equate it with Tau². Gould, on the other hand, noted that his G 72 was synonymous with BR 7220. He also noted that G 71 and G 72 appeared as one image to his naked eye, but he equated only his G 71 with Tau².

Hercules, Her

Hercules, Kneeler

Hercules (figures 41 and 42) is a Ptolemaic constellation that Bayer lettered from Alpha to Omega and from A to z. It is the only constellation where Bayer's lettering reached the letter z. Argo Navis has more stars, but Bayer duplicated many letters because of their close proximity.

The constellation Hercules has mystified students of astronomy for thousands of years. Who, they ask, does the kneeling figure in the heavens represent? Why is he kneeling? Some scholars have equated the demigod Hercules with the Mesopotamian epic hero Gilgamesh since both performed heroic deeds and superhuman feats of strength; both were descended from deities; both fought lions; and both wore lion skins. But as far as is known, the Mesopotamians did not honor Gilgamesh with a place in the heavens. In that sector of the sky now occupied by Hercules stood the Sumerian constellation UR.KU, Akkadian Kalbu, the Dog. According to Greek mythology, however, Hercules is ultimately rewarded by his father Zeus with immortality and a place among the stars.

The Greeks originally called this constellation the Kneeler, because it depicted the figure of a man on bended knee with upraised hands. In describing this constellation in his *Phaenomena*, the poet Aratus wrote:

> *The ghostly figure of a toiling hero*
> *rolls through heaven close by, on what labor bent*
> *no one can say, but men call him simply*
> *THE MAN ON HIS KNEES: an image of weariness*
> *sunk on his knees, his arms uplifted*
> *from both of his shoulders, and he holds his hands out*
> *a fathom apart (64-70).*

Although Eratosthenes was supposedly among the first astronomers to identify this constellation with Hercules, Ptolemy continued to refer to it as the Kneeler. Since classical times, scholars have puzzled over the placement of such a seemingly humble figure in the stars, but the solution to this puzzle is quite simple.

Belt-wrestling was among the major sports of the ancient Near East. In the Epic of Gilgamesh (II, 3), when Gilgamesh first encounters his future companion Enkidu, they begin to wrestle:

> *They seized each other (by their girdles);*
> *like experts*
> *they wrestled.*
> *They destroyed the doorpost,*
> *the wall shook,*
> *(yet) Gilgamesh and Enkidu*
> *(still) were holding each other('s girdle).*
> *Like experienced (wrestlers) they wrestled.*
> *(Eventually) Gilgamesh bent his one knee*
> *with his (other) foot (firmly) on the ground*
> *(and lifted up Enkidu)!*

In belt-wrestling, the victorious wrestler grasps his opponent by his belt and lifts him overhead while bending his own knee for added leverage, *knielaufstellung*. The celestial figure of the kneeling Hercules-Gilgamesh depicts a champion wrestler in his moment of glory with his defeated foe hoisted over his head. Similar images have been found carved on several seals and tablets from ancient Mesopotamia.

Although the constellation Hercules-Gilgamesh was unknown in Mesopotamia, it would appear that its origin can be traced back to the ancient Near East, perhaps to the West Semitic people of Syria, Canaan, and Phoenicia of the first millennium B.C., where belt-wrestling was considered both a popular sport and a method of fighting. As a matter of fact, the terms used in belt-wrestling were later applied to warfare. When Gilgamesh and Enkidu wrestled, the word used in the Akkadian (Semitic) edition of the epic to describe that action was *it-te-ig-ga-ru*, which is closely related to the Hebrew word for belt, *hagorah*, חגורה. Likewise, the Old Testament makes several references to "girding one's belt" or "girding one's loins" as a preparation for war (II Kings 3:21; II Kings 4:29).In Hebrew, a specific word for wrestling-belt, *halisah,* חליצה, was adopted as a military term and applied to soldiers who were well equipped (*halusim, halusi;* חלוצים, חלוצי) for battle with all their military accouterments (Numbers 32:30; Joshua 4:13). The Bible also mentions that a victorious soldier is entitled to the equipment (*halisatov,* חלצתו) of his fallen foe (II Samuel 2:21), just as the victorious belt-wrestler receives the girdle of his defeated rival. This tradition has persisted to the present day when a new champion boxer or wrestler is awarded the bejeweled belt worn previously by the former champion.

As previously noted, the Mesopotamians imagined a dog in that part of the sky later occupied by Hercules. Strange as it may seem, the Polish astronomer Hevelius two thousand years later devised the asterism Cerberus, the mythological watchdog of the Underworld, and placed it next to Hercules, who was reputed to have battled the beast and dragged it out of Hades. Was its placement next to Hercules merely coincidental or was Hevelius aware of some ancient astronomical tradition that associated this sector of the heavens with a canine figure?

See Allen, *Star Names,* pp. 238-41; Lum, *The Stars in Our Heaven,* pp. 78-88; Ridpath, *Star Tales,* pp. 72-75; Lombardo, *Sky Signs,* p. 3; van der Waerden, "Babylonian Astronomy. II. The Thirty-Six Stars," pp. 14-15; Gordon,"The Glyptic Art of Nazu," pp. 261-66; Oppenheim, "Mesopotamian Mythology II," pp. 29-30; Gordon, "Belt-Wrestling in the Bible World," pp. 131-36; Gordon, "Western Asiatic Seals," pp. 4-5; Wagman, "Hercules the Champion," pp. 134-36. For Hercules' early career, see Cancer; for Hercules and Cerberus, see Vulpecula.

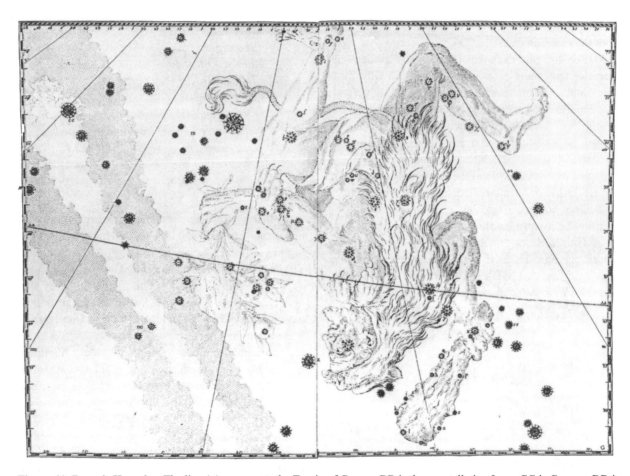

Figure 41. Bayer's Hercules. The line AA represents the Tropic of Cancer; BB is the constellation Lyra; CC is Cygnus; DD is the tail of Aquila; EE, to the left of Hercules' eyes, is the head of Ophiuchus; FF is the head of Serpens; and GG is the feet of Bootes. The 6th-magnitude star z, described by Bayer as *nebulosa*, at the top center of the chart, appears fuzzy and somewhat different from the other stars. This is the same way Bayer depicted the Double Cluster of Perseus on his chart of Cassiopeia. See Figure 20.

Lettered Stars - Hercules

Figure 42. Hevelius's Hercules with Cerberus. Hercules is depicted holding the three serpentine heads of Cerberus, the monstrous hound that guards the inmates of the Underworld. Approaching Hercules on the right is Vulpecula, the Little Fox, grasping Anser, the Goose, in its jaws (see Vulpecula). Cerberus, Vulpecula, and Anser were all devised by Hevelius. All the constellations in Hevelius's atlas are drawn backwards, as they would appear on a celestial globe.

Table 45. The lettered stars of Hercules

	MAGNITUDES			CATALOGUE NUMBERS				
Letter	Bayer	Lacaille	Visual	Flamsteed	Lacaille	Other	HR	HD
Alpha[1,2]+	3		3.08c	64			6406/7	156014/5
Beta	3		2.77	27			6148	148856
Gamma	3		3.75	20			6095	147547
Delta	3		3.14	65			6410	156164
Epsilon	3		3.92	58			6324	153808
Zeta	3		2.81	40			6212	150680
Eta	3		3.53	44			6220	150997
Theta	3		3.86	91			6695	163770
Iota	3		3.80	85			6588	160762
Kappa	4		5.02c	7			6008/9	145001/00
Lambda	4		4.41	76			6526	158899
Mu	4		3.42	86			6623	161797
Nu	4		4.41	94			6707	164136

(table continued on next page)

Table 45. The lettered stars of Hercules *(continued)*

	MAGNITUDES			CATALOGUE NUMBERS				
Letter	Bayer	Lacaille	Visual	Flamsteed	Lacaille	Other	HR	HD
Xi	4		3.70	92			6703	163993
Omicron	4		3.83	103			6779	166014
Pi	4		3.16	67			6418	156283
Rho	4		4.17c	75			6484/5	157778/9
Sigma	4		4.20	35			6168	149630
Tau	4		3.89	22			6092	147394
Upsilon	4		4.76	6			5982	144206
Phi	4		4.26	11			6023	145389
Chi	4		4.62	1			5914	142373
Psi+	4		{5.02 {5.02	{52 Boo {53 Boo			{5763 {5774	{138481 {138629
Omega+	5		4.57	24			6117	148112
A	5		4.97	104			6815	167006
b	5		5.04	99			6775	165908
c+	5		5.39				6377	155103
d	5		5.25	59			6332	154029
e	5		4.65	69			6436	156729
f	5		5.16	90			6677	163217
g	5		5.04	30			6146	148783
h	6		4.84	29			6159	149161
i+	6		5.15	43			6228	151217
k	6		5.49	47			6250	151956
l+	6		5.24	45			6234	151525
$m^{1,2}$+	6		5.40c	36/7			6194/5	150379/8
n+	6		5.63	28			6158	149121
o	6		5.85	21			6111	147869
p+	6							
q+	6		6.08	48 Ser			6035	145647
r	6		5.12	5			5966	143666
s+	6		5.25				6152	148897
t	6		5.12	107			6877	168914
u	6		4.82	68			6431	156633
w	6		5.39	72			6458	157214
x+	6		5.80	77			6509	158414
y+	6		5.37	82			6574	160290
z+	6		6.68	88			6664	162732

The Lost, Missing, or Troublesome Stars of Hercules

Alpha^1, Alpha^2, 64, HR 6406/7, 3.48V, 5.39V, comb. mag. 3.08V, *SA, HRP* as Alpha, *BS,* ADS 10418.

Bayer described Alpha as a single 3rd-magnitude star. Later astronomers added the indices to identify the components of this double star.

Psi, 52/3 Boo in Bootes, HR 5763, HR 5774, 5.02V, 5.02V, *HRP.*

Bayer noted that this star was synonymous with Nu^{1,2} Boo, two nearly adjacent stars. For Bayer's duplicate stars, see Alpha And; see also Nu Boo and 52 and 53 Boo in Part Two.

Omega, 24, HR 6117, 4.57V, *SA, HRP, BS,* HF 25*, BV 24*, H 29*.

Flamsteed's "Catalogus Britannicus" of 1725 includes two Omega's in Hercules: this star, 24, which is Bayer's

Lettered Stars - Hercules

Omega, and 66 (HR 6433, 5.03V, BV 38 in Ophiuchus as 66 Her, e Oph, B 247 as 66 Her, e Oph, BF 2377 as 66 Her, e Oph). Bevis, Bode, and Baily were all aware of this error and corrected it. In Halley's 1712 edition of Flamsteed's catalogue, only 24 is equated with Omega. 66 Her in Ophiuchus is synonymous with e Oph.

A, 104, HR 6815, 4.97V, *SA, HRP.*

b, 99, HR 6775, 5.04V, *SA, HRP.*

c, HR 6377, 5.39V, *SA, HRP,* H 97*.

    Bayer described c as a 5th-magnitude star, and Flamsteed designated his 61 as c (HR 6346, 6.69V, BV 68 as c, BF 2354 as c, BAC 5763 as c). Bevis and Baily agreed although its magnitude is beyond the range of the human eye. Argelander, who could not see 61, felt that c was HR 6377, about 1° to the northeast of 61 (*UN,* p. 33). See Hoffleit, "Discordances," p. 51. For the possibility that Bayer could indeed see 61, see Chi Per.

d, 59, HR 6332, 5.25V, *SA, HRP.*

e, 69, HR 6436, 4.65V, *SA, HRP.*

f, 90, HR 6677, 5.16V, *SA, HRP.*

g, 30, HR 6146, 5.04V, *SA, HRP.*

h, 29, HR 6159, 4.84V, *SA, HRP.*

i, 43, HR 6228, 5.15V, *SA, HRP.*

    Flamsteed mistakenly designated two i's in Hercules. This star, his 43, which is Bayer's i, and his 100 (HR 6781/2, 5.86V, 5.90V, BV 115, BF 2475). Bevis and Baily removed the i from 100. Bayer had included 100 in his atlas but left it unlettered since he considered it an *informis* outside the border of Hercules.

j

    Bayer did not use this letter.

k, 47, HR 6250, 5.49V, *SA, HRP.*

l, 45, HR 6234, 5.24V, *SA, HRP,* HF 43*, BV 71*, B 141*, BF 2313*.

    Flamsteed's 1725 catalogue designates this star, 45, as e, a typographical error that Bevis and Bode corrected. It is properly identified in Halley's 1712 edition of Flamsteed's catalogue.

m^1,m^2, 36/7, HR 6194/5, 6.93V, 5.77V, comb. mag. 5.40V, *SA, HRP* as m, BF 2290/1*, ADS 10149.

    Bayer described m as a single 6th-magnitude star. Flamsteed, using a telescope, resolved this double star and designated its components m-1 and m-2.

n, 28, HR 6158, 5.63V, *SA, HRP,* HF 28*, BV 51*, B 82*, BF 2271*.

    Although Halley's edition of Flamsteed's catalogue equates 28 with n, the letter is missing from Flamsteed's 1725 catalogue. Bevis and Bode corrected the omission. This star is synonymous with 11 Oph (*BF,* Table A, p. 645).

o, 21, HR 6111, 5.85V, *SA, HRP.*

p

    This may be one of Bayer's lost stars. Although Halley's pirated edition of Flamsteed's catalogue equates 13 with p (HF 16), Flamsteed's catalogue of 1725 removes p from 13. Baily at first felt p was synonymous with 13 (HD 146279, 7.4V, BF 2229 as p, BAC 5422), but he later changed his mind, probably because he realized that 13 was too faint for Bayer to have seen with his naked eye. Argelander suggested that either 13 or 15 (HD 146452, 7.4V) would conform to Bayer's chart, but he felt it was quite unlikely that either one was p because of its faintness (*Fide Uran.,*

p. 11). Bevis, though, suggested an interesting possibility. Although his catalogue labels 13 (BV 35) as p and leaves 15 (BV 38) unlettered, his atlas (Tabula VII) shows the two stars nearly adjacent — they are only about 15' apart — implying that the combined light of both is synonymous with Bayer's p. For the possibility that Bayer may have used more than his naked eye in preparing his *Uranometria,* see Chi Per.

q, 48 Ser in Hercules, HR 6035, 6.08V, *SA, HRP,* HF 13*, BV 30*, H 13*.

In Halley's 1712 edition of Flamsteed's catalogue, HF 13 Her is equated with Bayer's q Her. The star is also listed in Serpens as HF 44 Ser and is equivalent to Flamsteed's 48 Ser in his 1725 catalogue. In the latter work, q Her is equated with 8 Her (HR 6013, 6.14V, BF 2211 as q, 8 Her, BAC 6013 as q, 8 Her), an obvious error since q on Bayer's chart is south of Kappa and 8 is north. Baily commented on this but retained q as being synonymous with 8 Her. Bevis and Argelander (*UN,* p. 32) caught the error and corrected it by equating q Her with 48 Ser. What happened, apparently, was that in editing Flamsteed's 1725 catalogue, the editors discovered that HF 13 Her was the same as HF 44 Ser (48 Ser). To avoid duplication, the star HF 13 was removed from Hercules, but in the process the letter q was mistakenly assigned to 8 Her, the star (HF 12) immediately before HF 13 in the 1712 catalogue. See Hoffleit, "Discordances," p. 51.

r, 5, HR 5966, 5.12V, *SA, HRP.*

s, HR 6152, 5.25V, *SA, HRP,* BV 44*, BF 2269*, BAC 5527, H 36*.

Although Flamsteed observed this star, it was among 458 stars not included in his printed catalogue (*BF*, p. 392). Bevis and Baily felt it was synonymous with Bayer's s, so too did Argelander (*UN,* p. 32) and Heis.

t, 107, HR 6877, 5.12V, *SA, HRP.*

u, 68, HR 6431, 4.82V, *SA, HRP.*

v

Bayer did not use this letter.

w, 72, HR 6458, 5.39V, *SA, HRP.*

x, 77, HR 6509, 5.80V, *SA, HRP,* HF 72*, BV 80*, B 285*, BF 2400*, H 128*.
y, 82, HR 6574, 5.37V, *SA, HRP,* HF 73*, BV 93*, BF 2420*, H 142*.
z, 88, HR 6664, 6.68V, *SA, HRP,* HF 76*, BV 101*, BF 2440*, H 167*.

Flamsteed's catalogue of 1725 designates 77 as Kappa, a typographical error for x — both letters are similar in appearance, x, κ. Bevis and Bode caught and corrected the error. It is properly identified in Halley's 1712 edition of Flamsteed's catalogue.

All the standard authorities, such as Baily, Argelander (*UN,* p. 34), Heis, and Pickering agreed with Flamsteed that x, y, and z are 77, 82, and 88. There are, however, some problems with z. Its magnitude is so faint it is virtually invisible to the naked eye. Moreover, Bayer, like Tycho Brahe, described it as a patch of nebulosity and modern charts show no nebulosity in the area. Toomer (*Ptolemy's Almagest,* p. 349) suggested that Flamsteed and the others were possibly in error. He proposed that x, y, and z should be 74 (HR 6464, 5.59V), 77, and 82. This suggestion, though, would not agree with Bayer's chart, but the stars are at least all visible to the naked eye, and 82 is adjacent to a much fainter star, HD 160291, 7.6V. Is it possible that the combined, blurred image of both is the *nebulosa* that both Bayer and Brahe observed? For the possibility that Bayer could have observed objects invisible to the naked eye, see Chi Per.

In his *Uranometria,* Bayer mentioned several other nebulosities besides z Her: Omega Cyg; M 44, Praesepe, the Beehive Cluster in Cancer; Nu Sgr; and the two Magellanic Clouds. Although he also cited the Pleiades and Hyades (Seculae), he did not describe them as *nebulosae.* Oddly enough, he described Chi and h Per as stars, yet his atlas shows them as nebulous objects. See Chi and h Per.

Horologium, Hor

Clock, Pendulum Clock

Lacaille devised and lettered the constellation Horologium (figures 2c and 2d) from Alpha to Lambda. In his preliminary catalogue he described it as a clock with a pendulum and a hand indicating seconds *(l'Horloge à pendule & à secondes)*, but on his chart he labeled it simply l'Horloge. He translated this into Latin as Horologium in his catalogue of 1763. His chart of the southern skies shows a large pendulum clock with two weights.

Thanks to the pioneering work of Galileo and Christiaan Huygens, the invention of an accurate pendulum clock by about 1700 was of enormous importance not only to mankind in general but especially to Lacaille, Flamsteed, and the generations of astronomers who followed them. With an accurate timepiece, an astronomer could determine a star's coordinates by setting his clock to Earth's daily rotation and timing the passage of stars across the lens of his telescope. With the introduction of the clock, the practice of measuring stellar right ascension in degrees, minutes, and seconds of arc gave way to hours, minutes, and seconds of time.

See Boorstin, *The Discoverers,* Bk. I; Chapman, *The Preface to Flamsteed's Historia,* pp. 4-5, 151.

Table 46. The lettered stars of Horologium

	MAGNITUDES			CATALOGUE NUMBERS				
Letter	Bayer	Lacaille	Visual	Flamsteed	Lacaille	Other	HR	HD
Alpha		5	3.86		323		1326	26967
Beta+		5	4.99		229 ?		909	18866
Gamma		6	5.74		211		833	17504
Delta		6	4.93		320		1302	26612
Epsilon+		6	6.77		181		762	16226
Zeta		6	5.21		187		802	16920
Eta		6	5.31		183		778	16555
Theta+		6	6.15		197		821	17254
Iota		6	5.41		192		810	17051
Kappa+		6			223			
Lambda		6	5.35		169		714	15233
Mu+			5.11			G 33	934	19319
Nu+			5.26			G 21	852	17848

The Lost, Missing, or Troublesome Stars of Horologium

Beta, HR 909, 4.99V, *SA, HRP, BS,* CA 229*, BR 462, L 956*, BAC 956, G 32*.

Some confusion surrounds this star. Lacaille designated his CA 229 as Beta. In preparing the *BAC*, Baily included CA 229 and designated it BAC 931 but noted that it was probably non-existent since there was no star in the heavens at the coordinates Lacaille listed for it. He suggested that Lacaille probably meant BAC 956, but Baily did not specifically mark the latter as Beta in the *BAC*. He equated BAC 956 with Brisbane's BR 462. When Baily edited the expanded edition of Lacaille's catalogue, he identified L 956 as Beta. Like Baily, Gould not only noted that Lacaille's Beta coordinates were in error, but he also designated Brisbane's BR 462 as Beta, suggesting it was probably the star Lacaille meant. Brisbane, incidentally, did not include CA 229 in his catalogue, nor did he identify any star as Beta Hor.

Epsilon, HR 762, 6.77V, CA 181*, BR 373*, L 812, BAC 805, G 8.
Theta, HR 821, 6.15V, CA 197*, BR 403*, L 874, BAC 862, G 18.

Baily removed the letters from these two stars because of their dimness.

Kappa, CA 223*.

This is one of Lacaille's lost stars. In his catalogue of 1763, he designated CA 223, a 6th-magnitude star, as Kappa, but in the expanded catalogue, the nearest star, L 948, has a difference of 6' in declination, and consequently, Baily left it unlettered. Moreover, none of the leading astronomers of the southern skies — Brisbane, Gould, Maclear, Rumker, Stone, or Taylor — could positively identify Kappa. If L 948 (BR 439, BAC 919, G 25) is accepted as being synonymous with Kappa, CA 223, its modern equivalent is HD 18292, 6.4V.

Mu, HR 934, 5.11V, *SA, HRP, BS,* CA 238, BR 476, L 989, G 33*.

Although Lacaille observed and catalogued this star in Horologium, he left it unlettered since he considered it 6th magnitude. Gould designated it Mu since he felt a star of its brightness merited a letter.

Nu, HR 852, 5.26V, *SA, HRP, BS,* G 21*.

Lacaille did not letter this star. Gould designated this star Nu because he felt a star of its brightness merited a letter.

Hydra, Hya

Female Water Snake, Large Water Snake

Hydra (Figure 43) is a Ptolemaic constellation that Bayer lettered from Alpha to Omega and from A to b. Both Lacaille and Gould added additional Roman letters.

Flamsteed divided Hydra into three parts. The first or western part contained stars 1 to 44. The second or middle part he called Hydra and Crater, in effect creating a new constellation since he included here all the stars in Crater and those in Hydra south of Crater. He numbered these 1 to 31. The last or eastern part was a continuation of the first part, and he numbered its stars from 45 to 60.

Hydra was first noted in the astronomical literature of Mesopotamia in the latter half of the second millennium B.C. It was called in Sumerian MUS, the Snake. It probably owes its position in the heavens to the god Marduk. As recounted in Canis Major, Marduk agrees to serve as champion of the gods and do battle with Tiamat, the goddess of the sea and the symbol of primordial chaos. Among the monsters Tiamat enlists on her side are the Viper, the Mad-Dog, the Scorpion-Man, the Sphinx, and the Dragon. Marduk not only kills Tiamat but captures her allies, including the Dragon. In honor of his victory, Marduk was pictured on Mesopotamian tablets and boundary stones standing over a four-legged, horned dragon, the Mushhushshu. In fact, this monster became one of the symbols associated with him. It was usually depicted carrying on its scaly back the *marru,* an arrowhead or spear point, another symbol of Marduk.

In Mesopotamian mythology, it was not unusual for a deity to assume or absorb the identity or attributes of a defeated foe. Soon after the creation of the universe, for instance, the Earth god, Enki battles Abzu (Akkadian Apsu), the god of fresh water and husband of Tiamat, the goddess of salt water, the sea. After defeating him, Enki himself becomes the god of fresh water and lives in his sacred city of Eridu (see Eridanus) in a sacred chamber surrounded by subterranean fresh water that he calls the Abzu. So, too, when Marduk kills Abzu's wife, Tiamat, and captures her monstrous allies, he becomes associated with one of them, the Dragon, Mushhushshu.

Perched atop the constellation MUS, the Serpent or Snake, was another Mesopotamian constellation, UGA, the Raven, positioned exactly like the modern Hydra and Corvus. As noted in Corvus, the Raven is the messenger of Ishtar, while the Dragon is the symbol of Marduk. The Babylonians believed that these two gods, the principal deities in their pantheon, were closely associated. They represented the brightest "stars" in the night sky. Marduk was Nebiru, the planet Jupiter. Ishtar, the Sumerian Inanna, the mistress of heaven, was the Morning and Evening Star, Dilbat, the planet Venus. It would appear that the positioning of Hydra and Corvus, the symbols of Marduk and Ishtar, next to each other was intentional, not accidental. Marduk and Ishtar, who on several occasions save mankind from destruction and chaos, may have been the prototypes of the Biblical Mordecai (Marduk) and Esther — whose name is derived from either Ishtar or the Persian *stara* or star — who also saved their people from death and destruction.

See van der Waerden, "Babylonian Astronomy. II. The Thirty-Six Stars," pp. 11, 13-14; Reiner and Pingree, *Baby. Plan. Omens,* Part 2, p. 13; Jacobsen, *Toward the Image of Tammuz,* pp. 4-5, 8, 32-33, 35-36; Pritchard, *Anc. Near East. Texts,* pp. 61, 111, 514-17; Pritchard, *Anc. Near East Picts.,* plates 454, 519-21, 523; Heimpel, "A Catalogue of Near Eastern Venus Deities," pp. 9-15; Reiner and Pingree, *Baby. Plan. Omens,* Part 2, pp. 4, 8, 13; for the Greek astral legends of Corvus, Crater, and Hydra, see Crater.

Lost Stars

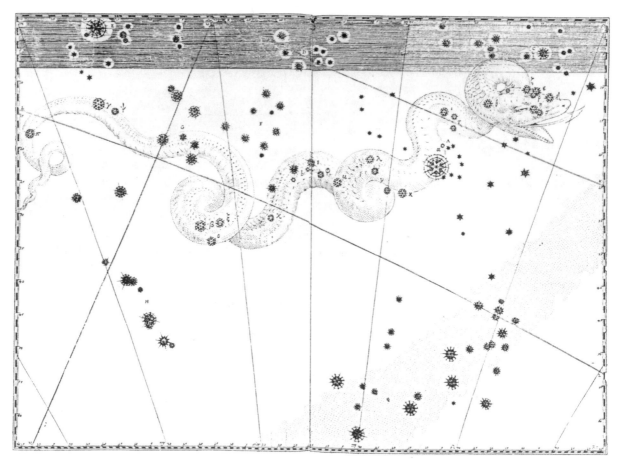

Figure 43. Bayer's Hydra. The letter C is Cancer; D is Leo; E is the star Spica, Alpha Vir; F is Crater; G is Corvus; H is Centaurus; and I is Argo Navis. The letter I, among the stars at the bottom right of the chart, is missing on some copies of the *Uranometria*. The lines K, L, and M are, respectively, the celestial equator, the Tropic of Capricorn, and the equinoctial colure. At the top left sector of the chart, the equator and the equinoctial colure meet on the ecliptic at the First Point of Libra, ♎, the time of the autumnal equinox. The shaded area across the top is the zodiac.

Table 47. The lettered stars of Hydra

	MAGNITUDES			CATALOGUE NUMBERS				
Letter	**Bayer**	**Lacaille**	**Visual**	**Flamsteed**	**Lacaille**	**Other**	**HR**	**HD**
Alpha	2		1.98	30			3748	81797
Beta	3		4.28	28 Crt			4552	103192
Gamma	3		3.00	46			5020	115659
Delta	4		4.16	4			3410	73262
Epsilon	4		3.38	11			3482	74874
Zeta	4		3.11	16			3547	76294
Eta	4		4.30	7			3454	74280
Theta	4		3.88	22			3665	79469
Iota	4		3.91	35			3845	83618
Kappa	4		5.06	38			3849	83754
Lambda	4		3.61	41			3994	88284
Mu	4		3.81	42			4094	90432
Nu	4		3.11	4 Crt			4232	93813
Xi	4		3.54	19 Crt			4450	100407

(table continued on next page)

Table 47. The lettered stars of Hydra (continued)

	MAGNITUDES			CATALOGUE NUMBERS				
Letter	Bayer	Lacaille	Visual	Flamsteed	Lacaille	Other	HR	HD
Omicron	4		4.70	25 Crt			4494	101431
Pi	4		3.27	49			5287	123123
Rho	5		4.36	13			3492	75137
Sigma	5		4.44	5			3418	73471
Tau^1+	5		4.60	31			3759	81997
Tau^2	5		4.57	32			3787	82446
Upsilon^1+	5		4.12	39			3903	85444
Upsilon^2	5		4.60	40			3970	87504
Phi^1+			7.6	43				91369
Phi^2			6.03	1 Crt			4156	91880
Phi^3	5		4.91	2 Crt			4171	92214
Chi^1+	5		4.94	9 Crt			4314	96202
Chi^2			5.71				4317	96314
Psi	5		4.95	45			4958	114149
Omega	6		4.97	18			3613	77996
A+	6		5.56	33			3814	82870
b^1+	6		5.42	3 Crt			4214	93397
b^2			6.6	5 Crt				94046
b^3	6		5.24	6 Crt			4251	94388
a+		5	5.00c	17 Crt	1016		4443/4	100286/7
b+		6	5.44	10 Crt	994		4334	96819
c		5	6.17		1048		4553	103266
d		5	5.45		1094		4803	109799
e		6	5.48		1102		4839	110666
f		6	5.78		1153		5120	118349
g		6	5.81		1163		5167	119752
h		6	5.48		1187		5265	122430
i+		6	5.08	50	1199		5312	124206
k+		6	4.77	51	1208		5381	125932
l+		6	4.97	52	1213		5407	126769
m^1+		5	4.94	54	1241		5497	129926
m^2		5	5.63	55	1242		5514	130158
m^3		5	5.24	56	1243		5516	130259
m^4		6	5.77	57	1244		5517	130274
C+			3.90	30 Mon		G 19	3314	71155
D			4.32	12		G 60	3484	74918
E+			4.41	58		G 387	5526	130694
F+			4.62	31 Mon		G 51	3459	74395
G			4.69			G 141	3749	81799
H+			4.68	2 Sex		G 168	3834	83425
I			4.77			G 174	3858	83953
N+			5.00c	17 Crt		G 284/5	4443/4	100286/7
P			4.80	27		G 124	3709	80586

The Lost, Missing, or Troublesome Stars of Hydra

Tau^1, 31, HR 3759, 4.60V, *SA, HRP, BS.*
Tau^2, 32, HR 3787, 4.57V, *SA, HRP, BS.*

Bayer described Tau as a couple of 5th-magnitude stars, and Flamsteed designated his 31 and 32 as Tau-1 and Tau-2.

Lost Stars

Upsilon¹, 39, HR 3903, 4.12V, *SA, HRP, BS.*
Upsilon², 40, HR 3970, 4.60V, *SA, HRP, BS.*

Bayer described Upsilon as a couple of 5th-magnitude stars, and Flamsteed designated his 39 and 40 as Upsilon-1 and Upsilon-2.

Phi¹, 43, HD 91369, 7.6V, HF 42, BF 1502*, BAC 3611*, LL 20438*.
Phi², 1 Crt in Hydra, HR 4156, 6.03V, *SA, BS,* HF 44, BF 1513*, BAC 3632*, LL 20525*, T 4742*, G 234.
Phi³, 2 Crt in Hydra, HR 4171, 4.91V, *SA* erroneously as Phi¹, *HRP* as Phi, *BS,* HF 45 as Phi, BF 1517*, BAC 3646*, LL 20526*, H 106 as Phi, G 242 as Phi.

Bayer noted that Phi was a single 5th-magnitude star. Although Halley's 1712 edition of Flamsteed's catalogue designates only 2 Crt as Phy Hya, Flamsteed's 1725 catalogue designates these three neighboring stars, his 43 Hya, 1 Hya and Crt, and 2 Hya and Crt, as Phi-1, Phi-2, and Phi-3. Baily felt that only the brightest, 2 Crt, was synonymous with Phi, but he kept Flamsteed's designations. Argelander designated only 2 Crt as Phi Hya since he could not see the others with his naked eye (*UN,* p. 99). Gould, agreeing with Argelander, also lettered 2 Crt as Phi Hya. See Hoffleit, "Discordances," p. 52.

Chi¹, 9 Crt in Hydra, HR 4314, 4.94V, *SA, HRP, BS,* CA 989 as Chi, RAS 1311*, BR 3376*, L 4583, BAC 3793*, G 264*.
Chi², HR 4317, 5.71V, *SA, HRP, BS,* CA 990 as Chi, RAS 1312*, BR 3382*, L 4587, BAC 3794*, G 265*.

Bayer described Chi as a single 5th-magnitude star. It was Lacaille who first designated both stars of this close pair as Chi.

A, 33, HR 3814, 5.56V, *SA, HRP, BS,* BV 36, H 71*, G 163.

Bayer noted that A was a single 6th-magnitude star. Flamsteed thought his 28 (HR 3738, 5.59V, BV 26 as A, BF 1328 as A, G 134) was synonymous with A, and Bevis and Baily agreed. Argelander, though, changed A to 33 (*UN,* p. 99). Gould left both stars unlettered, probably because he considered them 6th magnitude. Hoffleit felt that, based on Bayer's chart, A could be 28, 33, or a brighter star, HR 3750, 5.38V ("Discordances," p. 52).

b¹, 3 Crt in Hydra, HR 4214, 5.42V, *SA, HRP,* HF 46 as b, BV 51*, BF 1537, BAC 3697*, H 109*, G 249*.
b², 5 Crt in Hydra, HD 94046, 6.6V, *SA, HRP,* HF 48, BV 49, BF 1547, BAC 3722*, H 113*, G 255*.
b³, 6 Crt in Hydra, HR 4251, 5.24V, *SA, HRP,* HF 49 as b, BV 54 as b-2, BF 1553, BAC 3733*, H 115*, G 257*.

Bayer described b as a couple of 6th-magnitude stars about 3° apart. Although Halley's edition of Flamsteed's catalogue designates two stars, 3 Crt and 6 Crt, as Bayer's pair of b's, Flamsteed's 1725 catalogue designates three stars, 3 Crt, 5 Crt, and 6 Crt, as b-1, b-2, and b-3. Baily at first considered these three stars to be in Crater, so he removed their letters. He later changed his mind, placed them in Hydra, and restored their letters. Agreeing with Halley's edition of Flamsteed's catalogue, Bevis labeled only the two brighter stars, 3 and 6, as b-1 and b-2. Argelander did the same since he was unable to see 5 with his naked eye (*UN,* p. 99). Heis, on the other hand, observed all three and lettered them as they appeared in Flamsteed's authorized catalogue of 1725.

Lacaille added lower-case Roman letters from a to m in Hydra, but Baily removed them. In fact, as he stated in the preface to the *BAC* (pp. 62-63), Baily intended to remove Lacaille's Roman letters from all the constellations save the three newly devised components of Argo Navis. He did, though, make some exceptions, as in the case of Eridanus. Gould restored Lacaille's letters to several of the brighter stars in Hydra.

a, 17 Crt in Hydra, HR 4443/4, 5.76V, 5.64V, comb. mag. 5.00V, CA 1016*, BR 3628*, BR 3629, L 4770, ADS 8202.

Gould relettered this star N (see N). *HRP* and *The Henry Draper Catalogue* erroneously list 6 Hya (HR 3431, HD 73840) as a Hya.

b, 10 Crt in Hydra, HR 4334, 5.44V, CA 994*, L 4615, BAC 3815, G 271.

This star should not be confused with Bayer's b, above.

c, HR 4553, 6.17V, CA 1048*, BR 3813*, L 4926, BAC 4016, G 302.

d, HR 4803, 5.45V, CA 1094*, BR 4123*, L 5225, BAC 4253, G 328.

e, HR 4839, 5.48V, CA 1102*, BR 4172*, L 5263, BAC 4278, G 330.

f, HR 5120, 5.78V, CA 1153*, BR 4556*, L 5608, BAC 4541, G 351.

g, HR 5167, 5.81V, CA 1163*, BR 4620*, L 5670, BAC 4581, G 357.

h, HR 5265, 5.48V, CA 1187*, BR 4738*, L 5788, BAC 4671, G 367.

i, 50, HR 5312, 5.08V, CA 1199*, BR 4809 erroneously as Iota, L 5856, BAC 4708, G 370.

Gould considered the magnitude of this star to be 5.5 and hence left it unlettered.

j

Lacaille did not use this letter.

k, 51, HR 5381, 4.77V, *SA, HRP,* CA 1208*, BR 4887*, L 5917, G 375*.
l, 52, HR 5407, 4.97V, *SA, HRP,* CA 1213*, BR 4925*, L 5949, G 379*.

Although Baily dropped the letters from these two stars, Gould restored them because of their brightness.

m^1, 54, HR 5497, 4.94V, *SA* as m, *HRP* as m, CA 1241 as m, L 6087, BAC 4865, G 382/3 as m, ADS 9375.
m^2, 55, HR 5514, 5.63V, CA 1242 as m, L 6097, BAC 4877, G 384.
m^3, 56, HR 5516, 5.24V, CA 1243 as m, BR 5060*, L 6102, BAC 4880, G 385.
m^4, 57, HR 5517, 5.77V, CA 1244 as m, BR 5061*, L 6104, BAC 4882, G 386.

Lacaille designated these four neighboring stars m. Baily dropped the letters from all four, but Gould restored m to the brightest of the group, 54, since he felt its brightness merited a letter. Brisbane's catalogue does not include either CA 1241 or CA 1242.

HR 5497 is component A of the double-star system Herschel III 97. Component B is HD 129926B, 7.20V. William Herschel first noted its duplicity in the 1780's. Gould designated both components as m.

Gould assigned the letters C through I and N and P to the following stars that he considered brighter than magnitude 5.2. He did not assign letters A and B since Bayer had already preempted them. Nor did he assign K, L, or M since they were Lacaille letters, albeit lower case, which he retained.

C, 30 Mon in Hydra, HR 3314, 3.90V, *SA, HRP,* BF 1180, BAC 2825, G 19*.

Baily included this star in Hydra.

D, 12, HR 3484, 4.32V, *SA, HRP,* G 60*.

E, 58, HR 5526, 4.41V, *SA, HRP,* CA(P) 1244, CA 1247, BR 5080, L 6116, BF 2019, BAC 4891, G 387*.

Although Lacaille observed and catalogued this star, he left it unlettered because of a copying error. He noted in his observations that it was a 5th-magnitude star — and he invariably lettered 5th-magnitude stars — but when he prepared his catalogues of 1756 and 1763, he mistakenly assigned it a magnitude of 6. This star is the same as 6 Lib.

F, 31 Mon in Hydra, HR 3459, 4.62V, *SA, HRP,* BF 1233, BAC 2954, G 51*.

Baily included this star in Hydra.

G, HR 3749, 4.69V, *SA, HRP,* G 141*.

H, 2 Sex in Hydra, HR 3834, 4.68V, BAC 3295, G 168*.

Gould included this star in Hydra.

I, HR 3858, 4.77V, *SA, HRP,* G 174*.

N, 17 Crt in Hydra, HR 4443/4, 5.76V, 5.64V, comb. mag. 5.00V, *SA, HRP, BS,* CA 1016 as a, BR 3628 as a, BR 3629, L 4770, BAC 3921/2, G 284/5*, ADS 8202.

Lacaille described this star as a 5th-magnitude *duplex* and designated it a Hya, but Gould relettered it N to avoid confusion with Bayer's A. See Baily's comments in the *BAC* on the confusion over which component is which of this double-star system.

P, 27, HR 3709, 4.80V, *SA, HRP,* G 124*.

Hydrus, Hyi

Male Water Snake, Small Water Snake

Keyzer and Houtman devised the constellation Hydrus (Figure 44). In his catalogue of 1603, Houtman called it De Waterslang (The Water Snake). Plancius's globe of 1598 refers to it as Waterslang Hydrus. Bayer depicted it on his chart of the southern skies simply as Hydrus. The difference in spelling between Hydrus and Hydra reflects their different sexes. Hydra, the Female Water Snake, is a first declension Latin noun whose nominative or subjective form ends in "a," and with some rare exceptions, most words in the first declension are of the feminine gender. On the other hand, Hydrus is a second declension noun whose nominative ending is "us," and words in this declension are masculine.

Lacaille lettered its stars from Alpha to Tau.
See Warner, *Sky Explored,* pp. 203-4.

Table 48. The lettered stars of Hydrus

	MAGNITUDES			CATALOGUE NUMBERS				
Letter	Bayer	Lacaille	Visual	Flamsteed	Lacaille	Other	HR	HD
Alpha		3	2.86		136		591	12311
Beta		3	2.80		15		98	2151
Gamma		3	3.24		309		1208	24512
Delta		4	4.09		166		705	15008
Epsilon		5	4.11		196		806	16978
Zeta		5	4.84		214		837	17566
Eta^1+		6	6.7		126			11733
Eta^2		5	4.69		132		570	11977
Theta		5	5.53		240		939	19400
Iota		6	5.52		267		1025	21024
Kappa		6	5.01		174		715	15248
Lambda		6	5.07		55		236	4815
Mu		6	5.28		202		776	16522
Nu		6	4.75		234		872	18293
Pi^1+		6	5.55		156		667	14141
Pi^2		6	5.69		158		678	14287
Rho+		6			283			
Sigma		6	6.16		143		593	12363
Tau^1+		6	6.33		121		516	10859
Tau^2		6	6.06		135		550	11604

Lettered Stars - Hydrus

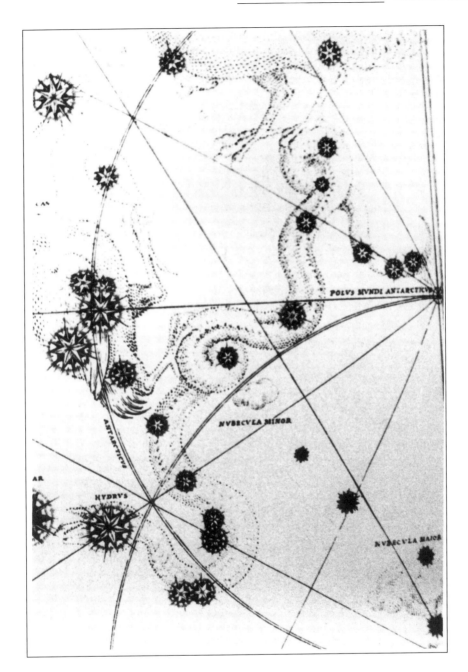

Figure 44. Bayer's Hydrus. This is an enlarged view of Hydrus from Bayer's Tabula XLIX. To the right of the Male Water Snake is the Small Magellanic Cloud, Nubecula Minor. At the bottom right corner of the chart is the Large Magellanic Cloud, Nubecula Major.

The Lost, Missing, or Troublesome Stars of Hydrus

Eta¹, HD 11733, 6.7V, *SA,* CA 126 as Eta, BR 276*, L 577*, G 19*.
Eta², HR 570, 4.69V, *SA, HRP, BS,* CA 132 as Eta, BR 283*, L 594*, G 21*.

Lacaille designated these two neighboring stars Eta.

Xi

Omicron

Lacaille did not use these two letters in Hydrus.

Pi¹, HR 667, 5.55V, *SA, HRP, BS,* CA 156 as Pi, BR 330*, L 701*, G 35*.
Pi², HR 678, 5.69V, *SA, HRP, BS,* CA 158 as Pi, BR 332*, L 706*, G 36*.

Lacaille designated these two nearly adjacent stars Pi.

Rho, CA 283*, L 1210, BAC 1116.

Lacaille observed this star on September 24, 1751, but apparently erred in copying its coordinates or in determining its time of transit across his lens. Neither Brisbane, Gould, Maclear, Rumker, Stone, nor Taylor could identify it. Baily placed it in Mensa but admitted in the *BAC* that he could neither locate it nor equate it with any star in any catalogue other than Lacaille's.

Tau¹, HR 516, 6.33V, *SA, HRP, BS,* CA 121 as Tau, BR 259*, L 551, G 17*.
Tau², HR 550, 6.06V, *SA, HRP, BS,* CA 135 as Tau, BR 279*, L 606, G 18*.

Lacaille designated these two neighboring stars Tau.

INDUS, IND

Indian

Keyzer and Houtman devised the constellation Indus (figures 45 amd 46). The latter's catalogue of 1603 refers to it as De Indiaen. Bayer depicted it on his chart of the southern skies as Indus, and Lacaille lettered its stars from Alpha to Rho. Although Keyzer and Houtman traveled to the Far East, the Indian they commemorated in their catalogue was an Amerindian, not an inhabitant of the subcontinent of India.

See Ridpath, *Star Tales,* p. 79; Allen, *Star Names,* p. 250.

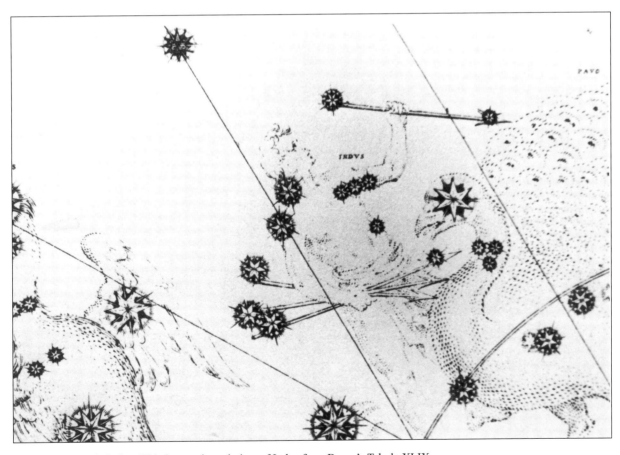

Figure 45. Bayer's Indus. This is an enlarged view of Indus from Bayer's Tabula XLIX.

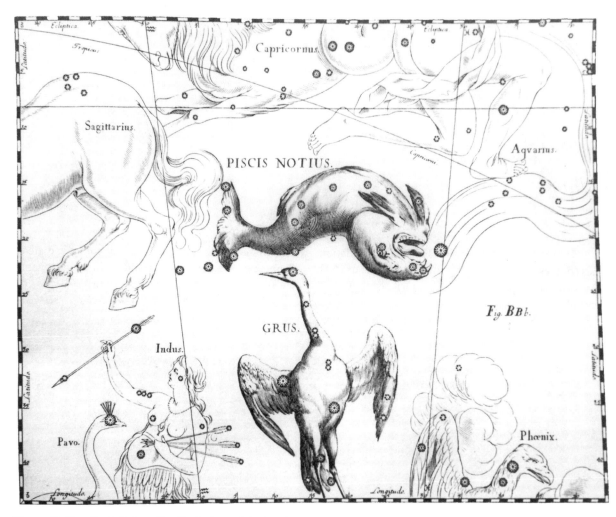

Figure 46. Hevelius's Indus. Hevelius's chart of the Southern Fish (Piscis Notius) and the Crane (Grus) depicts Indus as a full-breasted woman with long flowing hair but retains the masculine name for the constellation. All the constellations in Hevelius's atlas are drawn backwards, as they would appear on a celestial globe.

Table 49. The lettered stars of Indus

		MAGNITUDES			CATALOGUE NUMBERS			
Letter	Bayer	Lacaille	Visual	Flamsteed	Lacaille	Other	HR	HD
Alpha		3	3.11		1676		7869	196171
Beta		4	3.65		1691		7986	198700
Gamma		5	6.12		1731		8188	203760
Delta		5	4.40		1764		8368	208450
Epsilon		6	4.69		1769		8387	209100
Zeta		6	4.89		1684		7952	198048
Eta		6	4.51		1679		7920	197157
Theta		6	4.39		1721		8140	202730
Iota		6	5.05		1685		7968	198308
Kappa^1+		6	6.12		1765		8369	208496
Kappa^2		6	5.62		1775		8409	209529
Lambda^1+		6	5.42		1612		7531	186957
Lambda^2		6	5.13		1628		7625	189124
Mu		6	5.16		1702		8055	200365
Nu+		6	5.29		1793		8515	211998
Omicron		6	5.53		1748		8333	207241
Pi		6	6.19		1761		8362	208149
Rho		6	6.05		1838		8701	216437

The Lost, Missing, or Troublesome Stars of Indus

Kappa^1, HR 8369, 6.12V, *BS*, CA 1765 as Kappa, BR 7101*, L 8959*, G 61.
Kappa^2, HR 8409, 5.62V, *SA* as Kappa, *BS*, CA 1775 as Kappa, BR 7117*, L 9001*, G 65.

Lacaille designated these two neighboring stars Kappa, but Gould removed their letters because he considered them less than 6th-magnitude.

Lambda^1, in Pavo, HR 7531, 5.42V, CA 1612 as Lambda, BR 6733*, L 8207, BAC 6766, G 74 in Pavo.
Lambda^2, in Pavo, HR 7625, 5.13V, CA 1628 as Lambda, BR 6775 erroneously as Lambda, L 8269, BAC 6837, G 85 in Pavo.

Lacaille designated these two neighboring stars Lambda. Baily included them unlettered in Pavo.

Nu, HR 8515, 5.29V, *SA, HRP, BS*, CA 1793*, BR 7153*, L 9082*, G 70*, W 11*.

Lacaille inadvertently labeled two stars in his catalogue as Nu Ind, this star and CA 1673. Neither Baily, Brisbane, nor Maclear noticed this error. On Lacaille's chart, this star is located in the Indian's right foot, at the southern extremity of the constellation, and labeled Nu. CA 1673 is at the opposite end, just north of the arrow in the Indian's right hand. In fact, Lacaille's chart shows it as an unlabeled star within Sagittarius, about equidistant from Indus and Microscopium. Gould shifted CA 1673 to Microscopium and designated it Nu Mic. He retained CA 1793 as Nu Ind. See Nu Mic.

Xi

Lacaille did not use this letter in Indus.

Lacerta, Lac

Lizard

Hevelius devised the constellation Lacerta (Figure 47) and named it Lacerta sive Stellio, the Lizard or Newt. He noted that since its proposed home in the heavens would be a rather narrow, confined space between Andromeda and Cygnus, it would be inappropriate to place there a large or majestic beast. The slender Lizard or Newt, however, would fit in comfortably. Hevelius also pointed out that since this sector of the sky is sprinkled with many small stars, it would match the speckled spots that dot the hide of Lacerta or Stellio.

Flamsteed included Lacerta in his atlas and catalogue, and Baily lettered its two brightest stars Alpha and Beta. Both of these stars were included in the *BS*, and there are no missing or problem lettered stars in the constellation.

See Hevelius, *Prodromus Astronomiae,* p. 114; Allen, *Star Names,* p. 251.

Table 50. The lettered stars of Lacerta

	MAGNITUDES			CATALOGUE NUMBERS				
Letter	Bayer	Lacaille	Visual	Flamsteed	Lacaille	Other	HR	HD
Alpha			3.77	7		BAC 7855	8585	213558
Beta			4.43	3		BAC 7815	8538	212496

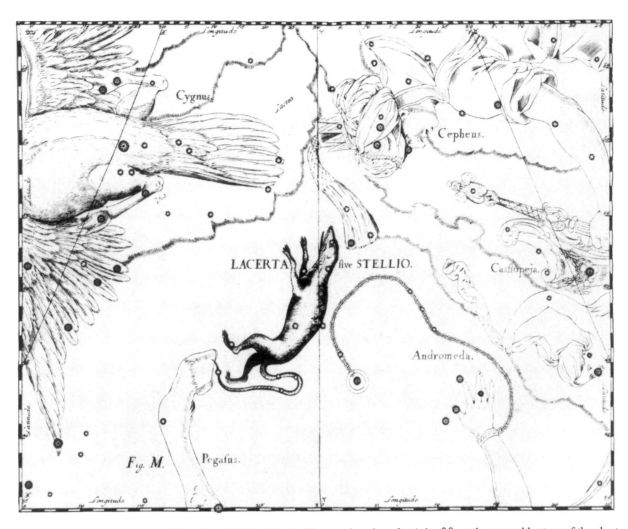

Figure 47. Hevelius's Lacerta (Lizard) or Stellio (Newt). The two tiny signs for Aries, ♈, at the top and bottom of the chart identify the vertical line down the center as 0° ecliptic longitude as well as 0° in the sign of Aries. (Each of the twelve constellations in the zodiac contains 30° of ecliptic longitude.) The lines of ecliptic longitude extend from the north ecliptic pole in Draco to the ecliptic and from there south to the south ecliptic pole in Dorado. The 0° line runs from the north ecliptic pole south to Cepheus, Lacerta, Pegasus, and Pisces, where it meets the junction of the ecliptic, the celestial equator, and the equinoctial colure at the First Point of Aries. When the sun reaches this point in the sky in March, it is the time of the vernal equinox.

Leo, Leo

Lion

Together with Taurus, Scorpius, and Ibex-Aquarius, Leo (Figure 48) was among the oldest constellations in Mesopotamian astronomy, dating back at least six thousand years to the fourth millennium B.C. when it was aligned with the summer solstice. The Sumerians called it UR.GU.LA, the Lion, a most appropriate name since it is one of the few constellations that actually resembles its namesake. The western or preceding half of the constellation, usually identified as the asterism the Sickle, accurately depicts a lion's majestic head and mane.

When the sun appeared in that part of the sky occupied by Leo, it was at its highest or most northerly point in its annual peregrination around the ecliptic. Because the sun was the brightest object in the sky, the Sumerians identified it with the most majestic of animals, the lion, the king of beasts. Utu (Akkadian Shamash), the sun god, was depicted symbolically on various wall sculptures, tablets, and boundary stones as a lion with star-shaped rosettes embedded in his shoulder, a winged lion, a winged solar disk, or simply a disk. The *lucida* of the constellation, Alpha (HR 3982, 1.35V), was appropriately enough dubbed LUGAL, the King, probably the source of its Latin name Regulus, Little King. A Ptolemaic constellation, Leo was included in Bayer's *Uranometria,* where its stars were lettered from Alpha to Omega and from A to p.

According to Greek and Roman mythology, Zeus sets the image of Leo in the heavens because he feels that the lion, as the king of beasts, merits a place among the stars. In another legend, Leo represents the Nemean lion, a great lion that is terrorizing the inhabitants of the town of Nemea, near Corinth, Greece. This is no ordinary beast, but one that cannot be killed with any weapons devised by humans. As the first of his twelve labors, Hercules is given the task of slaying this beast and he succeeds by killing him with his bare hands.

See Hartner, "The Earliest History of the Constellations in the Near East," pp. 1-16 and plates following the article; van der Waerden, "Babylonian Astronomy. II. The Thirty-Six Stars," pp. 11, 13-14; van der Waerden, "History of the Zodiac," pp. 219-20; O'Neil, *Time and the Calendars,* p. 57; Heuter, "Star Names," p. 243; Contenau, *Everyday Life in Babylon and Assyria,* p. 257; Pritchard, *Anc. Near East Picts.,* Plates 228, 453, 518-20, 529, 534, 855; Reiner and Pingree, *Baby. Plan. Omens,* Part 2, pp. 13, 16; Kunitzsch and Smart, *Short Guide to Modern Star Names,* pp. 40-41; Condos, *Star Myths,* pp. 125-27; Ridpath, *Star Tales,* pp. 80-81. For an account of the solstice and the four turning points of the year marking the seasons, see Aquarius; for a brief summary of the twelve labors of Hercules, see Vulpecula.

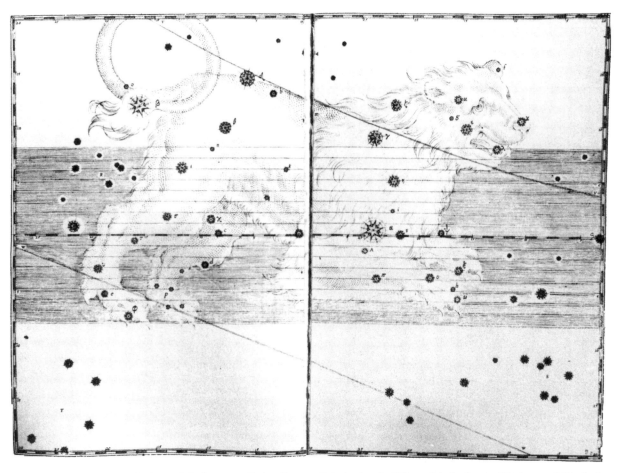

Figure 48. Bayer's Leo. The letter Q is the constellation Cancer; R is the head of Virgo; S is the head of Hydra; and T is Crater. The lines V and W are the Tropic of Cancer and the celestial equator. The shaded area is the zodiac.

Table 51. The lettered stars of Leo

	MAGNITUDES			CATALOGUE NUMBERS				
Letter	Bayer	Lacaille	Visual	Flamsteed	Lacaille	Other	HR	HD
Alpha	1		1.35	32			3982	87901
Beta+	1		2.14	94			4534	102647
Gamma[1,2]+	2		1.90c	41			4057/8	89484/5
Delta	2		2.56	68			4357	97603
Epsilon	3		2.98	17			3873	84441
Zeta	3		3.44	36			4031	89025
Eta	3		3.52	30			3975	87737
Theta	3		3.34	70			4359	97633
Iota	3		3.94	78			4399	99028
Kappa	4		4.46	1			3731	81146
Lambda	4		4.31	4			3773	82308
Mu	4		3.88	24			3905	85503
Nu	4		5.26	27			3937	86360
Xi	4		4.97	5			3782	82395
Omicron	4		3.52	14			3852	83808
Pi	4		4.70	29			3950	86663
Rho	4		3.85	47			4133	91316
Sigma	4		4.05	77			4386	98664
Tau	4		4.95	84			4418	99648
Upsilon	4		4.30	91			4471	100920
Phi	4		4.47	74			4368	98058
Chi	4		4.63	63			4310	96097
Psi	5		5.35	16			3866	84194
Omega	5		5.41	2			3754	81858
A	5		4.37	31			3980	87837
b	5		4.42	60			4300	95608
c	5		4.99	59			4294	95382
d	5		4.84	58			4291	95345
e	5		4.77	87			4432	99998
f	6		5.64	15			3861	84107
g	6		5.32	22			3900	85376
h	6		5.07	6			3779	82381
i+	6		5.46	46			4127	91232
k+	6		5.48	52			4209	93291
l	6		5.25	53			4227	93702
m	6		5.49	51			4208	93257
n	6		5.32	73			4365	97907
o	6		5.53	95			4564	103578
p[1]+	6		5.45				4253	94402
p[2]	6		4.74	61			4299	95578
p[3]	6		5.95	62			4306	95849
p[4]	6		5.52	65			4319	96436
p[5]	6		5.42	69			4356	97585

The Lost, Missing, or Troublesome Stars of Leo

Beta, 94, HR 4534, 2.14V, *SA, HRP, BS.*

Beta is unmistakenly a 2nd-magnitude star, but Ptolemy's catalogue describes it as somewhat less than 1st magnitude (Toomer, *Ptolemy's Almagest*, p. 368; Peters and Knobel, *Ptolemy's Catalogue*, p. 132). Tycho's catalogue of l000 stars lists it simply as 1st magnitude (Brahe, *Opera Omnia*, III, 348), and so too does Bayer's *Uranometria*.

Gamma^1, Gamma^2, 41, HR 4057/8, 2.61V, 3.80V, comb. mag. 1.90V, *SA, HRP* as Gamma, *BS*, ADS 7724.

Bayer described Gamma as a single 2nd-magnitude star. Later astronomers added the indices to identify the components of this double star.

A, 31, HR 3980, 4.37V, *SA, HRP.*

b, 60, HR 4300, 4.42V, *SA, HRP.*

c, 59, HR 4294, 4.99V, *SA, HRP.*

d, 58, HR 4291, 4.84V, *SA, HRP.*

e, 87, HR 4432, 4.77V, *SA, HRP.*

f, 15, HR 3861, 5.64V, *SA, HRP.*

g, 22, HR 3900, 5.32V, *SA, HRP.*

h, 6, HR 3799, 5.07V, *SA, HRP.*

i, 46, HR 4127, 5.46V, BF 1498*, BAC 3606*, B 176*, P X,97*, H 74, W 21*.

Bayer described i as a 6th-magnitude star, and Flamsteed designated his 46 as i. Baily would have preferred 34 (HR 3998, 6.44V, H 54) since he felt it was more in accord with Bayer's chart, but he went along with Flamsteed's designation. Argelander could not see 34 with his naked eye, and in his opinion 46 did not conform to Bayer's chart. Consequently, he omitted i from his catalogue (*Fide Uran.*, pp. 12-3; *UN*, p. 65). Heis, although he saw both stars, agreed with Argelander's decision not to designate any star as i Leo.

The confusion with 46 and i dates back to Tycho. In his catalogue, Tycho equated his 16 with Ptolemy's 16, but as Dreyer, the editor of Tycho's works, has pointed out, he erred in ascertaining its position. Bayer equated his i with Ptolemy's and Tycho's 16, but since he used Tycho's catalogue, he used the wrong coordinates, which were about 6° too far west and about 2½° too far south. Flamsteed equated his 46 with Ptolemy's and Tycho's 16, but he positioned his star correctly and equated it with Bayer's i. In other words, when Flamsteed realized that Bayer's i as it appeared on his chart was nonexistent, Flamsteed positioned it where he felt Bayer intended it to be. Bevis, apparently unaware of Tycho's error, left i where Bayer had originally placed it and in his catalogue equated it with Tycho's 16 (BV 34). He correctly equated Flamsteed's 46 (BV 51 and Tabula XXVI) with Ptolemy's 16. In sum, Bayer's i is synonymous with Flamsteed's 46 and Ptolemy's and Tycho's 16. It has nothing at all to do with Flamsteed's 34, except that the 34 just happens to be in the approximate position of Tycho's erroneous coordinates for his 16, which Bayer had copied. Although most astronomers omit i from Leo, Ptolemy's latest editor and translator, G. J. Toomer, has equated Ptolemy's 16 with Flamsteed's 46 and Bayer's i. See Toomer, *Ptolemy's Almagest*, p. 367; Brahe, *Opera Omnia*, III, 347, 412.

j

Bayer did not use this letter.

k, 52, HR 4209, 5.48V, *SA, HRP,* HF 51*, BV 61*, B 211*, BF 1532*.

As the result of a copying or typographical error, this letter appears as K in Flamsteed's 1725 "Catalogus Britannicus," but Bevis and Bode corrected it. The star is properly identified in Halley's 1712 edition of Flamsteed's catalogue.

l, 53, HR 4227, 5.25V, *SA, HRP.*

m, 51, HR 4208, 5.49V, *SA, HRP.*

n, 73, HR 4365, 5.32V, *SA, HRP.*

o, 95, HR 4564, 5.53V, *SA, HRP.*

p^1, HR 4253, 5.45V, *SA, HRP,* H 91*, G 21*.
p^2, 61, HR 4299, 4.74V, *SA, HRP,* H 100*, G 28*.
p^3, 62, HR 4306, 5.95V, *SA, HRP,* H 103*, G 31*.
p^4, 65, HR 4319, 5.52V, *SA, HRP,* H 106*, G 34*.
p^5, 69, HR 4356, 5.42V, *SA, HRP,* H 112*, G 37*.

Bayer noted that p was a group of five 6th-magnitude stars. Working without Bayer's star list, Flamsteed was not certain which or how many stars Bayer meant as p. Consequently, he designated only his 62 as g — a typographical error for p — because 62 was closest to the letter p on Bayer's chart. Baily corrected Flamsteed's error by removing the g and by assigning p to the five stars that seemed to him to correspond best to Bayer's chart:

p^1, 61, BF 1568.
p^2, 62, BF 1572.
p^3, 65, BF 1578.
p^4, 66, HD 96855, 6.8V, BF 1579, H 107.
p^5, 69, BF 1586.

Argelander was somewhat dissatisfied with Baily's designations, especially with equating 66 with p^4 since he himself could not see 66 (*UN,* p. 65-6). He decided, therefore, to drop p from 66, to add another star — one (HR 4253) not in Flamsteed's catalogue — as p^1, and to rearrange the indices as in the heading above. The sharp-eyed Heis saw 66, but he went along with Argelander's designations, and so too did Gould. It should be noted that although Heis's observations were made without telescopic assistance, he used a special device to block out extraneous light. This has led Allen to comment that his "observations, although by the naked eye, were not unaided" (*Star Names,* p. xiii, note). Gould, who also saw 66, noted that from his observatory high up in the Andes, he could observe stars as faint as 7th magnitude (*Uran. Arg.,* p. 6).

Other astronomers have made different selections for Bayer's p's. For example, Bevis and Bode, who designated only four p's, selected the following stars:

p-1, 62, BV 84, B 281.
p-2, 65, BV 85, B 290.
p-3, BV 91.

Bevis's chart (Tabula XXVI) does not show p-3. He noted in his catalogue that he was the first since Bayer to have observed this star. It is not in the catalogues of Ptolemy, Tycho, Flamsteed, or any other astronomer. From Bevis's coordinates, the closest star is HD 96694, 6.8V.

p-4, 66, BV 92, B 298 as p-3.
p-5, 69, BV 96, B 307 as p-4.

Despite the suggestions of Baily, Bevis, and Bode, most authorities have accepted Argelander's five stars primarily because they conform best to Bayer's chart.

Leo Minor, LMi

Lesser, Smaller, or Little Lion

Hevelius devised the constellation Leo Minor (Figure 49) and named it the Little Lion, in order, he said, not to offend astrologers or violate the rules of astrology. He noted that he formed the constellation from eighteen *informes* or unformed stars between the Great Bear and the Lion, two raging and ferocious beasts. He felt it would be most appropriate to name his new constellation Leo Minor or Leo Junior so that it would conform to the nature and character of its two neighbors.

Flamsteed included Leo Minor in his atlas and catalogue, and Baily assigned it only one letter, Beta. He did not use Alpha. As he stated in the preface to the *BAC*, he intended to assign Greek letters for those stars he estimated to be brighter than magnitude 4½ in the nine new constellations in Hevelius's catalogue that were also included in Flamsteed's catalogue: Camelopardalis, Canes Venatici, Coma Berenices, Lacerta, Leo Minor, Lynx, Monoceros, Sextans, and Vulpecula. There are only three stars in Leo Minor brighter than magnitude 4.5:

21, HR 3974, 4.48V, BAC 3446.
Beta, 31, HR 4100, 4.21V, *SA, HRP, BS,* BAC 3572*.
46, HR 4247, 3.83V, BAC 3728.

Baily designated 31 as Beta but he left 46 unlettered although it should have been assigned Alpha since it is the brightest star in the constellation. It is difficult to understand why Baily left it unlettered. But perhaps the fact that he died before he had a chance to correct the galley proofs for this constellation explains the omission. Baily left 21 unlettered since he considered it 5th magnitude. He estimated 31 and 46 to be stars of magnitude 4½.

See *Prodromus Astronomiae*, p. 114; Allen, *Star Names,* pp. 263-64; *BAC*, p. 70.

Table 52. The lettered stars of Leo Minor

	MAGNITUDES			CATALOGUE NUMBERS				
Letter	Bayer	Lacaille	Visual	Flamsteed	Lacaille	Other	HR	HD
Beta			4.21	31		BAC 3572	4100	90537
o+			3.83	46		B 122	4247	94264

Lost Stars

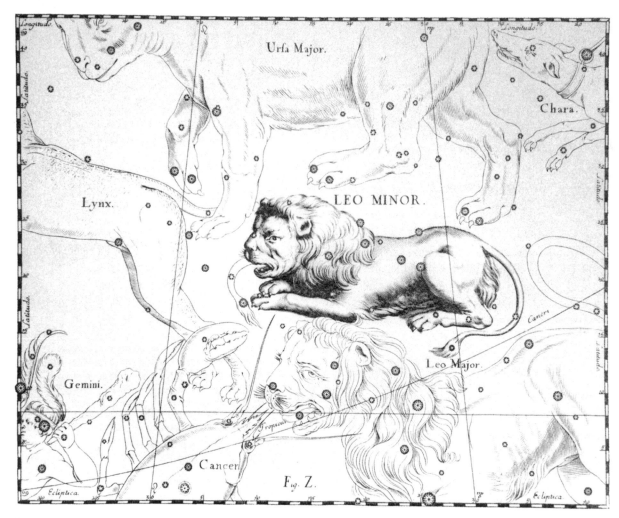

Figure 49. Hevelius's Leo Minor. In the upper right corner of the chart is Chara, one of the Hunting Dogs, one of the twelve constellations devised by Hevelius.

The Lost, Missing, or Troublesome Stars of Leo Minor

o, 46, HR 4247, 3.83V, *SA* erroneously as Omicron, B 122*.

This is lower-case o (oh). Baily did not letter this star. The letter was first introduced by Bode, who added additional letters in Leo Minor, none of which has survived in the literature except this one. It owes its survival to the fact that it is the *lucida* of the constellation.

Lepus, Lep

Hare

Bayer lettered the stars in the Ptolemaic constellation Lepus (Figure 50) from Alpha to Nu. All its lettered stars are listed in the *BS,* and there are no missing or problem lettered stars in the constellation.

Greek poets and writers have collected at least three myths describing how Lepus finds a place of honor among the stars. According to one account, Hermes (Roman Mercury), the messenger of the gods and the god of speed, sets the quick-footed Hare in the sky as the symbol of speed. Hermes is also considered the patron of merchants and thieves, both of whom depend on quick thinking and agility for success in their chosen fields. To some cynical observers, there is very little difference between the two professions. Another legend notes that the Hare in the heavens is part of a larger tableau featuring Orion, the Mighty Hunter, and his two hunting dogs, Canis Major and Canis Minor. According to this account, Orion is depicted in the heavens hunting Lepus with the aid of his two dogs. Some writers have wondered why the Mighty Hunter would be hunting such an insignificant little animal and why such a ludicrous scene should be honored with a place among the stars.

A third myth, suggests one scholar, may have a basis in historical fact. According to the Egyptian-born Greek grammarian Athenaeus, who lived sometime in the second century A.D., the island of Astypalaea was overrun with rabbits about the middle of the third century B.C. As the story goes, one of the inhabitants of the island of Leros — in the Eleusinian Gulf near Athens, about twenty miles northwest of Astypalaea — brought a pregnant female rabbit to the island. The islanders took an instant liking to the creature, and after the female gave birth to a huge brood, they began to breed them. In just a short time the rabbit population grew beyond control and began to devour the island's crops. Faced with possible famine, the inhabitants eventually rid themselves of the pesky little animals. Astronomers placed the image of a rabbit in the sky to remind humans that pleasure can have unpleasant consequences.

See Condos, *Star Myths,* pp. 129-31, 244; Ridpath, *Star Tales,* p. 83. According to Condos, the reference to the plague of rabbits can be found in *The Banquet,* 9.400.

Lost Stars

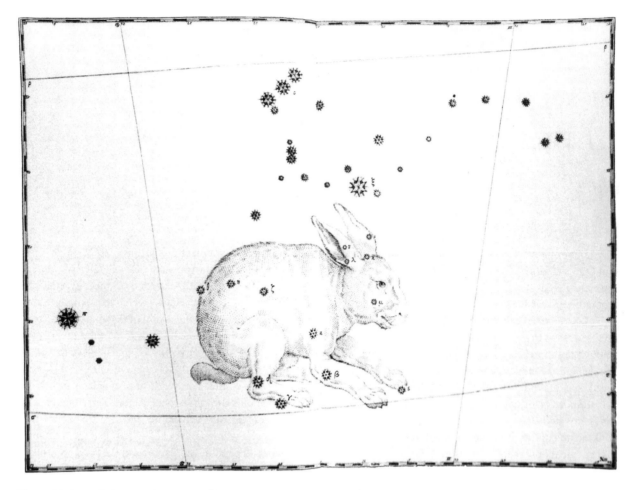

Figure 50. Bayer's Lepus. The letter Xi (ξ) is the 1st-magnitude star Rigel, Beta Ori; Omicron (o) is Orion's belt; and Pi (π) is the star Sirius, Alpha CMa, the next brightest star to the sun that is visible from Earth. The lines Rho (ρ) and Sigma (σ) are the celestial equator and the Tropic of Capricorn.

Table 53. The lettered stars of Lepus

	MAGNITUDES			CATALOGUE NUMBERS				
Letter	Bayer	Lacaille	Visual	Flamsteed	Lacaille	Other	HR	HD
Alpha	3		2.58	11			1865	36673
Beta	3		2.84	9			1829	36079
Gamma	3		3.60	13			1983	38393
Delta	3		3.81	15			2035	39364
Epsilon	4		3.19	2			1654	32887
Zeta	4		3.55	14			1998	38678
Eta	4		3.71	16			2085	40136
Theta	4		4.67	18			2155	41695
Iota	5		4.45	3			1696	33802
Kappa	5		4.36	4			1705	33949
Lambda	5		4.29	6			1756	34816
Mu	5		3.31	5			1702	33904
Nu	6		5.30	7			1757	34386

LIBRA, LIB

Scales, Balance

Libra (Figure 51) is a Ptolemaic constellation. Ptolemy referred to it as the (Scorpion's) Claws, but Latin authors used the name Iugum (Beam) or Libra (Balance or Pair of Scales). Bayer called it Libra and lettered its stars from Alpha to Omicron.

Libra was originally part of Scorpius, one of the four major constellations of Sumerian astronomy that were associated with the four turning points of the solar year. Libra was probably separated from the Scorpion sometime in the second millennium B.C. when, because of precession, the autumnal equinox shifted westward and away from the center of Scorpius. It appeared as a separate constellation in the latter half of the second millennium under the Sumerian name ZI.BA.AN.NA (Akkadian Zibanitu), Scales or Balance of Heaven.

Libra was so named not because the constellation resembled a pair of scales but because 3,500 years ago it was aligned with the autumnal equinox, an important event in the Mesopotamian religious calendar — the beginning of the New Year. In the ancient Near East, there were originally two calendars, each one closely tied to agriculture and the position of the sun as it crossed the celestial equator. One calendar began the New Year at the time of the vernal equinox, the beginning of spring planting, which occurred during the first month of the New Year, Nisanu. The other calendar began the New Year at the autumnal equinox, which coincided with the completion of the harvest and the end of the agricultural year in the month of Tashritu, the seventh month, which may at one time have actually been considered the first month. In short, there were two celebrations of the New Year in Mesopotamia. In Babylon, for instance, it was celebrated in Nisanu, and in the cities of Ur and Erech, it was celebrated in Tashritu. Some historians have suggested that there may have been two celebrations in these cities, one at each of the equinoxes. This Mesopotamian concept of two new years should be compared with the Jewish calendar. According to the Talmud, the New Year begins in the seventh month, Tishri (Tashritu). The rabbis assert that this is the religious New Year — when God created man — while the Old Testament refers to Nisan (Nisanu) as the first month, the beginning of the secular year, the New Year for kings.

The New Year Festival (Sumerian Zagmuk; Akkadian Akitu), whether in the first or seventh month, was the most important religious holiday for the people of Mesopotamia. Although there is some indication it originally lasted sixteen days, by the first millennium B.C. it was an eleven-day festival. Included among the activities was the *kuppuru*-ritual, a ceremony of atonement for past sins in which a ram was sacrificed to the gods. For it was at this time that the gods judge man and fix his destiny for the coming year. According to Sumerian beliefs, Nanshi, the goddess of social justice; her husband Haiia, the man of the tablets of destiny; and Nidabe, the goddess of writing and accounts, who holds the golden stylus, hold court on the first day of the New Year. These deities evaluate man's life and determine his fate. Centuries later, among the Babylonians, who replaced the Sumerians in ancient Mesopotamia, new deities took the place of the older gods, but the basic elements of the New Year Festival remained unchanged — atonement for past sins and determination of future destiny. It was the time for weighing man's fate.

As noted, the New Year Festival in the ancient Near East may have originally lasted longer than eleven days. The Jewish High Holy Days, for instance, extend for a period of twenty-two days. They begin with the New Year (Rosh Hashanah) on the first day of Tishri, which is followed by the Day of Atonement (Yom Kippur) on the 10th. On the 15th of Tishri there begins the eight-day celebration of Sukkot (Tabernacles), the Harvest Festival, which ends on the 22nd of Tishri with the holy day Shemini Atzeres (Holy Convocation of the Eighth Day), the Feast of Conclusion (Numbers 29: 1-38). As in the Jewish calendar, the Mesopotamian New Year began not exactly on the equinox, but on the evening of the first new moon nearest the equinox. The Jewish calendar, like that of the ancient Mesopotamians, is lunar-solar with occasional days and months intercalated, or added, to keep the lunar year (354+ days) synchronized with the solar (365+ days).

See Heuter, "Star Names," pp. 245-46; Hartner, "The Earliest History of the Constellations in the Near East," diagram 1; van der Waerden, "Babylonian Astronomy. II. The Thirty-Six Stars," pp. 11, 13-15; van der Waerden, "History of the Zodiac," pp. 218-20; O'Neil, *Time and the Calendars,* p. 58; Frankfort, *Kingship and the Gods,* chap. 22; Gaster, *Thespis,* pp. 35-37; Hooke, *Myth, Ritual, and Kingship,* pp. 37-40; Langdon, *Babylonian Menologies,* pp. 6-7, 68, 97-9, 110; Langdon, *The Babylonian Epic of Creation,* p. 28; Kramer, *History Begins at Sumer,* p. 265; Kramer, *The Sumerians,* pp. 124-25; Finegan, *Light from the Ancient Past,* pp. 552-98; Freedman and Simon, *Midrash Rabbah,* IV, 369-79; Epstein, *The Babylonian Talmud, Seder Mo'ed,* IV, "Rosh Hashanah;" Reiner and Pingree, *Baby. Plan. Omens,* Part 2, p. 16. For the turning points of the solar year, see Aquarius and Scorpius; for a description of the Babylonian New Year's Festival, see Aries.

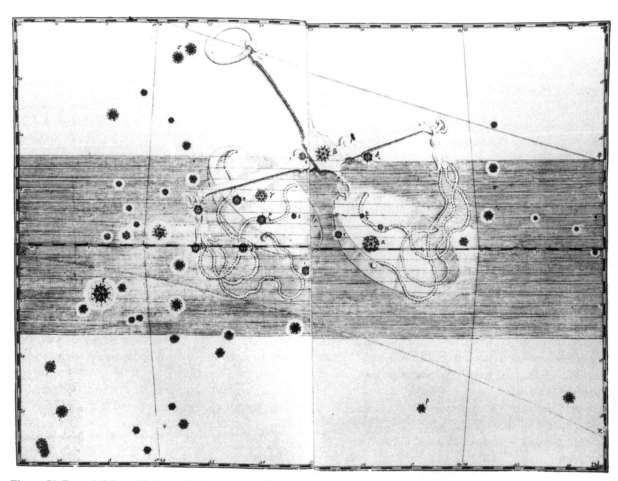

Figure 51. Bayer's Libra. The letter Pi (π) is the constellation Virgo; Rho (ρ) is Hydra's head; Sigma (σ) is the hand of Ophiuchus; Tau (τ) is the 1st-magnitude star Antares, Alpha Sco; and Upsilon (υ) is the head of Lupus. The lines Phi (φ) and Chi (χ) are the celestial equator and the Tropic of Capricorn. The shaded area is the zodiac.

Table 54. The lettered stars of Libra

	MAGNITUDES			CATALOGUE NUMBERS				
Letter	Bayer	Lacaille	Visual	Flamsteed	Lacaille	Other	HR	HD
Alpha^1+			5.15	8			5530	130819
Alpha^2	2		2.75	9			5531	130841
Beta	2		2.61	27			5685	135742
Gamma	3		3.91	38			5787	138905
Delta	4		4.92	19			5586	132742
Epsilon	4		4.94	31			5723	137052
Zeta^1+			5.64	32			5743	137744
Zeta^2			6.69	33				137949
Zeta^3			5.82	34			5750	138137
Zeta^4	4		5.50	35			5764	138485
Eta	4		5.41	44			5848	140417
Theta	4		4.15	46			5908	142198
Iota^1+	4		4.54	24			5652	134759
Iota^2			6.08	25			5656	134967
Kappa	4		4.74	43			5838	139997
Lambda	4		5.03	45			5902	142096
Mu	5		5.31	7			5523	130559
Nu+	5		5.20	21			5622	133774
Xi^1+	6		5.80	13			5554	131530
Xi^2	6		5.46	15			5564	131918
Omicron+	6		6.30	29			5703	136407
Sigma+			3.29	20		G 45	5603	133216
Tau+			3.66	40		G 98	5812	139365
Upsilon+			3.58	39		G 93	5794	139063

The Lost, Missing, or Troublesome Stars of Libra

Alpha^1, Alpha^2, 8, 9, HR 5530/1, 5.15V, 2.75V, *SA, HRP, BS,* G 25/6*.

Bayer noted that Alpha was a single 2nd-magnitude star. Although Flamsteed resolved this double star, h 186, with his telescope and catalogued the individual components separately, he assigned Alpha only to 9, the brighter of the two. Gould was among the first to designate 8 as well as 9 as Alpha and to add indices to both. This is a widely separated pair with almost 4' between the two stars.

Zeta^1, 32, HR 5743, 5.64V, *BS,* HF 27, BV 36, BF 2097*, H 35*, G 77.
Zeta^2, 33, HD 137949, 6.69V, HF 28, BV 37, BF 2099*, G 80.
Zeta^3, 34, HR 5750, 5.82V, *BS,* HF 29, BV 38, BF 2102*, H 36*, G 82.
Zeta^4, 35, HR 5764, 5.50V, *SA* as Zeta, *HRP* as Zeta, *BS,* HF 30 as Zeta, BV 39 as Zeta, BF 2104*, H 38*, G 87 as Zeta.

Bayer described Zeta as a single 4th-magnitude star. Halley's 1712 edition of Flamsteed's catalogue equates 35 with Zeta. Flamsteed's 1725 catalogue, however, designates these four neighboring stars, his 32, 33, 34, and 35, as Zeta-1, Zeta-2, Zeta-3, and Zeta-4. Bevis, though, went along with the 1712 catalogue, labeling only 35, the brightest star in the group, as Zeta. Argelander, who saw only 34 and 35, agreed with Bevis. He felt that only 35 was synonymous with Bayer's Zeta (*UN,* p. 106). Heis, on the other hand, saw 32 as well as 34 and 35. Consequently, he designated these three stars Zeta^1, Zeta^3, and Zeta^4. He omitted 33, Zeta^2, since its light was too faint even for his acute vision. See Heis's Corrigenda to his catalogue and Tabula XI of his atlas. Gould, from his observatory high atop the Andes at Cordoba, Argentina, saw all four stars — he could see, he claimed, stars to the 7th magnitude and even beyond — but he agreed with Argelander that only 35 should be considered Zeta.

Iota¹, 24, HR 5652, 4.54V, *SA* as Iota, *HRP* as Iota, *BS,* HF 19 as Iota, BV 27 as Iota, BF 2064*, H 26 as Iota, G 52 as Iota.
Iota², 25, HR 5656, 6.08V, *BS,* HF 20, BV 28, BF 2070*, G 57.

Bayer noted that Iota was a single star of the 4th magnitude. Halley's pirated edition of Flamsteed's catalogue designates his 24 as Iota, but the 1725 edition labels these two neighboring stars, his 24 and 25, as Iota-1 and Iota-2. Bevis, though, followed Halley's edition of Flamsteed's catalogue by labeling only the brighter star, 24, as Iota. Argelander agreed since he could not see 25 with his naked eye (*UN,* p. 105).

Nu¹, 21, HR 5622, 5.20V, *SA* as Nu, *HRP* as Nu, *BS* as Nu, HF 17 as Nu, BV 21 as Nu, BF 2054*, H 25 as Nu, G 48 as Nu.
Nu², 22, HD 133800, 6.39V, HF 18 as Nu, BV 22, BF 2055*, G 49.

Bayer described Nu as a single star of the 5th magnitude, but Flamsteed designated these two nearly adjacent stars, his 21 and 22, as Nu-1 and Nu-2. Bevis, though, labeled only the brighter star, 21, as Nu. Argelander, unable to see 22, agreed with Bevis that only the brighter star was synonymous with Bayer's Nu (*UN,* p. 105).

Xi¹, 13, HR 5554, 5.80V, *SA, HRP, BS,* BV 8 as Xi, BF 2031*, H 14*.
Xi², 15, HR 5564, 5.46V, *SA, HRP, BS,* BV 12, BF 2037*, H 15*.

Bayer described Xi as a single 6th-magnitude star, but Flamsteed designated these two neighboring stars, his 13 and 15, as Xi-1 and Xi-2.

Omicron¹, 29, HR 5703, 6.30V, *SA* as Omicron, *HRP* as Omicron, *BS* as Omicron, BF 2086*, H 30 as Omicron, G 70 as Omicron.
Omicron², 30, HD 136081, 6.47V, BF 2089*, G 74.

Bayer described Omicron as a single 6th-magnitude star, but Flamsteed designated these two neighboring stars, his 29 and 30, as Omicron-1 and Omicron-2. Argelander, unable to see 30 with his naked eye, assumed that only the brighter star, 29, was equivalent to Bayer's Omicron (*UN,* p. 105).

Rho

This letter is not used in Libra.

Sigma, 20, HR 5603, 3.29V, *SA, HRP, BS,* BF 2048, BAC 4950, H 1 as Gamma Sco, G 45*.
Tau, 40, HR 5812, 3.66V, *SA, HRP, BS,* BF 2115, BAC 5151, H 3 as Omicron Sco, G 98*.
Upsilon, 39, HR 5794, 3.58V, *SA, HRP, BS,* BF 2114, BAC 5138, H 2 in Scorpius, G 93*.

Bayer did not letter these stars in Libra. Ptolemy included these stars in Libra, but Bayer annexed them to Scorpius. As in other instances where Bayer switched Ptolemaic stars from one constellation to another (see Upsilon and Phi Per, for example), Flamsteed, the traditionalist, proceeded to set matters right. He returned to Libra the three stars that Bayer had removed: (1) Gamma Sco, (2) an unlettered *informis* on Bayer's chart about 1½° north and preceding Omicron Sco, and (3) Omicron Sco. These are Flamsteed's 20, 39, and 40 Lib. Argelander, faithful to Bayer, returned these three stars to Scorpius (*UN,* p. 106). Gould, however, decided to follow Flamsteed and Ptolemy by keeping them in Libra. He proposed to assign them letters because of their brightness, but to avoid confusion with Bayer's letters, he relettered them Sigma, Tau, and Upsilon. The International Astronomical Union, in one of the few instances where it altered a Bayer constellation (see Psi Gem and Upsilon and Phi Per), agreed with Gould. See Toomer, *Ptolemy's Almagest,* pp. 371-72; BF 2114/5, where Baily incorrectly noted that Bayer called both 39 and 40 Omicron Sco, whereas Bayer so designated only 40.

Phi
Chi

Phi and Chi are not used in Libra.

Psi, 48, HR 5941, 4.88V, BF 2176, H 50, G 119.

Bayer did not letter this star. Although he included it in his atlas, he considered it an *informis* outside the border of Libra and left it unlettered. Flamsteed lettered it Psi, but Baily removed the letter as superfluous (see i Aql). Gould left it unlettered since he considered it dimmer than 5th magnitude. See Phi Sco.

Lupus, Lup

Wolf

Lupus (Figure 52) is a Ptolemaic constellation that Bayer lettered from Alpha to Upsilon. Since it is mostly below the horizon in Central Europe, Bayer's chart does not conform to the constellation's actual appearance in the heavens. Consequently, Lacaille redesigned and relettered it from Alpha to Psi and from a to h. As Baily proposed in the preface to the *BAC* (pp. 62-63), he removed all of Lacaille's Roman letters. Gould restored most of them since they designated stars brighter than magnitude 5.5.

In Greek and Roman mythology, this constellation was considered part of Centaurus, which depicted the wise and pious centaur Chiron carrying a beast (Lupus) impaled on a pole to the Altar (Ara) as an offering to the gods. Ptolemy considered it a separate constellation and referred to it as the Wild Beast (θηριον, Therion). Some scholars have suggested that the constellation can be traced back to the Middle East. The Mesopotamians had a constellation called the Wolf, but it was located far away in Triangulum, in the northern sky. Another Mesopotamian constellation, the Mad Dog, was located in Lupus's sector of the heavens. According to myth, The Mad Dog is one of the minions of Tiamat, goddess of the sea and chaos. When Tiamat battles Marduk, the champion of the gods, she brings to the battlefield all her allies: the Viper, the Sphinx, the Dragon, the Scorpion-Man, and the Mad Dog. But as far as is presently known, no Mesopotamian astral myths or legends have been uncovered relating to the Mad Dog or connecting it to Lupus. Some might speculate, however, that the Centaur's sacrifice of the Wild Beast to the gods may somehow allegorically parallel Marduk's defeat of Tiamat and her allies — the victory of order, intelligence, and piety over the forces of savagery, brute force, and evil.

It is difficult to determine when the constellation was first called Lupus. As late as the sixteenth century it was labeled Fera (Wild Beast) on sky maps. By the late seventeenth century, some stellar cartographers were referring to it as Fera lupus (Wild Beast the Wolf).

See *BAC,* pp. 62-63; Condos, *Star Myths,* pp.79-82; Ridpath, *Star Tales,* 85-86; Staal, *New Patterns in the Sky,* pp. 170-71; Toomer, *Ptolemy's Almagest,* p. 396; Hunger and Pingree, *Mul.Apin,* pp. 137-38; Warner, *Sky Explored,* pp. 11, 17, and *passim.* See Hydra for an account of Marduk's battle with Tiamat.

Lost Stars

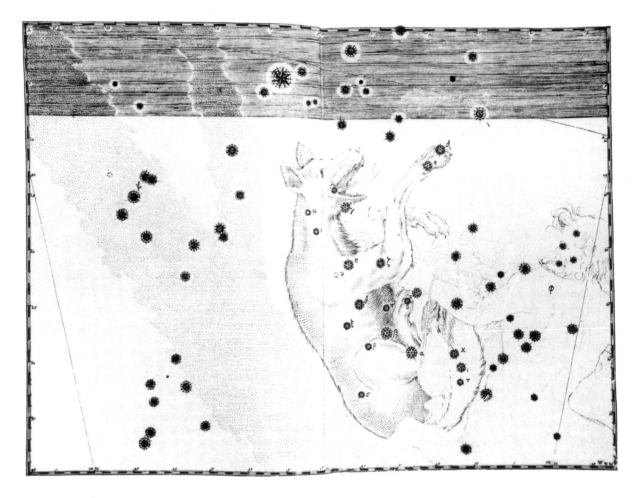

Figure 52. Bayer's Lupus. The letter Phi (φ) is the Centaur; Chi (χ), above the Wolf's ears and missing from some copies of the *Uranometria,* is Antares in the Scorpion; Psi (ψ) is the tail of the Scorpion; and Omega (o) is the Altar. The line A, immediately below the zodiac on the right of the chart, is the Tropic of Capricorn.

Table 55. The lettered stars of Lupus

	MAGNITUDES			CATALOGUE NUMBERS				
Letter	Bayer	Lacaille	Visual	Flamsteed	Lacaille	Other	HR	HD
Alpha		3	2.30		1231		5469	129056
Beta		3	2.68		1254		5571	132058
Gamma		3	2.78		1293		5776	138690
Delta		4	3.22		1283		5695	136298
Epsilon		4	3.37		1285		5708	136504
Zeta		4	3.41		1265		5649	134505
Eta		4	3.41		1325		5948	143118
Theta		5	4.23		1335		5987	144294
Iota+		5	3.55		1201		5354	125238
Kappa[1,2]+		5	3.97c		1266		5646/7	134481/2
Lambda		5	4.05		1263		5626	133955
Mu		5	4.27		1274		5683	135734
Nu[1]+		5	5.00		1281		5698	136351
Nu[2]		6	5.65		1284		5699	136352
Xi[1,2]+		6	4.78c		1321		5925/6	142629/30

(table continued on next page)

Table 55. The lettered stars of Lupus (continued)

	MAGNITUDES			CATALOGUE NUMBERS				
Letter	Bayer	Lacaille	Visual	Flamsteed	Lacaille	Other	HR	HD
Omicron		5	4.32		1246		5528	130807
Pi		5	3.88c		1258		5605/6	133242/3
Rho		5	4.05		1223		5453	128345
Sigma		5	4.42		1216		5425	127381
Tau^1+		5	4.56		1209		5395	126341
Tau^2		5	4.35		1210		5396	126354
Upsilon		6	5.37		1288		5719	136933
Phi^1+		5	3.56		1286		5705	136422
Phi^2		6	4.54		1287		5712	136664
Chi		5	3.95	5	1312		5883	141556
Psi^1+		6	4.67	3	1301		5820	139521
Psi^2		6	4.75	4	1303		5839	140008
Omega+			4.33			G 117	5797	139127
a^1+		6	5.55		1221		5444	128068
a^2		6	5.41		1222		5450	128266
a^3		6	6.07		1228		5457	128582
b		6	5.21		1236		5495	129893
c		6	5.38		1251		5556	131562
d+			4.54			B 63	5781	138769
e		6	4.82		1268		5651	134687
f+		5	4.34	2	1277		5686	135758
g+		6	4.64		1300		5825	139664
h		6	5.24		1304		5837	139980
i+			4.91	1		G 69	5660	135153
k+			4.60			G 97	5724	137058

The Lost, Missing, or Troublesome Stars of Lupus

Iota, HR 5354, 3.55V, *SA, HRP, BS,* CA 1201*, BR 4848*, L 5881*, BAC 4734*, G 1*.

Lacaille inadvertently included two Iota's in Lupus, this star and CA 1297. See Omega.

Kappa^1, HR 5646, 3.87V, *SA, HRP* as Kappa, *BS,* CA 1266 as Kappa, BR 5205*, L 6246 as Kappa, BAC 4986 as Kappa, G 62 as Kappa.

Kappa^2, HR 5647, 5.69V, *SA, BS,* BR 5207*, BAC 4988, G 63.

Lacaille described Kappa as a single star of the 5th magnitude. Later astronomers added the indices to distinguish the components of this double star, Dunlop 177. Although Gould assigned Kappa only to his G 62, he noted that both G 62 and G 63 appeared as one image to his naked eye with a combined magnitude of 4.1.

Nu^1, HR 5698, 5.00V, *SA, HRP, BS,* CA 1281 as Nu, BR 5286*, L 6322*, G 86*.
Nu^2, HR 5699, 5.65V, *SA, HRP, BS,* CA 1284 as Nu, BR 5288*, L 6324*, G 85*.

Lacaille described these two neighboring stars as Nu. When he first observed these stars on March 19, 1752, Nu^1 preceded Nu^2, but as a result of Nu^2's rapid proper motion, 0.1615 of a second per year westward, it now precedes Nu^1. Nu^1's proper motion is only 0.0135 of a second westward.

Xi^1, HR 5925, 5.37V, *SA, HRP* as Xi, *BS,* CA 1321 as Xi, BR 5535*, L 6592 as Xi, G 138 as Xi.
Xi^2, HR 5926, 5.73V, *SA, HRP* as Xi, *BS,* BR 5536*, G 139.

Lacaille described Xi as a single star of the 6th magnitude. Later astronomers added indices to distinguish the components of this double star. Although Gould assigned Xi only to his G 138, he noted that both G 138 and G 139 appeared as one image to his naked eye with a combined magnitude of 5.2.

Tau¹, HR 5395, 4.56V, *SA, HRP, BS,* CA 1209 as Tau, BR 4902*, L 5928 as Tau², BAC 4768*, G 9*.
Tau², HR 5396, 4.35V, *SA, HRP, BS,* CA 1210 as Tau, BR 4903*, L 5927 as Tau¹, BAC 4770*, G 10*.

In preparing his catalogue of 1763, Lacaille reduced his observations to 1750 and showed CA 1209 preceding CA 1210 by about 1^s of right ascension. He described these two nearly adjacent stars as Tau. When Baily prepared the expanded edition of Lacaille's catalogue, also reduced to epoch 1750, he noted that CA 1210 actually preceded CA 1209 by 3/10 of a second. Consequently, he designated it Tau¹ and CA 1209 as Tau². Soon afterward, CA 1209 preceded CA 1210, not only because of the precession of the equinoxes but also because of the proper motion of these two stars — they are moving in opposite directions. CA 1209 is moving westward at .0012 of a second each year, while CA 1210 is moving eastward at .0014 of a second. See Gould, *Uran. Arg.,* p. 88.

Phi¹, HR 5705, 3.56V, *SA, HRP, BS,* CA 1286 as Phi, BR 5293*, L 6335*, G 88*.
Phi², HR 5712, 4.54V, *SA, HRP, BS,* CA 1287 as Phi, BR 5299*, L 6349*, G 92*.

Lacaille described these two neighboring stars as Phi.

Psi¹, 3, HR 5820, 4.67V, *SA, HRP, BS,* CA 1301 as Psi, BR 5410*, L 6463*, G 119*.
Psi², 4, HR 5839, 4.75V, *SA, HRP, BS,* CA 1303 as Psi, BR 5440, L 6489*, G 125*.

Lacaille described these two neighboring stars as Psi. Lacaille made a clock error in recording CA 1303's position, thus generating erroneous coordinates. Brisbane, therefore, did not equate his BR 5440 with CA 1303. Baily discovered the error and corrected it while editing the expanded version of Lacaille's catalogue.

Omega, HR 5797, 4.33V, *SA, HRP, BS,* CA 1297 as Iota, BR 5399 as Iota, L 6443, BAC 5139 as Iota, G 117*.

Lacaille did not use the letter Omega in Lupus. Gould added the letter noting that Lacaille had inadvertently included two Iota's in Lupus, this star and CA 1201. Since both stars are brighter than 5th magnitude, Gould felt they both merited letters. He left CA 1201 as Iota (see Iota) since he felt it was synonymous with Bayer's Iota and he relettered CA 1297 as Omega. Neither Brisbane nor Baily caught or noted the duplication.

a¹, HR 5444, 5.55V, CA 1221 as a, BR 4971*, L 5995, BAC 4815, G 18.
a², HR 5450, 5.41V, *SA* as a, *HRP* as a, CA 1222 as a, BR 4974*, L 6001, BAC 4819, G 24 as a.
a³, HR 5457, 6.07V, CA 1228 as a, BR 4987*, L 6018, BAC 4829, G 24.

Lacaille described these three neighboring stars as a. Baily removed their letters because, in his opinion, Lacaille should have restricted Roman letters only to the components of Argo Navis. Gould restored a to the brightest of the group, CA 1222, since he felt a star of its magnitude merited a letter. He noted that CA 1222, together with CA 1221 and two smaller stars (HD 127629, 7.3V, BR 4955, L 5978; and HD 127864, 6.89V, BR 4964, L 5987), appeared as one image to his naked eye. Lacaille had observed these two small stars but he neither lettered them nor included them in his catalogue of 1763.

b, HR 5495, 5.21V, *SA, HRP,* CA 1236*, BR 5044*, L 6070, G 30*.

c, HR 5556, 5.38V, *SA, HRP,* CA 1251*, BR 5103*, L 6132, G 36*.

d, HR 5781, 4.54V, *SA, HRP,* CA 1294 as f, BR 5384 as f, BAC 5123, B 63*, G 114*.

Lacaille did not letter this star d in Lupus. Bode, noting that Lacaille had mistakenly designated two stars as f Lup, relettered the fainter one, CA 1294, as d. Three quarters of a century later, Gould made the exact same correction without crediting Bode. See f.

e, HR 5651, 4.82V, *SA, HRP,* CA 1268*, BR 5219*, L 6257, G 66*.

f, 2, HR 5686, 4.34V, *SA, HRP,* CA 1277*, BR 5266, L 6304, RAS 1731*, BAC 5032, B 38*, G 78*.

Lacaille mistakenly designated two stars as f Lup, this star and CA 1294. Bode retained f for this star, the brighter of the two, and relettered the other d. See d.

g, HR 5825, 4.64V, *SA, HRP,* CA 1300*, BR 5416*, G 121*.

Lacaille mistakenly identified two stars as g Lup, this star and CA 1272. Gould retained g for this star, the brighter of the two, and relettered the other i. See i.

h, HR 5837, 5.24V, *SA, HRP,* CA 1304*, BR 5437*, L 6486, G 124*.

i, 1, HR 5660, 4.91V, *SA, HRP,* CA 1272 as g, BR 5237 as g, L 6277, BAC 5009, G 69*.

    Lacaille did not letter this star i in Lupus. The letter was added by Gould, who, noting that Lacaille had inadvertently lettered two stars g Lup, relettered the dimmer one, CA 1272, as i. See g.

j

    Lacaille did not use this letter.

k, HR 5724, 4.60V, *SA, HRP,* CA 1289, BR 5313, L 6361, G 97*.

    Lacaille did not letter this star. He observed this star and included it in his catalogue of 1763 but left it unlettered; he considered it 6th magnitude. Gould designated it k because he felt a star of its brightness merited a letter.

Lynx, Lyn

Lynx or Tiger

Lynx (Figure 53) is one of the twelve new constellations devised by Hevelius. He called it Lynx or Tiger and formed it from nineteen small stars of the 5th and 6th magnitude between the Great Bear and Auriga. He took pride in noting that he had observed all these stars only with his naked eye, not with the telescope. In fact, in a challenge to future astronomers, he asserted: *"Qui Lyncem contemplari velit, oportet, ut sit Lynceus"* (He who wishes to observe the Lynx must, of necessity, have the eyesight of a lynx).

Although Hevelius originally called this constellation Lynx or Tiger and used this nomenclature in his catalogue, he labeled it simply Lynx in his atlas. Flamsteed included it as Lynx in his atlas and catalogue, and so it has remained ever after.

In compiling the *BAC*, Baily found only one star in Lynx brighter than magnitude 4.5, which he labeled Alpha, the only lettered star in the entire constellation. It is listed in the *BS*, and there are no problem or missing lettered stars in the constellation.

See Hevelius, *Prodromus Astronomiae*, pp. 114-15; for Hevelius's disdain for the telescope for stellar observations, see Sextans.

Table 56. The lettered stars of Lynx

	Magnitudes			Catalogue Numbers				
Letter	Bayer	Lacaille	Visual	Flamsteed	Lacaille	Other	HR	HD
Alpha			3.13	40		BAC 3178	3705	80493

Figure 53. Hevelius's Lynx. This chart depicts parts of two other "modern" constellations, Camelopardalis and Leo Minor. The little Lion and Lynx were devised by Hevelius; the Giraffe was probably devised by Petrus Plancius.

LYRA, LYR

Lyre, Harp

Lyra (Figure 54) is a Ptolemaic constellation that Bayer lettered from Alpha to Nu.

According to Greek mythology, the god Hermes (Roman Mercury) fashions a lyre from the shell of a tortoise. Soon afterward, he is caught stealing cattle from the sun god, Apollo. Hermes was, after all, the patron of merchants and thieves. To avoid punishment, he agrees to give Apollo, who was also the god of music, his lyre. Apollo, in turn, gives the lyre to Orpheus of Thrace, the son of Calliope, the Muse of epic poetry. Orpheus improves the lyre by adding additional strings. He becomes such a great singer and musician that it is said he moves not only humans but animals, trees, and rocks. He marries Eurydice, who tragically dies from the bite of a venomous snake. So stricken is Orpheus by the death of his young bride that he journeys to the Underworld to seek her release. His music and song are so beautiful that Hades, god of the Underworld, permits Eurydice to leave on the condition that Orpheus not look back. As they approach daylight, Orpheus is unable to resist looking back to see if Eurydice is following him. As he does, he sees Eurydice slip back into the Underworld. Grief stricken at the second loss of his beloved, Orpheus refuses to have anything to do with women and turns his attention instead to young boys. Angered by his indifference to their advances and by his introduction of homosexual love, the women of Thrace tear him to pieces. Another account of the death of Orpheus relates that while he is in the Underworld, he praises all the gods but inadvertently forgets Dionysus (Bacchus). To avenge this slight to his honor, Dionysus sends his crazed followers, the Bacchantes, to tear Orpheus to pieces. Orpheus's aunts, the Muses, collect the remnants of his body and bury them. They also retrieve his lyre and, with the permission of Zeus, they place it among the stars.

The constellation is often depicted on star charts as a lyre being carried in the talons of an eagle or vulture. This stems from another variation of the legend of Orpheus. After Orpheus's death, Zeus himself sends an eagle or vulture to retrieve the lyre and bring it up to heaven. Islamic astronomers visualized this constellation as either an eagle or vulture.

Some scholars have suggested that the origin of the constellation can be traced back to the ancient Near East, but as far as is presently known, no Mesopotamian astral legends are associated with Lyra. In its place in the heavens, the Mesopotamians saw the constellation the She-Goat, and it is not known to be associated with Lyra.

See Lum, *The Stars in Our Heaven,* pp. 95-98; Condos, *Star Myths,* pp. 133-39; Ridpath, *Star Tales,* pp. 87-89; Staal, *New Patterns in the Sky,* pp. 183-87; Hunger and Pingree, *Mul.Apin,* p. 138.

Lettered Stars - Lyra

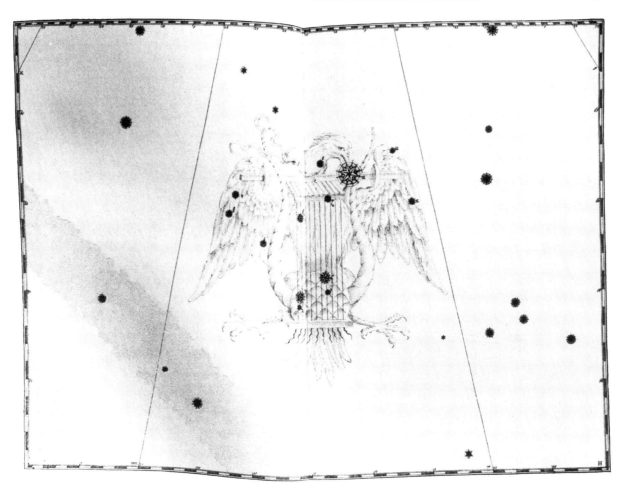

Figure 54. Bayer's Lyra. The letter Xi (ξ) is the beak of Cygnus, but the letter that should be next to the star below Lyra's right claw is missing on some copies of the *Uranometria*. The letter Omicron (o) is the right arm of Hercules and should be among the group of stars to the right of Lyra's left claw, but, like Xi, it is missing on some copies of the *Uranometria*.

Table 57. The lettered stars of Lyra

	MAGNITUDES			CATALOGUE NUMBERS				
Letter	Bayer	Lacaille	Visual	Flamsteed	Lacaille	Other	HR	HD
Alpha	1		0.03	3			7001	172167
Beta	3		3.45	10			7106	174638
Gamma	3		3.24	14			7178	176437
Delta^1+			5.58	11			7131	175426
Delta^2	4		4.30	12			7139	175588
Epsilon^1+			4.66c	4			7051/2	173582/3
Epsilon^2	5		4.59c	5			7053/4	173607/8
Zeta^1+	5		4.36	6			7056	173648
Zeta^2			5.73	7			7057	173649
Eta	5		4.39	20			7298	180163
Theta	5		4.36	21			7314	180809
Iota	5		5.28	18			7262	178475
Kappa	5		4.33	1			6872	168775
Lambda	6		4.93	15			7192	176670
Mu	6		5.12	2			6903	169702
Nu^1+			5.91	8			7100	174585
Nu^2	6		5.25	9			7102	174602

The Lost, Missing, or Troublesome Stars of Lyra

Delta¹, 11, HR 7131, 5.58V, *SA, HRP, BS,* BV 11, H 33*.
Delta², 12, HR 7139, 4.30V, *SA, HRP, BS,* BV 12 as Delta, H 34*.

Bayer noted that Delta was a single star of the 4th magnitude. Flamsteed designated these two nearly adjacent stars, his 11 and 12, as Delta-1 and Delta-2. Bevis, however, labeled only the brighter star, 12, as Delta. Argelander, seeing only 12, agreed with Bevis (*UN,* p. 37), but Heis saw both stars and restored Delta to 11. Heis noted that with good viewing conditions the pair could be split with the naked eye.

Epsilon¹, 4, HR 7051/2, 5.06V, 6.02V, comb. mag. 4.66V, *SA, HRP* as Epsilon, *BS,* BV 7 as Epsilon, BF 2533 as Epsilon, BAC 6390*, H 21 as Epsilon.
Epsilon², 5, HR 7053/4, 5.14V, 5.37V, comb. mag. 4.59V, *SA, HRP* as Epsilon, *BS,* BV 6 as Epsilon, BF 2534, BAC 6391*, H 22 as Epsilon.

Bayer described Epsilon as a single 5th-magnitude star. Although 4 and 5 are nearly adjacent, Flamsteed designated only 4 as Epsilon. Baily at first accepted Flamsteed's designation, but he later designated 4 as Epsilon¹ and 5 as Epsilon². Stars 4 and 5 appeared as one image to both Bevis and Argelander (*UN,* p. 37), but Heis noted that in a clear sky the pair could be split with the naked eye. The use of a telescope, however, reveals that each star in the pair is itself a double. This is the famous Double Double. The combined magnitude of all four stars is 3.83V.

Zeta¹, 6, HR 7056, 4.36V, *SA, HRP* as Zeta, *BS,* BV 4 as Zeta, BF 2535 as Zeta, BAC 6392*, H 23 as Zeta.
Zeta², 7, HR 7057, 5.73V, *SA, HRP* as Zeta, *BS,* BV 5, BF 2536, BAC 6394*, H 23 as Zeta.

Bayer described Zeta as a single star of the 5th magnitude. Although 7 is adjacent to 6, Flamsteed and Bevis designated only the brighter star, 6, as Zeta. Baily at first agreed, but he later designated 6 and 7 as Zeta¹ and Zeta². Both Argelander (*UN,* p. 37) and Heis noted that 6 and 7 appear as one image to the naked eye. They have a combined magnitude of 4.06V.

Nu¹, 8, HR 7100, 5.91V, *SA, BS,* HF 8, BV 9, BF 2545*.
Nu², 9, HR 7102, 5.25V, *SA, HRP* as Nu, *BS,* HF 9 as Nu, BV 8 as Nu, BF 2546*, H 27 as Nu.

Bayer described Nu as a single 6th-magnitude star. Halley's 1712 edition of Flamsteed's catalogue labels 9 as Nu, but Flamsteed's 1725 catalogue designates these two neighboring stars, his 8 and 9, as Nu-1 and Nu-2. Bevis, though, labeled only the brighter star, 9, as Nu. Argelander, unable to see 8, also equated 9 with Nu (*UN,* p. 37).

Xi

Omicron

These two letters are not used in Lyra to identify stars.

Pi, 13, HR 7157, 4.04V, BV 16, BF 2563.
Rho, 16, HR 7215, 5.01V, BV 17, BF 2579.

Bayer did not letter Pi and Rho. Although he included these stars in his atlas, he considered them *informes* between Lyra and Draco and left them unlettered. Flamsteed assigned them Pi and Rho, but Bevis and Baily removed these letters as superfluous (see i Aql). Flamsteed skipped Xi and Omicron because Bayer had used these two letters to describe two neighboring constellations, Cygnus and Hercules. For Bayer's use of Greek letters and upper-case Roman letters for miscellaneous objects on his charts, see P Cyg.

x, 1, HR 6872, 4.33V, HF 1 as Kappa, BV 1 as Kappa, B 11 as Kappa, BF 2498 as Kappa.

Flamsteed's 1725 "Catalogus Britannicus" designates this star as x, a copying or typographical error for Kappa — they are similar in appearance, κ, x — that Bevis and Bode corrected. The star is properly identified as Kappa in Halley's edition of Flamsteed's catalogue.

Mensa, Men

Table, Table Mountain

Lacaille devised and lettered the constellation Mensa (figures 2c and 2d) from Alpha to Lambda. He named it Mons Mensae in honor of Table Mountain, which overlooks Cape Town, where he set up his small, 8-power telescope to observe the southern skies. (Mensae is the Latin genitive or possessive of Mensa.) Lacaille drew Mons Mensae on his chart directly under the Large Magellanic Cloud. As he recounted in the introduction to his preliminary catalogue, whenever a turbulent southeast windstorm approached the Cape, a white cloud, like a tablecloth (*nappe*), settled over Table Mountain. He also noted that navigators often referred to the two Magellanic Clouds as Cape Clouds. Consequently, he felt it was most appropriate to honor Table Mountain with a place in the heavens underneath one of the Magellanic Clouds. Unlike the other constellations that Lacaille devised, Mons Mensae, with its overhanging cloud cover, resembles its namesake.

See Lacaille, "Table des Ascensions," p. 589.

Table 58. The lettered stars of Mensa

	Magnitudes			Catalogue Numbers				
Letter	Bayer	Lacaille	Visual	Flamsteed	Lacaille	Other	HR	HD
Alpha		6	5.09		515		2261	43834
Beta		6	5.31		400		1677	33285
Gamma		6	5.19		451		1953	37763
Delta		6	5.69		364		1426	28525
Epsilon		6	5.53		690		2919	60816
Zeta		6	5.64		606		2559	50506
Eta		6	5.47		395		1629	32440
Theta		6	5.45		638		2689	54239
Iota		6	6.05		469		1991	38602
Kappa+			5.47			G 32	2125	40953
Lambda		6	6.53		474		2062	39810
Mu+			5.54			G 14	1541	30612
Nu+			5.79			G 8	1456	29116
Xi+			5.85			G 21	1716	34172
Pi+			5.65			G 29	2022	39091

The Lost, Missing, or Troublesome Stars of Mensa

Kappa, HR 2125, 5.47V, *SA, HRP, BS,* CA 496, BR 1150, L 2210, G 32*.
Mu, HR 1541, 5.54V, *SA, HRP, BS,* CA 378, BR 801, L 1654, BAC 1502, G 14*.
Nu, HR 1456, 5.79V, *SA, HRP, BS,* CA 375, BR 764, L 1639, G 8*.
Xi, HR 1716, 5.85V, *SA, HRP, BS,* CA 429, BR 955, L 1921, G 21*.
Pi, HR 2022, 5.65V, *SA, HRP, BS,* CA 481, BR 1096, L 2138, G 29*.

Although Lacaille observed and catalogued these five stars in Mensa, he left them unlettered. Gould designated them Kappa, Mu, Nu, Xi, and Pi. He usually did not letter stars of their dimness, but he felt they merited letters because of their proximity to the southern celestial pole. A note to L 2210 erroneously states that it is synonymous with CA 495, which is Delta Pic. The note should probably refer to L 2205. Both Lacaille and Brisbane described CA 378 as being in the Large Magellanic Cloud, but this is a mistake. See the note to BAC 1502.

Omicron

This letter is not used in Mensa.

Microscopium, Mic

Microscope

Lacaille devised the constellation Microscopium (Figure 2b) and lettered its stars from Alpha to Iota. He described it as an ordinary microscope of his time.

The simple microscope — nothing more than a single lens, somewhat like a magnifying glass — had been known to the ancient Greeks. The compound microscope, on the other hand, consisted of two converging lenses, one convex and the other concave. Scholars believe it was developed by Sacharias Janssen, a lens maker of Middelburg, Holland. But it was Galileo who first used it for scientific purposes — to examine the bodies of insects. At about the same time, vast improvements were made in the construction of simple — or single lens — microscopes. Anton van Leeuwenhoek (1632-1723) used a simple microscope to systematically examine minute creatures, but his instruments were more than just magnifying glasses. They were constructed with brass screws, metal plates, and adjustable eyepieces. He built special microscopes for special purposes. For example, to examine a tadpole, he placed the animal in a glass tube containing water and attached a lens to the tube. Leeuwenhoek, who had no formal medical or scientific education, discovered single-cell creatures like Protozoa and was the first to see bacteria. He used his microscope to study the circulation of the blood and concluded, "an artery and a vein are one and the same vessel prolonged and extended." He and his contemporaries like Marcello Malpighi (1628-94) and Jan Swammerdam (1637-82) inaugurated the field of microbiology. Like the telescope that was invented at about the same time, the microscope marked another significant phase of the Scientific Revolution that was sweeping Western Europe in the seventeenth and eighteenth centuries.

See Wolf, *History of Science,* I, 71-75; II, 417-24; Boorstin, *The Discoverers,* chap. 42. For Janssen's reputed role in the development of the telescope, see Chi and h Per.

Table 59. The lettered stars of Microscopium

	Magnitudes			Catalogue Numbers				
Letter	Bayer	Lacaille	Visual	Flamsteed	Lacaille	Other	HR	HD
Alpha		5	4.90		1689		7965	198232
Beta		6	6.04		1693		7979	198529
Gamma+		6	4.67	1 PsA	1701		8039	199951
Delta		6	5.68		1712		8070	200718
Epsilon+		6	4.71	4 PsA	1722		8135	202627
Zeta		6	5.30		1704		8048	200163
Eta		6	5.53		1709		8069	200702
Theta^1+		6	4.82		1723		8151	203006
Theta^2		6	5.77		1729		8180	203585
Iota		6	5.11		1683		7943	197937
Nu+			5.11			G 6	7846	195569

The Lost, Missing, or Troublesome Stars of Microscopium

Gamma, 1 PsA in Microscopium, HR 8039, 4.67V, *SA, HRP, BS,* CA 1701*, BR 6957*, L 8639 in PsA, BAC 7280 in PsA, G 39*.

Epsilon, 4 PsA in Microscopium, HR 8135, 4.71V, *SA, HRP, BS,* CA 1722*, BR 7002*, L 8761 in PsA, BAC 7386 in PsA, G 62*.

Lacaille formed Microscopium from the stars between Sagittarius and Piscis Austrinus. Although Baily included these two stars, Gamma and Epsilon, in the Southern Fish as Flamsteed intended, Gould considered them to be in the Microscope. When definitive constellation boundaries were established in 1930, the International Astronomical Union confirmed the decision made by Lacaille and Gould.

Theta¹, HR 8151, 4.82V, *SA, HRP, BS,* CA 1723 as Theta, BR 7010*, L 8773*, G 65*.
Theta², HR 8180, 5.77V, *SA, HRP, BS,* CA 1729 as Theta, L 8793*, G 67*.

Lacaille designated these two neighboring stars Theta.

Nu, HR 7846, 5.11V, *SA, HRP, BS,* CA 1673 as Nu Ind, BR 6876 as Nu Ind, L 8472, G 6*.

Lacaille did not letter this star Nu in Microscopium. In the constellation Indus, he mistakenly designated two stars as Nu Ind, this star and CA 1793. Gould included this star in the Microscope and relabeled it Nu Mic. See Nu Ind.

Monoceros, Mon

Unicorn

Monoceros (Figure 55) is a modern constellation that appeared for the first time as Monoceros Unicornis on Petrus Kaerius's globe of 1613, which Warner felt was the work of Petrus Plancius. Plancius was a fervent Calvinist minister who bitterly assailed Arminians and others in the Netherlands who sought to liberalize the basic tenets of Calvinist theology. He was also a geographer and astronomer who tried to depaganize the heavens by introducing Biblical figures and symbols as constellations.

Like many of his contemporaries, including Jacob Bartsch, Plancius believed that the monoceros or unicorn is mentioned frequently in the Bible (Deuteronomy 33:17; Numbers 23:22; Numbers 24:8, and *passim*). As a matter of fact, it is not mentioned in the Bible at all. The references cited in the Old Testament are to the *re'em* (ראם), a now-extinct multi-horned beast of enormous proportions that roamed the Middle East. According to rabbinic tradition, the *re'em* was so huge it did not fit into Noah's Ark but had to be lashed to its side. When the Septuagint, the Greek translation of the Hebrew Bible, was prepared in the third century B.C., reputedly by seventy rabbis in seventy days, *re'em* was mistakenly translated into Greek as *monocerotos* (μονοκερωτος), single horned. It was this translation that led Plancius to believe that Monoceros was a Biblical creature.

Other editors of the Bible copied the mistaken translation in the Septuagint and interpreted *monocerotos* in their own way. When St. Jerome, for instance, translated the Old Testament into Latin in the famous Vulgate Edition, he referred to the *re'em* as *rhinoceros,* nose-horned, while the editors of the King James Version transformed the monstrous *re'em* into an elegant unicorn.

The apparent mistranslation in the Septuagint is traditionally ascribed to the haste of the rabbis in translating the Old Testament from Hebrew to Greek. They were required to finish their work in just seventy days. But another possibility exists for the difference between the authorized Hebrew version of the Old Testament and the Septuagint. The rabbis in Hellenistic Egypt may have been using non-Masoretic texts, which might have differed somewhat from the authorized or standard scriptures prepared by the Masoretes or scribes.

Although the history of the constellation Monoceros can be traced to the Biblical *re'em,* its origin may reach even farther back into the past. Among the Sumerians of ancient Mesopotamia, the *re'em* was known as the *am,* the mighty wild bull (Akkadian *remu* or *rimu).* In Sumerian mythology, *am* became $GU_4.AN.NA$, the Bull of Heaven, which, in turn, became Taurus. Strange as it may seem, the constellations Monoceros and Taurus share the distinction of being derived from the same source, the legendary Wild Bull or Ox of the ancient Near East.

Baily proposed to letter all of Monoceros's stars brighter than magnitude 4.5. He noted that three stars met this criterion:

5, HR 2227, 3.98V, BAC 2015.
22, HR 2714, 4.15V, BAC 2358.
26, HR 2970, 3.93V, BAC 2542 as Gamma.

But he lettered only one star, 26, Gamma. Gould noted there were five stars in Monoceros brighter than magnitude 5 and he lettered these from Alpha to Zeta in order of their magnitude. In doing so, he relettered Baily's Gamma to Alpha. All of Gould's lettered stars are listed in the *BS,* and there are no missing or problem lettered stars in the constellation.

See Warner, *Sky Explored,* p. 204; Bartsch, *Usus Astronomicus,* p. 64; Freedman and Simon, *Midrash Rabbah,* I, 247; Fish, "The Zu Bird," pp. 163-64 and especially n. 1, p. 164; Reiner and Pingree, *Baby. Plan. Omens,* Part 2, p. 12; Gould, *Uran. Arg.,* p. 95. For Plancius's efforts to depaganize the heavens, see Camelopardalis and Columba; for the possibility that the *re'em-remu* may have been a prehistoric rhinoceros, see Barton, "Traces of the Rhinoceros in Ancient Babylon," pp. 92-95, wherein the author bases his argument to a large extent on a disputed translation of the Gudea Cylinder found in Lagash (about 200 kilometers north of modern Basra) that dates from about the twenty-first century B.C.; for a possible explanation as to why Baily omitted Alpha and Beta, see Leo Minor.

Figure 55. Hevelius's Monoceros. The horizontal line across the center of the chart is the celestial equator; the line forming the top border of the chart is the ecliptic. The line 8° below the top border is the southern limit of the zodiac. The path extending from above Orion's club, through Monoceros, and beyond Canis Major is the Via Lactea, the Milky Way.

Table 60. The lettered stars of Monoceros

	MAGNITUDES			CATALOGUE NUMBERS				
Letter	Bayer	Lacaille	Visual	Flamsteed	Lacaille	Other	HR	HD
Alpha			3.93	26		G 149	2970	61935
Beta			3.74c	11		G 48	2356/7/8	45725/6/7
Gamma			3.98	5		G 16	2227	43232
Delta			4.15	22		G 118	2714	55185
Epsilon			4.30c	8		G 25	2298/9	44769/70
Zeta			4.34	29		G 164	3188	67594

Musca, Mus

Fly

Keyzer and Houtman devised the constellation Musca (Figure 56). In Houtman's catalogue of 1603, it is called De Vlieghe, the Fly, and in William Jansen Bleau's globe of the same year, it is translated into Latin as Musca, the Fly. Bayer depicted it on his chart of the southern skies as Apis, the Bee, but Bartsch called it Musca. Kepler's Rudolphine Tables of 1627 refer to it as "Apis, Musca." Lacaille called it Musca and lettered its stars from Alpha to Kappa. This constellation should not be confused with the asterism the Northern Fly located just above the Ram.

See Bartsch, *Usus Astronomicus*, p. 66. For the Northern Fly, see c Ari; for a possible explanation for the differences in nomenclature of the constellations devised by Keyzer and Houtman, see Apus.

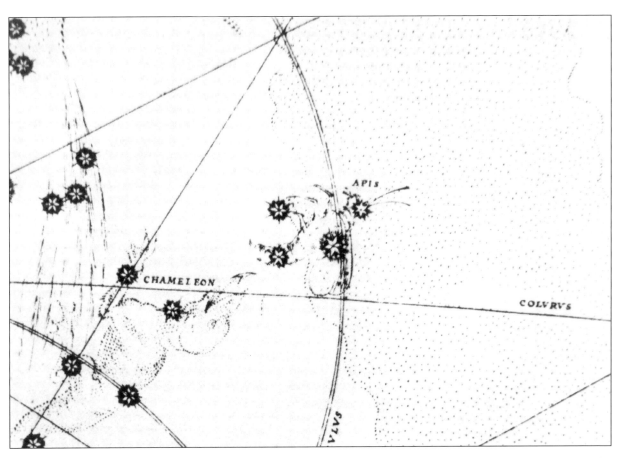

Figure 56. Bayer's Musca. This is an enlarged view of Musca from Bayer's Tabula XLIX, where it is drawn and labeled as Apis, the Bee.

Table 61. The lettered stars of Musca

	MAGNITUDES			CATALOGUE NUMBERS				
Letter	Bayer	Lacaille	Visual	Flamsteed	Lacaille	Other	HR	HD
Alpha		4	2.69		1092		4798	109668
Beta		4	3.05		1104		4844	110879
Gamma		4	3.87		1090		4773	109026
Delta		4	3.62		1119		4923	112985
Epsilon		6	4.11		1072		4671	106849
Zeta¹+		6	5.74		1078		4704	107567
Zeta²			5.15			BAC 4162	4703	107566
Eta		6	4.80		1135		4993	114911
Theta		6	5.51		1125		4952	113904
Iota¹+		6	5.05		1142		5042	116244
Iota²		6	6.63		1146		5051	116579
Kappa+		6	6.59		1152		5125	118522
Lambda+			3.64			G 15	4520	102249
Mu+			4.72			G 16	4530	102584

The Lost, Missing, or Troublesome Stars of Musca

Zeta¹, HR 4704, 5.74V, *SA, HRP, BS,* CA 1078 as Zeta, BR 4019 as Zeta, L 5113, BAC 4161*, CAP 2215*, G 39*.
Zeta², HR 4703, 5.15V, *SA, HRP, BS,* CA 1077, BR 4017, L 5112, BAC 4162*, CAP 2214*, G 38*.

Lacaille described only one star as Zeta, CA 1078. In the *BAC*, Baily designated these two stars, CA 1078 and CA 1077, as Zeta¹ and Zeta², despite the fact that they are far apart — about 1° in declination. He thought that Lacaille intended to designate the brighter star, CA 1077, as Zeta. In other words, Baily assigned Zeta with superscripts to both stars — to CA 1077 since it was brighter than CA 1078 and he thought Lacaille originally meant to designate this star as Zeta; and to CA 1078 since it was the star Lacaille actually designated as Zeta and he was hesitant to remove one of Lacaille's Greek letters.

In preparing the *BAC*, Baily erred about 2^s in calculating CA 1077's position. The *BAC* lists this star's right ascension as $12^h13^m52.05^s$ in 1850.0. Maclear's *Cape Catalogue of 1850* points out that it should be $12^h13^m50.12^s$. As a result of this error, Baily mistakenly figured that CA 1078 preceded CA 1077 and therefore designated the former as Zeta¹ and CA 1077 Zeta². Most later astronomers have adhered to these indices to avoid causing confusion between the two stars.

Iota¹, HR 5042, 5.05V, *SA, HRP, BS,* CA 1142 as Iota, BR 4423*, L 5486, BAC 4465*, G 66*.
Iota², HR 5051, 6.63V, *SA, HRP, BS,* CA 1146 as Iota, BR 4438*, L 5504, BAC 4476*, G 68*.

Lacaille designated these two nearly adjacent stars Iota.

Kappa, HR 5125, 6.59V, CA 1152*, BR 4552*, L 5595, BAC 4542*, R 349*.

Lacaille described Kappa as a star of the 6th magnitude, but Gould considered it fainter than 7th magnitude and did not include it in his *Uran. Arg.* Consequently, its letter has largely disappeared from the literature.

Lambda, HR 4520, 3.64V, *SA, HRP, BS,* CA 1037, BR 3756 in Centaurus, L 4883, G 15*.
Mu, HR 4530, 4.72V, *SA, HRP, BS,* CA 1040, BR 3778 in Centaurus, L 4899, G 16*.

Lacaille did not letter Lambda and Mu. Although he observed and catalogued these two stars, he considered them 6th-magnitude *informes,* unformed stars between Argo Navis and Centaurus. Baily included them in Musca, and Gould designated them Lambda and Mu because he felt that stars of their brightness merited letters.

Norma, Nor

Square, Architect's Square

Lacaille devised and lettered the stars in the constellation Norma (Figure 2b) from Alpha to Mu. In the first or preliminary edition of his catalogue, published in 1756, he called this constellation l'Équerre et la Règle (the Square and the Ruler). On the accompanying chart, he drew a square with a ruler and noted that these were the instruments of the architect. As a matter of fact, he dedicated this whole sector of the heavens to architecture. He went on to state that on his chart he purposely designed the figure of Triangulum Australe, a neighboring constellation devised earlier by Keyzer and Houtman, in the form of a level *(niveau)* with a lead weight dangling from its apex. This statement, coming right after his description of the Square and the Ruler, caused some astronomers to believe he meant to rename l'Équerre et la Règle to Niveau, the Level. On his chart accompanying his catalogue of 1763, the constellation's name is shortened in Latin simply to Norma, the Square.

See Lacaille, "Table des Ascensions," p. 589.

Table 62. The lettered stars of Norma

	Magnitudes			Catalogue Numbers				
Letter	Bayer	Lacaille	Visual	Flamsteed	Lacaille	Other	HR	HD
Alpha+		5	4.23		1371		6143	148703
Beta+		5	4.16			CA(P)1373	6166	149447
Gamma^1+		6	4.99		1350		6058	146143
Gamma^2		5	4.02		1351		6072	146686
Delta		5	4.72		1332		5980	144197
Epsilon		6	4.47		1365		6115	147971
Zeta		6	5.81		1339		6019	145361
Eta		6	4.65		1329		5962	143546
Theta		6	5.14		1349		6045	145842
Iota^1+		6	4.63		1324		5961	143474
Iota^2		6	5.57		1333		5994	144480
Kappa		6	4.94		1342		6024	145397
Lambda		6	5.45		1353		6071	146667
Mu		6	4.94		1373		6155	149038

The Lost, Missing, or Troublesome Stars of Norma

Alpha, in Scorpius, HR 6143, 4.23V, CA 1371*, BR 5747*, L 6859 in Scorpius, G 72 in Scorpius.

Although Lacaille lettered this star Alpha Nor, Baily included it in Scorpius and left it unlettered. Gould relettered it N Sco because he felt a star of its brightness merited a letter. See N Sco.

Beta, in Scorpius, HR 6166, 4.16V, CA(P) 1373*, CA 1376, BR 5767, L 6890 in Scorpius, G 76 in Scorpius.

Lacaille inadvertently omitted this letter from his 1763 catalogue. His map of the southern skies, however, shows Beta Nor, and it is also included as Beta in his earlier preliminary catalogue of 1756 ("Table des Ascensions," p. 573). Baily included it unlettered in Scorpius. Gould designated it H Sco because he felt a star of its brightness merited a letter. See H Sco.

Gamma^1, HR 6058, 4.99V, *SA, HRP, BS,* CA 1350 as Gamma, BR 5655*, L 6746*, G 46*.
Gamma^2, HR 6072, 4.02V, *SA, HRP, BS,* CA 1351 as Gamma, BR 5675*, L 6764*, G 49*.

Lacaille designated these two neighboring stars Gamma.

Iota^1, HR 5961, 4.63V, *SA, HRP, BS,* CA 1324 as Iota, BR 5559*, L 6615*, G 28*.
Iota^2, HR 5994, 5.57V, *SA, HRP, BS,* CA 1333 as Iota, BR 5590*, L 6665*, G 36*.

Lacaille designated these two neighboring stars Iota.

O*CTANS*, O*CT*

Octant

Lacaille devised and lettered the stars of the constellation Octans (figures 2c and 2d) from Alpha to Upsilon. Gould and other astronomers lettered additional stars, despite their faintness, because of their proximity to the southern celestial pole. In his planisphere of 1752, Lacaille designated this constellation l'Octans de Reflexion, but he later shortened it in Latin to Octans. It represents John Hadley's sea octant with its small telescope and two mirrors, one fixed and the other moveable, to enable a seaman "to shoot" a star or the sun and the horizon at the same time. The user adjusted the octant until the desired object was tangent to the horizon, at which time its latitude could be read from the scale engraved along the side of the instrument. This navigational instrument, the forerunner of the modern marine sextant, should not be confused with Sextans Uraniae, the large Astronomical Sextant devised by Hevelius. See Sextans.

Lost Stars

Table 63. The lettered stars of Octans

		Magnitudes			Catalogue Numbers			
Letter	Bayer	Lacaille	Visual	Flamsteed	Lacaille	Other	HR	HD
Alpha		5	5.15		1686		8021	199532
Beta		5	4.15		1811		8630	214846
Gamma^1+		5	5.11		1917		9032	223647
Gamma^2		5	5.73		1921		9061	224362
Gamma^3		5	5.28		1940		30	636
Delta		5	4.32		1190		5339	124882
Epsilon		6	5.10		1778		8481	210967
Zeta		6	5.42		876		3678	79837
Eta		6	6.19		999		4312	96124
Theta		6	4.78		1928		9084	224889
Iota		6	5.46		1105		4870	111482
Kappa		6	5.58		1140		5084	117374
Lambda		6	5.29		1732		8280	206240
Mu^1+		6	6.00		1661		7863	196051
Mu^2		6	6.55		1664		7864	196067
Nu+		6	3.76		1735/6		8254	205478
Xi+			5.35			G 77	8663	215573
Omicron+			7.21			R 12		1348
Pi^1+		6	5.65		1224		5525	130650
Pi^2		6	5.65		1229		5545	131246
Rho		6	5.57		1261		5729	137333
Sigma		6	5.47		1275		7228	177482
Tau		6	5.49		1828		8862	219765
Upsilon		6	5.77		1754		8505	211539
Phi+			5.47			G 33	6829	167468
Chi+			5.28			G 30	6721	164461
Psi+			5.51			G 69	8471	210853
Omega+			5.91			G 23	5557	131596
B+			6.57			CAP 4030	8294	206553

The Lost, Missing, or Troublesome Stars of Octans

Gamma^1, HR 9032, 5.11V, *SA, HRP, BS,* CA 1917 as Gamma, BR 7334*, L 9607*, G 86*.
Gamma^2, HR 9061, 5.73V, *SA, HRP, BS,* CA 1921 as Gamma, BR 7350*, L 9651*, G 87*.
Gamma^3, HR 30, 5.28V, *SA, HRP, BS,* CA 1940 as Gamma, BR 5*, L 9756*, G 1*.

Lacaille designated these three neighboring stars Gamma.

Mu^1, HR 7863, 6.00V, *SA, HRP, BS,* CA 1661 as Mu, BR 6870*, L 8435*, BAC 7068*, G 50*.
Mu^2, HR 7864, 6.55V, *SA, HRP, BS,* CA 1664 as Mu, L 8443*, BAC 7075, G 51*.

Lacaille designated these two neighboring stars Mu. Brisbane's catalogue does not include CA 1664, and the *BAC* asserts that the star is not at the coordinates listed by Lacaille. Gould, however, apparently had no difficulty in identifying this star.

Nu, HR 8254, 3.76V, *SA, HRP, BS,* CA 1735/6*, BR 7037*, L 8817/8 as Nu^1, Nu^2, BAC 7481*, G 60*.

Lacaille listed Nu as a very close pair, but Brisbane could find no evidence of duplicity, equating his BR 7037 to CA 1735. Nevertheless, Baily considered it double and equated BAC 7481 to CA 1735/6. A generation later, Gould also asserted that Nu had no duplicity. Oddly enough, in recent years it has been discovered to be a spectroscopic binary.

Xi, HR 8663, 5.35V, *SA, HRP, BS,* CA 1819, BR 7201, L 9202, G 77*.

Lacaille did not letter this star. Although he observed and catalogued it, he left it unlettered. Gould designated it Xi because of its brightness, greater than 6th magnitude, and because of its proximity to the south pole.

Omicron, HD 1348, 7.21V, *SA,* R 12*, BR 32*, L 260.

Although Lacaille observed and catalogued this star, he left it unlettered. Charles (Karl) Rumker designated it Omicron and noted it was the southern pole star at the time of Lacaille's observations (*Prel. Cat.,* pp. 14-15). In 1750, Omicron was at -89° 45' 28" while Sigma (HR 7228, 5.47V, CA 1275, BR 5912, L 6295, G 34), the current pole star, was at -89° 1' 37". As the result of an error in calculating their positions, Lacaille mistakenly designated Sigma as the pole star instead of Omicron. Gould did not include Omicron in his catalogue since it was dimmer than 7th magnitude, the limiting magnitude for his *Uran. Arg.* (p. 62).

Pi^1, HR 5525, 5.65V, *SA, HRP, BS,* CA 1224 as Pi, BR 5022, BAC 4859, CAP 2614, G 21*.
Pi^2, HR 5545, 5.65V, *SA, HRP, BS,* CA 1229 as Pi, BR 5046, BAC 4883, CAP 2626, G 22*.

Some confusion surrounds these two stars. Lacaille designated CA 1224 and CA 1229, about 30' apart, as Pi. Neither Baily, Brisbane, Maclear, Stone, nor Taylor was able to locate them. When Baily reduced Lacaille's observations, he nevertheless included these two stars (L 6009, L 6019) in Lacaille's expanded catalogue and even in the *BAC* (BAC 4866 and BAC 4871), but with the warning that they were probably nonexistent. Rumker suggested that CA 1224 was synonymous with his R 391, which he lettered Pi, but Maclear, in the *Cape Catalogue of 1850,* equated R 391 with Brisbane's BR 5046. Gould sought to end the confusion once and for all. He asserted that Lacaille's coordinates for CA 1224 and CA 1229 were obviously erroneous, but that his G 21 and G 22 were the two stars Lacaille intended to designate as Pi. Although these two stars are about 11' apart, in contrast to 30' for CA 1224 and CA 1229 as noted by Lacaille, most authorities have accepted Gould's suggestion.

Phi, HR 6829, 5.47V, *SA, HRP, BS,* CA 1487, BR 6337, L 7559, G 33*.
Chi, HR 6721, 5.28V, *SA, HRP, BS,* BR 6058, L 7001, BAC 5936, CAP 3332, G 30*.
Psi, HR 8471, 5.51V, *SA, HRP, BS,* CA 1782, BR 7137, L 9022, G 69*.
Omega, HR 5557, 5.91V, *SA, HRP, BS,* G 23*.

The lettering of these four stars did not originate with Lacaille. Although he observed all off these stars except Omega, he left them unlettered since he considered them 6th magnitude or less. Since they were all near the southern pole and relatively bright, greater than magnitude 6, Gould assigned them letters.

B, HR 8294, 6.57V, *SA, HRP,* BR 6644, L 6460, CAP 4030*, G 54*.

Although Lacaille observed and catalogued this star, he left it unlettered. Maclear designated it B and included it in the *Cape Catalogue of 1850.* Gould retained its letter despite its faintness because of its proximity to the pole (*Uran. Arg.,* p. 80). Maclear labeled several faint stars around the south celestial pole with Roman capitals, but none except B has survived in the literature. B has survived primarily because Gould included it in his catalogue.

OPHIUCHUS, OPH

Ophiuchus, Serpent Bearer

Ophiuchus ("Snake Handler;" Figure 57) is a Ptolemaic constellation that Bayer lettered from Alpha to Omega and from A to f. Bayer and the classical writers of antiquity referred to the constellation as Asclepius, the Greek god of healing and the reputed ancestor of Hippocrates, the father of modern medicine. Although the constellation was not, as far as we can determine, known to the peoples of the ancient Near East, certain aspects of it can be traced back to Mesopotamia.

In the first known epic of mankind, the Mesopotamian story of Gilgamesh, after the death of his companion Enkidu, Gilgamesh is unable to come to grips with the ultimate and irreversible fate of all humans. Frightened for the first time in his life by the prospect of dying, Gilgamesh embarks on a search for Utnapishtim, the Mesopotamian Noah, the only human known to have been granted immortality by the gods. Gilgamesh hopes that Utnapishtim, a distant relative, will pass on to him the secret of everlasting life. After a long, arduous journey, he ultimately finds Utnapishtim, who agrees to provide him with the secret of the gods, the plant of rejuvenation. On his way home, elated by the promise of eternal life, Gilgamesh relaxes his guard. He places the plant on the ground and refreshes himself in a pool of cool water. When he returns, he discovers that the plant has vanished, stolen by a snake that, having been rejuvenated by the plant, leaves behind a tell-tale clue of its crime, its old skin. Just as the serpent in the Garden of Eden deprives Adam and Eve of eternal life, its Mesopotamian counterpart deprives Gilgamesh of rejuvenation.

The serpent's association with rejuvenation, healing, and long life is probably based on the fact that many species can slough off their old skin and appear fresh and rejuvenated. Moreover, the snake was associated in Sumerian mythology with Ningizzida (Ningishzida), the snake god, whose name translates as Lord of the Good Tree. The tree was an apt symbol for the snake god, for of all living things, it is usually associated with longevity, and its entwined roots, twisted and half buried near the surface, bear an uncanny resemblance to snakes. This connection between snakes and longevity and healing probably accounts for the adoption of the caduceus, the wand with its two entwined serpents, as the symbol of the medical profession.

As noted, the Greeks believed that the Snake Handler in the sky was the physician Asclepius, son of Apollo, the god of medicine. Asclepius was closely associated with snakes and was often depicted as a serpent. According to Greek mythology, he learns how to restore life to the dead, possibly from his snake companions or from the goddess Athena. Naturally, this alarms Hades, the god of the dead, who fears that if Asclepius continues his practice, he will eventually be king of a kingdom with no subjects. He appeals to Zeus, who slays Asclepius with a bolt of lightning fashioned by the Cyclopes. In retaliation, Apollo — the god of archery as well as of medicine — slays the Cyclopes with his arrows. At Apollo's request, Zeus agrees to place Asclepius's image among the stars and to grant him divinity.

See Staal, *New Patterns in the Sky,* pp. 195-98; Allen, *Star Names,* pp. 297-98; Ridpath, *Star Tales,* pp. 94-95; Lum, *The Stars in Our Heaven,* pp. 88-91; Pritchard, *Anc. Near East. Texts,* p. 96; Pritchard, *Anc. Near East Picts.,* plate 511; Van Buren, *Symbols of the Gods,* pp. 40-42, 146; Jacobsen, *Treasures of Darkness,* pp. 7, 207-8, 214, 217; Condos, *Star Myths,* pp. 141-45, 247-48; Gayley, *Classic Myths,* pp. 38, 104; Reinhold, *Essentials of Greek and Roman Classics,* p. 334. See Sagitta.

Lettered Stars - Ophiuchus

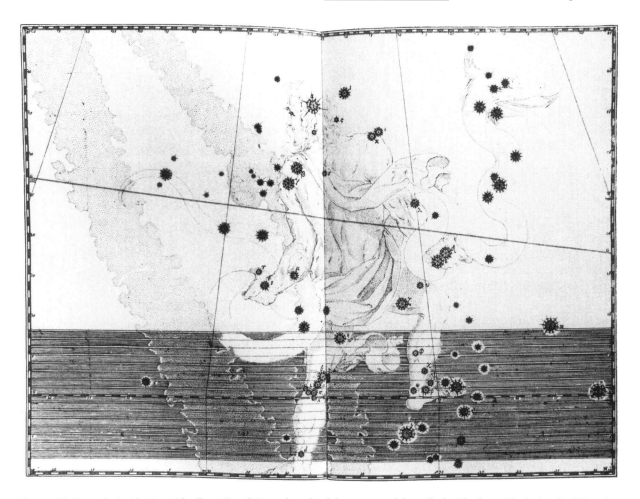

Figure 57. Bayer's Ophiuchus. The lines G and H are the celestial equator and the ecliptic. The letter I is the head of Hercules; K is the head of Sagittarius; L is the star Antares, Alpha Sco, the heart of the Scorpion; and M and N are the scales of Libra. The shaded area across the bottom is the zodiac.

Table 64. The lettered stars of Ophiuchus

	MAGNITUDES			CATALOGUE NUMBERS				
Letter	Bayer	Lacaille	Visual	Flamsteed	Lacaille	Other	HR	HD
Alpha	2		2.08	55			6556	159561
Beta	3		2.77	60			6603	161096
Gamma	3		3.75	62			6629	161868
Delta	3		2.74	1			6056	146051
Epsilon	3		3.24	2			6075	146791
Zeta	3		2.56	13			6175	149757
Eta	3		2.43	35			6378	155125
Theta	3		3.27	42			6453	157056
Iota	4		4.38	25			6281	152614
Kappa	4		3.20	27			6299	153210
Lambda	4		3.82	10			6149	148857
Mu	4		4.62	57			6567	159975
Nu	4		3.34	64			6698	163917
Xi	4		4.39	40			6445	156897
Omicron	4		5.15c	39			6424/5	156349/50
Pi+	4							
Rho	4		4.63c	5			6112/3	147933/4
Sigma	5		4.34	49			6498	157999
Tau	5		4.78c	69			6733/4	164764/5
Upsilon	5		4.63	3			6129	148367
Phi	5		4.28	8			6147	148786
Chi	5		4.42	7			6118	148184
Psi	5		4.50	4			6104	147700
Omega	5		4.45	9			6153	148898
A+	5		4.31c	36			6401/2	155885/6
b+	5		4.17	44			6486	157792
c+	5		4.81	51			6519	158643
d+	5		4.29	45			6492	157919
e+	6		5.03	66 Her			6433	156681
f+	6		5.81	53			6548	159480

The Lost, Missing, or Troublesome Stars of Ophiuchus

Pi

Bayer's stars in the legs of Ophiuchus are badly confused and do not conform to the constellation's actual appearance in the heavens, primarily because Bayer depended heavily on Tycho's expanded catalogue of 1,000 stars — actually 1,005 stars with one duplication. Unlike Tycho's earlier catalogue of 777 stars, the larger one was hastily put together with very little attention to detail or accuracy. As a result, there are numerous errors, especially in the southern section of Ophiuchus, where Tycho inserted a group of eight stars that, to this day, defy identification. See Baily, "Catalogues of Ptolemy *et al.*," pp. 161-62; *BF*, p. 402; Argelander, *Fide Uran.*, p. 14; Brahe, *Opera Omnia,* III, 364 and especially note, p. 416; Werner and Schmeidler, *Synopsis,* p. 355.

The most troublesome stars were those that Bayer lettered Theta, Xi, and Pi in the Serpent Bearer's left leg. Flamsteed and Baily tried to set matters right. They equated Theta with 42, Xi with 40, and eliminated Pi altogether since they felt that there was no star in the sky that conformed to its position in Bayer's *Uranometria* (BF 2362, BF 2372). On the other hand, Bode believed that Pi was 42, Theta was 40, and Xi, nonexistent. Gould agreed with Bode that Pi was 42, but for the sake of uniformity, he agreed to drop Pi, leave 42 synonomous with Theta, and 40 synonomous with Xi (G 124, G 126; *Uran. Arg.*, p. 93) as Flamsteed and Baily had proposed. (Flamsteed had erroneously designated 40 as Rho, but Baily relabeled it Xi.) In comparing the *Uranometria* with modern atlases, it appears that Bode's suggestions most nearly conform to Bayer's chart:

1. 42 should be Pi, not Theta (B 175 and his Tabula IX).
2. 40 should be Theta, not Xi (B 168).
3. Xi should be eliminated.

But most modern authorities have rejected Bode's suggestions and have adopted Flamsteed and Baily's proposals: Theta is 42, Xi is 40, and Pi is nonexistent.

It might be of interest to note that twenty-five years after completing his *Uranometria,* Bayer worked with Julius Schiller to prepare the latter's *Coelum Stellatum Christianum.* In this catalogue, Bayer made, with the help of Kepler, appropriate corrections in the southern section of Ophiuchus, which Schiller called St. Benedict. The problem stars were properly positioned and Xi eliminated, as Bode later suggested. See Schiller's combined constellation, Tabula XIII and XIV, St. Benedict and St. Benedict's Thorns, formerly Ophiuchus and his Serpent. Kepler helped to correct this constellation because of his interest in the Nova of 1604, Kepler's Star, which appeared in the southern section or legs of Ophiuchus. See Kepler, *De Stella Nova Serpentarii,* with its accurate map and star list of the constellation, in *Gesammelte Werke,* I, 225-26.

A, 36, HR 6401/2, 5.33V, 5.29V, comb. mag. 4.32V, *SA, HRP,* ADS 10417.

Bayer described A as a single star of the 5th magnitude.

b, 44, HR 6486, 4.17V, *SA, HRP,* B 195*, BF 2382*.

Flamsteed's catalogue mistakenly designates this star B, but Bode corrected the error.

c^1, 50, HD 158527, 7.4V, HF 47, B 216*, BF 2394*.
c^2, 51, HR 6519, 4.81V, HF 48 as C, *SA* as c, *HRP* as c, B 218*, BF 2395*, H 56, G 149 as c.

Bayer's *Uranometria* describes c as a single star of the 5th magnitude, and in Halley's 1712 edition of Flamsteed's catalogue, 51 is labeled C. In Flamsteed's 1725 catalogue, however, these two nearly adjacent stars, 50 and 51, are each designated e. This is one of the very few instances in which Flamsteed did not use indices; he joined 50 and 51 with a brace and equated them to Bayer's e, a typographical error, which Bode corrected by labeling them c-1 and c-2. Argelander saw only the brighter star, 51, but he left out its Bayer letter, as he did with 36, 39, 40, and 44 — A, Omicron, Xi, and b — because of the confusion surrounding the identities of these stars in Ophiuchus's left leg (*UN,* p. 73; see Pi). Gould, though, designated 51 as c. He omitted 50 from his *Uran. Arg.* since it was less than 7th magnitude.

d, 45, HR 6492, 4.29V, *SA, HRP,* BV 59*, B 197*, BF 2383*.

Flamsteed's catalogue omits d, but Bevis and Bode correctly equated d with Flamsteed's 45.

D, 58, HR 6595, 4.87V, BV 66, BF 2415.

Bayer did not letter this star. Flamsteed added the letter, but Bevis and Baily considered it superfluous and removed it. See i Aql.

e, 66 Her in Ophiuchus, HR 6433, 5.03V, *SA, HRP,* BV 38*, B 171*, BF 2377 as 66 Her, BAC 5841.

Flamsteed included this star in Hercules as 66, Omega Her, but Bevis switched it back to Ophiuchus. Baily also switched it back to Ophiuchus in the *BAC* without, though, designating it e. Bode included the star in both constellations. Bayer's Omega Her is not this star but 24 Her, HR 6117. See Omega Her.

f, 53, HR 6548, 5.81V, *SA, HRP,* BV 58*, B 241*, BF 2404*.

Flamsteed's catalogue omits f, but Bevis and Bode correctly equated f with Flamsteed's 53.

g, 5, HR 6112/3, 5.02V, 5.85V, comb. mag. 4.63V, BF 2245 as Rho, ADS 10049.

Bayer designated this star Rho. Flamsteed mistakenly designated it as g, but Baily corrected the error by relabeling it Rho.

K, 68, HR 6723, 4.45V, BF 2455.
n, 66, HR 6712, 4.64V, BF 2449.
o, 67, HR 6714, 3.97V, BF 2450.
p, 70, HR 6752, 4.03V, *SA,* BF 2465.
q, 73, HR 6795, 5.73V, BF 2472.
r, 74, HR 6866, 4.86V, BF 2491.
s, 71, HR 6770, 4.64V, BF 2467.
S, 72, HR 6771, 3.73V, BF 2468.

Bayer did not letter these eight stars. He included them, with the possible exception of 73, in his atlas, considered them *informes* outside the border of Ophiuchus, and left them unlettered. Flamsteed assigned them these letters, but Baily removed the letters as superfluous (see i Aql). Although all have disappeared from the literature, p has somehow managed to survive in the atlases of Becvar and Tirion, but not in the second edition of Tirion's *SA*.

ORION, ORI

Orion, Giant Hunter

Known in ancient Sumer as SIPA.ZI.AN.NA (Sibzi Anna), the True Shepherd of the Sky, Orion (Figure 58) was associated with Dumuzi or Tammuz, a shepherd god who represented the power of fertility in the mother's womb. Tammuz was also the god of bees and grain, and of sap that rises in trees and plants. In essence, he embodied the new life that was reborn in nature every spring. In the late spring, when the milking and lambing season drew to a close and the deadly summer sun began scorching plant life and turning green fields to desert brown, Tammuz was reputed to have perished, just as his image in the sky began sinking below the western horizon at sunset. This initiated a time of mourning, and women bemoaned the loss of Tammuz, who, as the god of the power of fertility, was so important in their lives. The cult of Tammuz spread throughout the Middle East and reached even to the very gates of the Holy Temple in Jerusalem, where the prophet Ezekiel (8:14) laments, "there sat women weeping for Tammuz."

The symbolic death of Tammuz heralded a period of general mourning throughout the ancient Near East, and the fourth month (June-July) was appropriately named in his memory, Duzu or Duuzu in Babylonian and Tammuz in Hebrew. During this period the astral image of Tammuz vanished completely from the night sky, neither visible in the west at sunset nor in the east at sunrise. It is perhaps not unrelated that Jewish tradition ascribes to the 17th of Tammuz numerous mournful events: Moses' destruction of the first set of Tablets on Mt. Sinai; and the Babylonian breach of the walls of Jerusalem in 586 B.C and later, the Roman breach in A.D. 70 with the concomitant destruction of the First and Second Temples. For Jews, the three weeks from the 17th of Tammuz to the 9th (Tisha B'Av) of the following month, Av (Babylonian Abu), is a time of sorrow and mourning. No marriages, parties, celebrations, or festivals are scheduled. Special rites and ceremonies are followed that express intense sorrow and grief. Marking numerous catastrophes and adversities, it is a period of sadness for the Jews, as it was for many of the peoples of the ancient Near East.

In addition to Tammuz, this constellation was also associated with the Canaanite legend of Aqhat, the hunter, whose story was found on several clay tablets uncovered at Ugarit (see Gemini). Aqhat obtains possession of a divine bow that was designed for the gods, perhaps one of the newly developed composite bows of wood and bone. Accidentally, Aqhat is killed by one of the followers of the virgin goddess Anat, queen of the chase, who desires his bow. Although this version of the story ends with the death of Aqhat, he is probably restored to life — at least for part of the year, during the growing season — just as his image in the heavens reappears each year. Near at hand in the sky to Aqhat, the ancients located his fateful bow, the Bow-Star, Sirius, in Canis Major.

The legends of Tammuz and Aqhat are both accounts of dying gods who, like their astral representation Orion, disappear in the late spring only to be resurrected a month or so later. This was a common theme throughout the ancient Near East and may even be the source of the Egyptian legend of the dying god Osiris, who is reborn annually along with the rising waters of the Nile that carry the hope of new life each year.

The peoples of Mesopotamia placed Orion in the heavens confronting Taurus. The Faithful Shepherd of the Sky with his upraised club seems ready to do battle with the Bull of Heaven, traditionally associated with plowing and farming. This heavenly struggle is perhaps meant to symbolize the eternal earthly struggle that pits Cain against Abel, the farmer against the herder, the homesteader against the cattleman

The classic Greek and Roman myths associated with Orion are closely related to the story of Aqhat. According to one account, Orion hunts in the company of Artemis (Roman Diana), goddess of the hunt and of virginity, and her mother Leto. In a moment of extreme hubris, Orion explains that he intends to slay all the beasts on Earth. When Mother Earth, Gaea, hears of his boast, she sends a deadly scorpion to kill him. The gods commemorate this event by placing both the Mighty Hunter and the Scorpion among the stars. In a variation of the legend, Artemis is so smitten with Orion that she plans to give up her virginity to marry him. Her brother Apollo, who is opposed to the union, tricks Artemis into killing Orion with her own bow and arrow.

Many other myths relate to Orion, including one that describes how he obtained his name. According to the poet Ovid, Hyrieus, son of Poseidon, god of the sea, entertains his father, Zeus, and Hermes. So pleased are the guests with the feast that they offer to grant their host one wish. Although unmarried, Hyrieus requests a son. The gods respond by spilling their semen on the skin of the animal upon which they had all just feasted. They direct Hyrieus to bury the skin in the earth for ten months, after which time a baby — Orion — will emerge. In a variation of this legend related by the poet Hyginus, the gods urinate on the hide, hence the name Urion ("urine-born"), later modified to Orion.

A Ptolemaic constellation, Orion is included in Bayer's *Uranometria* and lettered from Alpha to Omega and from A to p.

See van der Waerden, "Babylonian Astronomy. II. The Thirty-Six Stars," pp. 7, 13-14; van der Waerden, "History of the Zodiac," p. 219; Jacobsen, *Toward the Image of Tammuz,* pp. 24-25, 28-29, 73-103, 364 note 32; Gaster, *Thespis,* pp. 12-17, 65, 257-313; Pritchard, *Anc. Near East. Texts,* pp. 41-42, 149-55; Langdon, *Semitic Mythology,* p. 178; Langdon, *Babylonian Menologies,* pp. 87-88; Jacobsen, *Treasures of Darkness,* chap. 2; Reiner and Pingree, *Baby. Plan. Omens,* Part 2, p. 14; Condos, *Star Myths,* pp. 147-50; Grantz, *Early Greek Myth,* p. 273. For a possible connection between Tammuz and Osiris, see Frankfort, *Kingship and the Gods,* chap. 20; for the Bull's association with plowing and farming, see Taurus and Aries; for Orion's association with Canis Major, Canis Minor, and Lepus, see Lepus.

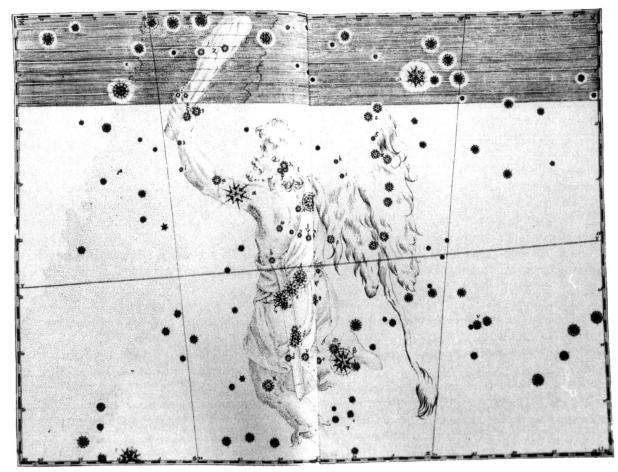

Figure 58. Bayer's Orion. The letter Q is the 1st-magnitude star Aldebaran, Alpha Tau, the Bull's Eye; R is the feet of Gemini; S, at the bottom left edge of the chart, is Canis Major; T is Lepus; V is Eridanus; and W, at the top left edge of the chart, is the star H Gem. The lines X and Y are the ecliptic and the celestial equator. The shaded area across the top is the zodiac.

Table 65. The lettered stars of Orion

		MAGNITUDES			CATALOGUE NUMBERS			
Letter	Bayer	Lacaille	Visual	Flamsteed	Lacaille	Other	HR	HD
Alpha	1		0.50	58			2061	39801
Beta	1		0.12	19			1713	34085
Gamma	2		1.64	24			1790	35468
Delta	2		2.46c	34			1851/2	36485/6
Epsilon	2		1.70	46			1903	37128
Zeta	2		1.76c	50			1948/9	37742/3
Eta	3		3.36	28			1788	35411
Theta^1+	3		4.85c	41			1893/4/5/6	37020/1/2/3
Theta^2			4.89c	43			1897	37041/2/62
Iota	3		2.77	44			1899	37043
Kappa	3		2.06	53			2004	38771
Lambda	4		3.39c	39			1879/80	36961/2
Mu	4		4.12	61			2124	40932
Nu	4		4.42	67			2159	41753
Xi	4		4.48	70			2199	42560
Omicron^1+	4		4.74	4			1556	30959
Omicron^2	4		4.07	9			1580	31421
Pi^1+	4		4.65	7			1570	31295
Pi^2	4		4.36	2			1544	30739
Pi^3	4		3.19	1			1543	30652
Pi^4	4		3.69	3			1552	30836
Pi^5	4		3.72	8			1567	31237
Pi^6	4		4.47	10			1601	31767
Rho+	4		4.46	17			1698	33856
Sigma	4		3.81	48			1931	37468
Tau	4		3.60	20			1735	34503
Upsilon	4		4.62	36			1855	36512
Phi^1+	5		4.41	37			1876	36822
Phi^2	5		4.09	40			1907	37160
Chi^1+	5		4.41	54			2047	39587
Chi^2	5		4.63	62			2135	41117
Psi^1+	5		4.95	25			1789	35439
Psi^2	5		4.59	30			1811	35715
Omega	5		4.57	47			1934	37490
A	5		4.20	32			1839	36267
b	5		4.91	51			1963	37984
c+	5		4.59	42			1892	37018
d	5		4.80	49			1937	37507
e	5		4.14	29			1784	35369
f^1+	6		4.95	69			2198	42545
f^2	6		5.30	72			2223	43153
g	6		5.19	6			1569	31283
h	6		5.43	16			1672	33254
i	6		5.34	14			1664	33054
k+	6		5.04	74			2241	43386
l	6		5.39	75			2247	43525
m	6		5.00	23			1770	35149
n^1+	6		5.46	33			1842	36351
n^2	6		5.36	38			1872	36777
o+	6		4.73	22			1765	35039
p+	6		5.08	27			1787	35410

The Lost, Missing, or Troublesome Stars of Orion

Theta¹, 41, HR 1893/4/5/6, 6.73V, 7.96V, 5.13V, 6.70V, comb. mag. 4.85V, *SA, HRP, BS,* BF 748*, H 64 as Theta, G 104*, ADS 4186, the Trapezium.

Theta², 43, HR 1897, HD 37042, HD 37062, 5.08V, 6.38V, 9.1V, comb. mag. 4.89V, *SA, HRP, BS,* BF 750*, H 64 as Theta, G 105*, ADS 4188.

Bayer considered Theta a single 3rd-magnitude star. Flamsteed, uncertain which of these two nearly adjacent stars Bayer meant or thinking that Bayer saw the combined light of both stars, designated them Theta-1 and Theta-2. Argelander considered the combined light of 41 and 43 synonymous with Theta (*UN,* p. 58). Gould also noted that the light of both stars appeared as one image to the naked eye, but he assigned them separate catalogue numbers.

Observed first by Christiaan Huygens in 1656, Theta¹, the Trapezium, so called because of the shape of its major components, is a complex multiple-star system with nine components ranging in magnitude from 5 to 16. Theta² has three components: component A, HD 37041 (HR 1897), 5.08V; component B, HD 37042, 6.38V; and component C, HD 37062, 9.1V. The combined magnitude of Theta¹ is 4.85V and of Theta² is 4.89V. Altogether, the twelve components of Theta as seen with the naked eye have a combined magnitude of about 4.2. They are all part of the Great Nebula, M 42, in Orion's sword. See *Burnham's Celestial Handbook,* II, 316-37; *BS Suppl.,* p. 116.

Although these two stars are at the heart of the Great Orion Nebula, neither Flamsteed nor Bayer described them as nebulous objects in their catalogues. For nebulosities in Bayer, see z Her; for Flamsteed's nebulosities, see 33 And in Part Two. Lacaille observed forty nebulous objects, of which he lettered three: Omega Cen, Kappa Cru, and Eta Tel.

Omicron¹, 4, HR 1556, 4.74V, *SA, HRP, BS,* H 8*.
Omicron², 9, HR 1580, 4.07V, *SA, HRP, BS,* H 15*.

Bayer described Omicron as two 4th-magnitude stars, and Flamsteed designated his 4 and 9 as Omicron-1 and Omicron-2.

Pi¹, 7, HR 1570, 4.65V, *SA, HRP, BS,* BV 8*, H 12*.
Pi², 2, HR 1544, 4.36V, *SA, HRP, BS,* BV 4*, H 5*.
Pi³, 1, HR 1543, 3.19V, *SA, HRP, BS,* BV 1*, H 4*.
Pi⁴, 3, HR 1552, 3.69V, *SA, HRP, BS,* BV 2*, H 6*.
Pi⁵, 8, HR 1567, 3.72V, *SA, HRP, BS,* BV 5*, H 10*.
Pi⁶, 10, HR 1601, 4.47V, *SA, HRP, BS,* BV 7*, H 17*.

Bayer described Pi as a group of six stars of the 4th magnitude, and placed the letter Pi between the upper two stars of the group. Since Flamsteed was working without Bayer's star list, he assumed Bayer meant only these two stars as Pi, so he designated his 2 and 7 as Pi-1 and Pi-2. Bevis sought to correct Flamsteed's omission by adding four additional stars from among those in Flamsteed's catalogue that he thought were equivalent to Bayer's Pi's. He arranged their indices in descending order, as Ptolemy had suggested and as in the heading above (Toomer, *Ptolemy's Almagest,* p. 383; Bevis's Tabula XXXV).

Baily accepted Bevis's selections, but he felt the indices should be arranged in strict order of right ascension, as had been Flamsteed's practice:

> Pi¹, 1, BF 613 and pp. 399-400.
> Pi², 2, BF 615.
> Pi³, 3, BF 618.
> Pi⁴, 7, BF 628.
> Pi⁵, 8, BF 629.
> Pi⁶, 10, BF 643.

Argelander agreed that these six stars were indeed synonymous with Bayer's Pi's, but he felt that the stars' indices should be arranged more logically in descending order, north to south, as Ptolemy and Bevis had suggested (*UN,* pp. 56-57). For another example of Argelander's opposition to Baily's rigid adherence to right ascension numbering of indices, see Upsilon Eri.

Rho, 17, HR 1698, 4.46V, *SA, HRP, BS,* HF 14*, BF 676*.

This star is properly identified as Rho in Halley's pirated edition of Flamsteed's catalogue. However, as the result of a copying or typographical error, Flamsteed's 1725 "Catalogus Britannicus" labels this star, 17, as Rho-1 and 27 as Rho-2. The latter should be p. The error arose because p and Rho are similar in appearance, p, ρ. Baily corrected the error in his revised edition of Flamsteed's catalogue. See p.

Phi^1, 37, HR 1876, 4.41V, *SA, HRP, BS,* H 62*.
Phi^2, 40, HR 1907, 4.09V, *SA, HRP, BS,* H 68*.

Bayer described Phi as a couple of 5th-magnitude stars, and Flamsteed designated his 37 and 40 as Phi-1 and Phi-2.

Chi^1, 54, HR 2047, 4.41V, *SA, HRP, BS,* HF 49 as Chi, H 95*.
Chi^2, 62, HR 2135, 4.63V, *SA, HRP, BS,* HF 58 as Chi, H 111*.

Bayer described Chi as a couple of stars of the 5th magnitude, and in Halley's edition of Flamsteed's catalogue, these two stars, 54 and 62, are labeled Chi. In Flamsteed's 1725 catalogue, however, five stars are designated Chi:

>Chi-1, 54.
>Chi-2, 57.
>Chi-3, 62.
>Chi-4, 64.
>Chi-5, 65.

In checking Flamsteed's manuscripts, Baily found that Flamsteed had made two mathematical errors: 65 was a nonexistent star that should be dropped from the catalogue (see 65 Ori in Part Two); and 64's right ascension was too great and the star should be moved westward with a lower index number. Baily, therefore, rearranged the group as follows:

>Chi^1, 54, BF 801.
>Chi^2, 57, HR 2052, 5.92V, BF 804.
>Chi^3, 64, HR 2130, 5.14V, BF 833.
>65, nonexistent, BF, Table B, p. 645.
>Chi^4, 62, BF 838.

Baily felt that either 54 or 57 was the first of the two stars in Bayer's Chi pair and either 62 or 64 was the second of the pair. Although Argelander observed all four stars, he felt Bayer must have meant only the two brightest, 54 and 62 (*UN,* p. 59), the very same two stars that were originally designated Chi in Halley's edition of Flamsteed's catalogue.

Psi^1, 25, HR 1789, 4.95V, *BS,* BV 28, BF 709*, H 44, G 71.
Psi^2, 30, HR 1811, 4.59V, *SA* as Pi, *HRP* as Psi, *BS,* BV 30 as Psi, BF 719*, H 48 as Psi, G 77 as Psi.

Bayer described Psi as a single star of the 5th magnitude, but Flamsteed designated these two neighboring stars, his 25 and 30, as Psi-1 and Psi-2. Bevis labeled only the brighter star, 30, as Psi. Like Bevis, Baily was certain that 30 was Bayer's Psi, but he kept Flamsteed's designations. Argelander designated only 30 as Psi (*UN,* p. 58).

A, 32, HR 1839, 4.20V, *SA, HRP.*

b, 51, HR 1963, 4.91V, *SA, HRP.*

c^1, 42, HR 1892, 4.59V, *SA* as c, *HRP* as c, BV 39 as c, BF 749 as c, H 65 as c, G 109 as c.
c^2, 45, HR 1901, 5.26V, BV 40, BF 752, H 65 as c, G 110.

Bayer described c as a single star of the 5th magnitude, but Flamsteed designated these two nearly adjacent stars, his 42 and 45, as c-1 and c-2. Bevis labeled only the brighter star, 42, as c, and Baily agreed. Argelander (*UN,* p. 58) and Heis, on the other hand, felt that the combined light of 42 and 45 was synonymous with c. Gould also noted that the combined light of these two stars appeared as one image to his naked eye, but he equated only 42 with Bayer's c.

d, 49, HR 1937, 4.80V, *SA, HRP.*

e, 29, HR 1784, 4.14V, *SA, HRP.*

f¹, 69, HR 2198, 4.95V, *SA, HRP,* H 121*.
f², 72, HR 2223, 5.30V, *SA, HRP,* H 127*.

     Bayer described f as two very close stars of the 6th magnitude, and Flamsteed designated his 69 and 72 as f-1 and f-2.

g, 6, HR 1569, 5.19V, *SA, HRP.*

h, 16, HR 1672, 5.43V, *SA, HRP.*

i, 14, HR 1664, 5.34V, *SA, HRP.*

j

     Bayer did not use this letter.

k¹, 73, HR 2229, 5.33V, HF 68, BV 75, BF 871*, H 128.
k², 74, HR 2241, 5.04V, *SA* as k, *HRP* as k, HF 69 as k, BV 76 as k, BF 874*, H 130 as k.

     Bayer described k as a single 6th-magnitude star, and Halley's 1712 edition of Flamsteed's catalogue labels his 74 as k. Flamsteed's 1725 catalogue, however, designates these two neighboring stars, his 73 and 74, as k-1 and k-2. Bevis and Argelander, though, agreed with the 1712 catalogue. They felt that only the brighter star, 74, was synonymous with k (*UN*, p.60).

l, 75, HR 2247, 5.39V, *SA, HRP.*

m, 23, HR 1770, 5.00V, *SA, HRP.*

n¹, 33, HR 1842, 5.46V, *SA, HRP,* BV 33*, BF 728*, H 56*.
n², 38, HR 1872, 5.36V, *SA, HRP,* BV 41*, BF 738*, H 61*.

     Bayer described n as two stars of the 6th magnitude and placed the letter n directly over the preceding star. Since Flamsteed did not have access to Bayer's star list, he designated only his 33 as n. Bevis and Baily corrected this by designating Flamsteed's 38 as the other half of Bayer's n pair and by labeling these stars n¹ and n².

o, 22, HR 1765, 4.73V, *SA, HRP,* HF 19*, B 83*, BF 700*, H 39*.

     Flamsteed's 1712 catalogue properly equates his 22 with o, but for some reason the letter is omitted from his "Catalogus Britannicus" of 1725. Bevis and Bode relettered 22 as o.

p, 27, HR 1787, 5.08V, *SA, HRP,* HF 23*, B 90*, BF 710*, H 42*.

     Although Halley's 1712 edition of Flamsteed's catalogue properly equates his 27 with p, Flamsteed's 1725 catalogue designates his 17 as Rho-1, which should be Rho, and his 27 as Rho-2. Since both Rho and p are similar in appearance — ρ, p — this is probably a copying or typographical error. Bode caught the mistake and relabeled 27 as p. See Rho.

y¹, 11, HR 1638, 4.68V, BV 11, BF 652.
y², 15, HR 1676, 4.82V, BV 18, BF 663.

     Bayer did not letter these stars. The letters were added by Flamsteed but removed by Bevis and Baily, who considered them superfluous. See i Aql.

Pavo, Pav

Peacock

Keyzer and Houtman devised the constellation Pavo (Figure 59). Houtman's catalogue of 1603 designates it De Pauw, the Peacock. Like several of the other constellations devised by Keyzer and Houtman — Flying Fish, Toucan, Bird of Paradise, Chameleon, and Dorado — the Peacock represented another of the exotic creatures Europeans were encountering in their voyages around the globe. Bayer included it on his chart of the southern skies and labeled it in Latin Pavo. Like the other constellations on this chart, Pavo's stars are unlettered. Lacaille assigned Pavo letters from Alpha to Omega.

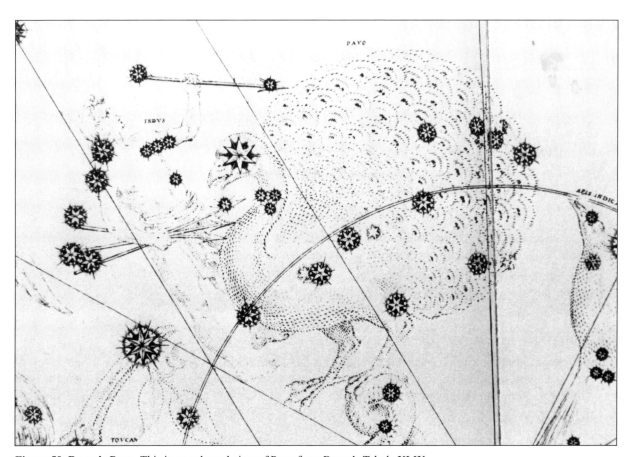

Figure 59. Bayer's Pavo. This is an enlarged view of Pavo from Bayer's Tabula XLIX.

Table 66. The lettered stars of Pavo

Letter	Bayer	Lacaille	Visual	Flamsteed	Lacaille	Other	HR	HD
Alpha		2	1.94		1657		7790	193924
Beta		3	3.42		1677		7913	197051
Gamma		3	4.22		1724		8181	203608
Delta		4	3.56		1635		7665	190248
Epsilon		4	3.96		1615		7590	188228
Zeta		4	4.01		1525		6982	171759
Eta		5	3.62		1449		6582	160635
Theta		5	5.73		1536		7036	173168
Iota		6	5.49		1485		6761	165499
Kappa		6	4.44		1547		7107	174694
Lambda		5	4.22		1541		7074	173948
Mu^1+		6	5.76		1620		7603	188584
Mu^2		6	5.31		1623		7612	188887
Nu		5	4.64		1512		6916	169978
Xi+			4.36			G 26	6855	168339
Omicron		6	5.02		1707		8092	201371
Pi		6	4.35		1476		6745	165040
Rho		6	4.88		1672		7859	195961
Sigma		6	5.41		1678		7934	197635
Tau		6	6.27		1572		7274	179009
Upsilon		5	5.15		1674		7881	196519
Phi^1+		6	4.76		1668		7848	195627
Phi^2		6	5.12		1675		7875	196378
Chi+		6	6.9		1551			175147
Omega		6	5.14		1554		7127	175329

The Lost, Missing, or Troublesome Stars of Pavo

Mu^1, HR 7603, 5.76V, *SA, HRP, BS,* CA 1620 as Mu, BR 6764*, L 8244*, G 82*.
Mu^2, HR 7612, 5.13V, *SA, HRP, BS,* CA 1623 as Mu, BR 6767*, L 8251*, G 83*.

Lacaille designated these two nearly adjacent stars Mu.

Xi, HR 6855, 4.36V, *SA, HRP, BS,* CA 1500, BR 6366, L 7638, G 26*.

Although Lacaille observed and catalogued this star, he left it unlettered since he considered it 6th magnitude. Gould designated it Xi because he felt a star of its brightness merited a letter.

Phi^1, HR 7848, 4.76V, *SA, HRP, BS,* CA 1668 as Phi, BR 6873*, L 8461*, G 104*.
Phi^2, HR 7875, 5.12V, *SA, HRP, BS,* CA 1675 as Phi, BR 6886*, L 8490*, G 109*.

Lacaille designated these two neighboring stars Phi.

Chi, HD 175147, 6.9V, CA 1551*, BR 6518*, L 7879, BAC 6425*, CAP 3693*.

Gould did not include this star in his *Uran. Arg.* (p. 81) because he considered its magnitude to be 7.2, beyond the 7.0 limiting magnitude of his catalogue. As a result, most modern astronomers and cartographers have dropped this star's letter from their catalogues and atlases.

Psi

Lacaille did not use this letter in Pavo.

Pegasus, Peg

Pegasus, Flying Horse

The ancient Mesopotamians knew the Great Square of Pegasus (Alpha, Beta, Gamma, and Delta Peg) (Figure 60) as Iku, Akkadian for the Square or Rectangular Field (Sumerian AS.GAN). The constellation represented a plowed, cultivated field. Its particular shape reflected the primitive nature of early farming. Since the first plows had no moldboard to turn a furrow — they could barely scratch the surface of the earth — farmers had to cross-plow the land in order to break the soil effectively. This usually produced square or rectangular fields. Iku's rising, near the time of the vernal equinox, reminded farmers that it was time to begin spring plowing. Formed about 4,000 years ago, it was one of the constellations that dominated Nisanu (March-April), the first month of the Babylonian New Year. The Field, the Hired Hand (LU.HUN.GA-Aries), the Plow (Triangulum), and the Ox (Taurus) were all part of a group of spring constellations associated with farming and agriculture.

The Field was first mentioned in Mesopotamian astronomical texts in the latter half of the second millennium B.C. and became one of the most prominent constellations in the ancient Near East. It is not surprising, therefore, that when Pegasus was later devised, only the forepart of the Flying Horse was depicted since its hind quarters, the Great Square, had already been preempted by the Field.

According to Greek mythology, Pegasus is the son of Poseidon, god of the sea, and Medusa, one of the Gorgons. The Gorgons live in the far west, near Oceanus, the river that surrounds Earth and the source of all water. When Perseus slays Medusa, Pegasus emerges — a full-grown steed with wings — from the blood of his mother. He is named for the place of his birth, pegai (πηγαι), the springs (of the Ocean). He is tamed by his half-brother, the Greek hero Bellerophon, who is also the son of Poseidon. With the help of Pegasus, Bellerophon succeeds in slaying the Chimaera, a monstrous fire-breathing beast with the head of a lion, the body of a goat, and the tail of a serpent. After a while, Bellerophon begins to think himself a god and tries to ascend to heaven on Pegasus. Zeus, offended that a mere mortal would dare think himself a god, sends a gadfly to sting Pegasus. The Flying Horse bucks and throws the arrogant Bellerophon down to Earth. But Zeus allows Pegasus to remain with the gods in heaven.

The Greek playwrite Euripides provided an entirely different account of the origin of the constellation Pegasus. According to Euripides in "Melanippe the Prisoner," Melanippe is the daughter of Thetis, who, in turn, is the daughter of the wise and pious centaur, Chiron. Seduced, impregnated, and ashamed, Thetis seeks to hide her condition from her father. As she labors to give birth to Melanippe, Thetis is about to be discovered by her father when she prays to the gods to be turned into a horse. Artemis hears her prayer and turns her into the mare, Hippe, and sets her among the stars. Through Melanippe, Euripides relates still another possible origin for this constellation. In "Melanippe the Wise," Melanippe claims that her mother, Thetis, was so successful in curing sickness, alleviating pain, and fortelling the future that Zeus feared she might reveal the secrets of the gods, so he placed her in the heavens as a horse.

Ptolemy called this constellation simply "the Horse (ιππος, hippos)." Bayer included Pegasus in his *Uranometria* and lettered its stars from Alpha to Psi.

See van der Waerden, "Babylonian Astronomy. II. The Thirty-Six Stars," pp. 9, 11, 13, 15, 21; Hartner, "The Earliest History of the Constellations in the Near East," pp. 8, 13; Langdon, *Babylonian Menologies,* pp. 3, 68; McNeill, *The Rise of the West,* pp. 40-44; Reiner and Pingree, *Baby. Plan. Omens,* Part 2, pp. 11-12; Condos, *Star Myths,* pp. 151-55; Ridpath, *Star Tales,* pp. 100-2; Staal, *New Patterns in the Sky,* pp. 27-32; Allen, *Star Names,* pp. 321-24; Gantz, *Early Greek Myth,* pp. 20, 313-16, 734-35; Bates, *Euripides,* pp. 258-61. For the possibility that Thetis-Hippe might be the Little Horse, see Equuleus; for an account of Chiron, see Centaurus; for the Gorgons and their family of monsters, see Cetus.

Lost Stars

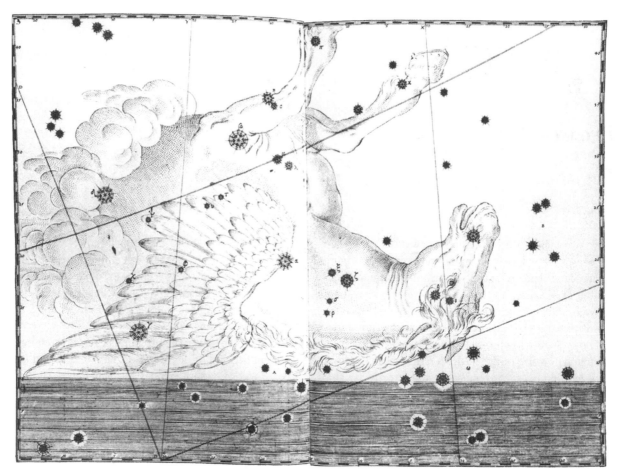

Figure 60. Bayer's Pegasus. Alpha (α), Beta (β), Gamma (γ), and Delta (δ) make up the Great Square of Pegasus. The letter Omega (o) is the constellation Aquarius; A is Pisces; and B is Equuleus. The lines C and D are the equator and the equinoctial colure. The unmarked line above the Flying Horse's wing is the Tropic of Cancer. The shaded area across the bottom is the zodiac.

Table 67. The lettered stars of Pegasus

	MAGNITUDES			CATALOGUE NUMBERS				
Letter	Bayer	Lacaille	Visual	Flamsteed	Lacaille	Other	HR	HD
Alpha	2		2.49	54			8781	218045
Beta	2		2.42	53			8775	217906
Gamma	2		2.83	88			39	886
Delta+	2		2.06	21 And			15	358
Epsilon	3		2.39	8			8308	206778
Zeta	3		3.40	42			8634	214923
Eta	3		2.94	44			8650	215182
Theta	4		3.53	26			8450	210418
Iota	4		3.76	24			8430	210027
Kappa	4		4.13	10			8315	206901
Lambda	4		3.95	47			8667	215665
Mu	4		3.48	48			8684	216131
Nu	4		4.84	22			8413	209747
Xi	5		4.19	46			8665	215648
Omicron	5		4.79	43			8641	214994
Pi^1+			5.58	27			8449	210354
Pi^2	5		4.29	29			8454	210459
Rho	6		4.90	50			8717	216735
Sigma	6		5.16	49			8697	216385
Tau	6		4.60	62			8880	220061
Upsilon	6		4.40	68			8905	220657
Phi	6		5.08	81			9036	223768
Chi	6		4.80	89			45	1013
Psi	6		4.66	84			9064	224427

The Lost, Missing, or Troublesome Stars of Pegasus

Delta, 21 And in Andromeda, HR 15, 2.06V, *SA, HRP, BS.*

Bayer noted that this star was synonymous with Alpha And. For Bayer's duplicate stars, see Alpha And. Delta Peg, and Alpha, Beta, and Gamma, form the Great Square of Pegasus.

Pi^1, 27, HR 8449, 5.58V, *SA, BS,* BV 41, B 128*, BF 3034*.
Pi^2, 29, HR 8454, 4.29V, *SA, HRP* as Pi, *BS,* BV 43 as Pi, B 133*, BF 3037*, H 55 as Pi.

Bayer noted that Pi was a single star, and Flamsteed designated his 29 as Pi. Bevis agreed, but Bode felt that Pi should apply to 27 and 29 since they are nearly adjacent and Bayer must have seen the combined light of both stars as one image. Consequently, he designated them Pi-1 and Pi-2. Although Baily agreed with Bode, Argelander equated only 29, the brighter of the two, with Bayer's Pi (*UN*, p. 82) since he could not see 27. Heis's catalogue and Corrigenda are confusing. They list H 54 as equivalent to Flamsteed's 29. It is not. They also list H 55 as equivalent to Flamsteed's 28. It is not. H 55 is equivalent to Flamsteed's 29, and H 56 to 28. H 54 is equivalent to HR 8458, 5.78V; it is not listed in Flamsteed's catalogue. Heis's atlas shows Pi as a single star. Neither his atlas nor his catalogue includes Flamsteed's 27.

Bayer did not seem certain of Pi's magnitude. On his star list, he classified Pi as 5th magnitude, but he noted that at times it was considered 4th magnitude. Twenty-five years later in Julius Schiller's *Coelum Stellatum Christianum,* which Bayer helped to prepare, the star was also shown as magnitude 4 (10 St. Gabriel). It is difficult to understand why Bayer considered it a 5th-magnitude star. Ptolemy classified it as brighter than 4th, and Peters and Knobel (*Ptolemy's Catalogue,* p. 129) noted that all the Ptolemaic manuscripts they consulted showed the same. Tycho listed it simply as 4th magnitude in both his catalogues. Is it possible that Bayer somehow managed to see both 27 and 29 and confused their magnitude? For another example of Bayer's seeming ability to see the unseeable, see Chi Per.

e, 1, HR 8173, 4.08V, BV 6, BF 2915.
f, 2, HR 8225, 4.57V, BV 9, B 14*, BF 2932.
g, 9, HR 8313, 4.34V, BV 11, B 52*, BF 2968.
l, 55, HR 8795, 4.52V, BV 48, B 295*, BF 3165.
m, 57, HR 8815, 5.12V, BV 49, B 300*, BF 3173.
n, 58, HR 8821, 5.39V, BV 51, BF 3174.
p, 59, HR 8826, 5.16V, BV 50, B 307*, BF 3175.
q, 70, HR 8923, 4.55V, *SA,* BV 60, B 363*, BF 3220.
r, 83, HD, 223792, 6.7V, BV 81, B 415*, BF 3271.
s, 75, HR 8963, 5.40V, BV 71, B 386*, BF 3236.
u, 87, HR 22, 5.53V, BV 86, B 446*, BF 3301.
y, 71, HR 8940, 5.32V, BV 75, B 371*, BF 3228.

The letters assigned to these twelve stars did not originate with Bayer. He included 1, 2, and 9 in his atlas but considered them *informes* outside the border of Pegasus and left them unlettered. Flamsteed labeled them as in the heading above. He started with e since Bayer's last letter in Pegasus was D, the vernal equinoctial colure (for Bayer's use of Roman capitals, see P Cyg). Bode used most of these letters in his works, but Bevis and Baily considered Flamsteed's letters superfluous and removed them all (see i Aql). The letter q has somehow managed to survive in the atlases of Becvar and Tirion, but not in the second edition of Tirion's *SA*.

Perseus, Per

Perseus, Champion

Perseus (Figure 61) is a Ptolemaic constellation that Bayer lettered from Alpha to Omega and from A to o.

Many myths and variations of myths and legends are associated with Perseus. One of the foremost Greek heroes, he is the son of Zeus and Danae, the daughter of the king of Argos. When Perseus grows to manhood, he provokes the enmity of the king of Seriphos, who sets him off on a mission to obtain the head of Medusa, one of the Gorgons. Besides a formidable appearance that includes a head of venomous snakes instead of hair, the Gorgons possess an even more dangerous characteristic. One look on their terrible faces turns a mortal to stone. Luckily for Perseus, the gods look favorably upon him and several offer to help. Athena provides him with a shiny shield and Hermes gifts him a pair of winged sandals that permit him to fly through the air. The Gorgons live far to the west, near the setting sun, protected by their sisters, the three Graiae, who share but one eye and one tooth among them. To achieve his goal, Perseus effectively blinds the Graiae by snatching their eye while they are passing it one to another. With their guardians thus disabled, the Gorgons are ready prey for Perseus. While the monsters sleep, Perseus uses Athena's shiny shield as a mirror to back up to the Gorgons without having to look directly upon their faces. He then slices off the head of Medusa. He places her head in his pouch and heads toward Atlas, who is holding up the sky. As he nears the Titan, he removes Medusa's head from his pouch and shows her face to Atlas, thus turning the mighty Titan into the Atlas Mountains of North Africa. Continuing on his journey homeward, Perseus rescues Andromeda from the sea monster Cetus, marries her, and fathers several children including a son Perses, who becomes the progenitor of the royal house of Persia. Ultimately, Perseus, Andromeda, her parents — Cassiopeia and Cepheus — and Cetus are placed among the stars.

See Condos, *Star Myths,* pp. 157-60; Ridpath, *Star Tales,* pp. 102-5; Staal, *New Patterns in the Sky,* pp. 19-26. For an account of the Gorgons and their sisters, the Graiae, see Cetus.

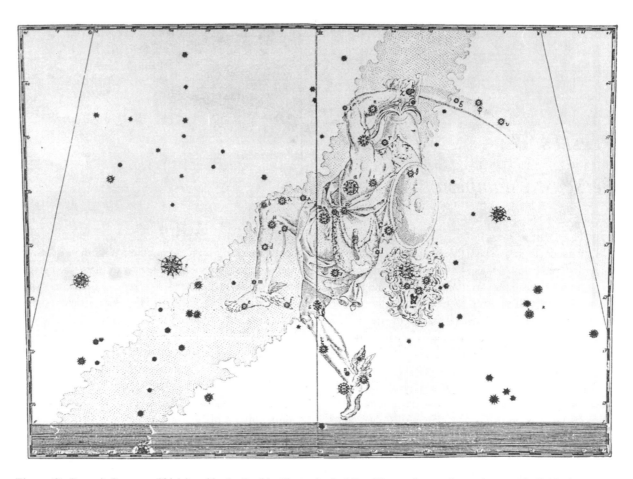

Figure 61. Bayer's Perseus. Chi (χ) and h, the Double Cluster in the hilt of Perseus's sword, are shown as individual stars, not as nebulous objects. However, see Bayer's chart of Cassiopeia, Figure 20, where they are shown as nebulous objects. The letter P is the 1st-magnitude star Capella, Alpha Aur; Q is the left foot of Andromeda; and R is Triangulum. The line S, near the bottom of the chart, marks the upper limit of the zodiac; Bayer's charts extend the path of the zodiac to 8° north and south of the ecliptic.

Table 68. The lettered stars of Perseus

	MAGNITUDES			CATALOGUE NUMBERS				
Letter	Bayer	Lacaille	Visual	Flamsteed	Lacaille	Other	HR	HD
Alpha	2		1.79	33			1017	20902
Beta	2		2.12	26			936	19356
Gamma	3		2.93	23			915	18925
Delta	3		3.01	39			1122	22928
Epsilon	3		2.89	45			1220	24760
Zeta	3		2.85	44			1203	24398
Eta+	4		3.76	15			834	17506
Theta	4		4.12	13			799	16895
Iota+	4		4.05				937	19373
Kappa	4		3.80	27			941	19476
Lambda	4		4.29	47			1261	25642
Mu	4		4.14	51			1303	26630
Nu	4		3.77	41			1135	23230
Xi	4		4.04	46			1228	24912
Omicron+	4		3.83	38			1131	23180
Pi	4		4.70	22			879	18411
Rho	4		3.39	25			921	19058
Sigma	4		4.36	35			1052	21552
Tau	5		3.95	18			854	17878
Upsilon+	5		3.57	51 And			464	9927
Phi+	5		4.07	54 And			496	10516
Chi+	5							
Psi	5		4.23	37			1087	22192
Omega	5		4.63	28			947	19656
A	5		5.28	43			1210	24564
b¹+	5		4.61				1324	26961
b²			5.55				1333	27192
c+	5		4.04	48			1273	25940
d	5		4.85	53			1350	27396
e	5		4.25	58			1454	29094
f	5		4.71	52			1306	26673
g+	6		5.04	4			590	12303
h+	6							
i	6		5.17	9			685	14489
k+	6		4.76				918	18970
l	6		4.95	32			1002	20677
m	6		6.09	57			1434	28704
n	6		5.11	42			1177	23848
o+	6		4.97	40			1123	22951

The Lost, Missing, or Troublesome Stars of Perseus

Eta, 15, HR 834, 3.76V, *SA, HRP, BS,* HF 17*, BF 348*, H 34*.

Although Halley's 1712 edition of Flamsteed's catalogue properly equates 15 with Bayer's Eta, the letter is omitted from Flamsteed's "Catalogus Britannicus" of 1725. Baily (*BF*, Table I, p. 655) labeled 15 as Eta in his revised edition of Flamsteed's catalogue. See Bevis's Remarks on his Tabula XI regarding Eta, Iota, and Omicron.

Iota, HR 937, 4.05V, *SA, HRP, BS,* HF 28*, BV 40*, B 115*, BF 391*, BAC 962*, H 56*, W 23*.

Halley's edition of Flamsteed's catalogue includes Iota, but it, together with several hundred other stars that Flamsteed observed, is omitted from his 1725 catalogue (*BF*, p. 392). Bevis and Bode properly equated Iota with this star. Baily was, at first, confused with Iota. In his revised edition of Flamsteed's catalogue, he labeled two stars Iota Per, this star and Flamsteed's 31 (HR 989, 5.03V, BF 410 as Iota, BAC 1011), about 1½° to the east. He later realized his mistake and dropped Iota from 31.

Omicron, 38, HR 1131, 3.83, *SA, HRP, BS,* HF 44*, BV 45*, BF 454*, H 94*.
o, 40, HR 1123, 4.97V, *SA, HRP,* HF 42*, BV 44*, BF 449*, H 91*.

Some confusion exists in regard to these two stars probably because on Bayer's chart the stars are in close proximity in Perseus's right foot, about 2° apart, and because their letters are very similar in appearance, the Greek letter Omicron, o, and the Roman o (oh). In Halley's edition of Flamsteed's catalogue, 38 is properly equated with Omicron and 40 with oh; and both stars' coordinates are properly listed. In Flamsteed's catalogue of 1725, however, 38 and 40 are mistakenly labeled Omicron-1 and Omicron-2 and 38's coordinates are erroneous; there is no star located at the position given in his catalogue. Bevis sought to correct the mistakes. He correctly equated oh with 40, but since the position of Bayer's Omicron did not coincide with 38's coordinates, he equated Omicron with Halley's HF 44. Baily, who had access to Flamsteed's manuscripts, finally set matters right. As Bevis had done, he equated oh with 40 and, upon discovering that Flamsteed had used an erroneous sine logarithm, he recalculated 38's coordinates and found that they corresponded exactly with Bayer's Omicron. Argelander (*UN*, p. 15) and Heis, however, did not identify any star as 38, probably because there was no star in the heavens at the coordinates originally calculated by Flamsteed in the authorized edition of his catalogue. They did include Omicron in their atlases and catalogues, though without identifying it as Flamsteed's 38. See Bevis's Remarks on his Tabula XI.

Upsilon, 51 And in Andromeda, HR 464, 3.57V, *HRP, BS,* H 1*.

Bayer included this star in Perseus although Ptolemy had placed it in Andromeda. More concerned with astronomical custom and tradition than Bayer, Flamsteed returned it to Andromeda, where it had been for over 1,500 years before Bayer switched it. Although Argelander (*UN*, p. 14) and Heis included Upsilon in Perseus, the International Astronomical Union, which established permanent constellation boundaries in 1930, vindicated Flamsteed by restoring Upsilon Per to Andromeda. For Bayer's switching stars between Ptolemaic constellations, see Sigma Lib.

Phi, 54 And in Perseus, HR 496, 4.07V, *SA, HRP, BS,* H 2*.

Bayer included this star in Perseus although Ptolemy had placed it in Andromeda. As he had done with Upsilon, above, Flamsteed returned Bayer's Phi Per to Andromeda, where he felt it belonged. But this time the IAU agreed with Bayer, Heis, and Argelander (*UN*, p. 14) and left Phi within the bounds of Perseus — but just barely since it is only 3½' from the border with Andromeda.

Chi and h

The identity of Chi and h Per has bedeviled astronomers from antiquity to the present day. Do the letters refer to specific stars or star clusters? It all began with Ptolemy, who described the first star in Perseus as a nebulous mass. Ptolemy's latest editor and translator, G. J. Toomer, felt that he was referring to the combined light of NGC 884 and NGC 868, the famous Double Cluster, which, Toomer asserted, appears "as a single hazy patch to the naked eye" (*Ptolemy's Almagest,* p. 352). Bayer, on the other hand, depicted two separate stars, Chi and h, in the area Ptolemy described, but without noting any nebulosity either in his star list or on his star chart. However, on his chart of the neighboring constellation Cassiopeia, which also shows Chi and h Per, Bayer depicted them as nebulous objects. They are clearly shown near the bottom of Cassiopeia's chart with the symbol Bayer reserved for nebulosities. Some twenty-five years later in Schiller's *Coelum Stellatum Christianum* that Bayer helped to prepare, the chart for St. Paul, formerly Perseus, shows two nebulous stars. The star list describes these as a *Nebulosa duplex* (a nebulous double star) and states specifically that they are Bayer's Chi and h. Bayer, it would seem, meant to describe Chi and h as nebulous patches.

Flamsteed, using a telescope, resolved the nebulosities and assigned Chi and h to specific stars around the clusters. In Halley's pirated edition of his catalogue, 7 (HR 662, 5.98V, HF 9) is equated with Bayer's Chi and 8 (HR 661, 5.75V, HF 10) with Bayer's h. In Flamsteed's 1725 catalogue, while 7 is still equated with Chi, his 6 is mistakenly equated with h, but Bevis and Baily switched it to 5, (HR 627, 6.36V, BV 19, BF 256 and Table I, p. 655). In each of these instances, Flamsteed was referring to particular stars, not to the Double Cluster, since he assigned to the stars specific stellar magnitudes and since he did not mention that they were associated with any nebulosity. In his revised edition of Flamsteed's catalogue and in the *BAC*, Baily did the same, describing 5 as synonymous with h (BF 256, BAC 658) and 7 as synonymous with Chi (BF 274, BAC 696). He noted in each of these references that these were stars, not patches of nebulosity. Baily assigned a separate designation, BAC 719, to NGC 884, the eastern half of the Double Cluster. In short, Flamsteed and Baily referred to Chi and h — 7 and 5 — as individual stars, not as the Double Cluster.

Argelander believed otherwise. Since he could not distinguish 5 and 7 as separate stars with his naked eye, he felt that they could not possibly be Bayer's h and Chi. He decided that the Double Cluster, with an integrated magnitude of about 4½, must have been what Bayer observed (*UN*, p. 14). Consequently, he designated h and Chi as the Double Cluster, as did Heis (H 15/6). The latter equated Chi with BAC 719 and h with BAC 700, noting they were synonymous with the Double Cluster. As stated above, BAC 719 is NGC 884; but since BAC 700 is a particular star, possibly 61 And (HD 14134, 6.55V) in Perseus (see 61 And in Part Two), Heis was mistaken in equating it with the preceding half of the Double Cluster. Like Argelander, Heis did not designate any star in his catalogue as 5 or 7 Per.

What did Bayer really observe? Based on his own work, *Uranometria* and *Coelum Stellatum Christianum*, it appears that Argelander was correct. Bayer intended Chi and h to refer to the Double Cluster, not to specific stars. As possibly one of the last astronomers to work before the proliferation of telescopes in 1608, he saw what Ptolemy and other naked-eye observers had seen, patches of nebulosities. See Epsilon Cnc; Hoffleit, "Discordances," pp. 57-58.

What is extremely puzzling in regard to Chi and h is how Bayer was able to see two separate patches of nebulosity. As Toomer and others have remarked, the Double Cluster appears to the naked eye as just one nebulous object. How could Bayer have possibly seen both clusters with his naked eye? Did he, or someone he knew, have access to some form of early telescope? It is generally accepted that the Dutch spectacle maker Hans Lipperhey of Middleburg displayed publicly the first successful telescope in 1608, but he is reported to have experimented with prototypes as early as 1600. Even earlier, in the 1580's, experiments were being conducted in Western Europe with concave and convex lenses. The only other spectacle maker in Middleburg, one Sacharias Janssen, was reported by his son to have developed a telescope as early as 1590 (King, *History of the Telescope*, pp. 29-33; Boorstin, *The Discoverers*, pp. 312-22; Sluitter, "The Telescope before Galileo," esp. note 7). Is it possible that Bayer, living some three hundred miles away at Augsburg, learned of this discovery and was able to split the Double Cluster using some primitive lens before publishing his *Uranometria* in 1603? See, for instance, Sigma Eri, p Her, and Pi Peg.

A, 43, HR 1210, 5.28V, *SA, HRP.*

b^1, HR 1324, 4.61V, *SA, HRP* as b, HF 58 as b, BF 515*, BAC 1301*, H 118 as b.
b^2, HR 1333, 5.55V, *SA,* HF 60 as b, BV 75, BF 521*, BAC 1314*, H 120.

Bayer described only one 5th-magnitude star as b, but Halley's edition of Flamsteed's catalogue designates three neighboring stars as b:

> HF 58, HR 1324, 4.61V.
> HF 59, HR 1330, 5.45V, BV 73 as b, BF 517, BAC 1307.
> HF 60, HR 1333, 5.55V.

For some inexplicable reason, these three stars, together with several hundred others, are not included in Flamsteed's 1725 catalogue (*BF*, p. 392). Bevis included two of the stars (HF 59 and HF 60) in his catalogue, labeling only one, HF 59, as b. Baily included all three stars in his revised edition of Flamsteed's catalogue, but he was not sure whether HF 58 or HF 60 was synonymous with Bayer's b, so he labeled them b^1 and b^2, as in the heading above; he left HF 59 unlettered. Argelander felt only the brighter of the two, HF 58, was equivalent to Bayer's b (*UN*, p. 16).

c, 48, HR 1273, 4.04V, *HRP,* the variable MX.

Tirion's *SA* erroneously labels 48 as Nu on chart 1 and as Upsilon on charts 4 and 5. Becvar also designated this star Upsilon. The error is corrected in Tirion's *Uranometria 2000.0* and in the second edition of *SA*, but the star is not identified as c in either atlas.

d, 53, HR 1350, 4.85V, *SA, HRP.*

e, 58, HR 1454, 4.25V, *SA, HRP.*

f, 52, HR 1306, 4.71V, *SA, HRP.*

g, 4, HR 590, 5.04V, *SA, HRP,* H 9, BKH 13*.

Flamsteed designated his 2 (HR 536, 5.79V, BF 213 as g, BAC 560 as g, H 6) as g. Baily went along with this designation but noted that 2 did not accord with Bayer's chart. Argelander was uncertain which star was g. On chart I of his atlas, he designated 2 as g, but on chart II he omitted the letter g, as he did in his catalogue (*UN,* p. 19). Heis did not include g in either his atlas or his catalogue. Pickering was among the first to equate g with 4 in *HRP*, but this designation also does not accord well with Bayer's chart. Bevis thought that his BV 4 (HR 529, 5.90V) best matched Bayer's chart and data. In magnitude, position, and distance from its neighbors, it agrees better than 2 or 4. Is it Bayer's g?

h

See Chi.

i, 9, HR 685, 5.17V, *SA, HRP.*

j

Bayer did not use this letter.

k, HR 918, 4.76V, *SA, HRP,* HF 26*, BV 48, B 109*, BF 378*, H 50*.

Although Flamsteed observed this star and Halley's 1712 edition of his catalogue labels it k, it, together with 457 other stars, is not included in his 1725 "Catalogus Britannicus" (*BF*, p. 392). Bode equated it properly with Bayer's k, and Baily included it as k in his revised edition of Flamsteed's catalogue. Bevis labeled this star k on his chart (Tabula XI) but omitted the letter in his catalogue.

l, 32, HR 1002, 4.95V, *SA, HRP.*

m, 57, HR 1434, 6.09V, *SA, HRP.*

n, 42, HR 1177, 5.11V, *SA, HRP.*

o, 40, HR 1123.

See Omicron.

p^1, 16, HR 840, 4.23V, BF 354.
p^2, 20, HR 855, 5.33V, BF 360.
q, 12, HR 788, 4.91V, BF 330.
r, 17, HR 843, 4.53V, BF 358.
s, 24, HR 882, 4.93V, BF 371.

Bayer did not letter these five stars. Although he included them in his atlas, he considered them *informes* outside the border of Perseus and left them unlettered. Flamsteed added these letters, but Baily removed them as superfluous. See i Aql.

PHOENIX, PHE

Phoenix

Keyzer and Houtman devised the constellation Phoenix (Figure 62). In Houtman's catalogue of 1603, it is called Den voghel Fenicx, the Bird Phoenix. Bayer depicted it on his chart of the southern skies as Phoenix, and Lacaille lettered its stars from Alpha to Omega. On some early charts, a cross is shown in this area of the heavens.

Traditionally, the Phoenix is depicted as a mythical bird that is consumed by fire only to be reborn from its own ashes. According to the Roman poet Ovid, the bird feeds only on frankincense and fragrant gums. After 500 years, it constructs a nest in an oak tree or atop a palm, builds a pile of various spices, and lies down and dies. From the body of the dead bird its offspring is born. A variation of this tale describes how a worm comes forth from the body of the dead bird, and with the passage of time, the worm becomes a full-grown Phoenix.

See Ridpath, *Star Tales,* pp. 105-6; Staal, *New Patterns in the Sky,* p. 57. See Crux for the cross that once occupied this sector of the sky.

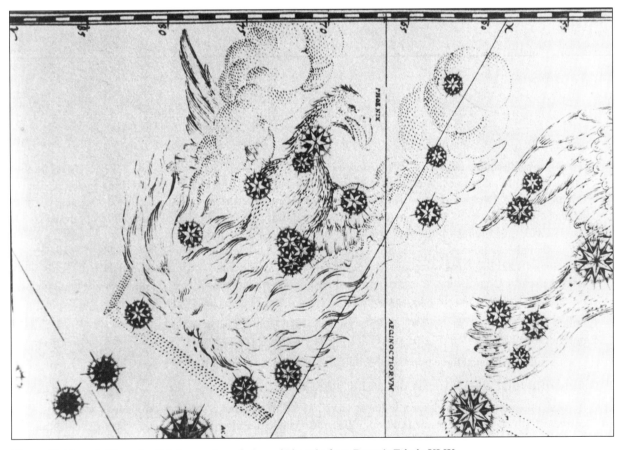

Figure 62. Bayer's Phoenix. This is an enlarged view of Phoenix from Bayer's Tabula XLIX.

Table 69. The lettered stars of Phoenix

		Magnitudes			Catalogue Numbers			
Letter	Bayer	Lacaille	Visual	Flamsteed	Lacaille	Other	HR	HD
Alpha		2	2.39		17		99	2261
Beta		4	3.31		68		322	6595
Gamma		3	3.41		89		429	9053
Delta		4	3.95		91		440	9362
Epsilon		4	3.88		1938		25	496
Zeta		5	3.92		71		338	6882
Eta		5	4.36		47		191	4150
Theta+		5	6.09c		1906		8966	222287
Iota		5	4.71		1901		8949	221760
Kappa		5	3.94		16		100	2262
Lambda^1+		5	4.77		25		125	2834
Lambda^2		6	5.51		34		147	3302
Mu		5	4.59		41		180	3919
Nu		6	4.96		78		370	7570
Xi		6	5.70		43		183	3980
Omicron+		6	6.97		100			10121
Pi		6	5.13		1924		9069	224554
Rho		6	5.22		54		242	4919
Sigma		6	5.18		1914		9006	223145
Tau		6	5.71		1927		9081	224834
Upsilon		6	5.21		69		331	6767
Phi		6	5.11		124		558	11753
Chi		6	5.14		137		602	12524
Psi+			4.41			G 126	555	11695
Omega+		6	6.11			L 288	295	6192

The Lost, Missing, or Troublesome Stars of Phoenix

Theta, HR 8966, 6.09V, *SA* as Theta^{1,2}, *BS,* CA 1906*, BR 7315 as Theta^2, L 9543*, G 12.

Some confusion has arisen over whether this star or HR 8959 (4.74V, CA 1904, BR 7309, L 9535, G ll) is Lacaille's Theta. Gould felt that Lacaille erred in designating CA 1906 as Theta and that he probably meant to select a brighter star, CA 1904, about 1° north. Regardless of what Gould said he intended, Lacaille did designate CA 1906 as Theta while leaving CA 1904 unlettered. Brisbane and Baily did the same. Gould himself left both stars unlettered — CA 1904 because of the uncertainty and CA 1906 because of its dimness.

Although Lacaille described Theta as a single 5th-magnitude star, it is a double, Dunlop 251. Component A is HR 8966, HD 222287, 6.6V; component B is SAO 231718, 7.2V. The combined magnitude of the two stars is 6.09V. Brisbane was among the first to assign indices to this double-star system. His BR 7314, Theta^1, is component B. See notes to HR 8959 and HR 8966 in the *BS*.

Lambda^1, HR 125, 4.77V, *SA, HRP, BS,* CA 25 as Lambda, BR 57*, L 115 as Lambda, G 54*.
Lambda^2, HR 147, 5.51V, *SA, HRP, BS,* CA 34 as Lambda, BR 69*, L 143, G 59*.

Lacaille designated these two neighboring stars Lambda.

Omicron, HD 10121, 6.97V, CA 100*, BR 236, L 478, BAC 505, G 116.

Lacaille made a 22^s clock error in his observation of this star that resulted in erroneous coordinates. Baily corrected the error in his expanded edition of Lacaille's catalogue and equated this star with Brisbane's BR 236. Both Baily and Gould dropped its letter because of its dimness.

Psi, HR 555, 4.41V, *SA, HRP, BS,* CA 123, BR 272, L 559, G 126*.

This is not the star Lacaille designated as Psi. He observed and catalogued this star but left it unlettered; he considered it 6th magnitude. Gould designated it Psi because of its brightness. Lacaille's Psi Phe is HR 494 (6.17V, CA 105 as Psi, BR 245 as Psi, L 496, BAC 526, G 127) in Sculptor. Gould included it in Sculptor and left it unlettered because of its dimness (*Uran. Arg.,* p. 84).

Omega, HR 295, 6.11V, *SA, BS,* CA 65, BR 136, L 288*, BAC 292*, G 80.

Lacaille designated this star Omega in his observations but left it unlettered in his catalogue. Baily lettered it Omega in editing the expanded version of Lacaille's catalogue and in the *BAC*. Gould dropped its letter because of its dimness.

Pictor, Pic

Painter's Easel

Lacaille devised the constellation Pictor (figures 2c and 2d) and lettered its stars from Alpha to Nu. In his preliminary catalogue, published in 1756, he called the constellation le Chevalet et la Palette (Easel and Palette) and noted that it represented the painter's easel with his palette. In his catalogue of 1763, he labeled the constellation in Latin simply as Equuleus Pictorius (Painting or Painter's Easel) although his star chart, like the one included in his preliminary catalogue, depicted both an easel and palette. Lacaille intended that Pictor, like the other constellations that he devised, should represent the arts and sciences that were flourishing during the Enlightenment that was spreading throughout Western Europe in the eighteenth century.

Table 70. The lettered stars of Pictor

Letter	Magnitudes			Catalogue Numbers				
	Bayer	Lacaille	Visual	Flamsteed	Lacaille	Other	HR	HD
Alpha		4	3.27		583		2550	50241
Beta		6	3.85		450		2020	39060
Gamma		6	4.51		459		2042	39523
Delta		6	4.81		495		2212	42933
Epsilon+		6	6.39		443		1984	38458
Zeta		6	5.45		408		1767	35072
Eta^1+		6	5.38		388		1649	32743
Eta^2		6	5.03		390		1663	33042
Theta		6	6.27		417		1818	35860
Iota		6	5.19c		376		1563/4	31203/4
Kappa		6	6.11		415		1801	35580
Lambda		6	5.31		365		1516	30185
Mu		6	5.70		549		2412	46860
Nu		6	5.61		520		2320	45229

The Lost, Missing, or Troublesome Stars of Pictor

Epsilon, HR 1984, 6.39V, CA 443*, BR 1029, L 1981*, BAC 1836, G 29.

Lacaille made a 1^m clock error in his observation of this star that resulted in the generation of erroneous coordinates. Baily corrected the mistake in his expanded edition of Lacaille's catalogue and equated this star with Brisbane's BR 1029. Gould dropped its letter because of its dimness.

Eta^1, HR 1649, 5.38V, *SA, HRP, BS,* CA 388 as Eta, BR 861*, L 1717*, G 10*.
Eta^2, HR 1663, 5.03V, *SA, HRP, BS,* CA 390 as Eta, BR 870*, L 1728*, G 11*.

Lacaille designated these two neighboring stars Eta.

Pisces, Psc

Fishes

A Ptolemaic constellation that Bayer lettered from Alpha to Omega and from A to l, Pisces (Figure 63) presently consists of two fish, the Northern Fish and the Western Fish, usually pictured tied together at their tails by a ribbon or chord. From Mesopotamian astronomical literature dating from the latter part of the second millennium B.C., the constellation that occupied the position of Pisces in the night sky was actually two separate constellations: Anunitum, or Anunitu, and SIM.MAH. The latter, located to the west, was the Sumerian constellation the Great Swallow; and the former, located to the east, was an Akkadian goddess whose attributes and nature were eventually merged with her Sumerian counterpart, Inanna, queen of heaven, and the Akkadian Ishtar. Anunitum can be translated as she of the skirmish, goddess of battle. She was originally the goddess of spring thunderstorms, the approach of which resembled the rumblings of war chariots (see Virgo). She was represented symbolically as a monstrous lion-headed thunderbird.

In the early half of the first millennium B.C., a third constellation was added to the group, Zibbati, Akkadian for Tails, referring perhaps to the tails of the Swallow and thunderbird, or, more likely, to the tails of fish. A few centuries later, the entire group was referred to in an Akkadian astronomical text as Rikis Nuni, the Band of the Fishes or the Fish Chord.

The Fishes came to be identified with this group through a number of associations and symbols. At Inanna's (Anunitum's) temple at Nippur — the modern village of Nuffar in Iraq — a large votive plaque was recently discovered that depicted, among several other animals, a large, oversized fish. In a land dominated by the Tigris and Euphrates and crisscrossed with innumerable irrigation canals, fish held a very special place in the life and folklore of the people. Fish became not only an important element in their economy, but because of the great mass of roe each fish produces, it also became a sacred symbol of fertility and reproduction. Fertility was also one of the main attributes of the goddess Inanna (Anunitum-Ishtar). When, in the epic poem "Descent of Ishtar to the Nether World," the goddess dies and is confined to the Nether World:

The bull springs not upon the cow, the ass impregnates not the jenny,
In the street the man impregnates not the maiden.

Inanna, or Ishtar, was closely associated with fish. On various cylinder seals, boundary stones, and wall carvings, fish were frequently shown as sacrificial offerings to the goddess, as depicted at her temple at Nippur. In one hymn to Inanna, she is portrayed clad in a fish mantle, with fish sandals on her feet, holding a fish scepter in her hand, and seated on a throne of fish.

While it may seem that there is no direct connection between fish (Inanna-Ishtar-Anunitum) and the other half of the Mesopotamian constellation, SIM.MAH, the Great Swallow, some archeological evidence suggests that the two may indeed be symbolically related. Mesopotamian carvings have been found that show the figure of a fish with the head of a swallow, suggesting, perhaps, a symbolic cyclical regeneration — the swallow rising from the depths of the waters as the phoenix would later be said to rise from its own ashes. Swallows would also have been seen feeding over bodies of water. In sum, fishes were associated in one way or another with both Anunitum and SIM.MAH, the Great Swallow.

Although the early Mesopotamians, as far as we currently know, had no constellation exactly equivalent to Pisces, several cylinder seals from Ur depict a heroic figure, possibly Gilgamesh, holding two fish threaded on a chord. On these and other seals and carvings depicting fish on chords, a goat is usually shown, one of the symbols of the water-god Enki. Could these fish on chords closely associated with divine figures have represented early prototypes of the modern constellation Pisces or is it merely coincidental?

It is also possible that this constellation evolved as Pisces because of its proximity in the sky to the Ibex (Aquarius), which was the favorite animal and symbol of Enki, the god of fresh water. Enki, or Ea as he is known in Akkadian, was often pictured on Mesopotamian seals and wall carvings with the Tigris and Euphrates rivers, filled with fish, flowing from his shoulders. When the giant Ibex, one of the four major constellations of ancient Sumer, was eventually reduced in size, all the new constellations that replaced it were in one way or another associated with Enki: Aquarius, who was Enki himself; Capricornus, the Goat-Fish, another symbol of Enki; Pisces; and Piscis Austrinus.

There are very few classical legends associated with Pisces. The poet Hyginus, however, relates that one day Venus and her son Cupid are near the Euphrates River when the hundred-headed, fire-breathing monster, Typhon, appears. Jumping into the river, Venus and Cupid transform themselves into fish and swim safely away. Two fish are later placed among the stars to commemorate their escape. The locale of this tale is an indication of the probable Mesopotamian origin of the constellation.

See van der Waerden, "Babylonian Astronomy. II. The Thirty-Six Stars," pp. 13-15; van der Waerden, "History of the Zodiac," pp. 219-20, 226; Reiner and Pingree, *Baby. Plan. Omens,* Part 2, pp. 10, 14; O'Neil, *Time and the Calendars,* p. 60; Jacobsen, "Mesopotamian Religions," pp. 458-59; Jacobsen, *Toward the Image of Tammuz,* pp. 34, 323-24; Pritchard, *Anc. Near East. Texts,* pp. 108, 111; Pritchard, *Anc. Near East Picts.,* plates 454, 685, 846; Kramer, *The Sumerians,* p. 110; Contenau, *Everyday Life in Babylon and Assyria,* pp. 47-48; Cirlot, *A Dictionary of Symbols,* pp. 106-7; Heuter, "Star Names," p. 245; Jacobsen, *Treasures of Darkness,* p. 111; Van Buren, "Fish Offerings in Ancient Mesopotamia," pp. 101-21; Hansen, "New Votive Plaques from Nippur," p. 162; Condos, *Star Myths,* pp. 161-62; Ridpath, *Star Tales,* pp. 107-8; Staal, *New Patterns in the Sky,* pp. 45-47.

Table 71. The lettered stars of Pisces

	MAGNITUDES			CATALOGUE NUMBERS				
Letter	Bayer	Lacaille	Visual	Flamsteed	Lacaille	Other	HR	HD
Alpha	3		3.82c	113			595/6	12446/7
Beta	4		4.53	4			8773	217891
Gamma	4		3.69	6			8852	219615
Delta	4		4.43	63			224	4656
Epsilon	4		4.28	71			294	9186
Zeta	4		5.18c	86			361/2	7344/5
Eta	4		3.62	99			437	9270
Theta	5		4.28	10			8916	220954
Iota	5		4.13	17			8969	222368
Kappa+	5		4.94	8			8911	220825
Lambda	5		4.50	18			8984	222603
Mu	5		4.84	98			434	9138
Nu	5		4.44	106			489	10380
Xi	5		4.62	111			549	11559
Omicron	5		4.26	110			510	10761
Pi	5		5.57	102			463	9919
Rho	5		5.38	93			413	8723
Sigma+	5		5.50	69			291	6118
Tau	5		4.51	83			352	7106
Upsilon	5		4.76	90			383	7964
Phi	5		4.65	85			360	7318
Chi	5		4.66	84			351	7087
Psi^1+	5		4.92c	74			310/1	6456/7
Psi^2	5		5.55	79			328	6695
Psi^3	5		5.55	81			339	6903
Omega	5		4.01	28			9072	224617
A	6		5.40	5			8807	218527
b	6		5.05	7			8878	220009
c+	6		5.63	32			9093	225003
d	6		5.37	41			80	1635

(table continued on next page)

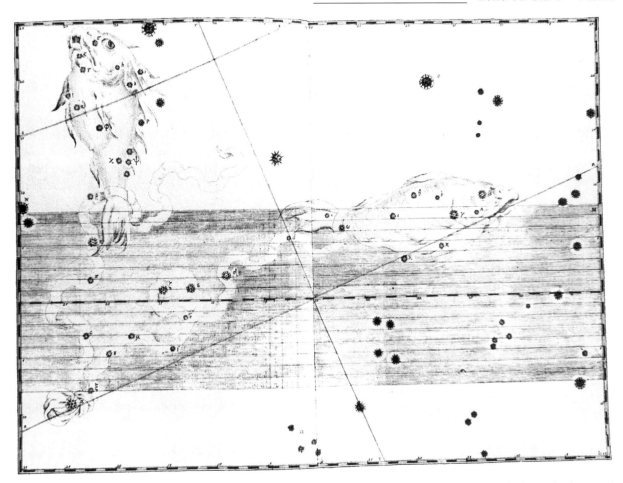

Figure 63. Bayer's Pisces. The center of the chart, where the line R (the celestial equator) crosses the ecliptic, marks the vernal equinox, the First Point of Aries, ♈, 0^h right ascension, where the sidereal day begins. The interval between two successive passages of the First Point of Aries across the meridian or zenith point marks a sidereal day, which is 3 minutes and 55.91 seconds shorter than a solar day.

Because of the wobble of Earth's axis, the First Point of Aries moves (precession) westward along the ecliptic about 1.4° every hundred years, thus slightly changing the position of the stars relative to the sun as seen by an observer on Earth. Four thousand years ago, the First Point of Aries was in the Ram, the Age of Aries. Today, as in Bayer's day, it is in the Fishes, the Age of Pisces, but several degrees farther west. In four hundred years, it will eventually move out of the Fishes and into the Water Bearer, thus ushering in the Age of Aquarius.

The letter M is the constellation Aquarius; N is the left horn of the Ram; O is the wing of Pegasus; P is the left elbow of Andromeda; and Q is the tail of Cetus. The lines R, S, and T are, respectively, the celestial equator, the Tropic of Cancer, and the equinoctial colure. The point where the equinoctial colure, the celestial equator, and the ecliptic meet is the First Point of Aries, which is marked by the letter V on Bayer's chart. When the sun reaches this point in March it is the time of the vernal equinox. The shaded area is the zodiac.

Table 71. The lettered stars of Pisces *(continued)*

	MAGNITUDES			CATALOGUE NUMBERS				
Letter	Bayer	caille	Visual	Flamsteed	caille	Other	HR	HD
e	6		5.52	80			330	6763
f	6		5.16	89			378	7804
g	6		5.16	82			349	7034
h	6		5.42	68			274	5575
i+	6		5.54c	65			230/1	4757/8
k	6		6.00	67			262	5382
l	6		5.23	91			389	8126

The Lost, Missing, or Troublesome Stars of Pisces

Kappa¹, 8, HR 8911, 4.94V, *SA* as Kappa, *HRP* as Kappa, *BS* as Kappa, HF 8 as Kappa, BV 9 as Kappa, BF 3214 as Kappa, H 14 as Kappa.
Kappa², 9, HR 8912, 6.25V, HF 9, BV 10, BF 3216.

Bayer described Kappa as a single star of the 5th magnitude, and Halley's edition of Flamsteed's catalogue labels his 8 as Kappa. Flamsteed's 1725 catalogue, though, designates these two nearly adjacent stars, his 8 and 9, as Kappa-1 and Kappa-2. Flamsteed estimated 9's magnitude as 7.6, which Baily interpreted to mean a bit brighter than 7 or about 6¾. Consequently, Baily dropped the letter from 9. He felt the difference in magnitude between Flamsteed's and Bayer's measurements was too great for the two stars to be synonymous. Eighty years earlier, Bevis had come to the same conclusion since he too equated only 8 with Kappa.

Sigma¹, 69, HR 291, 5.50V, *SA* as Sigma, *HRP* as Sigma, *BS* as Sigma, HF 65 as Sigma, BV 100 as Sigma, BF 99*, H 69 as Sigma.
Sigma², 76, HD 6476, 6.27V, HF 71, BV 105, BF 109*.

Bayer described Sigma as a single 4th-magnitude star, and Halley's pirated edition of Flamsteed's catalogue designates his 69 as Sigma. Flamsteed's 1725 "Catalogus Britannicus," however, designates these two neighboring stars, his 69 and 76, as Sigma-1 and Sigma-2. Bevis, though, labeled only the brighter star, 69, as Sigma, and Argelander agreed. He felt that Bayer must have meant the brighter star since he could not see 76 with his naked eye (*UN,* p. 48). The star 69 Psc is the same as 40 And.

Psi¹, 74, HR 310/1, 5.34V, 5.56V, *SA, HRP, BS,* H 74*.
Psi², 79, HR 328, 5.55V, *SA, HRP, BS,* H 78*.
Psi³, 81, HR 339, 5.55V, *SA, HRP, BS,* H 81*.

Bayer described Psi as a group of three 5th-magnitude stars, and Flamsteed designated his 74, 79, and 81 as Psi-1, Psi-2, and Psi-3.

A, 5, HR 8807, 5.40V, *SA, HRP.*

b, 7, HR 8878, 5.05V, *SA, HRP.*

c¹, 31, HR 9092, 6.32V, HF 30, BV 32, BF 3288*, H 37.
c², 32, HR 9093, 5.63V, *SA* as c, *HRP* as c, HF 31 as c, BV 31 as c, BF 3289*, H 38 as c.

Bayer described c as a single 6th-magnitude star, and Halley's 1712 edition of Flamsteed's catalogue designates his 32 as c. Flamsteed's 1725 catalogue, however, labels these two neighboring stars, his 31 and 32, as c-1 and c-2. Bevis dropped c from the dimmer star, 31. Argelander also dropped 31's letter since he could not see it without telescopic assistance (*UN,* p. 48). Although Heis observed both stars with his naked eye, he accepted Argelander's designation.

d, 41, HR 80, 5.37V, *SA, HRP.*

e, 80, HR 330, 5.52V, *SA, HRP.*

f, 89, HR 378, 5.16V, *SA, HRP.*

g, 82, HR 349, 5.16V, *SA, HRP.*

h, 68, HR 274, 5.42V, *SA, HRP.*

i, 65, HR 230/1, 7.0V, 7.1V, comb. mag. 5.54V, *SA, HRP,* B 152*, BF 82*, ADS 683.

Bayer described i as a single star of magnitude 6. Later astronomers discovered it to be a double star, Struve 61.

j

Bayer did not use this letter.

k, 67, HR 262, 6.00V, *SA, HRP.*

l, 91, HR 389, 5.23V, *SA, HRP.*

PISCIS AUSTRINUS, PsA
Southern Fish

Piscis Austrinus (Figure 64) is a Ptolemaic constellation that Bayer lettered from Alpha to Mu. Since the southern section is barely visible from Central Europe, virtually the entire constellation was redrawn and relettered by Lacaille, with some additional letters added by Bode. Most astronomers, however, felt that Lacaille's wholesale revisions in the upper half of the constellation were unwarranted and they restored Bayer's letters.

Various synonyms for south or southern have been employed in describing this Fish. Bayer, for instance, called the constellation Piscis Notius, Meridanus, or Austrinus while Lacaille labeled it Piscis Australis. Flamsteed designated it Piscis Austrinus, and most modern astronomers have accepted his designation.

Piscis Austrinus can be traced back in Mesopotamian astronomical literature at least to the latter half of the second millennium B.C., when it was considered one of the three reigning constellations of the twelfth month, Addaru (February-March). Its cuneiform symbol is KU_6, which is Akkadian for Nunu, Fish. The constellation evolved as a fish because of its location directly below the Ibex (Aquarius), the hindquarters of which were usually depicted as submerged in the marshes of southern Mesopotamia. The Ibex, a huge constellation and one of the four major constellations of Sumerian astronomy — together with Leo, Scorpius, and Taurus — was the favorite animal and symbol of Enki (Aquarius), the god of fresh water, who was often shown with fish swimming in streams of water that flowed from his shoulders.

The Greek and Roman legends associated with the Southern Fish, like those of its neighbor Pisces, take place in the Middle East. One such legend recounts the tale of the Syrian goddess Derceto — another form of Ishtar, Astarte, and Aphrodite — who falls into a lake not far from the Euphrates. She is saved by a big fish who is rewarded with a place in the heavens.

See van der Waerden, "Babylonian Astronomy. II. The Thirty-Six Stars," pp. 9, 11, 13-14; Hartner, "The Earliest History of the Constellations in the Near East," diagram 1; Langdon, *Babylonian Menologies,* p. 9; Pritchard, *Anc. Near East Picts.,* plate 685; Jacobsen, *Toward the Image of Tammuz,* p. 7; Reiner and Pingree, *Baby. Plan. Omens,* Part 2, pp. 13; Condos, *Star Myths,* pp. 163-65; Ridpath, *Star Tales,* pp. 108-9.

Lost Stars

Figure 64. Bayer's Piscis Austrinus. The letter Nu (ν) is the constellation Grus; Xi (ξ) is the belly of Capricornus; and Omicron (o) is the star Delta Aqr. The line Pi (π) is the Tropic of Capricornus. The shaded area across the top is the zodiac.

Table 72. The lettered stars of Piscis Austrinus

	MAGNITUDES			CATALOGUE NUMBERS				
Letter	**Bayer**	**Lacaille**	**Visual**	**Flamsteed**	**Lacaille**	**Other**	**HR**	**HD**
Alpha+	1		1.16	24			8728	216956
Beta	4		4.29	17			8576	213398
Gamma	4		4.46	22			8695	216336
Delta	4		4.21	23			8720	216763
Epsilon	4		4.17	18			8628	214748
Zeta+	4		6.43				8570	213296
Eta	4		5.42	12			8386	209014
Theta	4		5.01	10			8326	207155
Iota	4		4.34	9			8305	206742
Kappa+	4		3.01				8353	207971
Lambda	5		5.43	16			8478	210934
Mu	5		4.50	14			8431	210049
Nu+		6	6.47	13	1777		8405	209476
Pi+		6	5.11		1852		8767	217792
Tau+			4.92	15		B 27	8447	210302
Upsilon+			4.99			B 24	8433	210066

The Lost, Missing, or Troublesome Stars of Piscis Austrinus

Alpha, 24, HR 8728, 1.16V, *SA, HRP, BS,* Fomalhaut.

Ptolemy included Fomalhaut, the *lucida* of the constellation, in both Piscis Austrinus and Aquarius, but Bayer depicted it only in the Southern Fish. Flamsteed, the traditionalist, listed it in both constellations — as 24 PsA and as 79 Aqr.

Zeta, HR 8570, 6.43V, *SA, HRP, BS,* HF 11*, BF 3073*, CA 1808*, BR 7175*, L 9160*, BAC 7839*, H 9*, G 41*.

Although Flamsteed observed this star and Halley's 1712 edition of his catalogue labels it Zeta, it, together with several hundred other stars, is omitted from his authorized catalogue of 1725 (*BF,* p. 392). Both Bevis and Lacaille identified this star as Bayer's Zeta. Baily mentioned in his notes to the revised edition of Flamsteed's catalogue that this star was Bayer's Zeta, and he identified it as Zeta in the *BAC.* Although low on the horizon in Central Europe, it was observed without telescopic assistance by Argelander (*UN,* p. 115) and Heis.

There appears to be some problem with this star's reported magnitude. It is a relatively faint star, about magnitude 6½, but Ptolemy included it in his catalogue and described it as 5th magnitude (Toomer, *Ptolemy's Almagest,* p. 398). Although not observed by Tycho, it is included in Bayer's *Uranometria* and described as 4th magnitude. Almost a quarter of a century later, while preparing Schiller's *Coelum Stellatum Christianum,* Bayer drew the constellation more accurately. He reduced Zeta's magnitude to 5 and renamed it **7 Hydriae Farinae Sarepthanae Viduae** or the Vase of Grain of the Widow of Zarephath (I Kings 17:9-16), formerly Zeta PsA. In Halley's 1712 edition of Flamsteed's catalogue, Zeta's magnitude is given as 4.5, or somewhat fainter than 4, about 4½. But in Lacaille's catalogue it is listed as magnitude 6 and in Piazzi's catalogue (XIII, 118, erroneously as Xi) as magnitude 7. Argelander and Heis described it as 5½.

Gould was troubled by these discrepancies in Zeta's magnitude and paid special attention to it while observing the southern skies at Cordoba, Argentina. After several observations, he determined its magnitude at 6.7 without any sign of variability (*Uran. Arg.,* p. 296). More recently, the *BS* lists Zeta as a suspected variable, but even this would not account for the wide range of reported magnitudes. If HR 8570 is indeed the star that Ptolemy and Bayer observed — and its coordinates match almost exactly — it is one of the faintest stars in their catalogues, just at the limit of naked eye visibility.

Kappa, in Grus, HR 8353, 3.01V, CA 1762 as Gamma Gru, BR 7094 as Gamma Gru, L 8951 as Gamma Gru, G 18 as Gamma Gru.

Lacaille shifted the lower section of Piscis Austrinus to Grus and relettered Bayer's Kappa PsA as Gamma Gru. See Ptolemy's star 1022 in Baily, "Catalogues of Ptolemy *et al.*"; Toomer, *Ptolemy's Almagest,* p. 309.

Nu, 13, HR 8405, 6.47V, CA 1777*, BR 7118*, L 9009, BAC 7673, G 23.

Bayer did not letter this star. The letter was added by Lacaille but dropped by Baily because of the star's dimness.

Xi
Omicron

These letters are not used in Piscis Austrinus.

Pi, HR 8767, 5.11V, *SA, HRP, BS,* CA 1852*, BR 7237*, L 9350, G 72*.

Bayer did not letter this star in Piscis Austrinus. Pi was added by Lacaille.

Rho
Sigma

These letters are not used in Piscis Austrinus.

Tau, 15, HR 8447, 4.92V, *SA, HRP, BS,* CA 1784 as Delta, BR 7135 as Delta, L 9037, B 27*, G 30*.
Upsilon, HR 8433, 4.99V, *SA, HRP, BS,* BR 7132, L 9030, B 24*, G 26*.

The letters of these two stars did not originate with either Bayer or Lacaille. Lacaille observed these two stars, and lettered the first Delta, although Bayer had already assigned Delta to another star, HR 8720, 4.21V. He left the second unlettered. Tau and Upsilon were assigned by Bode, who also added additional letters, none of which has survived in the literature except for these two. They owe their survival primarily to Gould, who included them in his catalogue because he felt stars of their brightness merited letters.

In redrawing and revising the Southern Fish, Lacaille relettered several of Bayer's stars:

Gamma, 22, HR 8695, 4.46V, *SA, HRP, BS,* CA 1839 as Epsilon, BR 7218 as Epsilon, L 9287 as Gamma, G 62 as Gamma.
Delta, 23, HR 8720, 4.21V, *SA, HRP, BS,* CA 1842 as Eta, BR 7224 as Eta, L 9304 as Delta, G 63 as Delta.
Epsilon, 18, HR 8628, 4.17V, *SA, HRP, BS,* CA 1821 as Gamma, BR 7193 as Gamma, L 9206 as Epsilon, G 52 as Epsilon.

Lacaille relettered these three stars as Epsilon, Eta, and Gamma, but Baily and Gould restored Bayer's letters. In addition, Lacaille called his CA 1845 (HR 8732, 6.13V, BR 7226 as Kappa, L 9316, G 67) Kappa, but Baily and Gould removed the letter, not only because of the star's dimness but also because of its potential to become confused with Bayer's Kappa, which Lacaille had relettered as Gamma Gru. See Kappa.

Lacaille's relettering of Piscis Austrinus resulted from his efforts to revise some of Bayer's southern constellations that bore little or no resemblance to the star's actual appearance in the heavens. Lacaille proposed to redesign Ara, Argo Navis, Centaurus, Lupus, and Piscis Austrinus ("Table des Ascensions," p. 589). Although it was necessary to revise the first four, the Southern Fish required no such extensive relettering since Bayer's chart was quite accurate. Consequently, most astronomers, like Baily and Gould, sought to undo Lacaille's work and to restore Bayer's original letters.

Puppis, Pup

Ship's Stern

Puppis (Figure 65) is part of the former Ptolemaic constellation Argo Navis. Lacaille broke it up into three separate constellations — Carina (Keel), Puppis (Stern), and Vela (Sails). Bayer lettered Argo Navis from Alpha to Omega and from A to s. Lacaille eliminated Bayer's letters and relettered Puppis from a to z and from A to Z.

On star charts and in stellar catalogues, Argo Navis is shown only from the stern to amidships. The whole front of the ship appears to be missing. This has led to speculation as to why only half of the vessel is pictured. Some have suggested that it should remind seafarers not to fear shipwrecks; that is, that damaged ships can still be beached, repaired, and relaunched. But a recent and more logical explanation has suggested that Argo Navis may have been modeled after Phoenician warships of the sixth and seventh century B.C. These vessels, unlike Egyptian, Greek, and Roman ships with their curved or pointed prows, ended in a strait vertical line, giving the impression that their bows or forward parts had been sheered off.

See Condos, *Star Myths,* pp. 39-42. For an account of the legends associated with the constellation and a detailed analysis of its Greek letters, see Carina.

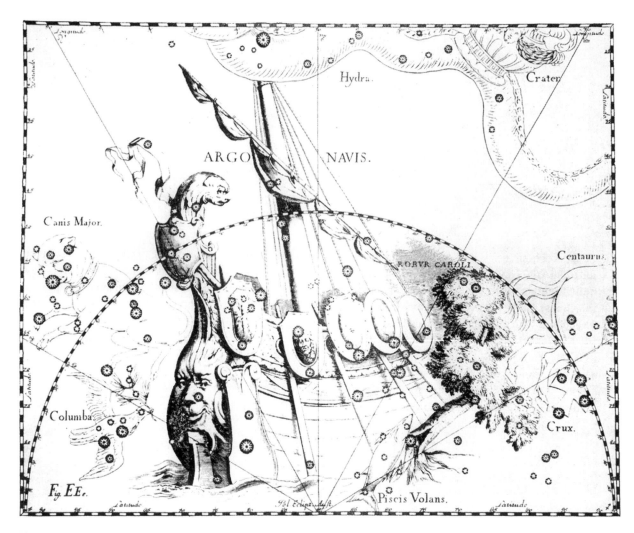

Figure 65. Hevelius's Argo Navis with the asterism Robur Caroli, the Oak of Charles II (see Figure 31). All the lines of longitude below the ecliptic come together at the south ecliptic pole at the bottom center of the chart.

Table 72. The lettered stars of Puppis

		MAGNITUDES			CATALOGUE NUMBERS			
Letter	Bayer	Lacaille	Visual	Flamsteed	Lacaille	Other	HR	HD
Zeta		2	2.25		729		3165	66811
Nu		3	3.17		557		2451	47670
Xi		4	3.34	7 Nav	691		3045	63700
Pi		3	2.70		627		2773	56855
Rho		4	2.81	15 Nav	731		3185	67523
Sigma		4	3.25		655		2878	59717
Tau		4	2.93		579		2553	50310
a		5	3.73		704		3080	64440
b+		5	4.49		706		3084	64503
c+		5	3.61		683		3017	63032
d^1+		6	4.84		669		2961	61831
d^2		6	5.73		670		2963	61878

(table continued on next page)

Table 72. The lettered stars of Puppis (continued)

	MAGNITUDES			CATALOGUE NUMBERS				
Letter	Bayer	Lacaille	Visual	Flamsteed	Lacaille	Other	HR	HD
d³		6	5.76		671		2964	61899
d⁴		6	6.00		672		2968	61925
e+		6	5.80		667		2954	61641
f		6	4.53		664		2937	61330
g+		5	6.65		658		2912	60629
h¹+		6	4.45		739		3225	68553
h²		6	4.44		746		3243	69142
i+		5	6.44		736		3219	68450
j+			4.20	11 Nav		G 222	3102	65228
k¹+		5	4.50		665		2948	61555
k²			4.62				2949	61556
l+		6	3.96	3 Nav	678		2996	62623
m+		6	4.70		663		2944	61429
n+		5	5.09c		656		2909/10	60584/5
o+		6	4.50		686		3034	63462
p+		6	4.64		661		2922	60863
q		5	4.45		750		3270	70060
r+		6	4.78		743		3237	68980
s+		6	5.43		640		2819	58155
t		6	5.06		587		2619	52092
u+		6	5.96		576		2558	50445
v¹+			4.66			G 83	2787	57150
v²			5.11			G 84	2790	57219
w		5	4.83		753		3282	70555
x+		6	5.26		570		2518	49591
y+		6	5.43		653		2875	59635
z		6	5.54		659		2911	60606
A		5	4.83		607		2702	54893
B+		6	4.40		581		2554	50337
C		6	5.20		595		2666	53704
D+		6	5.79		601		2691	54475
E		6	5.31		610		2727	55719
F		6	5.25		632		2791	57240
G		6	5.76		523		2348	45572
H		6	4.93		598		2672	53811
I		5	4.49		615		2740	55892
J+			4.24			G 218	3090	64760
K+		6	6.26		735		3205	68242
L¹+		5	4.89		616		2746	56022
L²		6	5.10		617		2748	56096
M+		6	5.85		633		2789	57197
N		6	5.09		718		3116	65551
O		6	5.17		720		3121	65685
P		5	4.11		698		3055	63922
Q		6	4.71		697		3046	63744
S+		6	7.2		693			63451

The Lost, Missing, or Troublesome Stars of Puppis

a, HR 3080, 3.73V, *SA, HRP,* CA 704*, BR 1799*, L 3044*, G 213*.

b, HR 3084, 4.49V, *SA, HRP,* CA 706*, BR 1801*, L 3049*, G 214*.
c, HR 3017, 3.61V, *SA, HRP,* CA 683*, BR 1735*, L 2958*, G 175*.

Hoffleit confused Lacaille's designations for these two stars in the *BS* (pp. 398-99). Lacaille's catalogue of 1763, his *Coelum Australe,* correctly refers to these stars as b and c Pup.

d^1, HR 2961, 4.84V, *SA, HRP,* CA 669 as d, BR 1692*, L 2909*, G 143*.
d^2, HR 2963, 5.73V, *SA, HRP,* CA 670 as d, BR 1695*, L 2912*, G 144*.
d^3, HR 2964, 5.76V, *SA, HRP,* CA 671 as d, BR 1697*, L 2913*, G 145*.
d^4, HR 2968, 6.00V, *SA,* CA 672 as d, BR 1698*, L 2914*, G 146.

Lacaille designated these four neighboring stars d. Gould removed CA 672's letter because it stood somewhat apart from the other three and because of its dimness.

e, HR 2954, 5.80V, *SA, BS,* CA 667*, BR 1687*, L 2903, BAC 2536*, G 138.

SA shows two e's in Puppis, this star and HR 3102 (see j). Gould dropped this star's letter because of its dimness; he considered it 6th magnitude.

f, HR 2937, 4.53V, *SA, HRP,* CA 664*, L 2890*, G 127*.

g, HR 2912, 6.65V, CA 658*, BR 1650*, L 2854*, BAC 2500*, G 114.

Gould dropped this star's letter because of its dimness.

h^1, HR 3225, 4.45V, *SA, HRP,* CA 739 as h, BR 1925*, L 3191*, G 267*.
h^2, HR 3243, 4.44V, *SA, HRP,* CA 746 as h, BR 1943*, L 3223*, G 279*.

Lacaille designated these two neighboring stars h.

i, HR 3219, 6.44V, CA 763*, BR 1922*, L 3183*, BAC 2758, G 265.

Gould dropped this star's letter because of its dimness.

j, 11, HR 3102, 4.20V, *HRP, BS,* H 32 as e, G 222*.

Lacaille did not letter nor observe this star. Gould assigned j to Flamsteed's 11 since he felt a star of its brightness merited a letter. In assigning j to this star, Gould noted it was Bayer's original e Nav, but he changed the star's letter to j to avoid confusing it with Lacaille's e (see e, CA 667; *Uran. Arg.,* p. 89). Other astronomers, like Flamsteed, Argelander (*UN,* p. 97), and Heis, retained Bayer's e for this star. As a result, some modern works, like Becvar's atlases and Tirion's *SA,* show two e's in Puppis: Lacaille's e, CA 667, and this star. This is one of the rare instances in which one of Gould's proposals has not been universally accepted. In the second edition of *SA,* both e's have been removed.

k^1, HR 2948, 4.50V, *SA, HRP* as k, *BS,* HF 1 as Kappa, CA 665 as k, BR 1679*, L 2896 erroneously as k^3, BAC 2530*, BF 1074 as Kappa, H 14 as Kappa, G 133 as k.
k^2, HR 2949, 4.62V, *SA, BS,* BR 1680*, BAC 2531*, G 134.

Lacaille described k as a single star of the 5th magnitude. Brisbane was among the first astronomers to use indices to distinguish the components of this double star, ADS 6255. Gould (G 133), and Pickering in the *HRP,* designated only HR 2948 as k but they noted that its light combined with the light of HR 2949 (G 134) appears as one image to the naked eye with a combined magnitude of 3.82V. Although this star is included in Halley's 1712 edition of Flamsteed's catalogue, it is omitted from Flamsteed's later catalogue (*BF,* p. 392). It was originally Bayer's Kappa Nav before Lacaille relettered it k Pup, and both Argelander (*UN,* p. 96) and Heis referred to it as Kappa. Baily designated it Kappa in his revised edition of Flamsteed's catalogue and k in the *BAC* (see his note to BAC 2530). It should not be confused with Lacaille's k^1, k^2 CMa, which Baily included in Puppis:

k¹, in Puppis, HR 2873, 5.77V, CA 650 as k CMa, BR 1624 as k CMa, L 2823 erroneously as k¹ Pup, BAC 2478 in Puppis, G 96 in Puppis.

k², in Puppis, HR 2881, 4.65V, CA 654 as k CMa, BR 1634 as k CMa, L 2834 erroneously as k² Pup, BAC 2484 in Puppis, G 100 in Puppis.

In his expanded edition of Lacaille's catalogue, Baily included k¹ and k² CMa in Puppis erroneously as k¹ and k² Pup. He lettered HR 2948/9 as k³ Pup (L 2896) to follow his designation of HR 2873 and HR 2881 as k¹ and k² Pup. Gould left the latter two stars in Puppis but he removed their letters. See Gould, *Uran. Arg.*, p. 64; Hoffleit, "Discordances," p. 60; *BS,* p. 398.

l, 3, HR 2996, 3.96V, *HRP, BS,* CA 678*, BR 1717*, L 2938*, BAC 2562, BF 1088 as Tau, H 19 as Tau, G 157*.

This star was originally Bayer's Tau Nav before Lacaille relettered it l Pup. Both Argelander (*UN,* p. 96) and Heis referred to it as Tau. See *BS,* p. 398.

m, HR 2944, 4.70V, *SA, HRP, BS,* CA 663*, L 2888*, BAC 2525*, H 13 as Pi, G 128*.

This star was originally Bayer's Pi Nav before Lacaille relettered it m Pup. Both Argelander (*UN* p. 96) and Heis referred to it as Pi. See *BS,* p. 398.

n, HR 2909/10, 5.83V, 5.87V, comb. mag. 5.09V, *SA,* CA 656*, BR 1648*, L 2849*, BAC 2497 as n¹, BAC 2498 as n², G 111/2*, ADS 6190.

Lacaille described n as a single star of the 5th magnitude. Later astronomers added indices to distinguish the components of this double star. Brisbane noted it was a double but referred to it simply as n. Gould split the pair and designated each component as n. Although the *HRP* notes that the combined light of both stars appears as one image to the naked eye, it assigns n only to HR 2909.

o, HR 3034, 4.50V, *SA,* CA 686*, BR 1750*, L 2981*, G 183*.

Both the *HRP* and the *BS* erroneously designate this star Omicron. It should be o (oh). See Omicron in Carina.

p, HR 2922, 4.64V, *SA, HRP,* CA 661*, BR 1657, L 2867, BAC 2508*, G 119*.

Lacaille made a clock error in his observation of this star that resulted in erroneous coordinates. Baily corrected the error in his expanded edition of Lacaille's catalogue and equated this star with Brisbane's BR 1657.

q, HR 3270, 4.45V, *SA, HRP,* CA 750*, BR 1968*, L 3259*, G 289*.

r, HR 3237, 4.78V, *SA, HRP,* CA 743*, BR 1938, L 3212, BAC 2774*, CAP 1463*, G 274*.

Lacaille described r, CA 743, as 6th magnitude, and Brisbane equated his BR 1930, also 6th magnitude, with CA 743. Baily, in his expanded edition of Lacaille's catalogue and in the *BAC,* described BR 1938 as synonymous with r, and Maclear in the *Cape Catalogue of 1850* agreed. Neither Baily nor Maclear could locate Brisbane's BR 1930.

s, in Canis Major, HR 2819, 5.43V, CA 640*, BR 1581*, L 2769, BAC 2449*, G 161 in Canis Major.

Gould included this star in Canis Major and left it unlettered.

t, HR 2619, 5.06V, *SA, HRP,* CA 587*, BR 1421*, L 2554*, G 49*.

u, HR 2558, 5.96V, CA 576*, BR 1382*, L 2493*, BAC 2258*, G 40.

Gould dropped this star's letter because of its dimness; he considered it less than 6th magnitude.

v¹, HR 2787, 4.66V, *SA, HRP,* CA 629, BR 1544, L 2733, G 83*.
v², HR 2790, 5.11V, *SA, HRP,* CA 631, BR 1548, L 2736, G 84*.

Lacaille observed and catalogued these two nearly adjacent stars but left them unlettered; he considered them 6th magnitude. Gould designated them v¹ and v², since he felt stars of their brightness merited letters.

w, HR 3282, 4.83V, *SA, HRP,* CA 753*, BR 1979*, L 3277*, G 294*.

Lost Stars

x, HR 2518, 5.26V, *SA, HRP,* CA 570*, BR 1359 erroneously as Chi, L 2455*, G 31*.

Brisbane's use of Chi was probably the result of a typographical or copying error since Chi and x are similar in appearance, χ, x.

y, HR 2875, 5.43V, *SA* erroneously as y¹, *HRP,* CA 653*, BR 1628*, L 2832, BAC 2479*, G 98*.

See Y.

z, HR 2911, 5.54V, *SA, HRP* erroneously as Z, CA 659*, BR 1653*, L 2860, BAC 2502*, G 115*.

A, HR 2702, 4.83V, *SA, HRP,* CA 607*, BR 1486*, L 2649*, G 67*.

B, in Carina, HR 2554, 4.40V, CA 581*, BR 1388*, L 2511, BAC 2259 erroneously as B Car, G 18 as A Car.

Baily included this star in Carina and erroneously lettered it B Car. Gould relettered it A Car. See A Car.

C, HR 2666, 5.20V, *SA, HRP,* CA 595*, BR 1462*, L 2607*, G 59*.

D, HR 2691, 5.79V, *SA,* CA 601*, BR 1479*, L 2638*, G 66.

Gould dropped this star's letter because of its dimness; he considered it less than 6th magnitude.

E, HR 2727, 5.31V, *SA, HRP,* CA 610*, BR 1504*, L 2672*, G 70*.

F, HR 2791, 5.25V, *SA, HRP,* CA 632*, BR 1552*, L 2739*, G 87*.

G, HR 2348, 5.76V, *SA, HRP,* CA 523*, BR 1245*, L 2297*, G 15*.

H, HR 2672, 4.93V, *SA, HRP,* CA 598*, BR 1467*, L 2624*, G 61*.

I, HR 2740, 4.49V, *SA, HRP,* CA 615*, BR 1512*, L 2687*, G 71*.

J, HR 3090, 4.24V, *SA, HRP, BS,* CA 713 as R, BR 1812 as R, L 3068 as R, BAC 2644 as R, G 218*.

Lacaille did not use this letter for this star. This star was originally Lacaille's R. Since Argelander had started the practice of assigning Roman capitals beginning with R to variable stars, Gould dropped R to avoid confusion with variables and relettered this star J because of its brightness. See *BS,* p. 399.

K, HR 3205, 6.26V, CA 735*, BR 1914*, L 3179, BAC 2753*, G 264.

Gould dropped this star's letter because of its dimness.

L¹, HR 2746, 4.89V, *SA, HRP,* CA 616 as L, BR 1516*, L 2690*, G 72*.
L², HR 2748, 5.10V, *SA, HRP,* CA 617 as L, BR 1520*, L 2691*, G 73*.

Lacaille designated these two neighboring stars L.

M, HR 2789, 5.85V, *SA,* CA 633*, BR 1553*, L 2742*, G 85.

Gould dropped this star's letter because of its dimness; he considered it less than 6th magnitude.

N, HR 3116, 5.09V, *SA, HRP,* CA 718*, BR 1831*, L 3089, BAC 2661*, G 228*.

O, HR 3121, 5.17V, *SA, HRP,* CA 720*, BR 1836 erroneously as o, L 3099*, G 231*.

P, HR 3055, 4.11V, *SA, HRP,* CA 968*, BR 1778*, L 3022*, G 199*.

Q, HR 3046, 4.71V, *SA, HRP,* CA 697*, BR 1772*, L 3017*, G 196*.

R, HR 3090, 4.24V, CA 713*.

Gould relettered this star J. See J.

Lettered Stars - Puppis

S, HD 63451, 7.2V, *SA,* CA 693*, BR 1759*, L 2999, BAC 2597*.

Gould retained the letter S for this star since he considered it variable, but, because of its dimness, he did not include it in his catalogue (*Uran. Arg.,* p. 89). Although S Pup is no longer considered variable, it still bears Lacaille's designation and is included in Kholopov's *Gen. Cat. Var. Stars,* III, 50-51. See also Hirschfeld and Sinnott, *Sky Cat. 2000.0,* II, 233.

T, HR 2998, 5.06V, CA 682*, BR 1722*, L 2950*, BAC 2566*, G 159.

Gould dropped this star's letter to avoid confusion with variables.

U

Lacaille did not use this letter.

V, HR 2462, 4.93V, *BS,* CA 559*, BR 1321 erroneously as Upsilon, L 2402*, BAC 2193*, G 21.

Gould dropped this star's letter to avoid confusion with variables. Becvar's atlases, as well as Tirion's *SA* and *Uranometria 2000.0,* erroneously designate this star as Y. It should not be confused with the semiregular variable Y shown on *SA* at approximately 8^h12^m, -35°. Brisbane's mistake in lettering this star Upsilon was probably the result of a copying or typographical error since V and Upsilon are similar in appearance, V, υ. This may also have been at the root of Becvar's error since the letter Y is derived from the Greek Upsilon. See *BS,* p. 394. Tirion's second edition of *SA* leaves the star unlettered.

W, HR 3002, 5.17V, *BS,* CA 681*, BR 1723*, L 2945*, BAC 2570*, G 162.

Gould dropped this star's letter to avoid confusion with variables.

X, HR 2548, 5.14V, *BS,* CA 575*, BR 1379 erroneously as x, L 2492*, BAC 2253*, G 38.

Gould dropped this star's letter to avoid confusion with variables.

Y^1, HR 2940, 5.72V, CA 668 as Y, BR 1681*, L 2904, BAC 2529*, G 130.
Y^2, HR 2957, 5.68V, CA 673 as Y, BR 1694*, L 2918, BAC 2541*, G 142.

Lacaille described these two adjacent stars as Y, but Gould dropped their letters to avoid confusion with variables. Becvar's atlases, as well as Tirion's *SA* and *Uranometria 2000.0,* erroneously designate these stars as y^2 and y^3, and Lacaille's y, CA 653, HR 2875, as y^1 although it is over 10° to the north. The second edition of *SA* leaves these two stars (HR 2940 and HR 2957) unlettered and properly designates HR 2857 as y. See y and V.

Z, HR 2384, 5.27V, CA 538*, BR 1267 erroneously as z, L 2333 erroneously as z, BAC 2137*, G 16.

Gould dropped this star's letter to avoid confusion with variables. L 2333 is corrected in the Errata, p. 299. The *HRP* erroneously designates Lacaille's z, CA 659, as Z. See z.

Pyxis, Pyx

Ship's Compass

Lacaille devised and lettered the stars in the constellation Pyxis (Figure 2d) from Alpha to Lambda. He originally called it la Boussole (Compass) and described it as the Marine Compass. On the chart accompanying his catalogue of 1763 it is labeled Pixis [Pyxis] Nautica, the Nautical Box Compass. Many astronomers believe that this is one of the constellations Lacaille formed out of the old Argo Navis. This is not quite accurate. While formed from some of the stars of the Ship, Pyxis was meant to represent *a* ship's compass, not *the* compass of Argo Navis. This generic presentation of a compass can readily be understood by examining Lacaille's chart of the southern skies, where an archetypical compass is shown that is not in anyway related to the picture of the Ship. Moreover, Pyxis's letter designations are different from the three constellations formed from Argo Navis: Carina, Puppis, and Vela. It has, for instance, an independent set of Greek letters, whereas the others have a single alphabet divided among all three. Finally, as Lacaille himself noted in regard to the Ship: *"je l'ai ensuite partagée en trois parties, favoir, la Pouppe, le Corps & la Voilure* (I have divided it into three parts: the Stern, the Hull, and the Sails)." Baily renamed the constellation Malus, or the Ship's Mast, because several of the stars that Lacaille included in Pyxis were originally placed by Ptolemy in the Ship's Mast, especially Bayer's four o's. However, Gould restored Lacaille's original designation, Pyxis.

See Lacaille, "Table des Ascension," p. 590; *BAC*, pp. 62-63; Allen, *Star Names*, p. 348.

Table 73. The lettered stars of Pyxis

	Magnitudes			Catalogue Numbers				
Letter	Bayer	Lacaille	Visual	Flamsteed	Lacaille	Other	HR	HD
Alpha		5	3.68		788		3468	74575
Beta		5	3.97		780		3438	74006
Gamma		6	4.01		802		3518	75691
Delta		6	4.89		809		3556	76483
Epsilon		6	5.59		831		3644	78922
Zeta		6	4.89		777		3433	73898
Eta		6	5.27		773		3420	73495
Theta		6	4.72		846		3718	80874
Kappa+			4.58			G 49	3628	78541
Lambda		6	4.69		850		3733	81169

The Lost, Missing, or Troublesome Stars of Pyxis

Iota

Lacaille did not use the letter Iota in Pyxis.

Kappa, HR 3628, 4.58V, *SA, HRP, BS,* CA 828, BR 2338, L 3685, G 49* and Errata.

Although Lacaille observed and catalogued this star, he left in unlettered since he considered it 6th magnitude. Gould designated it Kappa because he felt a star of its brightness merited a letter.

RETICULUM, RET

Reticle

Isaac Habrecht II devised a constellation called Rhombus, which he depicted on his globe of 1621 as a simple quadrilateral. It became quite popular with stellar cartographers and was copied on numerous other globes and star charts. Lacaille drew his Reticle (figures 2c and 2d) in the approximate location of Habrecht's Rhombus and lettered its stars from Alpha to Iota.

Translated literally as the Net by countless astronomers, Reticulum actually referred to the reticle of Lacaille's rhomboidal micrometer, which was composed of fine lines for measuring stellar positions. A picture of the reticle eyepiece, with its rhomboidal diaphragm, is depicted on his chart of the southern skies. In the first edition of his catalogue, Lacaille called it Le Réticule rhomboïde and described it as a "small astronomical instrument which has helped in the preparation of this catalogue; it is made by the intersection of four straight chords from each angle of a square to the middle of the two opposite sides." In his catalogue of 1763, this constellation is called simply Reticulus [Reticulum].

See Warner, *Sky Explored,* pp. 104-5 and *passim*; Bartsch, *Usus Astronomicus,* p. 66; Lacaille, "Table des Ascensions," p. 588. For a detailed account of how Lacaille used this instrument to determine a star's coordinates, see the Preface to Lacaille's expanded catalogue that Baily edited, p. iii; and Owen Gingerich's article on Lacaille in the *Dict. Sci. Biog.,* VII, 542-45.

Table 74. The lettered stars of Reticulum

Letter	Magnitudes			Catalogue Numbers				
	Bayer	Lacaille	Visual	Flamsteed	Lacaille	Other	HR	HD
Alpha		3	3.35		329		1336	27256
Beta		4	3.85		292		1175	23817
Gamma		5	4.51		317		1264	25705
Delta		5	4.56		313		1247	25422
Epsilon		5	4.44		331		1355	27442
Zeta^1+		6	5.54		255		1006	20766
Zeta^2		6	5.24		256		1010	20807
Eta		5	5.24		340		1395	28093
Theta		5	5.87		336		1372	27657
Iota		6	4.97		316		1266	25728
Kappa+			4.72			G 6	1083	22001

The Lost, Missing, or Troublesome Stars of Reticulum

Zeta¹, HR 1006, 5.54V, *SA, HRP, BS,* CA 255 as Zeta, BR 536*, L 1074*, G 3*.
Zeta², HR 1010, 5.24V, *SA, HRP, BS,* CA 256 as Zeta, BR 537*, L 1077*, G 4*.

Lacaille designated these two nearly adjacent stars Zeta.

Kappa, HR 1083, 4.72V, *SA, HRP, BS,* CA 270, BR 566, L 1143, G 6*.

Although Lacaille observed and catalogued this star, he left it unlettered since he considered it only 6th magnitude. Gould designated it Kappa since he felt a star of its brightness merited a letter.

SAGITTA, SGE

Arrow

Sagitta (Figure 66) is a Ptolemaic constellation that Bayer lettered from Alpha to Theta.

There are several astral legends relating to Sagitta. In one account Sagitta represents the arrow used by Apollo, the god of medicine and archery, to slay the Cyclopes. The physician Asclepius, son of Apollo and the mortal Coronis, is such a skilled healer that he is even able to restore life to the dead. This angers Hades, god of the Underworld, who fears that Asclepius's power is so great that it will eventually lead to the depopulation of his realm. Hades appeals to Zeus, who destroys Asclepius with a burst of lightning. Angered by the death of his son, Apollo vents his rage on the Cyclopes, who had fashioned Zeus's lightning bolts, by slaying them with his bow and arrows. At Apollo's request, Zeus places Asclepius (Ophiuchus) among the stars.

In another legend, the Arrow in the heavens commemorates the arrow that Hercules uses to slay the eagle that continually torments the Titan Prometheus. Prometheus ("Forethought" or Intelligence) creates man and gives him the gift of fire to advance civilization. Prometheus discovers that when men make sacrifices to the gods, they offer the entire body of the animal, thereby depriving themselves of much needed food. Hoping to help the creatures he created, Prometheus tricks Zeus into accepting only the skin and bones of sacrifices, leaving the meat for humans. Upon discovering the deception, Zeus is so angered that he orders the god Hephaestus to create woman to plague man. In addition, Zeus takes back the gift of fire from mankind, but Prometheus restores it. As punishment for daring to defy him, Zeus has Prometheus chained to a rock on Mount Caucasus where he is tortured daily by an eagle that comes to gnaw away at his liver. This torture lasts for over thirty thousand years until Hercules slays the eagle with his bow and arrow.

See Condos, *Star Myths,* pp. 175-81; Ridpath, *Star Tales,* pp. 111-12; Reinhold, *Essentials of Greek and Roman Classics,* p. 335.

Table 75. The lettered stars of Sagitta

	Magnitudes			Catalogue Numbers				
Letter	Bayer	Lacaille	Visual	Flamsteed	Lacaille	Other	HR	HD
Alpha	4		4.37	5			7479	185758
Beta	4		4.37	6			7488	185958
Gamma	4		3.47	12			7635	189319
Delta	5		3.82	7			7536	187076
Epsilon	6		5.66	4			7463	185194
Zeta	6		5.00	8			7546	187362
Eta	6		5.10	16			7679	190608
Theta+	6		6.48	17			7705	191570

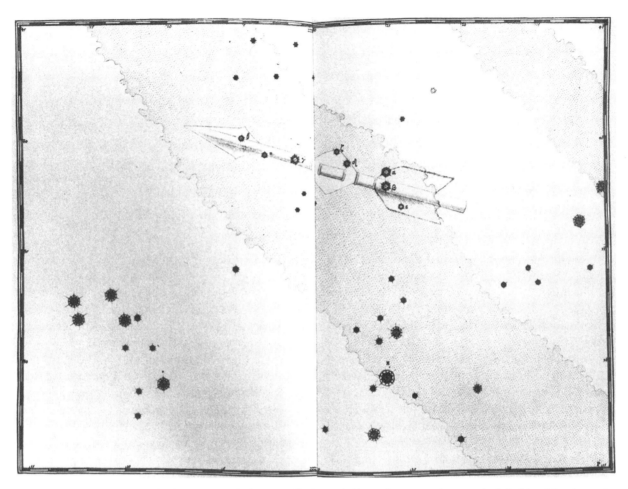

Figure 66. Bayer's Sagitta. The letter Iota (ι) is the constellation Delphinus, the group of stars at the left side of the chart. The letter Kappa (κ) is Aquila. On some copies of the *Uranometria* the letter Iota is missing.

The Lost, Missing, or Troublesome Stars of Sagitta

Theta, 17, HR 7705, 6.48V, *SA, HRP, BS,* H 17*, ADS 13442.

Although just at the limit of naked eye visibility, Theta was seen by Bayer, Argelander (*UN,* p. 79), and Heis. What they observed was the light of Theta, a multiple-star system, combined with that of several small stars just a few arc seconds away, all located in a dense sector of the Milky Way. HR 7705 is component A of the system, which consists of two additional stars: B and D. Component C, an optical double, is not part of the system (*Burnham's Celestial Handbook,* III, 1526; Hirschfeld and Sinnott, *Sky Cat 2000.0,* II, 131).

Chi, 13, HR 7645, 5.37V, HF 18 as x, BV 12, BF 2713, the variable VZ.
y, 14 Sge in Aquila, HR 7664, 5.67V, BV 14, BF 2730.
z, 15, HR 7672, 5.80V, BV 15, BF 2731.

Bayer did not letter these three stars. The letters were added by Flamsteed but removed by Bevis and Baily as superfluous (see i Aql). The Chi in Flamsteed's "Catalogus Britannicus" of 1725 is either a typographical or copying error for x since Chi and x are similar in appearance, χ, x; in Halley's 1712 pirated edition 13 is labeled x.

SAGITTARIUS, SGR

Archer

Sagittarius (Figure 67) is a Ptolemaic constellation that Bayer lettered from Alpha to Omega and from A to h. Lacaille introduced Roman capitals in the southern sector of the Archer that is mostly below the horizon in Europe and used them to replace several of Bayer's Greek letters. Baily removed them all since he felt only the three new components of Argo Navis should be designated with Lacaille's Roman letters. Moreover, since Bayer's original Greek lettered stars were, by and large, correctly positioned in the heavens, Baily made sure they were all properly restored. A generation later, Gould confirmed what Baily had done.

Sagittarius is a relatively new constellation that appeared for the first time in Mesopotamian astronomical literature in the eighth century B.C. It was considered one of the three reigning constellations of Kislimu (Kislev), the ninth month (November-December). Its cuneiform ideogram, PA.BIL.SAG, has been translated by some scholars as archer or arrow shooter, but its meaning is uncertain. It was usually depicted on Mesopotamian monuments and boundary stones as a centaur armed with a drawn bow and arrow, and probably was placed in the heavens to symbolize the new elements of warfare that revolutionized the battlefield in the second millennium B.C. — the horse and compound bow.

The horse and wild ass, or onager, had been used in warfare in the Middle East as early as the third millennium, but primarily to pull four-wheeled and later two-wheeled chariots. With the passage of time, however, the men of the steppes just north of Mesopotamia learned to ride horseback without the use of saddle, stirrups, or bridle. They controlled their mounts with their thighs, thus freeing their hands to use bows and arrows, like the Plains Indians of nineteenth-century North America. Their bows were radically different from those used by infantrymen since they had to be carried on horseback and aimed over the necks of their mounts. Their bows were therefore much shorter. To compensate for the shorter draw without compromising force, these horsemen used the compound bow, one made of wood and strengthened with bone and sinew (see Canis Major and Orion).

As a result of their masterful horsemanship, the cavalrymen of the latter half of the second millennium appeared to their enemies as centaurs — half-man, half-horse — as though man and beast were wedded together. The cavalrymen were fearsome warriors, terrorizing their opponents with their extraordinary skill as horsemen, their maneuverability on the field of battle, and their ability to rain a deadly shower of arrows on the ranks of massed infantrymen. These nomadic horsemen, like the Scythians and Cimmerians of Central Asia, invaded Asia Minor and Assyria in the eighth century, just about the time the constellation Sagittarius first appeared in Mesopotamian astronomical literature.

Sagittarius and Centaurus are often confused in Greek mythology since both are depicted as half-man and half-horse. Some classical poets, however, have suggested that Sagittarius is really a satyr, a man with a goat's hindquarters. They believed that Sagittarius represented the satyr Crotus, who was reputed to have invented archery.

See van der Waerden, "Babylonian Astronomy. II. The Thirty-Six Stars," p. 15; van der Waerden, "History of the Zodiac," p. 219; O'Neil, *Time and the Calendars*, pp. 58-59; Langdon, *Babylonian Menologies*, p. 8; McNeill, *The Rise of the West*, pp. 256-59; Reiner and Pingree, *Baby. Plan. Omens*, Part 2, p. 14; Condos, *Star Myths*, pp. 183-86.

Lettered Stars - Sagittarius

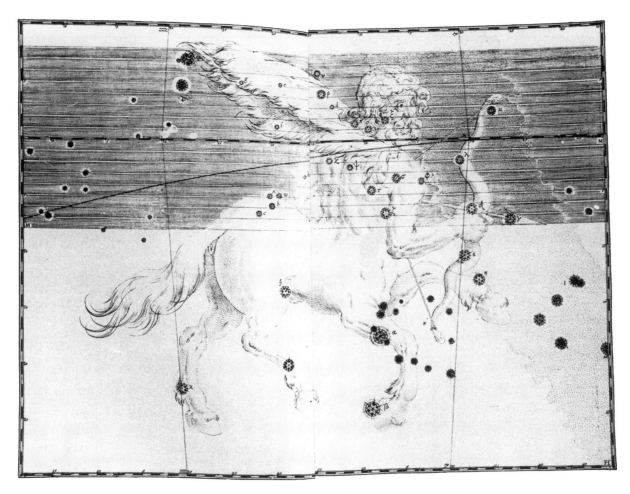

Figure 67. Bayer's Sagittarius. The circlet of stars in the Archer's front legs designated K is Corona Australis, the Southern Crown. The letter I is the tail of Scorpius, and L, beneath the Archer's cape, is the head of Capricornus. M is the Tropic of Capricorn. The point N, where the latter meets the ecliptic, marks the winter solstice. This solstice occurs in December when the ecliptic is 23½° south of the celestial equator and marks the sun's lowest point in the sky in its annual peregrination around the celestial globe. The shaded area is the zodiac.

Table 76. The lettered stars of Sagittarius

	MAGNITUDES			CATALOGUE NUMBERS				
Letter	Bayer	Lacaille	Visual	Flamsteed	Lacaille	Other	HR	HD
Alpha+	2		3.97				7348	181869
Beta^1+	2		4.01				7337	181454
Beta^2			4.29				7343	181623
Gamma^1+			4.69				6742	164975
Gamma^2	3		2.99	10			6746	165135
Delta	3		2.70	19			6859	168454
Epsilon	3		1.85	20			6879	169022
Zeta	3		2.60	38			7194	176687
Eta+	3		3.11				6832	167618
Theta^1+	3		4.37				7623	189103
Theta^2			5.30				7624	189118
Iota+	3		4.13				7581	188114
Kappa^1+	3		5.59				7779	193571
Kappa^2			5.64				7787	193807
Lambda	4		2.81	22			6913	169916
Mu+	4		3.86	13			6812	166937
Nu^1+	4		4.83	32			7116	174974
Nu^2			4.99	35			7120	175190
Xi^1+			5.08	36			7145	175687
Xi^2	4		3.51	37			7150	175775
Omicron	4		3.77	39			7217	177241
Pi	4		2.89	41			7264	178524
Rho^1+	4		3.93	44			7340	181577
Rho^2			5.87	45			7344	181645
Sigma	4		2.02	34			7121	175191
Tau	4		3.32	40			7234	177716
Upsilon	5		4.61	46			7342	181615
Phi	5		3.17	27			7039	173300
Chi^1+	5		5.03	47			7362	182369
Chi^2			7.1	48				182391
Chi^3			5.43	49			7363	182416
Psi	5		4.85	42			7292	179950
Omega	5		4.70	58			7597	188376
A+	5		4.83	60			7618	189005
b+	5		4.52	59			7604	188603
c	5		4.58	62			7650	189763
d	6		4.96	43			7304	180540
e^1+			6.2	54			7476	185644
e^2	6		5.06	55			7489	186005
f	6		4.86	56			7515	186648
g	6		5.02	61			7614	188899
h^1+			5.65	51			7431	184552
h^2	6		4.60	52			7440	184707

The Lost, Missing, or Troublesome Stars of Sagittarius

Alpha, HR 7348, 3.97V, *SA, HRP, BS,* BV 47*, CA 1590*, BR 6650*, L 8087*, G 177*.

Like many of the stars in the southern sector of Sagittarius, Alpha is below the horizon in Central Europe. Neither Tycho nor Bayer actually observed it, but the latter classified it nonetheless as a 2nd-magnitude star. He probably based this description on Ptolemy's catalogue where it is listed as somewhat fainter than 2nd magnitude (Toomer, *Ptolemy's Almagest,* p. 374). Both Bevis and Lacaille identified this star as Bayer's Alpha despite the discrepancy in magnitude. Both of their catalogues correctly list it as a star of the 4th magnitude.

Beta^1, HR 7337, 4.01V, *SA, HRP, BS,* BV 44 as Beta-2, CA 1587 as Beta, L 8075*, G 168/9*.
Beta^2, HR 7343, 4.29V, *SA, HRP, BS,* BV 45 as Beta-1, CA 1589 as Beta, L 8079*, G 172*.

Like Ptolemy, Bayer described Beta as a star of the 2nd magnitude. Toomer (*Ptolemy's Almagest,* p. 374) suggested that Ptolemy may have seen the combined light of these two closely placed stars. Bevis was among the first to label them Beta-1 and Beta-2. Lacaille saw both and also designated each as Beta. Since they are below the horizon in Central Europe, neither Tycho, Bayer, Flamsteed, Argelander, nor Heis actually observed them. See Halley's comments on the magnitude discrepancies in Sagittarius in Baily, "Catalogues of Ptolemy *et al.,*" p. 169. HR 7337 is a double star, Dunlop 226.

Abraham Sharp, Flamsteed's assistant and loyal friend who helped publish his three-volume *Historia Coelestis Britannica* after his death, prepared an appendix to the "Catalogus Britannicus" that included some stars in the southern skies that had been observed by Halley during his expedition to Saint Helena in 1677 but that had not been included in the body of Flamsteed's catalogue. Although Halley's original work, *Catalogus Stellarum Australium* (London, 1679), did not identify any stars by their Bayer letters, Sharpe included Bayer letters in his abbreviated version, erroneously lettering as Alpha the two stars that would later be identified as Beta. Moreover, both Halley and Sharpe listed HR 7343 before HR 7337. When Bevis prepared his catalogue, he correctly identified these two stars as Beta, not Alpha; but since HR 7343 was listed by Halley and Sharpe before HR 7337, he labeled it Beta-1 and the latter as Beta-2. Later astronomers, like Brisbane and Baily, reversed the indices to reflect the fact that HR 7337 precedes HR 7343 in order of right ascension. See Bevis's comments on the confusion between Alpha and Beta in his Remarks, Tabula XXX.

Gamma^1, HR 6742, 4.69V, *SA* as W, *HRP* as W, *BS,* CA 1484 as Gamma, L 7552*, H 8 as W, G 26 as W, the variable W.
Gamma^2, 10, HR 6746, 2.99V, *SA* as Gamma, *HRP* as Gamma, *BS,* CA 1486 as Gamma, L 7557*, H 9 as Gamma, G 28 as Gamma.

Bayer described Gamma as a single 3rd-magnitude star, and so too did Flamsteed, but Lacaille designated these two neighboring stars as Gamma. Since Argelander could see only the brighter star with his naked eye, he felt it alone was synonymous with Bayer's Gamma (*UN,* p. 108). Heis saw both stars, but agreed with Argelander's designation. When Roman capitals were introduced by Argelander in the nineteenth century to distinguish variables, Gamma^1 became W and Gamma^2 became simply Gamma.

Eta, HR 6832, 3.11V, *SA, HRP, BS,* BV 17*, CA 1503 as Beta Tel, G 46*.

Bevis was among the first to identify this star with Bayer's Eta. A few years later, Lacaille relabeled it Beta Tel, but Baily switched it back to Sagittarius with its original Bayer designation. See Beta Tel.

Theta^1, HR 7623, 4.37V, *SA, HRP, BS,* CA 1632 as K, L 8291, G 256*.
Theta^2, HR 7624, 5.30V, *SA, HRP, BS,* CA 1633 as K, L 8292, G 257*.

Bayer described Theta as a star of the 3rd magnitude. Lacaille, noting there were two neighboring stars in the area, designated each as K even though one of these had already been assigned a Greek letter by Bayer. Baily removed Lacaille's K's as he did with all Lacaille's Roman letters except those in the three new components of Argo Navis. For some reason, he failed to relabel them Theta. It was Gould who restored Bayer's designation by labeling them Theta^1 and Theta^2 since he felt stars of their brightness merited letters. See K.

Iota, HR 7581, 4.13V, *SA, HRP, BS,* CA 1624 as E, G 241*.

Lacaille relettered this star E. See E.

Kappa¹, HR 7779, 5.59V, *SA, HRP, BS,* CA 1656 as I, L 8415, G 292*.
Kappa², HR 7787, 5.64V, *SA, HRP, BS,* CA 1658 as I, L 8417, G 294*.

Bayer described Kappa as a single 3rd-magnitude star. Lacaille, noting there were two neighboring stars in the area, designated each as I. As he had done with Theta, Baily removed Lacaille's letters and left these two stars unlettered. Gould restored Bayer's letter by designating these two stars Kappa¹ and Kappa² (*Uran. Arg.*, p. 90). See I.

Mu¹, 13, HR 6812, 3.86V, *SA* as Mu, *HRP* as Mu, *BS* as Mu, BV 13 as Mu, BF 2477 as Mu, H 13 as Mu, G 41 as Mu.
Mu², 15, HR 6822, 5.38V, BV 15, BF 2481, H 14, G 43.

Although Bayer described Mu as a single star of the 4th magnitude, Flamsteed designated these two neighboring stars, his 13 and 15, as Mu-1 and Mu-2. Bevis, though, labeled only the brighter star, 13, as Mu. Baily, Argelander (*UN,* p. 108), Heis, and Gould all agreed.

Nu¹, 32, HR 7116, 4.83V, *SA, HRP, BS,* H 37*, G 111*.
Nu², 35, HR 7120, 4.99V, *SA, HRP, BS,* H 39*, G 113*.

Bayer, like Ptolemy, noted that Nu was a 4th-magnitude nebulous *duplex* or double star. Consequently, Flamsteed designated these two closely positioned stars, his 32 and 35, as Nu-1 and Nu-2. He also noted their nebulosity. Although neither of these stars has proven to be associated with any nebulosity, they are in a particulry rich section of the Milky Way next to the globular cluster NGC 6717. For Flamsteed's other nebulosities, see 33 And in Part Two; for Bayer's nebulosities, see z Her.

Xi¹, 36, HR 7145, 5.08V, *SA, HRP, BS,* HF 21, BV 39, BF 2548*, H 43*, G 119*.
Xi², 37, HR 7150, 3.51V, *SA, HRP, BS,* HF 22 as Xi, BV 40 as Xi, BF 2550*, H 44*, G 120*.

Bayer described Xi as a single 4th-magnitude star, and Halley's 1712 edition of Flamsteed's catalogue labels his 37 as Xi. Flamsteed's 1725 catalogue, however, designates these two neighboring stars, his 36 and 37, as Xi-1 and Xi-2. Bevis, on the other hand, labeled only the brighter star, 37, as Xi. Argelander agreed. Seeing only 37 with his naked eye, he felt it alone was synonymous with Bayer's Xi (*UN,* p. 108), but Heis observed both stars and restored Xi to 36. Since Gould also observed both stars and since they were both brighter than magnitude 6, he accepted Heis's designations, as have most modern astronomers.

Rho¹, 44, HR 7340, 3.93V, *SA, HRP, BS,* HF 29 as Rho, BV 56 as Rho, BF 2601*, H 60 as Rho, G 174*.
Rho², 45, HR 7344, 5.87V, *SA, HRP, BS,* HF 30, BV 54, BF 2602*, G 175*.

Bayer described Rho as a single 4th-magnitude star. Halley's 1712 edition of Flamsteed's catalogue labels his 44 as Rho, but Flamsteed's 1725 catalogue designates these two neighboring stars, his 44 and 45, as Rho-1 and Rho-2. Bevis, though, labeled only the brighter star, 44, as Rho. Argelander (*UN,* p. 109) and Heis agreed. Since they saw only 44, they felt that it alone was synonymous with Bayer's Rho. Gould saw both stars and restored Rho to 45 although he considered its magnitude 6.5

Chi¹, 47, HR 7362, 5.03V, *SA, HRP, BS,* HF 32 as Chi, BF 2611*, H 63 as Chi, G 181*, GC 26879*.
Chi², 48, HD 182391, 7.1V, HF 33, BF 2612*, GC 26793.
Chi³, 49, HR 7363, 5.43V, *SA* erroneously as Chi², *HRP, BS,* HF 34 as Chi, BF 2613*, H 64, G 182*, GC 26801 erroneously as Chi².

Bayer described Chi as a single star of the 5th magnitude, but Halley's 1712 edition of Flamsteed's catalogue labels two neighboring stars, his 47 and 49, as Chi. Flamsteed's 1725 "Catalogus Britannicus" designates these two stars Chi-1 and Chi-3; and adds his 48, almost adjacent to 47, as Chi-2. Argelander, seeing only the brightest one, 47, felt it alone was Bayer's Chi (*UN,* p. 109). Although Heis observed both 47 and 49, he kept Argelander's designation. Gould, too, saw only 47 and 49, but he lettered them Chi¹ and Chi³, as Flamsteed had proposed in his authorized catalogue. Gould erroneously credited Bode with first describing 49 as Chi³ (*Uran. Arg.,* pp. 290-91). Like Tirion in *SA* and *Uranometria 2000.0,* several astronomers have mistakenly identified 49 as Chi². Becvar, for example, in his atlases equated 49 with Chi² and left 48 unlettered and unnumbered. The mistake can possibly be traced to Boss's *General Catalogue* where 49 is also designated as Chi². The second edition of *SA* properly designates 47 as Chi¹ and 49 as Chi³, but omits Chi² from 48. See Hoffleit, "Discordances," p. 62.

Lettered Stars - Sagittarius

A, 60, HR 7618, 4.83V, *SA, HRP,* HF 45*, BV 72*, BF 2704, G 255*.

Bayer described A as a single 5th-magnitude star. Although it is properly identified in Halley's pirated edition of Flamsteed's catalogue, in the authorized edition of 1725, it is mistakenly designated as a. Bevis and Baily corrected the error. It should not be confused with Lacaille's A, HR 7223, CA 1570. See A, below.

b, 59, HR 7604, 4.52V, *SA, HRP.*

This star should not be confused with Flamsteed's b, 4, HR 6700. Flamsteed labeled two stars b, 59 and 4. See a, b, i, and p, below.

c, 62, HR 7650, 4.58V, *SA, HRP.*

d, 43, HR 7304, 4.96V, *SA, HRP.*

e¹, 54, HR 7476, 6.2V, *HRP,* BF 2660*, H 76*, G 221*.
e², 55, HR 7489, 5.06V, *SA* as e, *HRP,* BF 2666*, H 77*, G 225*.

Bayer described e as a single 6th-magnitude star, but Flamsteed designated these two neighboring stars, his 54 and 55, as e-1 and e-2. Although Argelander saw both stars, he felt that only the brighter one, 55, was synonymous with Bayer's e (*UN,* p. 109). Like Argelander, Heis also saw both stars, but he proposed that both stars should be designated as e.

f, 56, HR 7515, 4.86V, *SA, HRP.*

g, 61, HR 7614, 5.02V, *SA, HRP.*

h¹, 51, HR 7431, 5.65V, *SA, HRP,* HF 36, BV 61, BF 2638*, H 70*, G 204*.
h², 52, HR 7440, 4.60V, *SA, HRP,* HF 37 as h, BV 62 as h, BF 2639*, H 71*, G 207*.

Bayer described h as a single star of the 6th magnitude, and Halley's edition of Flamsteed's catalogue labels his 52 as h. Flamsteed's authorized catalogue, however, designates these two nearly adjacent stars, his 51 and 52, as h-1 and h-2. Bevis, though, labeled only the brighter star, 52, as h. Seeing only 52 with his naked eye, Argelander also felt it alone was synonymous with Bayer's h (*UN,* p. 109). Heis, who saw both stars, restored h to 51.

a, 7, HR 6724, 5.34V, BV 7, CA 1481*, BF 2446, G 20.
b, 4, HR 6700, 4.76V, BV 4, CA 1477*, BF 2435, H 4, G 15.
i, 5, HD 164031, 6.7V, BV 5, BF 2437.
p, 3, HR 6616, 4.54V, BV 3, CA 1466*, BF 2418, H 1 as X, G 2 as X, the variable X Sgr.

Bayer did not assign letters to these four stars. The letters were added by Flamsteed but removed by Bevis and Baily as superfluous. Bayer included p in his atlas but considered it an *informis* outside the border of Sagittarius and left it unlettered (see i Aql). Lacaille copied Flamsteed's a, b, and p, noting they were *informes* between Scorpius and Sagittarius. HR 6700, b, 4, should not be confused with Bayer's b, HR 7604. See also p Sco.

Lacaille lettered additional stars from A to T even though some had earlier been assigned Greek letters by Bayer. None of Lacaille's Roman letters in Sagittarius has survived in the literature because Baily removed them when he edited both Lacaille's expanded catalogue of 9,766 stars and the *BAC.* Lacaille's Roman letters might have survived, but Gould followed Baily's lead in eliminating them when he compiled his own atlas and catalogue of the southern skies. Gould also shifted some of Lacaille's Roman-lettered stars from Sagittarius to neighboring constellations.

A, in Telescopium, HR 7223, 5.97V, CA 1570*, BR 6569*, L 7973, BAC 6506, G 47 in Telescopium.

Gould included this star unlettered in Telescopium. It should not be confused with Bayer's A, HR 7618, above.

B, in Telescopium, HR 7289, 5.40V, CA 1583*, BR 6607*, L 8037, BAC 6570, G 51 in Telescopium.
C, in Telescopium, HR 7416, 5.61V, CA 1602*, L 8129, BAC 6675, G 63 in Telescopium.

Gould included these two stars unlettered in Telescopium.

D, in Telescopium, HR 7424, 4.90V, CA 1603*, BR 6689*, L 8137, BAC 6689, G 64 in Telescopium as Iota.

Gould included this star in Telescopium and relettered it Iota because he felt a star of its brightness merited a letter.

E, HR 7581, 4.13V, CA 1624*, L 8255, BAC 6812 as Iota, G 241 as Iota

This is Bayer's Iota. Lacaille relettered it E, but Baily restored Iota. See Iota.

F, HR 7630, 5.81V, CA 1631*, L 8285, BAC 6846, G 259.

G, HR 7552, 5.33V, CA 1619*, L 8239, BAC 6790, G 235.

H, HR 7652, 4.77V, CA 1639*, L 8310, BAC 6872, G 268.

I, HR 7779, 5.59V, CA 1656*, G 292 as Kappa[1].
I, HR 7787, 5.64V, CA 1658*, G 294 as Kappa[2].

These two stars are Bayer's Kappa. Lacaille relettered them I, but Gould restored Kappa. See Kappa.

J

Lacaille did not use this letter.

K, HR 7623, 4.37V, CA 1632*, G 256 as Theta[1].
K, HR 7624, 5.30V, CA 1633*, G 257 as Theta[2].

This is Bayer's Theta. Lacaille relettered it K, but Gould restored Theta. See Theta.

L, HR 7659, 4.99V, CA 1643*, L 8322, BAC 6877, G 270.
L, HR 7668, 6.53V, CA 1644*, L 8330, BAC 6886, G 272.

Lacaille designated these two neighboring stars L.

M, HR 7585, 6.46V, CA 1625*, L 8260, BAC 6816, G 244.

N, HR 7507, 5.52V, CA 1613*, L 8211, BAC 6755, G 228.

O, HR 7380, 5.67V, CA 1598*, L 8107, BAC 6639, G 186.

P, HR 7355, 6.04V, CA 1593*, L 8097, BAC 6628, G 179.

Q, HR 7398, 5.52V, CA 1600*, L 8123, BAC 6666, G 193.

R, in Capricornus, HR 7722, 5.73V, CA 1654*, L 8381, BAC 6947, G 5 in Capricornus.

Gould included this star unlettered in Capricornus. It should not be confused with the variable R Cap at $20^h11.3^m, -14°16'$.

S

Lacaille did not use this letter in Sagittarius.

T, HR 7029, 4.87V, CA 1540*, L 7830, BAC 6362, G 91.

This star should not be confused with the variable T Sgr at $19^h16.3^m, -16°59'$.

Scorpius, Sco

Scorpion

Scorpius (Figure 68) is a Ptolemaic constellation that Bayer lettered from Alpha to Omega and from A to c. Lacaille and Gould added additional letters, especially in the southern section of the constellation, which is below the horizon in Central Europe. Although Bayer called this constellation Scorpio, he also referred to it as Scorpius, as have most authorities including Eugène Delporte for the IAU in 1930. Both spellings are correct in Latin. Scorpius is a second declension noun whose possessive is Scorpii while Scorpio is a third declension noun whose possessive is Scorpionis.

Together with the Bull, the Lion, and the Ibex (Aquarius), Scorpius is one of the oldest and most important constellations dating back almost 6,000 years to ancient Sumer, where it was known as GIR.TAB, the Scorpion. The appearance of the sun in that portion of the sky occupied by each of these constellations signaled the beginning of a new season, one of the four turning points of the year. Like the other three, Scorpius actually resembles its namesake. It was originally a much larger constellation, with its claws extending westward to include what is now Libra. When it was first devised in about the fourth millennium B.C., its apparent heliacal rising corresponded with the autumnal equinox. By the second millennium, however, precession shifted the equinox away from the center of the constellation and toward its western section, its claws. The Mesopotamians adjusted to this move by splitting Scorpius in two. They retained the eastern part as Scorpius but broke off its claws to create a new constellation, which they dubbed the Scales (Libra).

Although Scorpius occupied an important place in the calendar, few references to scorpions occur in Mesopotamian mythology. One of the most significant occurs in Tablet IX of the Gilgamesh Epic, where the protagonist meets the scorpion people who guard the gates leading to the Garden of the Gods, where Gilgamesh hopes to find immortality. They warn Gilgamesh that the road is long, harrowing, and fraught with danger:

Never has a mortal man done that, Gilgamesh.
Over the mountain, no one has traveled the remote path,
for twelve double-hours it takes to reach its center,
and thick is the darkness; there is no light.

The Scorpion-Man is mentioned as one of the monsters who ally themselves with Tiamat, goddess of the sea and of chaos, in her battle with Marduk. Images of scorpion people — half-scorpion, half-man — have been uncovered on Mesopotamian boundary stones and monuments, but none seem to relate to stellar mythology.

The Greeks believed that the scorpion in the sky was associated with one of the myths relating to Orion. On one occasion, Orion is heard to boast that he is such a great hunter that he will kill all the beasts on Earth. This so upsets Gaea, goddess of the Earth, that she sends a poisonous scorpion to sting Orion's foot, killing the boastful hunter.

See Hartner, "The Earliest History of the Constellations in the Near East," p. 3 and diagram 1; van der Waerden, "Babylonian Astronomy. II. The Thirty-Six Stars," pp. 9, 11, 14-15; van der Waerden, "History of the Zodiac," pp. 219-20; O'Neil, *Time and the Calendars,* p. 58; Langdon, *Babylonian Menologies,* p. 7; Krupp, *Beyond the Blue Horizon,* pp. 137-38; Gardner and Maier, *Gilgamesh,* pp. 196-208; Reiner and Pingree, *Baby. Plan. Omens,* Part 2, p. 12; Pritchard, *Anc. Near East Picts.,* plates 192, 519. For the four turning points of the year, see Aquarius; for Marduk's battle with Tiamat and her allies, see Hydra; for the legend of the Scorpion and Orion, see Orion.

Lost Stars

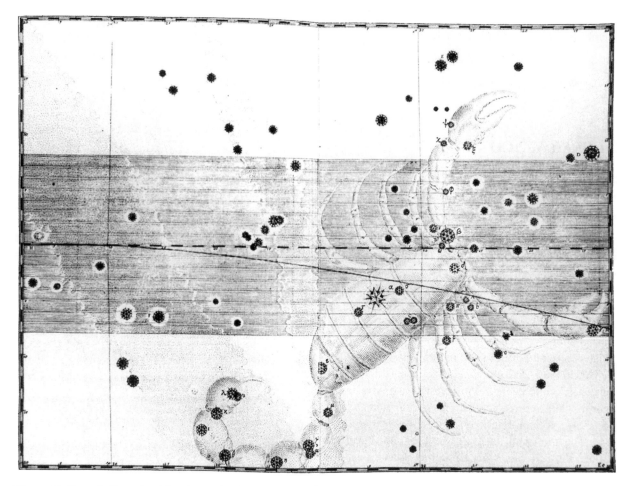

Figure 68. Bayer's Scorpius. The close pair of stars, Lambda (λ) and Upsilon (υ), in the Scorpion's tail is known as the Cat's Eyes. The letter D is the constellation Libra; E is the right hand of Ophiuchus; F is the arrow in the bow of Sagittarius; and G is the head of Lupus. The line H is the Tropic of Capricorn. The shaded area is the zodiac.

Table 77. The lettered stars of Scorpius

	Magnitudes			Catalogue Numbers				
Letter	Bayer	Lacaille	Visual	Flamsteed	Lacaille	Other	HR	HD
Alpha	1		0.96	21			6134	148478
Beta[1,2]+	2		2.76c	8			5984/5	144217/8
Gamma+	3		3.29	20 Lib			5603	133216
Delta	3		2.32	7			5953	143275
Epsilon	3		2.29	26			6241	151680
Zeta[1]+			4.73				6262	152236
Zeta[2]	3		3.62				6271	152334
Eta	3		3.33				6380	155203
Theta	3		1.87				6553	159532
Iota[1]+	3		3.03				6615	161471
Iota[2]			4.81				6631	161912
Kappa	3		2.41				6580	160578
Lambda+	3		1.63	35			6527	158926
Mu[1]+	4		3.08				6247	151890
Mu[2]	4		3.57				6252	151985
Nu	4		4.16c	14			6026/7	145501/2
Xi[1,2]+	4		4.16c	51 Lib			5977/8	144069/70
Omicron+	4		3.66	40 Lib			5812	139365
Pi	4		2.89	6			5944	143018
Rho	4		3.88	5			5928	142669
Sigma	4		2.89	20			6084	147165
Tau	4		2.82	23			6165	149438
Upsilon	4		2.69	34			6508	158408
Phi+	5							
Chi+	5		5.22	17			6048	145897
Psi+	5		4.94	15			6031	145570
Omega[1]+	5		3.96	9			5993	144470
Omega[2]			4.32	10			5997	144608
A+	5		4.59	2			5904	142114
b	5		4.64	1			5885	141637
c[1]+	5		5.67	12			6029	145483
c[2]	5		4.59	13			6028	145482
d+		6	4.78		1354		6070	146624
e+		5	3.58	39 Lib	1298		5794	139063
f		6	5.42		1317		5907	142184
i+		6	4.79	22	1370		6141	148605
k+		6	4.87		1410		6334	154090
l		6	5.08		1412		6371	154948
m		6	5.38		1337		6001	144690
o+		6	4.55	19	1359		6081	147084
p+		6	4.54	3 Sgr	1466		6616	161592
G+			3.21			G 172	6630	161892
H+			4.16			G 76	6166	149447
N+			4.23			G 72	6143	148703
Q+			4.29			G 159	6546	159433

The Lost, Missing, or Troublesome Stars of Scorpius

Beta¹, Beta², 8, HR 5984/5, 2.62V, 4.92V, comb. mag. 2.76V, *SA, HRP* as Beta, *BS,* ADS 9913.

Bayer described Beta as a single star of the 2nd magnitude. Later astronomers added the indices to identify the components of this multiple-star system.

Gamma, 20 Lib in Libra, HR 5603, 3.29V, *HRP, BS.*

Gould relettered this star Sigma Lib. See Sigma Lib.

Zeta¹, HR 6262, 4.73V, *SA, HRP, BS,* CA 1394 as Zeta, BR 5873*, L 7016*, G 103*.
Zeta², HR 6271, 3.62V, *SA, HRP, BS,* CA 1398 as Zeta, BR 5881*, L 7025*, G 104*.

Bayer described Zeta as a single 3rd-magnitude star. Zeta² was the star Bayer designated as Zeta; Zeta¹ was the more southerly of the two stars, which Bayer originally designated as Mu. For an explanation of the confusion that has arisen over Mu and Zeta, see Mu.

Iota¹, HR 6615, 3.03V, *SA, HRP, BS,* CA 1462 as Iota, BR 6198*, L 7425*, G 169*.
Iota², HR 6631, 4.81V, *SA, HRP, BS,* CA 1467 as Iota, BR 6213*, L 7447*, G 173*.

Bayer described Iota as a single star of the 3rd magnitude, but Lacaille, noting a smaller star less than ½° following, labeled them both Iota.

Lambda, 35, HR 6527, 1.63V, *SA, HRP, BS,* HF 47*, BV 54*, B 196*, BF 2396*, H 41*.

Flamsteed designated this star as Bayer's Lambda in his manuscripts, and it is included as Lambda in Halley's pirated edition of his catalogue, but the letter is inadvertently omitted from his "Catalogus Britannicus" of 1725. Bevis and Bode restored Lambda to 35.

Mu¹, HR 6247, 3.08V, *SA, HRP, BS,* CA 1392 as Mu, BR 5860*, L 7006*, BAC 5638*, G 98*.
Mu², HR 6252, 3.57V, *SA, HRP, BS,* CA 1393 as Mu, BR 5864*, L 7009*, BAC 5640*, G 99*.

Bayer designated two 4th-magnitude stars as Mu — one in the Scorpion's first tail-joint and the other about 5° to the south in its second joint. Bayer labeled each of these stars with the letter Mu on his chart. (For Bayer's practice of labeling stars in the same constellation with the same letter, see A Ser.) The first of these stars, the northern one, is this double star, Mu¹ and Mu², which Bayer saw as one star. Bayer's second Mu is the star currently designated Zeta¹, HR 6262 (see Zeta). The confusion started with Lacaille. He assigned Mu to each element of the double star HR 6247 and HR 6252, a wide pair, which Brisbane and others would later designate as Mu¹ and Mu². He then assigned Zeta to HR 6262 and HR 6271, two closely placed stars. The easterly was Bayer's Zeta, but the westerly, HR 6262, was originally Bayer's second Mu. Baily noted Lacaille's mistake but commented that since the designation had become widely accepted, it was too late to correct it (*BAC,* p. 422), and so it has remained to the present day. To summarize, Zeta¹ is the more southerly of Bayer's two Mu's; Zeta² is Bayer's Zeta; and Mu¹, Mu², a double star, is the more northerly of Bayer's two Mu's.

Xi¹, Xi², 51 Lib in Scorpius, HR 5977/8, 5.07V, 4.77V, comb. mag. 4.16V, *SA, HRP* as Xi, *BS* as Xi, H 13 as Xi, ADS 9909.

Bayer described Xi as a single 4th-magnitude star. Later astronomers added the indices to identify the components of this double star. Although Ptolemy included this star in Libra, Bayer switched it to Scorpius, but Flamsteed returned it to Libra where it remained until Argelander restored it to Scorpius (*UN,* p. 107), where it is today. For Bayer's switching stars from Libra to Scorpius, see Sigma Lib.

Omicron, 19, HR 6081, 4.55V, *SA, HRP, BS,* CA 1359 as o, L 6798, H 30, G 59 as o.

This is not Bayer's Omicron. His Omicron is the current Tau, 40 Lib, HR 5812 (see Lib). This is Lacaille's o (oh) that has been mistakenly copied as Omicron by several generations of astronomers and stellar cartographers. Although Gould correctly noted that it was o (*Uran. Arg.,* p. 90), it is still described as Omicron by various modern atlases and catalogues, such as *HRP, BS, SA,* Hirshfeld and Sinnott's *Sky Cat.,* and Tirion's *Uranometria 2000.0.*

Phi

This is one of Bayer's lost stars. Bayer noted that Phi was a 5th-magnitude star in the lower part of the Scorpion's northern claw, but it seems to have vanished. Flamsteed did not assign Phi to any star in Scorpius, but his 48 Lib (HR 5941, 4.88V) is erroneously designated as Psi (BF 2176) and is in the general area of Bayer's Phi Sco. Is it possible that Flamsteed's Psi was a typographical or copying error — there are many in his catalogue — for Phi (they are similar in appearance, ψ, φ) and that he thought his 48 Lib was synonymous with Phi Sco? Bode, on the other hand, felt that 49 Lib (HR 5954, 5.47V, B 208) was Phi Sco. But while both 48 and 49 Lib are in the general area where Phi Sco should be, neither conforms well with Bayer's chart. See the note to Phi on Bevis's Tabula XXIX.

Chi, 17, HR 6048, 5.22V, *SA, HRP, BS,* BV 18*, B 86*, BF 2216*, H 25*, G 44*.
Psi, 15, HR 6031, 4.94V, *SA, HRP, BS,* BV 11*, B 80*, BF 2212*, H 23*, G 39*.

Flamsteed's 1725 catalogue erroneously designates 15 as Chi although his manuscript catalogue describes it correctly as Psi; moreover, the 1725 catalogue leaves 17 unlettered. Bevis and Bode corrected these errors by relettering 15 as Psi and 17 as Chi.

Omega^1, 9, HR 5993, 3.96V, *SA, HRP, BS,* HF 9 as Omega, BV 17 as Omega, BF 2196*, H 15*, G 20*.
Omega^2, 10, HR 5997, 4.32V, *SA, HRP, BS,* HF 10 as Omega, BV 19, BF 2200*, H 16*, G 21*.

Bayer noted that Omega was a single 5th-magnitude star, but Flamsteed designated these two nearly adjacent stars, his 9 and 10, as Omega-1 and Omega-2. Bevis, though, labeled only the brighter star, 9, as Omega. Argelander agreed since he could see only 9 and therefore felt it alone was synonymous with Bayer's Omega (*UN*, p. 107). Heis, observing both stars, restored Omega to 10.

A^1, 2, HR 5904, 4.59V, *SA* as A, *HRP* as A, HF 2 as A, BV 5 as A, BF 2160 as A, G 3 as A.
A^2, 3, HR 5912, 5.87V, HF 3, BV 6, BF 2164, G 7.

Bayer described A as a single 5th-magnitude star, and Halley's 1712 edition of Flamsteed's catalogue labels his 2 as A. Flamsteed's 1725 catalogue, however, designates these two closely placed stars, his 2 and 3, as A-1 and A-2. Bevis, though, labeled only the brighter one, 2, as A. Baily agreed and as a result, he dropped 3's letter.

b, 1, HR 5885, 4.64V, *SA, HRP.*

c^1, 12, HR 6029, 5.67V, *SA, HRP,* HF 12 as c, BV 23*, CA 1347 as c, BR 5636*, L 6729, BF 2204*, G 36*.
c^2, 13, HR 6028, 4.59V, *SA, HRP,* HF 13, BV 22, CA 1348 as c, CAP 2957*, L 6730, BF 2206*, H 21 as c^1, G 37*.

Bayer described c as a *binae* or pair of 5th-magnitude stars, and Flamsteed's 1725 catalogue designates these two stars, his 12 and 13, as c-1 and c-2. Argelander, unable to see the fainter of the two, 12, relettered 13 as c^1 and Lacaille's d (see d) as c^2 (*UN*, p. 107). Heis was also unable to see 12. He agreed with Argelander that 13 was c^1, but he omitted any reference to c^2 either in his catalogue or atlas. Gould, on the other hand, observing both stars, agreed with Flamsteed's 1725 catalogue that 12 and 13 were c^1 and c^2, with the result that these designations became fixed in astronomical literature, one of the rare instances in which one of Argelander's proposals was rejected. See Hoffleit's "Discordances," p. 62.

The confusion surrounding c^1, c^2, and d dates back to Ptolemy. As Toomer has pointed out (*Ptolemy's Almagest,* p. 372), Ptolemy's 10 Sco was equivalent to Flamsteed's 13 and Bayer's c^2, and his 11 to Lacaille's d. Bayer affirmed that his pair of c's was synonymous with Ptolemy's 10, and in depicting his c's on his chart, he drew them next to each other on an east-west line. More than twenty year later in Schiller's *Coelum Stellatum Christianum,* which Bayer helped prepare, Bayer drew a more detailed diagram of this area of the heavens. The chart of St. Bartholomew, formerly Scorpius, depicts three very closely placed stars — two on an east-west line and the third to the northwest (11, 10, and 35 St. Bart). These three reflect the positions of Lacaille's d and Flamsteed's 12 and 13. In Schiller's accompanying star list, 10 and 11 St. Bart, the two stars on the east-west line, are equated with Bayer's c pair, and 35 St. Bart is left without any designation, indicating it has no Bayer equivalent and is not therefore included on Bayer's chart of Scorpius.

Schiller's 10 and 11 St. Bart are equated on his star list with Tycho's 10, which is synonymous with Ptolemy's 10, but the position of these two stars does not conform to Ptolemy's 10. They reflect instead the position of Lacaille's d and Flamsteed's 12, while 35 St. Bart is in the position of Flamsteed's 13, which is the same as Ptolemy's 10.

Ninety years later in Halley's 1712 edition of Flamsteed's catalogue, the two east-west stars that Schiller referred to as 10 and 11 St. Bart are both labeled as Bayer's c, while 35 St. Bart, Flamsteed's 13, is left unlettered. The exact same nomenclature is used in Bevis's atlas and catalogue. In Flamsteed's authorized catalogue of 1725, however, Flamsteed's 12 is labeled c-1, his 13 is labeled c-2, and the easterly of the three stars, Lacaille's d (BF 2227), is inadvertently omitted.

In sum, it appears that the stars Bayer designated as his pair of c's were Lacaille's d and Flamsteed's 12, c-1, although Flamsteed's 13, c-2, and Ptolemy's 11, may very well be what Bayer intended. Flamsteed's designation of his 12 and 13 as c-1 and c-2 was later adopted by Lacaille in his catalogue of the southern skies and by most modern authorities.

$$* \ 13, c^2, 35 \text{ St. Bart., Ptolemy's 10}$$
$$* \ 12, c^1, 10 \text{ St. Bart.}$$
$$* \ d, 11 \text{ St. Bart., Ptolemy's 11}$$

The relative position of c^1, c^2, and d with their Flamsteed, Schiller, and Ptolemy numbers.

Although Bayer's letters ended at c, Lacaille added lower-case Roman letters from d to p. Baily removed all of them in his expanded edition of Lacaille's catalogue and in the *BAC*, but Gould restored the letters to some of the brighter stars.

d, HR 6070, 4.78V, *SA, HRP,* HF 18 as c, BF 2227, BV 25 as c-2, CA 1354*, BR 5681*, L 6777, BAC 5429, H 28*, G 53*.

Halley's pirated edition of Flamsteed's catalogue labels this star c, and Bevis and Argelander (*UN*, p. 107) agreed that it was indeed Bayer's c^2. Heis, however, referred to it as d, the only time he ever cited one of Lacaille's letters either in his catalogue or atlas. Gould also designated this star as d. Although Baily had removed its letter, Gould restored Lacaille's d because he felt a star of its magnitude merited a letter. Hoffleit ("Discordances," p. 62) mistakenly asserted that Heis later called this star c^2 as Argelander had proposed (see c). Although included in Halley's edition of Flamsteed's catalogue, this star was inadvertently omitted from the authorized edition of 1725 (*BF*, p. 392).

e, 39 Lib in Libra, HR 5794, 3.58V, CA 1298*, BR 5400*, L 6445, G 93 as Upsilon Lib.

Gould included this star in Libra and relettered it Upsilon Lib because of its brightness. For the account of how this star was switched between Libra and Scorpius, see Upsilon Lib.

f, HR 5907, 5.42V, CA 1317*, L 6579, BAC 5254, G 5.

g
h

Lacaille did not use these two letters in Scorpius.

i, 22, HR 6141, 4.79V, *SA, HRP,* CA 1370*, L 6858, G 70*.

Gould restored this star's letter because he felt a star of its brightness merited a letter.

j

Lacaille did not use this letter.

k, HR 6334, 4.87V, *HRP,* CA 1410*, BR 5950*, L 7109, G 120*.

Gould restored this star's letter because he felt a star of its brightness merited a letter.

l, HR 6371, 5.08V, CA 1412*, BR 5975*, L 7147, BAC 5772, G 125.

m, HR 6001, 5.38V, CA 1337*, BR 5605*, L 6702, BAC 5347, G 25.

n

Lacaille did not use this letter in Scorpius.

o, 19, HR 6081, 4.55V, *SA* erroneously as Omicron, *HRP* erroneously as Omicron, *BS* erroneously as Omicron, CA 1359*, L 6798, RAS 1871*, G 59*.

Gould restored this star's letter because he felt a star of its brightness merited a letter. This letter, o (oh), has frequently been confused with Omicron. See Omicron.

p, 3 Sgr in Sagittarius, HR 6616, 4.54V, BV 3 in Sagittarius, CA 1466*, L 7440 in Sagittarius, BF 2418 in Sagittarius, H 1 as X Sgr, G 2 as X Sgr, the variable X Sgr.

Flamsteed labeled this star p Sgr, but Bevis and Baily removed the letter as superfluous (see i Aql). Lacaille copied Flamsteed's p and noted it was an *informis* between Scorpius and Sagittarius. It was relettered X in the nineteenth century because of its variability. See p Sgr.

Gould added upper-case Roman letters to four stars he considered brighter than 5th magnitude. He purposely selected only those letters Lacaille had skipped.

G, HR 6630, 3.21V, *SA, HRP,* CA 1468 as Gamma Tel, BR 6214 as Gamma Tel, L 7449, G 172*.

This star was originally Lacaille's Gamma Tel, which Baily included unlettered in Scorpius. Gould relettered it G Sco because he felt a star of its brightness merited a letter. See Gamma Tel.

H, HR 6166, 4.16V, *SA, HRP,* CA 1376, BR 5767, L 6890, G 76*.

This was originally Lacaille's Beta Nor, which Baily included unlettered in Scorpius. Gould relettered it H Sco because of its brightness. See Beta Nor.

N, HR 6143, 4.23V, *SA, HRP,* CA 1371 as Alpha Nor, BR 5747 as Alpha Nor, L 6859, G 72*.

This was originally Lacaille's Alpha Nor, which Baily included unlettered in Scorpius. Gould relettered it N Sco because of its brightness. See Alpha Nor.

Q, HR 6546, 4.29V, *SA, HRP,* CA 1445, BR 6133, L 7350, G 159*.

Although Lacaille observed and catalogued this star, he left it unlettered; he considered it 6th magnitude. Gould lettered it Q because he felt a star of its brightness merited a letter.

Sculptor, Scl

Sculptor, Sculptor's Implements

Lacaille devised and lettered the stars of the constellation Sculptor (Figure 2c) from Alpha to Tau. In his preliminary catalogue, written in French and published in 1756, he called it the Sculptor's Workshop (l'Atelier du Sculpteur) and depicted a bust resting on a three-legged platform next to a block of marble on which were set a pair of chisels and a mallet. In his catalogue of 1763, written in Latin, he labeled the constellation Apparatus Sculptoris (Sculptor's Implements). Like the other constellations that he devised, Sculptor represented the arts and sciences of Western Europe.

Table 78. The lettered stars of Sculptor

		Magnitudes			Catalogue Numbers			
Letter	Bayer	Lacaille	Visual	Flamsteed	Lacaille	Other	HR	HD
Alpha		5	4.31		61		280	5737
Beta		5	4.37		1898		8937	221507
Gamma		5	4.41		1879		8863	219784
Delta		5	4.57		1915		9016	223352
Epsilon		5	5.31		113		514	10830
Zeta		6	5.01		1931		9091	224990
Eta		6	4.81		18		105	2429
Theta		6	5.25		1942		35	739
Iota		6	5.18		13		84	1737
Kappa^1+		6	5.42		1937		24	493
Kappa^2		6	5.41		1941		34	720
Lambda^1+		6	6.06		44		185	4065
Lambda^2		6	5.90		45		195	4211
Mu		6	5.31		1907		8975	222433
Nu+		6	6.70		98		474	10161
Xi+			5.59			G 94	288	6055
Pi		6	5.25		107		497	10537
Rho+		6	6.65		84			8487
Sigma		6	5.50		64		293	6178
Tau		6	5.69		95		462	9906

The Lost, Missing, or Troublesome Stars of Sculptor

Kappa¹, HR 24, 5.42V, *SA, HRP, BS,* CA 1937 as Kappa, BR 2*, L 9741, G 51*.
Kappa², HR 34, 5.41V, *SA, HRP, BS,* CA 1941 as Kappa, BR 7*, L 9758, G 52*.

Lacaille designated these two neighboring stars Kappa.

Lambda¹, HR 185, 6.06V, *SA, HRP, BS,* CA 44 as Lambda, BR 89*, L 183*, G 83*.
Lambda², HR 195, 5.90V, *SA, HRP, BS,* CA 45 as Lambda, BR 93*, L 192*, G 84*.

Lacaille designated these two neighboring stars Lambda.

Nu, HR 474, 6.70V, CA 98*, L 475, BAC 504, G 123.

Baily removed the letter from this star because of its dimness.

Xi, HR 288, 5.59V, *SA, HRP, BS,* L 277, G 94*.

Although Lacaille observed and catalogued this star, he did not include it in his catalogue of 1763 nor did he assign it a letter. Baily placed it unlettered in Phoenix, but Gould switched it back to Sculptor and designated it Xi because he felt a star of its brightness merited a letter.

Omicron

Lacaille did not use this letter in Sculptor.

Rho, in Cetus, HD 8487, 6.65V, CA 84*, L 381, BAC 418, G 134 in Cetus.

Baily dropped this star's letter because of its dimness. Gould included it unlettered in Cetus.

Scutum, Sct

Shield

Hevelius devised the constellation Scutum (Figure 69) in 1684 and included it in his star atlas and catalogue, which were bound together and published posthumously in 1690 as *Prodromus Astronomiae*. He called the constellation Scutum Sobiescianum, Sobieski's Shield, and placed it in the heavens in honor of John III Sobieski, the elected King of Poland (1674-96), who was also his monarch and patron. He noted with considerable pride that Sobieski personally lead a charge of the Polish cavalry against the Ottomans and broke the Turkish siege of Vienna, the capital of the Holy Roman Empire, on the 12th of September 1683, saving Europe from the infidels and freeing thousands of Christians from the yolk of Islam. Since Sobieski had no coat of arms and fought with a bare shield, Hevelius decided to devise for him a special coat of arms and to place it in a very special place in the heavens. He set it next to Aquila and the asterism Antinous and above Sagittarius and Capricornus. He formed it symbolically from seven stars: four around the rim of the shield to represent the four royal princes and three in the center to represent the king, queen, and royal princess. In honor of the king's successful battles for Christendom, he emblazoned the center of the shield with a cross bearing the letters INRI, Iesus Nazarenus Rex Iudaeorum, Jesus of Nazareth, King of the Jews, the very same words that Pontius Pilate had inscribed on Jesus' cross (John 19:19).

Hevelius further explained that the neighboring constellation Aquila represented for him the Polish eagle. The two constellations below Aquila and Scutum Sobiescianum — Sagittarius and Capricornus — represented, as he learned from astrologers, the countries of Dalmatia, Slavonia, Hungary, Arabia Felis (Yemen), Macedonia, Illyria, Thrace, Bosnia, Albania, Bulgaria, and Greece. Consequently, he felt that Scutum Sobiescianum, together with Aquila, the Polish Eagle, would defend the Poles against these nations, their traditional enemies. Moreover, in his atlas, Hevelius depicted Scutum warding off Sagittarius while Aquila, holding a pugnacious Antinous armed with a bow and arrow, attacked Capricornus.

Hevelius's selection of seven stars for this constellation was, of course, not fortuitous. Since ancient times, seven has always been considered a significant number, representing for early man the seven classical planets — the Sun, Moon, Mercury, Venus, Mars, Jupiter, and Saturn — that dominate the heavens and the seven-day periods of the Moon as it passes through its four phases in the completion of its twenty-eight day cycle. These factors undoubtedly led to the seven-day week, the Biblical Seven Days of Creation, and the sacredness of the seventh day, the Sabbath. The word Sabbath is itself derived from the Hebrew word for seven, which some scholars believe is derived, in turn, from the Akkadian word *sapattu*, meaning full moon, at which time Mesopotamians were advised against undertaking any type of activity. There are many other examples of the importance of seven: the Jewish Festival of the First Fruits, Shavuot, commemorating the Revelation on Mt. Sinai, celebrated seven weeks from the first day of Passover; Noah's seven pairs, male and female, of all the creatures of the world; the seven fat and lean years of Pharaoh's dream; the seven plagues inflicted on the Egyptians; the seven priests with their seven trumpets marching around the walls of Jericho, which finally collapsed on the seventh circuit of the seventh day; the seven kings of ancient Rome; the seven hills of Rome; the seven wonders of the ancient world; the seven sacraments; the seven Liberal Arts; the 777 stars in Tycho's first catalogue, and so forth. As the Jewish sages noted: "All sevenths are favorites in the world." For other significant numbers, see Apus and Canes Venatici.

Although Flamsteed did not recognize Scutum as a separate constellation — he included its stars in Aquila — he did acknowledge that Hevelius had introduced Scutum and had been the first to observe and catalogue its stars: 1, 2, 6, 9 Aql. Actually, as noted above, Hevelius observed seven stars in Scutum: the four just noted; 3 Aql, which Flamsteed inadvertently forgot to attribute to him; and two additional stars that Flamsteed could not locate. Gould lettered all of Hevelius's stars from Alpha to Eta.

See *Prodromus Astronomiae,* pp. 115-16; Warner, *Sky Explored,* p. 113; Allen, *Star Names,* pp. 373-74; Dvornick, *The Slavs in European History and Civilization,* pp. 480-82; Freedman and Simon, *Midrash Rabbah,* IV, 377-78; Pritchard, *Anc. Near East. Texts,* p. 68 and note 84; O'Neil, *Time and the Calendars,* pp. viii, 6.

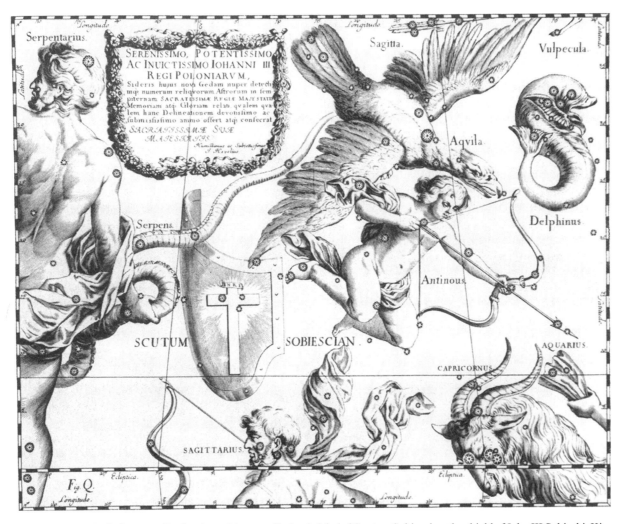

Figure 69. Hevelius's Scutum. On the chart, this constellation is labeled Scutum Sobiescian, the shield of John III Sobieski, King of Poland and patron of Hevelius. Hevelius was not the only astronomer seeking to honor his patron. Halley devised Charles II's Oak (see figures 31 and 65) to honor his sovereign. Bode created the asterism Friedrichs Ehre, Fredrick's Honor, north of Pegasus, for Frederick the Great of Prussia. And Maximilian Hell devised Psalterium Georgianum, George III's Harp, between Taurus and Eridanus. Probably the most famous astronomer to court favor was Galileo. He dubbed his newly discovered satelites of Jupiter the "Medicean stars" in honor of Cosmo de Medici, Grand Duke of Tuscany. For additional long-forgotten asterisms honoring political figures, see Warner, *Sky Explored,* p. xii.

Table 79. The lettered stars of Scutum

	MAGNITUDES			CATALOGUE NUMBERS				
Letter	Bayer	Lacaille	Visual	Flamsteed	Lacaille	Other	HR	HD
Alpha+			3.85	1 Aql		G 14	6973	171443
Beta+			4.22	6 Aql		G 24	7063	173764
Gamma+			4.70			G 6	6930	170296
Delta+			4.72	2 Aql		G 19	7020	172748
Epsilon+			4.90	3 Aql		G 21	7032	173009
Zeta+			4.68			G 3	6884	169156
Eta+			4.83	9 Aql		G 33	7149	175751

The Lost, Missing, or Troublesome Stars of Scutum

Alpha, 1 Aql in Scutum, HR 6973, 3.85V, *SA, HRP, BS,* HEV 2, H 5, G 14*, W 4*.
Beta, 6 Aql in Scutum, HR 7063, 4.22V, *SA, HRP, BS,* HEV 7, H 9, G 24*, W 7*.
Gamma, HR 6930, 4.70V, *SA, HRP, BS,* HEV 5, H 2, G 6*, W 2*.
Delta, 2 Aql in Scutum, HR 7020, 4.72V, *SA, HRP, BS,* HEV 4, H 6, G 19*, W 5*.
Epsilon, 3 Aql in Scutum, HR 7032, 4.90V, *SA, HRP, BS,* HEV 3, H 7, G 21*, W 6*.
Zeta, HR 6884, 4.68V, *SA, HRP, BS,* HEV 6, H 1, G 3*, W 1*.
Eta, 9 Aql in Scutum, HR 7149, 4.83V, *SA, HRP, BS,* HEV 1, H 11, G 33*, W 10*.

Argelander (*UN,* p. 110) and Heis observed all of these seven stars. Neither Flamsteed nor Baily could locate HEV 5 and 6 (Gamma and Zeta). Gould assumed that HR 6930 and HR 6884 were the stars Hevelius meant, although these stars did not conform to his catalogue or atlas. It should be noted that Gould did not have access to Hevelius's original works. He used Flamsteed's edition of Hevelius's catalogue and Gould's references to Hevelius are to Flamsteed's catalogue, not to the original. Flamsteed rearranged the stars in Hevelius's catalogue in order of right ascension. The references to HEV are to Baily's edition ("Catalogues of Ptolemy *et al.,*" p. 229) of Hevelius's catalogue, in which the stars are arranged in their original order.

Serpens, Ser

Serpent

Serpens (Figure 70) is a Ptolemaic constellation that Bayer lettered from Alpha to Omega and from A to e. Gould subdivided the Serpent into Caput — the Head, for that part preceding Ophiuchus, and Cauda — the Tail, for that part following. Eugène Delporte, for the IAU, recognized these subdivisions in 1930 by referring to them as Serpens Caput and Serpens Cauda.

This is the Serpent held by Ophiuchus ("Serpent Handler"). The Greeks believed that the Serpent Handler in the heavens was Asclepius, the physician and son of the god of medicine, Apollo. The association of snakes with long life and healing can be traced to ancient Mesopotamia.

See Gould, *Uran. Arg.*, p. 94; Delporte, *Atlas Céleste*. See Ophiuchus for the association of snakes with medicine.

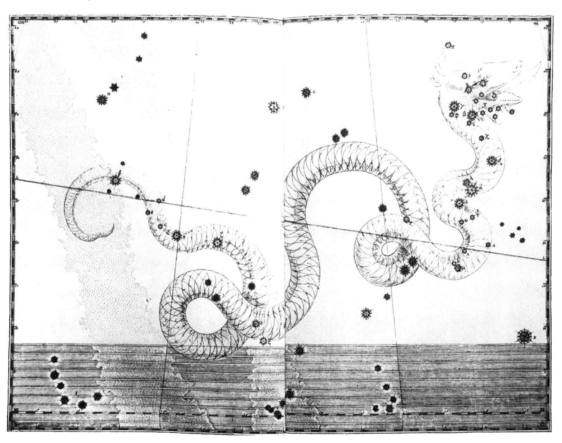

Figure 70. Bayer's Serpens. The letter F, to the left of G in the upper-middle section of the chart, is the head of Ophiuchus; G is head of Hercules; H is the tail of Aquila; I is the head of Sagittarius; and K is Libra. The lines L and M are the celestial equator and the ecliptic. The shaded area across the bottom is the zodiac.

Table 80. The lettered stars of Serpens

		MAGNITUDES		CATALOGUE NUMBERS				
Letter	Bayer	Lacaille	Visual	Flamsteed	Lacaille	Other	HR	HD
Alpha	2		2.65	24			5854	140573
Beta	3		3.67	28			5867	141003
Gamma	3		3.85	41			5933	142860
Delta	3		3.8c	13			5788/9	138917/8
Epsilon	3		3.71	37			5892	141795
Zeta	3		4.62	57			6710	164259
Eta	3		3.26	58			6869	168723
Theta[1,2]+	3		4.10c	63			7141/2	175638/9
Iota	4		4.52	21			5842	140159
Kappa	4		4.09	35			5879	141477
Lambda	4		4.43	27			5868	141004
Mu	4		3.53	32			5881	141513
Nu	4		4.33	53			6446	156928
Xi	4		3.54	55			6561	159876
Omicron	4		4.26	56			6581	160613
Pi	5		4.83	44			5972	143894
Rho	5		4.76	38			5899	141992
Sigma	5		4.82	50			6093	147449
Tau[1]+	6		5.17	9			5739	137471
Tau[2]	6		6.22	12			5770	138527
Tau[3]	6		6.12	15			5795	139074
Tau[4]	6		6.65	17				139216
Tau[5]	6		5.93	18			5804	139225
Tau[6]	6		6.01	19			5840	140027
Tau[7]	6		5.81	22			5845	140232
Tau[8]	6		6.14	26			5858	140729
Upsilon	6		5.71	31			5870	141187
Phi+	6		5.54				5940	142980
Chi	6		5.33	20			5843	140160
Psi	6		5.88	23			5853	140538
Omega	6		5.23	34			5888	141680
A[1]+	6		5.51	11			5772	138562
A[2]	6		5.40	25			5863	140873
b	6		5.11	36			5895	141851
c	6		5.39	60			6935	170474
d	6		5.21	59			6918	169985
e+	6		5.75				6993	171978

The Lost, Missing, or Troublesome Stars of Serpens

Theta[1], Theta[2], 63, HR 7141/2, 4.62V, 4.98V, comb. mag. 4.10V, *SA, HRP* as Theta, *BS,* H 80 as Theta, G 71 as Theta, ADS 11853.

Bayer considered Theta a single star of the 3rd magnitude. Argelander noted it was a double star but referred to it simply as Theta (*UN,* p. 72), as did Heis and Gould. Later astronomers added the indices to identify its components.

Tau¹, 9, HR 5739, 5.17V, *SA, HRP, BS,* B 14*, BF 2098*, H 8*.
Tau², 12, HR 5770, 6.22V, *SA, HRP, BS,* B 27*, BF 2111*, H 12*.
Tau³, 15, HR 5795, 6.12V, *SA, HRP, BS,* B 36*, BF 2127*, H 17*.
Tau⁴, 17, HD 139216, 6.65V, *SA, BS Suppl.,* B 38*, BF 2130*, H 19*.
Tau⁵, 18, HR 5804, 5.93V, *SA, HRP, BS,* B 39*, BF 2131*, H 20*.
Tau⁶, 19, HR 5840, 6.01V, *SA, HRP, BS,* B 49*, BF 2138*, H 22*.
Tau⁷, 22, HR 5845, 5.81V, *SA, HRP, BS,* B 53*, BF 2142*, H 25*.
Tau⁸, 26, HR 5858, 6.14V, *SA, HRP, BS,* B 63*, BF 2148*, H 28*.

Bayer described Tau as a group of eight 6th-magnitude stars and placed the letter Tau in the middle of the cluster on his star chart. Working without Bayer's star list, Flamsteed was uncertain which of the stars Bayer meant and consequently designated only three as Tau (*BF*, pp. 399-400). He chose those nearest the letter Tau on Bayer's chart — his 12, 18, and 19 — and designated them Tau-1, Tau-2, and Tau-3. Bode corrected this by assigning indices, in order of right ascension, to the eight stars, as in the heading above, that he believed were synonymous with Bayer's Tau's. A few years before, Bevis had selected almost the exact same stars (BV 7, 8, 11, 13, 15, 16, 19, 26) except he included 33 (BV 26) instead of 9. But, as Baily has pointed out, 33 is the result of a mathematical error committed by Flamsteed and does not actually exist (*BF*, Table B, p. 645; see also 33 Ser in Part Two).

Tau⁴, 17, is an irregular variable. Kholopov's *Gen. Cat. Var. Stars* (III, 238-39) gives its range as 5.89V to 7.07V. Argelander was able to view it with his naked eye, assigning it a magnitude of 6 (*UN*, p. 70).

Phi, HR 5940, 5.54V, *SA, HRP, BS,* B 104*, BF 2183*.

Flamsteed observed this star and referred to it as Phi in his manuscripts, but it, along with several hundred others that he had observed, was omitted from his printed catalogue (*BF*, p. 392). Bevis noted the omission, as did Bode and Baily. Baily included Phi in his revised edition of Flamsteed's catalogue. See Bevis's comments on this star, Remarks, Tabula XIV, where he noted that Phi's position in his catalogue was taken from Royer's charts.

Augustin Royer, a French architect and amateur astronomer, was the first to employ Bayer's letters to identify individual stars. In fact, Warner (*Sky Explored,* pp. 213-16) calls him "Bayer's first disciple." In 1679, Royer published a set of four charts that covered the entire sky along with tables of stellar longitude and latitude.

A¹, 11, HR 5772, 5.51V, *SA, HRP,* H 13*, G 15*.
A², 25, HR 5863, 5.40V, *SA, HRP,* H 30*, G 24*.

Bayer designated two 6th-magnitude stars, about 3° apart, as A Ser and placed individual letter A's next to each of them. Bayer rarely did this. He usually placed only one letter between two stars or in the midst of a group of stars as in the case of Tau Ser, above. Flamsteed designated his 14 and 25 as A-1 and A-2. But since 14 (HR 5799, 6.51V) was invisible to Argelander's naked eye, he felt it was not the star Bayer meant and consequently switched A¹ to 11 (*UN*, p. 71).

b, 36, HR 5895, 5.11V, *SA, HRP.*

c, 60, HR 6935, 5.39V, *SA, HRP.*

d, 59, HR 6918, 5.21V, *SA, HRP.*

e, HR 6993, 5.75V, *SA, HRP,* H 78*, G 57*.

Bayer noted that e was a star of the 6th magnitude, and Flamsteed designated his 61 (HR 6957, 5.94V, BF 2507 as e, G 48) as e. Argelander, however, switched e to this star, HR 6993, since he could not see 61. In addition, he felt that HR 6993 conformed better to Bayer's chart than 61 (*UN*, p. 71). Heis, who also was unable to see 61 with his naked eye, agreed with Argelander. Although Gould observed 61 and included it in his *Uran. Arg.,* he accepted Argelander's designation that HR 6993 was synonymous with e. See Hoffleit, "Discordances," p. 63, where it is argued that either 61 or HR 6993 could be Bayer's e.

SEXTANS, SEX

Sextant, Astronomical Sextant

Hevelius devised the constellation Sextans (Figure 71) in honor of his own sextant that had faithfully served him for over twenty years, from September 1658 to September 1679, when it was destroyed in a fire that ravaged his observatory at Danzig. He noted that it was this instrument that had helped him compile his stellar catalogue. He placed it in the heavens between Leo and Hydra because, after consulting with astrologers, he learned that these two creatures were considered fiery beasts, and, therefore, what more fitting place could be found for an object that had been consumed by flames!

Hevelius originally called this constellation Sextans Uraniae, Astronomical Sextant. It should not be confused with the mariner's sextant, a small, hand-held instrument with telescopic sights that evolved from the octant, which was honored with a place in the heavens by Lacaille a half century later. Hevelius's sextant, with a radius of six feet, was made of brass and equipped with plain sights *(nudis oculis)* rather than telescopic ones. As a matter of fact, Hevelius disdained telescopes for stellar observations and this involved him in a long-standing dispute with the English astronomer Robert Hooke. He argued with Hooke and Halley, who visited him at Danzig to check the accuracy of his observations, that while telescopes were necessary for lunar and planetary work, they were unnecessary in stellar observations. He felt, in fact, that the passage of light through their lenses would distort the resulting image. Incidentally, Halley later reported back to Hooke and his colleagues at the Royal Society that he had checked many of Hevelius's naked-eye observations with a telescope and found that in each instance they differed by less than 1'.

Flamsteed included Sextans in his atlas and catalogue as a separate constellation and credited Hevelius with having devised it. Baily had planned to letter its stars, but he found none brighter than magnitude 4.5, so he left them unlettered. Gould assigned to its brighter stars letters from Alpha to Epsilon. All of Gould's lettered stars are listed in the *BS,* and there are no missing or problem lettered stars in the constellation.

See Hevelius, *Prodromus Astronomiae,* p. 116; Baily, "Catalogues of Ptolemy *et al.,*" pp. 41-48.

Table 81. The lettered stars of Sextans

	MAGNITUDES			CATALOGUE NUMBERS				
Letter	Bayer	Lacaille	Visual	Flamsteed	Lacaille	Other	HR	HD
Alpha			4.49	15		G 27	3981	87887
Beta			5.09	30		G 56	4119	90994
Gamma			5.05	8		G 13	3909	85558
Delta			5.21	29		G 54	4116	90882
Epsilon			5.24	22		G 36	4042	89254

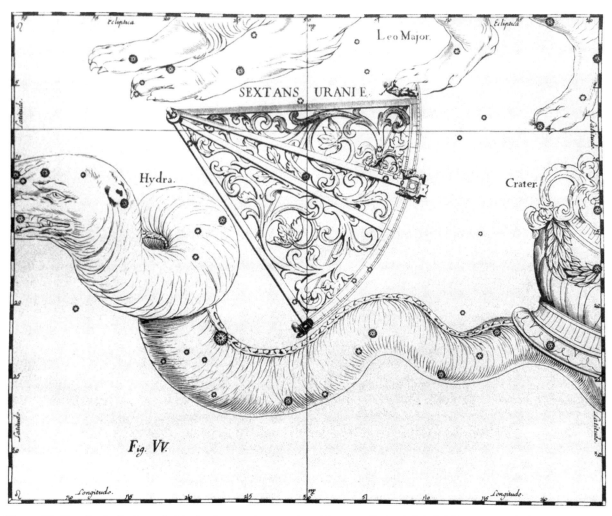

Figure 71. Hevelius's Sextans. In his *Prodromus Astronomiae*, this constellation is labeled Sextans Urani[a]e, Astronomical Sextant. The horizontal line across the chart marks the southern limit of the zodiac, 8° below the ecliptic, which is the upper border of the chart.

Taurus, Tau

Bull

Taurus (Figure 72) is a Ptolemaic constellation that Bayer lettered from Alpha to Omega and from A to u. It is one of the oldest constellations, dating back some 6,000 years to ancient Mesopotamia, where the Sumerians considered it the most important constellation in the heavens. The sun's appearance in this constellation coincided with the vernal equinox, the first month of the New Year, and the beginning of spring planting. Like the harbingers of other seasons — the Lion, the Scorpion, and the Ibex (Aquarius) — the Bull of Heaven actually resembles its earthly namesake. The vee-shaped Hyades bears an uncanny resemblance to the head of a bull or ox. It was called in Sumerian GU_4.AN.NA, Bull or Ox of Heaven and in Akkadian Is le, the Jaw of the Bull.

The bull or ox was extremely important in the life of ancient man. Its two slim horns resembled the crescent moon, the symbol of rebirth and the passage of the seasons. Of greater importance, though, was the use to which the animal's horns were put. It was in the ancient Near East that man first learned to attach a plow to the bull or ox's horns, enabling him to cultivate the land more effectively and thus to break the cycle of hunger, starvation, and death that had plagued humankind. Prior to this time, farmers were forced to turn the soil with a primitive spade or hand plow — difficult, backbreaking work that produced a limited quantity of food. The development of the plow permitted man to settle down, produce surplus food, found cities, and start on the long road from barbarism to civilization. No wonder the people of ancient Mesopotamia honored the bull, especially its horned head, with a place in the sky marking the coming of spring, the time for plowing.

The importance of the bull to early man can be seen in the religious ceremonies of the ancient Near East. They were the animals most frequently selected as sacrificial offerings to the gods. In Mesopotamia, they were sacrificed during the New Year Festival and at various other holidays. At the city of Uruk or Erech, the modern village of Warka in southern Iraq, they were offered to the deities almost on a daily basis. Similar sacrifices are found in the Old Testament (Leviticus 4:3, 13, etc.). Later, when the vernal equinox shifted by precession to Aries, the ram, or male sheep, replaced the bull as the sacrificial animal of choice.

Since the Bull signified the beginning of the New Year from about the fourth to the second millennium B.C., it is not at all surprising that when the people of the Middle East developed the prototype of the modern alphabet, the first letter was called Aleph, a variation of the Akkadian *alpu,* or ox. The letter itself, from which the modern "A" evolved, is a stylized pictograph of an ox's head with horns, ∀. In Hebrew, *aloof* (אלוף) can be translated as either ox or leader, a reminder that its heavenly counterpart once led the parade of stars across the sky as the first constellation of the zodiac.

Taurus is pictured in the heavens with only his head, shoulders, and forefeet; his hind part is missing. Some scholars have attributed this to an incident in the Mesopotamian Epic of Gilgamesh (VI, 158-59) when the hero's companion Enkidu kills the Bull of Heaven, which had been sent by the goddess Ishtar to destroy them. In an act of open defiance, Enkidu rips off the dead Bull's thigh and hurls it at the face of the goddess. To the people of Mesopotamia, this helped to explain why the Bull's hindquarters were missing from the sky. A more likely explanation, however, is that only the Bull's horned head seemed worthy of a place in the heavens.

Although the Pleiades are now considered part of Taurus, the Sumerians knew them as a separate constellation, MUL.MUL, the Stars. Mul is the Sumerian determinative for star, and its repetition indicates the importance that was attached to the group. It was the star par excellence, the star of stars. In Akkadian, the Pleiades were known as Zappu, the Bristle. They were depicted on Mesopotamian tablets and boundary stones as a group of seven stars or round balls, usually in association with the symbols of other heavenly bodies: a disk (sun), a crescent (moon), and an eight-pointed star (Venus-Ishtar). The seven balls may have represented Il Sibittum, Akkadian for the Seven Gods, the seven most important deities in the Mesopotamian pantheon, those who had the power to determine fate or destiny. Or, the seven balls may have represented the seven sacred pellets that were used in casting lots. In Greek mythology, the Pleiades, a matronymic, represent the seven daughters of the Titan Atlas and Pleione. There are several classical legends associated with them. In one account, Pleione and her beautiful daughters are spotted by Orion, who chases after them for seven years. Zeus finally takes pity on them and places them in the heavens, forever safe from the lecherous Hunter.

The Hyades are another group of stars in Taurus. As already noted, they form a vee-shaped cluster in the Bull's face. According to classic mythology this group represents the daughters of Atlas and Aethra, daughter of Oceanus. According to one account, the Hyades derive their name from their brother Hyas, who is killed while hunting. The sisters cry so much for their lost sibling that they die of grief and are honored with a place in the heavens. Another legend suggests that the sisters nurse the infant Dionysus (Bacchus), and as a reward, they are transformed into stars. This asterism is usually associated with rain and some historians have suggested that the name Hyades is derived from the Greek *hein* (υειν), to rain, since the group rises and sets in the sky in the fall and spring, the rainy seasons of the year. This association with rain may also be related to the tears shed by the sisters for their dead brother. The Romans, however, sometimes referred to them as Suculae, piglets, perhaps from the Latin *sucus,* moisture; or, as one scholar has postulated, from a mistranslation of the Greek υαδες, Hyades, as συς or υς, swine or pig.

Ancient Greek poets and writers relate two legends associated with Taurus. Zeus happens one day to spy Europa, the beautiful daughter of the king of Tyre. He changes into a handsome white bull, induces Europa to mount his back, and carries her across the sea to Crete, where he seduces her. She gives her name to her new homeland, Europe, and gives birth to a son, Minos, who later becomes the king of Crete and establishes bull-worship on the island. Another legend regarding Taurus relates how the philandering Zeus seduces Io, daughter of King Inachus of the Pelasgians. Zeus changes Io into a white heifer to hide her from his jealous wife, Hera, although another account claims that Hera herself changes Io into a heifer to thwart the relationship between her husband and his new-found love.

See Kramer, *The Sumerians,* pp. 305-6; Oates, *Babylon,* p. 17; McNeill, *The Rise of the West,* pp. 19-78; Hartner, "The Earliest History of the Constellations in the Near East," pp. 3-4, 8; van der Waerden, "History of the Zodiac," pp. 219-20; van der Waerden, "Babylonian Astronomy. II. The Thirty-Six Stars," pp. 9, 11, 15; O'Neil, *Time and the Calendars,* p. 55; Langdon, *Babylonian Menologies,* pp. 3-4, 110-11; Pritchard, *Anc. Near East. Texts,* pp. 85, 334-38, 343-45; Pritchard, *Anc. Near East Picts.,* plates 286, 534, 658; Jacobsen, *Toward the Image of Tammuz,* pp. 165-66, 404-5; Heimpel, "A Catalogue of Near Eastern Venus Deities," p. 9; van Buren, *Symbols of the Gods,* pp. 74-82; Reiner and Pingree, *Baby. Plan. Omens,* Part 2, pp. 12-13, 16; Gantz, *Early Greek Myth* p. 213; Ridpath, *Star Tales,* pp. 118-22; Condos, *Star Myths,* pp. 171-73, 191-94; Staal, *New Patterns in the Sky,* pp. 74-77; Allen, *Star Names,* pp. 386-91. For the significance of the number seven, see Scutum; for other constellations involved in farming and agriculture, see Aries, Pegasus, and Triangulum; for Io's fate as a heifer, see Cepheus. It is interesting to note that the development of the alphabet, as noted above, with only twenty-two letters was an important step in the democratization of learning since it replaced thousands of cuneiform and hieroglyphic signs and symbols. Henceforth, reading and writing would no longer be restricted to a small group of professional scribes, as was the case in ancient Mesopotamia and Egypt.

Lost Stars

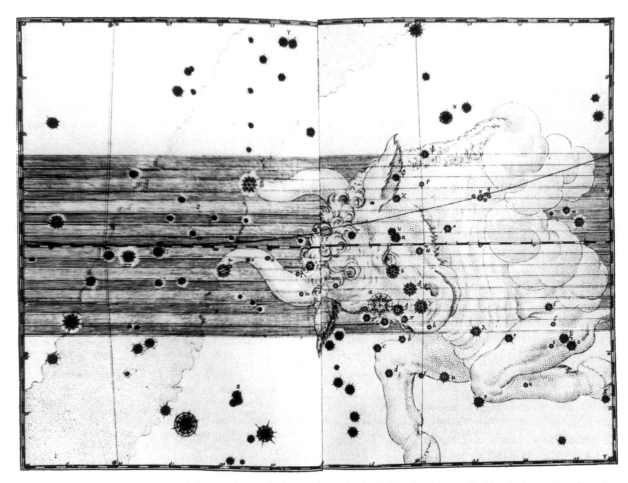

Figure 72. Bayer's Taurus. The Pleiades are the small cluster of stars in the Bull's shoulder marked by the letters Eta (ζ) and q. The letter W is the tail of Aries; X is the right foot of Perseus; Y is the group of kids in the hands of Auriga; Z is the head of Orion; and ZZ is the foot of one of the Twins, Castor. The unmarked line on the right side of the chart passing through the Pleiades is the Tropic of Cancer. The shaded area is the zodiac.

Table 82. The lettered stars of Taurus

	MAGNITUDES			CATALOGUE NUMBERS				
Letter	Bayer	Lacaille	Visual	Flamsteed	Lacaille	Other	HR	HD
Alpha	1		0.85	87			1457	29139
Beta+	2		1.65	112			1791	35497
Gamma	3		3.65	54			1346	27371
Delta¹+	3		3.76	61			1373	27697
Delta²			4.80	64			1380	27819
Delta³			4.29	68			1389	27962
Epsilon	3		3.53	74			1409	28305
Zeta	3		3.00	123			1910	37202
Eta+	3		2.87	25			1165	23630
Theta¹+			3.84	77			1411	28307
Theta²	4		3.40	78			1412	28319
Iota	4		4.64	102			1620	32301
Kappa¹+	4		4.22	65			1387	27934
Kappa²			5.28	67			1388	27946
Lambda	4		3.47	35			1239	25204
Mu	4		4.29	49			1320	26912
Nu	4		3.91	38			1251	25490
Xi	4		3.74	2			1038	21364
Omicron	4		3.60	1			1030	21120
Pi	5		4.69	73			1396	28100
Rho	5		4.65	86			1444	28910
Sigma¹+			5.07	91			1478	29479
Sigma²	5		4.69	92			1479	29488
Tau	5		4.28	94			1497	29763
Upsilon+	5		4.28	69			1392	28024
Phi	5		4.95	52			1348	27382
Chi	5		5.37	59			1369	27638
Psi	5		5.23	42			1269	25867
Omega¹+			5.50	43			1283	26162
Omega²	5		4.94	50			1329	27045
A¹+	5		4.36	37			1256	25604
A²			5.90	39			1262	25680
b	5		5.03	79			1414	28355
c¹+	5		4.27	90			1473	29388
c²			5.46	93			1484	29589
d	5		4.25	88			1458	29140
e	5		5.07	30			1174	23793
f	5		4.11	5			1066	21754
g+	5		5.56				958	19926
h+	6		5.59	57			1351	27397
i	6		5.10	97			1547	30780
k	6		5.81	98			1590	31592
l+	6		5.29	106			1658	32977
m	6		5.00	104			1656	32923
n	6		4.94	109			1739	34559
o	6		4.88	114			1810	35708
p	6		5.41	44			1287	26322
q+	6		4.30	19			1145	23338
r	6		5.12	66			1381	27820
s	6		5.14	4			1061	21686
t	6		5.77	6			1079	21933
u+	6		5.35	29			1153	23466

The Lost, Missing, or Troublesome Stars of Taurus

Beta, 112, HR 1791, 1.65V, *SA, HRP, BS,* El Nath.

Bayer, like Ptolemy, noted that this star was the same as Gamma Aur. For Bayer's duplicate stars, see Alpha And.

Delta¹, 61, HR 1373, 3.76V, *SA* as Delta, *HRP* as Delta, *BS,* BV 63 as Delta, BF 543*, H 91*.
Delta², 64, HR 1380, 4.80V, *BS,* BV 65, BF 547*, H 94*.
Delta³, 68, HR 1389, 4.29V, *BS,* BV 69, BF 552*, H 99*.

Although Bayer described Delta as a single 3rd-magnitude star, Flamsteed designated these three closely placed stars, his 61, 64, and 68, as Delta-1, Delta-2, and Delta-3. Bevis, though, labeled only the brightest star, 61, as Delta, and Argelander agreed (*UN,* p. 54). Heis felt otherwise. He restored Delta to all three stars as Flamsteed had originally proposed.

Eta, 25, HR 1165, 2.85V, *SA, HRP, BS,* Alcyone, one of the Pleiades.

Bayer included several stars of the Pleiades Cluster in his *Uranometria.* He noted that Eta was the *lucida* or the brightest of the group and assigned her a magnitude of 3. At the other extreme, he noted that q was the least bright, or as he described her *Pleiadum minima,* with a magnitude of only 6. Bayer placed four Pleiades on his star list between Omega and A, without Greek or Roman letter designations, and assigned them a magnitude of 5. All in all, he listed six Pleiades, thus preserving the tradition that there were originally seven, but one has been lost. See q, below.

Several Greek legends seek to explain why only six Pleiades are visible to the naked eye. One legend notes that six of the seven consort with the gods, but the seventh, Merope, marries a mere mortal. As a consequence, her image in the heavens appears much fainter than that of her sisters. A variation says that the faint Pleiad is Electra. Another legend suggests that one of the seven original sisters is struck by a bolt of lightening, thus rendering her invisible (Condos, *Star Myths,* pp. 171-73).

Theta¹, 77, HR 1411, 3.84V, *SA, HRP, BS,* HF 72 as Theta, H 112*.
Theta², 78, HR 1412, 3.40V, *SA, HRP, BS,* HF 73 as Theta, H 113*.

Bayer noted that Theta was a 4th-magnitude *duplex* or double star. Halley's 1712 pirated edition of Flamsteed's catalogue also refers to these two stars, his 77 and 78, as *duplex* and designates each of them Theta, while his authorized "Catalogus Britannicus" of 1725 designates them Theta-1 and Theta-2. This is a widely separated double, Struve I 10, in the Hyades Cluster.

Kappa¹, 65, HR 1387, 4.22V, *SA* as Kappa, *HRP* as Kappa, *BS,* BF 549*, H 97 as Kappa.
Kappa², 67, HR 1388, 5.28V, *BS,* BF 551*, H 98.

Although Bayer described Kappa as a single star of the 4th magnitude, Flamsteed designated these two nearly adjacent stars, his 65 and 67, as Kappa-1 and Kappa-2. Argelander, unable to see 67 with his naked eye, felt that the brighter star, 65, must have been Bayer's Kappa (*UN,* p. 54). Heis saw both stars, but he accepted Argelander's designation.

Sigma¹, 91, HR 1478, 5.07V, *SA, HRP, BS,* BV 101, BF 591*, H 132*.
Sigma², 92, HR 1479, 4.69V, *SA, HRP, BS,* BV 102 as Sigma, BF 592*, H 132*.

Although Bayer described Sigma as a single star of the 5th magnitude, Flamsteed designated these two nearly adjacent stars, his 91 and 92, as Sigma-1 and Sigma-2. Bevis, though, labeled only the brighter star, 92, as Sigma. Argelander agreed since he could not see 91 (*UN,* p. 55). Heis saw both and felt that their combined light was equivalent to Bayer's Sigma. Consequently, he restored Sigma to 91 as Flamsteed had originally proposed. This is the multiple-star system, Struve I 11, where a third component of magnitude 7.4V has recently been discovered.

Upsilon¹, 69, HR 1392, 4.28V, *SA* as Upsilon, *HRP* as Upsilon, *BS* as Upsilon, BV 90 as Upsilon, BF 554*, H 101 erroneously as Upsilon².
Upsilon², 72, HR 1399, 5.53V, BV 92, BF 558*, H 105.

Although Bayer described Upsilon as a single star of the 5th magnitude, Flamsteed designated these two closely placed stars, his 69 and 72, as Upsilon-1 and Upsilon-2. Bevis, though, labeled only the brighter star, 69, as Upsilon, and Argelander agreed (*UN*, p. 54). Heis, however, restored Upsilon to both stars, as Flamsteed had originally proposed. There is an error in Heis's catalogue where Flamsteed's 70, about 7° to the south, is equated with Upsilon^1 and his 69 with Upsilon^2. Hoffleit mistakenly noted that Heis's Corrigenda corrects the error ("Discordances," p. 64). Heis's erroneous catalogue listings probably account for the fact that most modern authorities have dropped Upsilon's indices and have labeled only 69 as Upsilon. On Heis's atlas, the stars are properly labeled.

Omega^1, 43, HR 1283, 5.50V, *BS*, BF 509*, H 65.

Omega^2, 50, HR 1329, 4.94V, *SA* as Omega, *HRP* as Omega, *BS,* BF 526*, H 78 as Omega.

Bayer described Omega as a single 5th-magnitude star. Apparently uncertain which of these two stars, his 43 or 50, about 3° apart, Bayer meant, Flamsteed designated them Omega-1 and Omega-2. Baily was puzzled by this since he felt it was quite obvious from Bayer's chart that only 50 was in any way synonymous with Omega, but he nevertheless retained both 43 and 50 as Omega. Argelander, though, had no hesitation in designating only 50 as Omega (*UN,* pp. 53-54).

A^1, 37, HR 1256, 4.36V, *SA, HRP,* B 170*, BF 499*, H 59*.

A^2, 39, HR 1262, 5.90V, *SA, HRP,* B 174*, BF 502*, H 61*.

Bayer noted that A was a single star of the 5th magnitude, and Flamsteed designated his 37 as A. Bode, however, was uncertain which of these two nearly adjacent stars, 37 or 39, Bayer meant, so he designated them A-1 and A-2. Argelander, on the other hand, had no doubt whatsoever which star was Bayer's A. Since he could see only 37, the brighter of the pair, with his naked eye, he felt certain it alone was synonymous with Bayer's A (*UN,* p. 53), but Heis saw both stars and, agreeing with Bode, he restored A to 39.

b, 79, HR 1414, 4.27V, *SA, HRP.*

c^1, 90, HR 1473, 4.27V, *SA, HRP* as c, BV 96 as c, BF 586*, H 130*.

c^2, 93, HR 1484, 5.46V, *SA,* BV 98, BF 595*, H 134*.

Although Bayer described c as a single 5th-magnitude star, Flamsteed designated these two neighboring stars, his 90 and 93, as c-1 and c-2. Both Bevis and Argelander, though, felt that only the brighter star, 90, was Bayer's c (*UN,* p. 55), but Heis disagreed and restored c to 93.

d, 88, HR 1458, 4.25V, *SA, HRP.*

e, 30, HR 1174, 5.07V, *SA, HRP.*

f, 5, HR 1066, 4.11V, *SA, HRP.*

g, in Cetus, HR 958, 5.56V, *SA, HRP, BS,* H 1*.

Both Bevis and Baily felt that Bayer's g Tau was synonymous with Kappa Cet (HR 996, 4.83V), not with this star. Argelander was among the first to equate this star with g (*UN,* p. 52). See Kappa Cet.

h, 57, HR 1351, 5.59V, *SA, HRP,* HF 53*, BV 52*, B 213 as h-1, BF 537*, H 84*.

Bayer described h as a single 6th-magnitude star, and Halley's 1712 edition of Flamsteed's catalogue designates his 57 as h. Oddly enough, Flamsteed's 1725 catalogue switches h, probably owing to a copying or typographical error, to his 58 (HR 1356, 5.26V, HF 54, BV 56, B 216 as h-2, BF 538, H 86), about 1° north. Bevis realized that this was erroneous since 58 did not conform to Bayer's chart. He felt that 57 was synonymous with h, as Halley's edition originally noted. Bode, possibly in deference to Flamsteed, designated both 57 and 58 as h-1 and h-2. Baily felt only 57 could possibly be h, so he dropped h from 58. Heis and Argelander agreed; the latter could not even see 58 (*UN,* p. 54).

i, 97, HR 1547, 5.10V, *SA, HRP.*

j

Bayer did not use this letter.

k, 98, HR 1590, 5.81V, *SA, HRP.*

l¹, 106, HR 1658, 5.29V, *SA* as l, *HRP* as l, BV 116 as l, BF 657 as l, H 149 as l.
l², 107, HD 33121, 6.5V, BV 118, BF 659.

Although Bayer noted that l was a single star of the 6th magnitude, Flamsteed designated these two neighboring stars, his 106 and 107, as l-1 and l-2. Bevis, though, labeled only the brighter one, 106, as l. Baily agreed.

m, 104, HR 1656, 5.00V, *SA, HRP.*

n, 109, HR 1739, 4.94V, *SA, HRP.*

o, 114, HR 1810, 4.88V, *SA, HRP.*

p, 44, HR 1287, 5.41V, *SA, HRP.*

q, 19, HR 1145, 4.30V, *SA, HRP,* BV 23*, H 26*, Taygeta, one of the Pleiades.

Although Bayer designated this star q, Flamsteed called it e. Bevis was among the first to label 19 as q. Argelander (*UN,* p. 53), Heis, and most modern authorities have accepted his designation.

Flamsteed listed in numerical order by right ascension fourteen stars in the Pleiades and except for Eta, 25, he lettered them with his own set of lower-case Roman letters. Baily removed them all as being superfluous (see i Aql; *BF,* p. 397). Listed below are Flamsteed's fourteen Pleiades with the letters he assigned them:

    15, n, nonexistent, BF Table B, p. 645.
    16, g, HR 1140, 5.46V, BF 457, Celaeno.
    17, b, HR 1142, 3.70V, BF 458, Electra.
    18, m, HR 1144, 5.64V, BF 459.
    19, e, HR 1145, 4.30V, BF 460, q, Taygeta.
    20, c, HR 1149, 3.87V, BF 461, Maia.
    21, k, HR 1151, 5.76V, BF 462, Asterope.
    22, l, HR 1152, 6.43V, BF 463, Asterope (combined light with 21).
    23, d, HR 1156, 4.18V, BF 464, Merope.
    24, p, HD, 23629, 6.29V, BF 467.
    25, Eta, HR 1165, 2.87V, *SA, HRP, BS,* BF 468 as Eta, Alcyone.
    26, s, HD 23822, 6.47V, BF 472.
    27, f, HR 1178, 3.63V, BF 473, Atlas, father of the Pleiades.
    28, h, HR 1180, 5.09V, BF 474, Pleione, mother of the Pleiades, the variable BU.

r, 66, HR 1381, 5.12V, *SA, HRP.*

s, 4, HR 1061, 5.14V, *SA, HRP.*

t, 6, HR 1079, 5.77V, *SA, HRP.*

u¹, 29, HR 1153, 5.35V, *SA* as u, *HRP* as u, BV 9 as u, BF 470*, H 29 as u.
u², 31, HR 1199, 5.67V, BV 14, BF 483*, H 41.

Although Bayer described u as a single 4th-magnitude star, Flamsteed designated these two stars, about 2° apart, his 29 and 31, as u-1 and u-2. Bevis and Argelander, however, felt that Bayer must have meant the brighter star, 29 (*UN,* p. 53), so they removed u from 31.

TELESCOPIUM, TEL

Telescope

Lacaille devised the constellation Telescopium (figures 2a and 2b) and lettered its stars from Alpha to Tau. He depicted it on his planisphere of the southern sky as a huge instrument suspended from a pole that stretched almost 40° across the sky into Corona Australis, Sagittarius, Scorpius, and Ophiuchus. Warner suggested it was meant to represent the aerial telescopes used by Giovanni Cassini at the Royal Observatory at Paris. Baily and Gould later reduced the size of the constellation to a neat quadrilateral and in the process switched several of its stars back to the neighboring constellations from which they had been taken by Lacaille.

Like the microscope, the telescope was first developed in the Netherlands by Dutch lens makers about the year 1608. Its first use was in warfare against the Spanish. It is generally believed that Galileo was the first to use the telescope in astronomy in 1609. Among his first discoveries were the four largest satelites of Jupiter, which he dubbed the Medici stars; the moon-like phases of Venus, thus proving that both Venus and Mercury rotate around the sun; the rings of Saturn; the myriad of stars in the Milky Way, which astronomers had earlier thought was a solid band across the sky; the craters and seas (*mares*) of the moon; and sunspots, which eventually caused Galileo's blindness. Within just a few short years, other astronomers made new and startling discoveries with the telescope, and a new era dawned for the science of astronomy in Western Europe.

See Warner, *Sky Explored,* p. 143; King, *History of the Telescope,* chap. III. For the discovery of the telescope, see Chi and h Per.

Table 83. The lettered stars of Telescopium

	MAGNITUDES			CATALOGUE NUMBERS				
Letter	Bayer	Lacaille	Visual	Flamsteed	Lacaille	Other	HR	HD
Alpha		4	3.51		1513		6897	169467
Beta+		4	3.11		1503		6832	167618
Gamma+		4	3.21		1468		6630	161892
Delta^1+		5	4.96		1521		6934	170465
Delta^2		5	5.07		1524		6938	170523
Epsilon		5	4.53		1496		6783	166063
Zeta		5	4.13		1517		6905	169767
Eta+			5.05			G 55	7329	181296
Theta+		5	4.29	45 Oph	1435		6492	157919
Iota+			4.90			G 64	7424	184127
Kappa		6	5.17		1550		7087	174295
Lambda		6	4.87		1556		7134	175510
Mu		6	6.30		1595		7393	183028
Nu		6	5.35		1611		7510	186543
Xi+			4.94			G 78	7673	190421
Rho		6	5.16		1567		7213	177171
Sigma+		6	5.25		1509		6875	168905
Tau+		6	5.70		1518		6922	170069

The Lost, Missing, or Troublesome Stars of Telescopium

Beta, in Sagittarius, HR 6832, 3.11V, CA 1503*, BR 6360*, L 7643 as Eta Sgr, G 46 as Eta Sgr.

This was originally Bayer's Eta Sgr that Lacaille relettered as Beta Tel. Baily switched it back to Sagittarius, removed Lacaille's Beta, and restored Bayer's original designation.

Gamma, in Scorpius, HR 6630, 3.21V, CA 1468*, BR 6214*, L 7449, G 172 as G Sco.

Baily switched this star to Scorpius and left it unlettered. Gould relettered it G Sco because of its brightness. See G Sco.

Delta¹, HR 6934, 4.96V, *SA, HRP, BS,* CA 1521 as Delta, BR 6419*, L 7729*, G 16*.
Delta², HR 6938, 5.07V, *SA, HRP, BS,* CA 1524 as Delta, BR 6420*, L 7734*, G 17*.

Lacaille designated these two nearly adjacent stars Delta.

Eta, HR 7329, 5.05V, *SA, HRP, BS,* CA 1586, BR 6629, L 8062, G 55*.

Lacaille did not letter this star Eta. He observed and catalogued it but left it unlettered; he considered it 6th magnitude. His Eta is the open cluster M 7, NGC 6475 (CA 1470 as Eta, BR 6241 as Eta, L 7478 in Scorpius, G 179 in Scorpius), which Baily switched to Scorpius. Since the letter Eta was therefore no longer in use in Telescopium, Gould used it to designate this star. He felt a star of its brightness merited a letter. For Lacaille's nebulous objects, see Xi Tuc.

Theta, 45 Oph in Ophiuchus, HR 6492, 4.29V, CA 1435*, L 7293 as d Oph, G 139 as d Oph.

This was originally Bayer's d Oph that Lacaille relettered as Theta Tel. Baily switched it back to Ophiuchus, removed Lacaille's Theta, and restored Bayer's original designation.

Iota, HR 7424, 4.90V, *SA, HRP, BS,* CA 1603 as D Sgr, BR 6689 as D Sgr, L 8137, G 64*.

Lacaille did not use this letter in Telescopium. He observed this star and lettered it D Sgr. Gould switched it to Telescopium and relettered it Iota because he felt a star of its brightness merited a letter.

Xi, HR 7673, 4.94V, *SA, HRP, BS,* CA 1642, BR 6794, L 8321, G 78*.

Although Lacaille observed and catalogued this star, he left it unlettered since he considered it a 6th-magnitude *informis* on the border between Sagittarius and Pavo. Baily considered it in Pavo, but Gould switched it to Telescopium and designated it Xi because of its brightness.

Omicron
Pi

Lacaille did not use these two letters in Telescopium.

Sigma, in Corona Australis, HR 6875, 5.25V, CA 1509*, BR 6386*, L 7680, BAC 6228, G 8 in Corona Australis.

Baily switched this star to Corona Australis and left it unlettered.

Tau, HR 6922, 5.70V, CA 1518*, BR 6411*, L 7713, BAC 6262*, G 15.

Gould removed this star's letter because of its dimness. He considered its magnitude 6.5.

TRIANGULUM, TRI

Triangle

Triangulum (Figure 73) is a Ptolemaic constellation that Bayer lettered from Alpha to Epsilon. Although a small, relatively insignificant constellation today with no star brighter than 3rd magnitude, Triangulum was most important to the ancient Sumerians. Since its shape resembled the point of a plow, they named the constellation APIN (Akkadian Epinnu), the Plow, the implement invented in the Middle East that revolutionized the life of primitive man.

The Plow was closely associated with several neighboring constellations: the Bull or Ox (Taurus), the Hired Farm-Worker (LU.HUN.GA-Aries), the Field (Iku-Pegasus), and Anunitum-SIM.MAH (Pisces). All of these constellations rose near the time of the vernal equinox and reminded farmers to prepare the land for cultivation. In an expansive celestial scenario, the Plow was fastened to the horns of the Ox, which was guided by the Hireling with his goad across the Field while Anunitum (Ishtar), goddess of spring rainstorms and fertility, would hopefully lend her assistance. In ancient Mesopotamia, APIN was among the reigning constellations of Nisanu (Nisan), the first month (April-May) of the Babylonian calendar, the beginning of the new year. MUL.APIN (Plow Star) are the first words of a surviving seventh century B.C. cuneiform text that has provided historians with a glimpse into the astronomical knowledge of ancient Mesopotamia.

In contrast to the very rich Mesopotamian tradition, relatively few Greek astral myths or legends are associated with Triangulum. Some poets have suggested that, because of its shape, the constellation may have represented to the Greeks the letter Delta, Δ; or, perhaps, Sicily, once called Trinacria, three-cornered; or, possibly, the Nile Delta. Greek mythology suggests that the constellation owes its place in the heavens to Hermes, who is assigned by the gods to set the stars in the sky. Hermes places Triangulum just above Aries, which due to precession replaced Taurus as the first sign in the zodiac during classical times. Since Delta represents the initial of the genitive or possessive form of Zeus's name, Dios (Διος, god), Hermes honors his sovereign by placing the symbol with Aries, the sign marking the beginning of the new year.

See Hartner, "The Earliest History of the Constellations in the Near East," p. 10; van der Waerden, "Babylonian Astronomy. II. The Thirty-Six Stars," pp. 9, 11, 13, 15; Contenau, *Everyday Life in Babylon and Assyria,* p. 183; Oates, *Babylon,* pp. 17, 120; Reiner and Pingree, *Baby. Plan. Omens,* Part 2, p. 10; Hunger and Pingree, *Mul.Apin,* Introduction; Ridpath, *Star Tales,* pp. 124; Condos, *Star Myths,* pp.195-96; Staal, *New Patterns in the Sky,* pp. 48-49. See also Aries, Pegasus, Pisces, and Taurus.

Table 84. The lettered stars of Triangulum

	MAGNITUDES			CATALOGUE NUMBERS				
Letter	Bayer	Lacaille	Visual	Flamsteed	Lacaille	Other	HR	HD
Alpha	4		3.41	2			544	11443
Beta	4		3.00	4			622	13161
Gamma	4		4.01	9			664	14055
Delta	5		4.87	8			660	13974
Epsilon	6		5.50	3			599	12471
Eta+			5.28	7			655	13869
Iota+			4.94	6			642	13480

Lost Stars

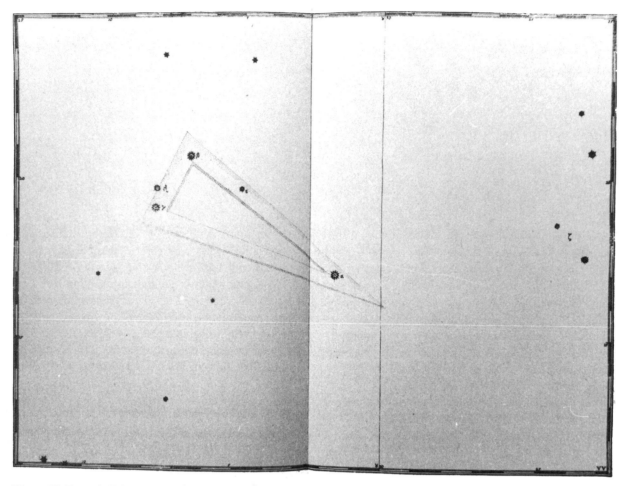

Figure 73. Bayer's Triangulum. The letter Zeta (ζ) at the right edge of the chart represents the stars in Pisces.

The Lost, Missing, or Troublesome Stars of Triangulum

Eta, 7, HR 655, 5.28V, BV 8, BF 277, H 15.
Iota, 6, HR 642, 4.94V, *SA,* BV 4, BF 267, RAS 233*, H 13.
a, 10, HR 675, 5.03V, BV 6, BF 286, H 18.
c, 12, HR 717, 5.29V, BV 11, BF 303, H 22.
d, 1, HD 10407, 7.4V, BV 1, BF 200.
d, 11, HR 712, 5.54V, BV 13, BF 301, H 21.

Bayer did not letter these six stars. The letters were assigned by Flamsteed but removed by Bevis and Baily, who considered them superfluous (see i Aql). Only Iota has managed to survive in the literature, possibly because Baily included it in the "General Catalogue" of the Royal Astronomical Society of 1826.

Triangulum Australe, TrA

Southern Triangle

Keyzer and Houtman devised the constellation Triangulum Australe (Figure 74) as Den Zuyden Trianghel (Southern Triangle); Bayer depicted it on his chart of the southern skies; and Lacaille lettered its stars from Alpha to Lambda.

The Reverend Petrus Plancius had drawn a Triangulus Antarcticus below Argo Navis on his globe of 1589. Although its position does not conform to that of the Southern Triangle, its inclusion on Plancius's globe may have influenced his pupil Petrus Keyzer to include a triangle in his own catalogue. When Lacaille prepared his chart of the southern skies, he purposely redesigned Keyzer and Houtman's Triangle to represent a level *(niveau)*, with the intention of having it conform to its two neighboring constellations, Compass Dividers (Circinus) and Square and Ruler (Norma). These were the instruments of the architect, whom he was honoring with a place in the heavens. Unlike its modern counterpart with its tiny water vials and bubbles, this level was shaped like a triangle, with legs extending beyond the base and a plumb line fixed to the triangle's apex so that when its legs were set on a perfectly level surface, the line would exactly bisect its base.

See Warner, *Sky Explored,* pp. 201-4; Lacaille, "Table des Ascensions," p. 589.

Table 85. The lettered stars of Triangulum Australe

	MAGNITUDES			CATALOGUE NUMBERS				
Letter	Bayer	Lacaille	Visual	Flamsteed	Lacaille	Other	HR	HD
Alpha		2	1.92		1381		6217	150798
Beta		3	2.85		1311		5897	141891
Gamma		3	2.89		1267		5671	135382
Delta		5	3.85		1338		6030	145544
Epsilon		5	4.11		1292		5771	138538
Zeta		6	4.91		1352		6098	147584
Eta[1]+		6	5.91		1372		6172	149671
Eta[2]		6	6.7		1379			150550
Theta		6	5.52		1367		6151	148890
Iota		6	5.27		1358		6109	147787
Kappa		6	5.09		1307		5891	141767
Lambda+		6	5.75		1314		5920	142514

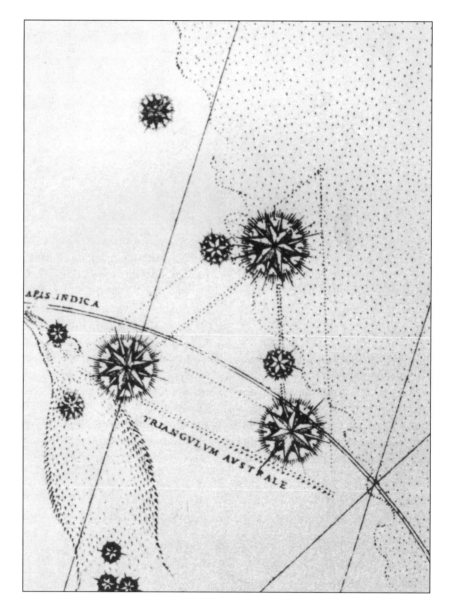

Figure 74. Bayer's Triangulum Australe. This is an enlarged view of Triangulum Australe from Bayer's Tabula XLIX.

The Lost, Missing, or Troublesome Stars of Triangulum Australe

Eta¹, HR 6172, 5.91V, *SA* as Eta, *HRP, BS,* CA 1372 as Eta, BR 5756*, L 6865 as Eta, BAC 5536*, G 37*.
Eta², HD 150550, 6.7V, CA 1379 as Eta, BR 5797*, L 6900, BAC 5565*, G 40*.

    Lacaille designated these two neighboring stars Eta. Baily was inconsistent in dealing with these stars. When he edited Lacaille's expanded catalogue, he designated only the brighter star, CA 1372, as Eta. However, when editing the *BAC* at about the same time, he designated both stars as Eta. Gould considered them fainter than 6th magnitude, but he kept their letters since they were less than 25° from the southern celestial pole and he felt it was important to highlight the stars in this area.

Lambda, HR 5920, 5.75V, CA 1314*, BR 5519*, L 6559, BAC 5256*, G 20.

    Gould dropped the letter from this star because of its dimness; he considered it less than 6th magnitude.

Tucana, Tuc

Toucan

Keyzer and Houtman devised the constellation Tucana (Figure 75). In his catalogue of 1603, Houtman called it Den Indiaenschen Exster (Indian Magpie) or Lang[bek] (Snipe or Long-beak). Petrus Plancius labeled it Toucan on his globe of 1598, as did Bayer on his chart of the southern skies. Lacaille included it in his catalogue and lettered its stars from Alpha to Rho. Like many of the other constellations devised by Keyzer and Houtman, the Toucan represented one of the exotic creatures first encountered by Europeans during the Age of Exploration.

See Warner, *Sky Explored,* p. 204. For a possible explanation of the differences in nomenclature in the constellations devised by Keyzer and Houtman, see Apus.

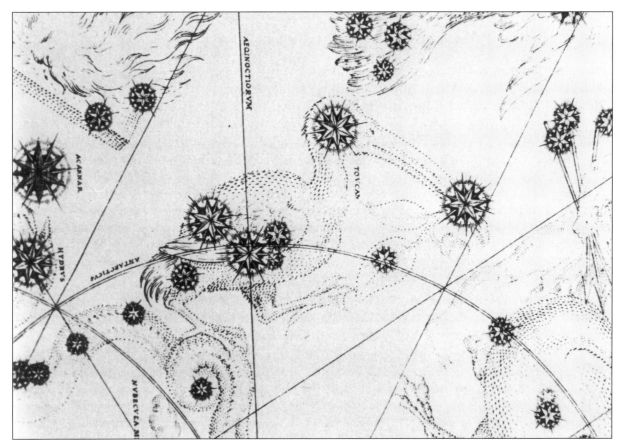

Figure 75. Bayer's Tucana. This is an enlarged view of Tucana from Bayer's Tabula XLIX. The bright 1st-magnitude star marked Acarnar [Achernar] near Hydrus's mouth at the left edge of the chart is Alpha Eri, the end of the River Eridanus.

Table 86. The lettered stars of Tucana

		MAGNITUDES			CATALOGUE NUMBERS			
Letter	Bayer	Lacaille	Visual	Flamsteed	Lacaille	Other	HR	HD
Alpha		3	2.86		1789		8502	211416
Beta¹+		4	4.37		26		126	2884
Beta²		4	4.54		27		127	2885
Beta³		5	5.09		28		136	3003
Gamma		4	3.99		1877		8848	219571
Delta		5	4.48		1799		8540	212581
Epsilon		5	4.50		1926		9076	224686
Zeta		5	4.23		11		77	1581
Eta		5	5.00		1923		9062	224392
Theta		6	6.13		32		139	3112
Iota		6	5.37		70		332	6793
Kappa¹+		7	7.22c					7693
Kappa²		6	4.86c		79		377	7788
Lambda¹+		6	6.22		57		252	5190
Lambda²		6	5.45		60		270	5457
Mu+		6	6.34		1795		8509	211726
Nu		6	4.81		1806		8582	213442
Xi+			Neb			G 51	95	
Pi		6	5.51		12		83	1685
Rho		6	5.39		46		187	4089

The Lost, Missing, or Troublesome Stars of Tucana

Beta¹, HR 126, 4.37V, *SA, HRP, BS,* CA 26 as Beta, BR 58*, L 119*, G 52*.
Beta², HR 127, 4.54V, *SA, HRP, BS,* CA 27 as Beta, BR 59*, L 120*, G 53*.
Beta³, HR 136, 5.09V, *SA, BS,* CA 28 as Beta, BR 61*, L 123, G 54.

Lacaille designated these three stars, all part of a multiple-star system, as Beta. Baily dropped CA 28's letter because of its dimness. Gould also dropped CA 28's letter, noting it was a faint 7½-magnitude companion of Beta¹ and Beta² (*Uran. Arg.,* p. 81). In the body of his catalogue, however, he correctly listed CA 28 as a 5th-magnitude star, as did Lacaille. Gould very likely observed one of the fainter elements in this system and somehow confused it with CA 28.

Beta¹ and Beta² comprise the double star Innes 260. Beta¹ consists of components A, magnitude 4.4V, and B, magnitude 13.5V. Beta² consists of components C, magnitude 4.8V, and D, magnitude 6.0V. Beta³ — the double star, Bos 8 — consists of two components with magnitudes of 5.8V and 6.0V. All three Beta's have the same common proper motion; that is, they are traveling through space together. See Hirshfeld and Sinnott, *Sky Cat.,* II, 5, 163; *Burnham's Celestial Handbook,* III, 1907.

Kappa¹, HD 7693, 7.8V, 8.2V, comb. mag. 7.22V, *SA,* BR 176, L 353, GC 1535*, Herschel 3423.
Kappa², HR 377, 5.1V, 7.3V, comb. mag. 4.86V, *SA, HRP* as Kappa, *BS* as Kappa, CA 79 as Kappa, BR 178 as Kappa, L 356 as Kappa, G 78 as Kappa, GC 1536*, Innes 27.

Kappa is a quadruple-star system consisting of two pairs of double stars moving together through space. Lacaille saw both pairs but designated only the brighter, L 356 (CA 79) as Kappa; he considered it a single 6th-magnitude star. He was unaware that stars L 353 and L 356 were actually both doubles. Later astronomers, like Boss in his *General Catalogue* of 1936, assigned indices to the pair: Kappa¹, HD 7693, consisting of component C, magnitude 7.8V, and D, magnitude 8.2V; and Kappa², HR 377, consisting of components A, magnitude 5.1V, and B, magnitude 7.3V. Gould saw both pairs but succeeded in splitting only the brighter one, L 356. He noted that the two pairs, L 353 and L 356, appeared as one star to his naked eye and assigned Kappa only to L 356. See Hirshfeld and Sinnott, *Sky Cat.,* II, 10.

Lambda¹, HR 252, 6.22V, *HRP, BS,* CA 57 as Lambda, BR 114*, BR 116*, L 250, G 68*, Dunlop 2.
Lambda², HR 270, 5.45V, *SA* as Lambda, *HRP, BS,* CA 60 as Lambda, BR 122*, L 262 as Lambda, G 70*.

Lacaille designated these two neighboring stars Lambda. Baily dropped the letter from the fainter star, CA 57, but Gould restored it because of its proximity to the southern celestial pole. Brisbane noted that Lambda¹ was a double star and designated each of its components as Lambda¹.

Mu, HR 8509, 6.34, CA 1795*, L 9092, BAC 7780, CAP 4470, G 4.

Baily removed the letter from this star because of its dimness.

Xi, HR 95, *HRP, BS,* CA 33, BR 38, L 80, B 47, G 51*, 47 Tucanae.

Lacaille did not use this letter in Tucana. Gould assigned Xi to Bode's 47, the globular cluster NGC 104. Although Lacaille observed and catalogued the cluster, he left it unlettered since he considered it nebulous. All in all, Lacaille observed forty nebulous objects, but he lettered only three: Omega Cen, Kappa Cru, and Eta Tel. Tirion's *SA* shows another Xi Tuc, about 30° to the west, a star (HD 215562, 6.4V, L 9227, BAC 7933, G 14) that Lacaille observed but left unlettered. Tirion copied this latter Xi from Becvar's atlases. The letter may have originated with Bode, who listed a Xi Tuc (B 24) at this star's approximate declination but at 1^h greater in right ascension. Tirion removed this Xi from his *Uranometria 2000.0* and from the second edition of *SA*.

Omicron

Lacaille did not use this letter in Tucana.

URSA MAJOR, UMa

Larger, Greater, or Big Bear

Some scholars have speculated that Ursa Major (Figure 76), Ursa Minor, and Bootes are among the oldest constellations of all, dating back at least 10,000 years. They suggest that since the legends surrounding these three constellations were familiar to Europeans, Siberians, and Amerindians, they must have been created before the Bering Strait land bridge was submerged thus cutting off America from the Eurasian continent shortly after the last Ice Age.

Ursa Major is a Ptolemaic constellation that Bayer lettered from Alpha to Omega and from A to h. Its more famous component, an asterism known to many people in the West as the Big Dipper, was known to the ancient Sumerians as MAR.GID.DA, the Wagon or Wain, a title that has been associated with the asterism from that time to the present. In Babylon on the 5th day of the first month, Nisanu (March-April), during the Akitu Festival celebrating the New Year, the urigallu-priest, recited a series of prayers to the goddess Ishtar, queen of heaven. He named the various stars sacred to Ishtar including "The star Margidda [the Wagon], the bond of heaven, whose name is My Lady...." The phrase "bond of heaven" refers perhaps to the asterism's faithful appearance each night of the year as it rotates or wheels around the celestial pole. This may also explain why the asterism is referred to as a wagon.

Although the Mesopotamians specifically devised a constellation, GIS.GIGIR (Akkadian Narkabtu, Narkabti), the Chariot, to occupy an area of the sky currently held by Perseus and Taurus, it has been suggested that MAR.GID.DA, the Wagon, is a heavenly representation of a four-wheeled war wagon that was developed in Mesopotamia in the middle of the third millennium B.C. The four stars in the Dipper's bowl supposedly represent the cart, which held a driver and a fighting-man, while the three stars in the handle represent the shaft, to which were yoked two to four onagers. This proto-chariot, while slow and clumsy with its four solid-wood wheels, was a significant innovation in warfare until it, in turn, was replaced several centuries later by a spoked, two-wheeled version pulled by a team of horses.

While the people of the Middle East placed the image of a wagon in the polar region of the sky, the people of Europe saw something quite different. They saw an animal that could be found in their more northerly regions — a great bear, Ursa Major.

The Great Bear was the source of several Greek legends. In one account, Callisto, daughter of the king of Arcadia, pledges herself to Artemis (Diana), goddess of the hunt and virginity. Zeus, however, ignores Calisto's pledge and rapes her. As a result, Callisto gives birth to a son, Arcas. When Artemis learns that Callisto has lost her virginity, the goddess vents her anger not on the rapist, but on the hapless victim, Callisto, whom she turns into a bear. Years later, Arcas comes upon his mother while hunting and just as he is about to slay her, Zeus intervenes and places both mother and son in the heavens — Callisto as Ursa Major and Arcas as Ursa Minor. In a variation of this tale, it is Zeus's wife, Hera, who, angered by her husband's infidelity, turns Callisto into a bear. While hunting in the forest, Arcas comes across Callisto. Because he does not recognize his mother, he prepares to kill her. Zeus intervenes and sets both in the sky — Callisto as Ursa Major and Arcas as Arctophylax, Protector of the Bear (Bootes).

Since the most visible parts of Ursa Major and its neighbor Ursa Minor are similarly shaped — in the west many see them as dippers — the myths and legends that surround the constellations are also similar. The Romans referred to the seven stars of the asterism as the Septem Triones, the Seven Plowing, or Threshing, Oxen. The Romans believed that these seven oxen perpetually circled the heavenly north pole in their never-ending task of separating the grain from the chaff on the celestial threshing floor. Other observers have suggested that the seven stars form a heavenly plow: the four stars in the bowl form the plow-blade and the three stars in the handle form the shaft.

See Schaefer, "The Southern Greek Constellations," pp. 334-35; van der Waerden, "Babylonian Astronomy. II. The Thirty-Six Stars," pp. 9, 11, 13-14; Langdon, *Babylonian Menologies,* p.5; Pritchard, *Anc. Near East. Texts,* p. 333; Allen, *Star Names,* pp. 426-37; Krupp, *Beyond the Blue Horizon,* pp. 230-31; Pritchard, *Anc. Near East Picts.,* plate 303; Reiner and Pingree, *Baby. Plan. Omens,* Part 2, pp. 11-13; Hunger and Pingree, *Mul.Apin,* p. 110; Ridpath, *Star Tales,* pp. 126-29; Condos, *Star Myths,* pp.197-200; Staal, *New Patterns in the Sky,* pp. 121-38. See Bootes for a somewhat different version of the legend surrounding Ursa Major; see also Ursa Minor.

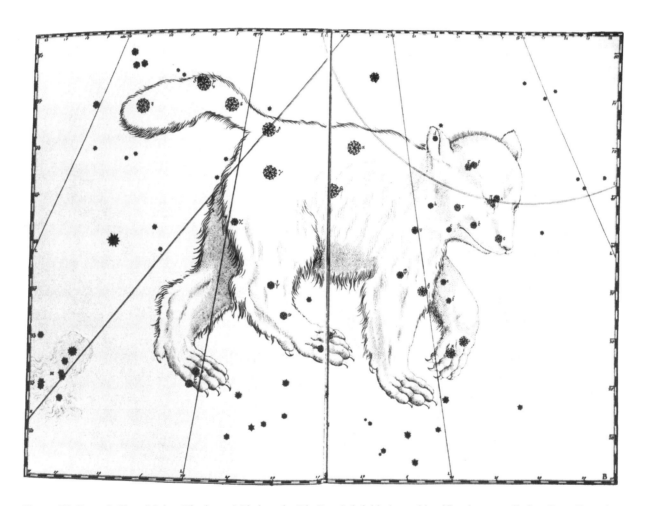

Figure 76. Bayer's Ursa Major. The letter M below the Big Bear's left hind paw identifies the constellation Coma Berenices (Berenice's Hair), which Bayer considered merely an asterism and thus not worthy of a separate chart. On his chart of Bootes (Figure 11), it is depicted as Azimeth, a sheaf of wheat. The letters I and K are the Arctic Circle and the equinoctial colure. The letter L located at the bottom of each line of ecliptic longitude is to remind readers that these lines are drawn every 30° around the celestial globe starting from the First Point of Aries for each of the twelve signs of the zodiac. Bayer noted that these signs affect one's character. He called them *dodecatemoria,* an old Greek astrological term derived from the two words meaning twelve parts. Bayer inserted this reminder only on the first two charts of his atlas—this chart, which is the second in his atlas, and his chart for Ursa Minor, which is the first (Figure 77). *Dodecatemoria,* or as it is sometimes called *dodecatemorion,* has taken on various meanings in astrology. As noted above, it was originally meant to refer to the 30° sectors of each of the twelve signs of the zodiac, but it was later interpreted to mean one-twelfth of a sign or 2½° of ecliptic longitude, and each of these sections was, in turn, assigned a zodiacal sign. See Tester, *History of Western Astrology,* pp. 27-28 and *passim.*

Table 87. The lettered stars of Ursa Major

	Magnitudes			Catalogue Numbers				
Letter	Bayer	Lacaille	Visual	Flamsteed	Lacaille	Other	HR	HD
Alpha	2		1.79	50			4301	95689
Beta	2		2.37	48			4295	95418
Gamma	2		2.44	64			4554	103287
Delta+	2		3.31	69			4660	106591
Epsilon	2		1.77	77			4905	112185
Zeta	2		2.17c	79			5054/5	116656/7
Eta	2		1.86	85			5191	120315
Theta	3		3.17	25			3775	82328
Iota	3		3.14	9			3569	76644
Kappa	3		3.60	12			3594	77327
Lambda	4		3.45	33			4033	89021
Mu	4		3.05	34			4069	89758
Nu	4		3.48	54			4377	98262
Xi	4		3.79c	53			4374/5	98230/1
Omicron	4		3.36	1			3323	71369
Pi^1+			5.64	3			3391	72905
Pi^2	4		4.60	4			3403	73108
Rho	4		4.76	8			3576	76827
Sigma^1+			5.14	11			3609	77800
Sigma^2	4		4.80	13			3616	78154
Tau	4		4.67	14			3624	78362
Upsilon	4		3.80	29			3888	84999
Phi	4		4.59	30			3894	85235
Chi	4		3.71	63			4518	102224
Psi	4		3.01	52			4335	96833
Omega	4		4.71	45			4248	94334
A	5		5.47	2			3354	72037
b+	5		5.72	5			3505	75486
c	5		5.13	16			3648	79028
d	5		4.56	24			3771	82210
e	5		4.83	18			3662	79439
f	5		4.48	15			3619	78209
g+	5		4.01	80			5062	116842
h	5		3.67	23			3757	81937

The Lost, Missing, or Troublesome Stars of Ursa Major

Delta, 69, HR 4660, 3.31V, *SA, HRP, BS*.

Bayer described Delta as a 2nd-magnitude star and this description has puzzled generations of astronomers. Some have concluded that Delta is variable or that it has decreased significantly in brightness since Bayer's day. Argelander, however, suggested a simpler explanation. He showed, first of all, that Bayer and his contemporaries, like Bartsch and Schiller, knew that Delta was 3rd magnitude. Even Ptolemy listed it as 3rd magnitude in his catalogue of the stars. Bayer described it as 2nd magnitude, Argelander argued, for the sake of esthetics. According to this argument, since all the other stars in the Big Dipper are 2nd magnitude, this "flawed" star should not be allowed to detract from the others. In other words, Argelander felt Bayer purposely overlooked Delta's faintness in order not to detract from the beauty of the entire asterism (*Fide Uran.*, pp. 18-20).

Tycho may also have influenced Bayer to describe Delta as 2nd magnitude. In both of Tycho's catalogues, the original of 777 stars and his expanded version of just over 1,000, he noted that Delta (Tycho's 19) was 2nd magnitude. And as Argelander has observed, Bayer was "seduced by Tycho's catalogue" (*Fide Uran.*, p. 21).

Pi¹, 3, HR 3391, 5.64V, *SA, HRP, BS,* HF 73, BF 1185, H 7*.
Pi², 4, HR 3403, 4.60V, *SA, HRP, BS,* HF 74 as Pi, BF 1186 as Pi, H 8*.

    Bayer noted that Pi was a single star of the 4th magnitude, and Halley's 1712 edition of Flamsteed's catalogue labels his 4 as Pi. Flamsteed's 1725 "Catalogus Britannicus," however, designates these two neighboring stars, his 3 and 4, as Pi-1 and Pi-2. Baily, though, felt that only the brighter star, 4, should be Pi. In fact, he suggested that since 3 was closer to A than to 4, it might more appropriately be called A^2. Argelander, unable to see 3 with his naked eye, agreed with Baily that only 4 was synonymous with Pi (*UN*, p. 21). Heis, though, saw both stars and restored Pi to 3.

Sigma¹, 11, HR 3609, 5.14V, *SA, HRP, BS,* HF 82, BV 14, BF 1266*, H 27*.
Sigma², 13, HR 3616, 4.80V, *SA, HRP, BS,* HF 84 as Sigma, BV 15 as Sigma, BF 1271*, H 29*.

    Bayer noted that Sigma was a single 4th-magnitude star, and Halley's pirated edition of Flamsteed's catalogue labels his 13 as Sigma. Flamsteed's authorized "Catalogus Britannicus", however, designates these two stars, his 11 and 13, as Sigma-1 and Sigma-2.

A, 2, HR 3354, 5.47V, *SA, HRP.*

b, 5, HR 3505, 5.72V, *SA, HRP,* BV 16*, B 31*, BF 1229*, H 13*.

    Flamsteed's catalogue designates his 7 as Bayer's b, but Bevis, noting that 7's position did not conform to Bayer's chart, instead designated 5 as b. Baily agreed. He explained that Flamsteed had erred in his observation of 7, that the star he observed was actually 5, and that 7 (BV 18, B 33, BF 1232 and Table C, p. 646) was probably nonexistent. As Baily pointed out, Flamsteed's manuscripts indicate that he was very uncertain about 7 and called the star "the companion of b," not b. Fifty years later, C. H. F. Peters took exception to 7's nonexistence and claimed that 7 was synonymous with HD 73745, 7.3V ("Flamsteed's Stars," p. 78). See 7 UMa in Part Two.

c, 16, HR 3648, 5.13V, *SA, HRP.*

d, 24, HR 3771, 4.56V, *SA, HRP.*

e, 18, HR 3662, 4.83V, *SA, HRP.*

f, 15, HR 3619, 4.48V, *SA, HRP.*

g, 80, HR 5062, 4.01V, *SA, HRP,* Alcor.

    Despite its prominent position next to Mizar (Zeta UMa) in the Big Dipper, Alcor was not included in the works of either Ptolemy or Tycho. It was observed by Persian and Arab astronomers in the Middle Ages and first appeared in Western Europe in the colophon of Peter Apian's *Cosmographicus* of 1524 where it was specifically identified as Alcor. See Allen, *Star Names,* pp. 445-46; Warner, *Sky Explored,* p. 8; Kunitzsch and Smart, *Short Guide to Modern Star Names,* pp. 56, 58.

h, 23, HR 3757, 3.67V, *SA, HRP.*

Ursa Minor, UMi

Lesser, Smaller, or Little Bear

Ursa Minor (Figure 77) is a Ptolemaic constellation that Bayer lettered from Alpha to Theta. Bode added more letters from Iota to Phi, but only Lambda and Pi have survived in the literature, probably because of their close proximity to the celestial north pole. The constellation is sometimes referred to unofficially as the Little Dipper because of its dipper-like appearance.

In ancient Mesopotamia, the constellation was known as MAR.GID.DA.AN.NA, the Wagon of Heaven, probably because of its continuous wheeling around the celestial pole star. As might be expected, the Greeks had several legends that related to Ursa Minor. One legend claims that Ursa Minor represents Arcas, son of Callisto. When Zeus places Callisto in the heavens as Ursa Major, he also honors her son by placing him near her as Ursa Minor. Another story suggests that Ursa Minor represents Zeus's nurse, Cynosura, whom he rewards with a place in the heavens. According to this explanation, Ursa Major represents his other nurse, Helice.

Ursa Minor was sometimes called Phoenice, representing yet another maiden seduced by Zeus, turned into an animal, and eventually honored with a place among the stars. The constellation was also called the Phoenician, an indication, certainly, of a possible origin in the Levant although no Near Eastern astral myths or legends have thus far been associated with it. The poet Hyginus suggested that it was called the Phoenician since Phoenician mariners sailed by it and Greek mariners sailed by Ursa Major.

Because the most prominent stars of both Ursa Major and Ursa Minor, Big and Little Dipper, are similar in appearance and name, the legends associated with these two constellations are also similar and intertwined. As noted in Ursa Major, the story of Callisto and Arcas relates to both constellations. Similarly, the ancient Romans called the seven stars that make up each of the dippers, Septem Triones, the Seven Plowing, or Threshing, Oxen.

See Condos, *Star Myths,* pp. 201-4; Ridpath, *Star Tales,* pp. 129-31; Allen, *Star Names,* p. 431. See Ursa Major for the myths surrounding Callisto and the Septem Triones.

Table 88. The lettered stars of Ursa Minor

	Magnitudes			Catalogue Numbers				
Letter	Bayer	Lacaille	Visual	Flamsteed	Lacaille	Other	HR	HD
Alpha	2		2.02	1			424	8890
Beta	2		2.08	7			5563	131873
Gamma	3		3.05	13			5735	137422
Delta	4		4.36	23			6789	166205
Epsilon	4		4.23	22			6322	153751
Zeta	4		4.32	16			5903	142105
Eta	5		4.95	21			6116	148048
Theta	6		4.96	15			5826	139669
Lambda+			6.38			B 85	7394	183030
Pi^1+			6.58			B 53	5829	139777
Pi^2			6.9	18		B 58		141652

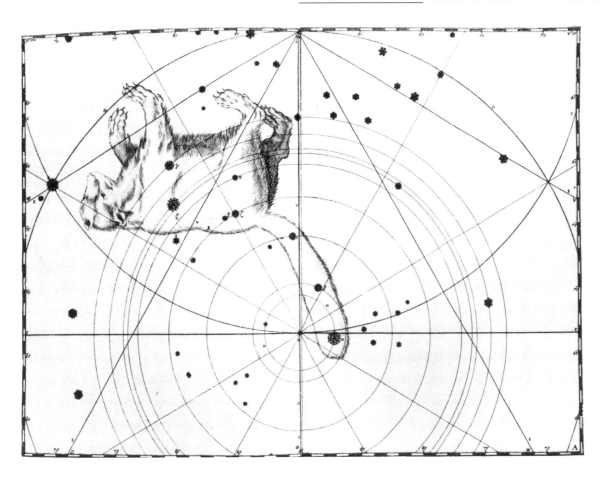

Figure 77. Bayer's Ursa Minor. Letters A and B are the north ecliptic pole and the celestial north pole; D is the Arctic Circle and C is the north ecliptic circle. The lines F and G are the equinoctial colure and the solstitial colure. The latter is a great circle that extends from the north celestial pole around the celestial globe, touching the two solstitial points on the ecliptic. When the sun reaches these two points in its journey across the heavens, on June 21 and on December 21, it is the time of the summer and winter solstice, the sun's highest and lowest points in the sky at noon. The letter H represents the miscellaneous stars around the celestial north pole. The letter E, located at the end of every line of ecliptic longitude, represents the *dodecatemoria*. See Figure 76, letter L.

The Lost, Missing, or Troublesome Stars of Ursa Minor

Lambda, HR 7394, 6.38V, *HRP, BS,* B 85*, H 54, GC 25111*.

Bode designated this star Lambda in his atlas and catalogue of 1801. Although Tirion did not identify Lambda in his *SA,* he did include it later in his *Uranometria 2000.0* and in the second edition of *SA.*

Pi¹, HR 5829, 6.58V, *SA,* B 53*, BAC 5205, BKH 31.
Pi², 18, HD 141652, 6.9V, *SA,* B 58*, BF 2220, BAC 5274, BKH 33.

Bode lettered these two stars Pi-1 and Pi-2. Heis claimed (H 31) that he observed 18, but Backhouse has suggested that Heis was in error since the coordinates listed in Heis's catalogue are not those of 18. Although Tirion identified both Pi¹ and Pi² in his *SA,* he omitted their letters and HD 141652's Flamsteed number in his more recent *Uranometria 2000.0.* In the second edition of *SA* he also removed their letters but properly identified HD 141652 as 18. Both of these stars with their Greek letters are included in Webb's *Celestial Objects for Common Telescopes* (II, 248-49) and *Burnham's Celestial Handbook* (III, 2006). See also 18 UMi in Part Two and Backhouse, *Catalogue of 9842 Stars*, p.182.

a, 5, HR 5430, 4.25V, BF 2006, H 15.
b, 4, HR 5321, 4.82V, BF 1974, H 13.

Neither Bayer nor Bode lettered these two stars. Although Bayer included these stars in his atlas, he considered them *informes* outside the border of the Little Bear and left them unlettered. Flamsteed designated them a and b, but Baily removed these letters as superfluous. See i Aql.

Vela, Vel

Ship's Sails

Vela (Figure 2d) is part of the former Ptolemaic constellation Argo Navis. Bayer originally lettered Argo Navis from Alpha to Omega and from A to s. Lacaille split the Ship into three separate constellations — Carina (Keel), Puppis (Stern), and Vela (Sails). He dropped Bayer's letters and lettered Vela a to z and from A to Z. Vela (Sails) is the Latin plural of velum (sail).

For the legend associated with Argo Navis and a detailed analysis of its Greek letters, see Carina; see also Puppis.

Table 89. The lettered stars of Vela

	MAGNITUDES			CATALOGUE NUMBERS				
Letter	Bayer	Lacaille	Visual	Flamsteed	Lacaille	Other	HR	HD
Gamma^1			4.27				3206	68243
Gamma^2		2	1.78		737		3207	68273
Delta		3	1.96		796		3485	74956
Kappa		3	2.50		851		3734	81188
Lambda		3	2.21		830		3634	78647
Mu		3	2.69		970		4216	93497
Omicron		4	3.62		786		3447	74195
Phi		4	3.54		901		3940	86440
Psi		4	3.60		864		3786	82434
a		5	3.91		794		3487	75063
b		5	3.84		783		3445	74180
c		5	3.75		827		3614	78004
d		6	4.07		792		3477	74772
e		6	4.14		776		3426	73634
f		6	5.10		807		3527	75821
g		6	4.93		805		3520	75710
h+		6	5.48		804		3514	75630
i+			4.39			G 239	4293	95370
k^1+		6	5.86		836		3677	79807
k^2		6	4.62		840		3684	79940
l		6	4.94		839		3682	79917
m+			4.58			G 163	3912	85622
n+			4.77			G 58	3452	74272
p		5	3.84		949		4167	92139
q		4	3.85		914		4023	88955
r		5	4.83		925		4080	89998
s+		6	5.16c		940		4135/6	91355/6
t		6	5.02		942		4143	91504
u		6	5.08		891		3898	85355

(table continued on next page)

Table 89. The lettered stars of Vela (continued)

	MAGNITUDES			CATALOGUE NUMBERS				
Letter	Bayer	Lacaille	Visual	Flamsteed	Lacaille	Other	HR	HD
w+			4.45			G 91	3591	77258
x+			4.28			G 225	4180	92449
y		6	5.50		877		3842	83548
z+		6	5.25		837		3674	79735
A		6	5.33		765		3358	72108
B		6	4.82		757		3294	70930
C		6	5.01		772		3407	73155
D		6	5.16		793		3476	74753
E+		6	5.80		775		3413	73340
F		6	5.09		764		3350	71935
G+		6	6.5		767			72303
H		6	4.69		816		3574	76805
I+		6	5.11		857		3753	81848
J+			4.50			G 203	4074	89890
K		6	5.26		845		3703	80456
L		6	5.01		871		3819	83058
M		6	4.35		875		3836	83446
N		5	3.13		868		3803	82668
O		6	5.56		886		3875	84461
P+		6	4.66		934		4110	90772
Q		6	4.86		909		3990	88206

The Lost, Missing, or Troublesome Stars of Vela

a, HR 3487, 3.91V, *SA, HRP,* CA 794*, BR 2198*, L 3526*, G 66*.

b, HR 3445, 3.84V, *SA, HRP,* CA 783*, BR 2141*, L 3470*, G 53*.

c, HR 3614, 3.75V, *SA, HRP,* CA 827*, BR 2326*, L 3677*, G 97*.

d, HR 3477, 4.07V, *SA, HRP,* CA 792*, BR 2179*, L 3508*, G 64*.

e, HR 3426, 4.14V, *SA, HRP,* CA 776*, BR 2114*, L 3446*, G 48*.

f, HR 3527, 5.10V, *SA, HRP,* CA 807*, BR 2241*, L 3572, BAC 3020*, G 78*.

g, HR 3520, 4.93V, *SA, HRP,* CA 805*, BR 2234*, L 3565, BAC 3014*, G 76*.

h, HR 3514, 5.48V, *SA,* CA 804*, BR 2228*, L 3556, BAC 3009*, G 74.

Gould dropped this star's letter since he considered it dimmer than 6th magnitude.

i, HR 4293, 4.39V, *SA, HRP,* CA 985, BR 3323, L 4550, G 239*.

Although Lacaille observed and catalogued this star, he noted it was in Centaurus and left it unlettered since he considered it 6th magnitude. Gould switched it to Vela and designated it i Vel because he felt a star of its brightness merited a letter.

j

Lacaille did not use this letter.

k¹, HR 3677, 5.86V, *SA,* CA 836 as k, BR 2401*, L 3748*, G 113.

k², HR 3684, 4.62V, *SA, HRP* as k, CA 840 as k, BR 2408*, L 3755*, G 117 as k.

    Lacaille designated these two closely positioned stars k, but Gould dropped the letter from CA 836 since he considered it dimmer than 6th magnitude.

l, HR 3682, 4.94V, *SA, HRP,* CA 839*, BR 2407*, L 3756, BAC 3163*, G 115*.

m, HR 3912, 4.58V, *SA, HRP,* BR 2704, L 4057, G 163*.

    Although Lacaille observed and catalogued this star, he left it unlettered since he considered its magnitude 6½. Gould designated it m because of its brightness.

n, HR 3452, 4.77V, *SA, HRP,* CA 785, BR 2154, L 3478, G 58*.

    Although Lacaille observed and catalogued this star, he left it unlettered since he considered it 6th magnitude. Gould designated it n because of its brightness.

o

    Lacaille did not use this letter in Vela.

p, HR 4167, 3.84V, *SA, HRP,* CA 949*, BR 3114*, L 4378*, G 222*.

q, HR 4023, 3.85V, *SA, HRP,* CA 914*, BR 2904*, L 4212*, G 191*.

r, HR 4080, 4.83V, *SA, HRP,* CA 925*, BR 2974*, L 4271*, G 204*.

s, HR 4135, 5.74V, *SA,* CA 940*, BR 3058*, L 4334, BAC 3613*, CAP 1912*, G 216*.

s, HR 4136, 6.09V, *SA, HRP,* BR 3059, BAC 3615, CAP 1913, G 217*.

    Lacaille described s as a single star of the 6th magnitude. HR 4135/6 are components A and B of the double star Dunlop 88. Although some astronomers sought to assign s to one or another of its components, Gould felt the combined light of both A and B, 5.16V, was synonymous with s.

t, HR 4143, 5.02V, *SA, HRP,* CA 942*, BR 3069*, L 4344, BAC 3618*, G 219*.

u, HR 3898, 5.08V, *SA, HRP,* CA 891*, BR 2688*, L 4047, BAC 3370*, G 160*.

v

    Lacaille did not use this letter.

w, HR 3591, 4.45V, *SA, HRP,* CA 820, BR 2300, L 3638, G 91*.

    Although Lacaille observed and catalogued this star, he left it unlettered since he considered it 6th magnitude. Gould designated it w because he felt a star of its brightness merited a letter.

x, HR 4180, 4.28V, *SA, HRP,* CA 953 as X, BR 3135 as X, L 4398 as X, G 225*.

    Lacaille lettered this star X, which Gould relettered as x to avoid confusion with variables. See R Cen.

y, HR 3842, 5.50V, *SA, HRP,* CA 877*, BR 2579*, L 3956, BAC 3302*, G 151*.

z, HR 3674, 5.25V, *SA, HRP,* CA 837*, BR 2400, L 3749, BAC 3156*, G 112*.

    See Z.

A, HR 3358, 5.33V, *SA, HRP,* CA 765*, BR 2056*, L 3367, BAC 2865*, G 33*.

B, HR 3294, 4.82V, *SA, HRP,* CA 757*, BR 2003*, L 3308*, G 26*.

C, HR 3407, 5.01V, *SA, HRP,* CA 772*, BR 2099*, L 3428*, G 46*.

D, HR 3476, 5.16V, *SA, HRP,* CA 793*, BR 2180*, L 3514*, G 63*.

E, HR 3413, 5.80V, *SA,* CA 775*, BR 2106*, L 3443*, G 47.

Gould dropped this star's letter since he considered it dimmer than 6th magnitude.

F, HR 3350, 5.09V, *SA, HRP,* CA 764*, BR 2047*, L 3359, BAC 2857*, G 30*.

G, in Carina, HD 72303, 6.5V, CA 767*, BR 2065*, L 3380, BAC 2873*, G 93 in Carina.

Gould switched this star to Carina and left it unlettered because of its dimness.

H, HR 3574, 4.69V, *SA, HRP,* CA 816*, BR 2280*, L 3620, BAC 3066*, G 88*.

I, HR 3753, 5.11V, *HRP,* CA 857*, BR 2495*, L 3854, BAC 3236*, G 136*.

Tirion erroneously designated this star as J in his *SA*, as did Becvar in his atlases. Tirion corrected the error in his *Uranometria 2000.0* and in the second edition of *SA*.

J, HR 4074, 4.50V, *HRP,* CA 926 as T, BR 2972 as T, L 4272 as T, G 203*.

Lacaille did not letter this star J. This is Lacaille's T, which Gould relettered as J to avoid confusion with variables. Tirion's *SA* erroneously designates this star I, and so too do Becvar's atlases. Tirion corrected the error in his *Uranometria 2000.*0 but again erroneously identified it as I in the second edition of *SA,* thus designating two stars as I in Vela, this star and HR 3753, above.

K, HR 3703, 5.26V, *SA, HRP,* CA 845*, BR 2428*, L 3786*, G 122*.

L, HR 3819, 5.01V, *SA, HRP,* CA 871*, BR 2555*, L 3925*, G 146*.

M, HR 3836, 4.35V, *SA, HRP,* CA 875*, BR 2577*, L 3952*, G 148*.

N, HR 3803, 3.13V, *SA, HRP,* CA 868*, BR 2535*, L 3910*, G 144*.

O, HR 3875, 5.56V, *SA, HRP,* CA 886*, BR 2637*, L 4003, BAC 3338*, G 154*.

P, in Carina, HR 4110, 4.66V, *SA* erroneously as P Car, CA 934*, BR 3023*, L 4310*, G 195 in Carina.

Gould switched this star to Carina and left it unlettered since he considered it dimmer than 5th magnitude (see P Car). Becvar's atlases, *SA,* and *Uranometria 2000.0* all erroneously label P Vel as P Car. Tirion has also carried over the error to the second edition of *SA*.

Q, HR 3990, 4.86V, *SA, HRP,* CA 909*, BR 2860*, L 4172*, G 186*.

R, HR 4017, 5.28V, CA 913*, BR 2895*, L 4206*, BAC 3499*, G 189.

Gould dropped this star's letter to avoid confusion with variables. See R Cen.

S

Lacaille did not use this letter in Vela.

T, HR 4074, 4.50V, CA 926*.

Gould dropped this star's letter to avoid confusion with variables. He relettered it J because of its brightness. See J.

U

Lacaille did not use this letter.

V, HR 4063, 4.57V, CA 923*, BR 2952*, L 4263*, BAC 3536*, G 201, SAO 23796, the variable GZ.

Gould dropped this star's letter to avoid confusion with variables. Since he considered it dimmer than 5th magnitude, he did not reletter it. Oddly enough, astronomers have recently discovered it to be variable and have relettered it GZ. There appears to be some difference of opinion about its magnitude and its range of variability. Gould listed its magnitude as 5.4 with no variability. Scovil's *AAVSO Variable Star Atlas* also does not list it as variable; nor does the *SAO*. Hoffleit in the *BS* and Hirshfeld and Sinnott in *Sky Cat. 2000.0,* 2nd ed. (I, 295; II, 236) both gave its magnitude as 4.57V with a range of 6.37V to 6.44V. Most recently, Kholopov's *Gen. Cat. Var. Stars* (III, 296-97) notes it is an irregular variable with a range of 3.43V to 3.81V. Which source is correct? Gould was especially interested in stellar magnitudes and variability. Since he and his colleagues left this star unlettered, it seems highly probable that when they observed it, it must indeed have been fainter than 5th magnitude. Even taking into account differences that arise from visual observations in comparison to modern photometry, it does not seem possible that Gould could have misjudged a 3.4 or even a 3.8 magnitude star for one fainter than 5th magnitude.

W

Lacaille did not use W in Vela.

X, HR 4180, 4.28V, CA 953*.

Gould relettered this star lower-case x to avoid confusion with variables. See x.

Y, HR 4134, 4.89V, *SA, BS,* CA 941*, BR 3062*, L 4336, BAC 3614*, G 215.

Gould dropped this star's letter to avoid confusion with variables. Since he considered it dimmer than 5th magnitude, he did not reletter it. Both this star and the variable Y (9^h29^m, -52°11') are labeled Y on Becvar's *Atlas Australis* and Tirion's *Uranometria 2000.0*. In Tirion's *SA,* the variable is omitted but this star, HR 4134, is labeled Y. The second edition of *SA* removes the letter from HR 4134 and identifies the variable as Y.

Z, HR 4221, 5.23V, CA 973*, BR 3211*, L 4468*, BAC 3705*, G 230.

Gould dropped this star's letter to avoid confusion with variables. Both Tirion's *SA* and his *Uranometria 2000.0* erroneously designate it z, thus labeling two stars z Vel — this star and Lacaille's z, CA 837, HR 3674. Becvar's *Atlas Australis* erroneously labels both stars Z. The second edition of *SA* removes the letter froom HR 4221. See z and Hoffleit, "Discordances," p. 65.

VIRGO, VIR

Virgin

Virgo (Figure 78) is a Ptolemaic constellation that Bayer lettered from Alpha to Omega and from A to q. It first appeared in Mesopotamian astronomical literature of the first millennium B.C., but it is certainly at least a thousand years older. The *lucida* of the constellation, Spica, was known by its Sumerian name AB.SIN, the Furrow. The word AB.SIN is probably a derivitive of Ezinu, which in Sumerian referred not only to the grain growing in the furrow but also to the grain goddess herself, Ezinu. Spica was also known as "the Corn [Wheat] Ear of the Goddess Shala." The Akkadian equivalent of the goddess Ezinu, Shala personified ripening grain. Her husband, Adad (Hadad), was the West Semitic god of thunderstorms and spring rains whose sweet waters impregnated the furrows of the plowed fields, giving life to the newly planted seeds. Unlike the inhabitants of Sumer or southern Mesopotamia who relied on ground water and irrigation for their crops, the people of Akkad or northern Mesopotamia, Syria, and Palestine were dependent on rainwater. Hence, while Enki (Ea), the personification of subterranean water, became an important figure in the Sumerian pantheon, Adad, the god of rainstorms, dominated in the lands north and west of Sumer, and his name became an eponym for many of the kings of ancient Syria.

Since the peoples of the ancient Near East believed that a bountiful harvest depended heavily upon winning the favor of both Adad and Shala, these deities were especially revered. Their images and symbols appeared on monuments and they occurred together on cylinder seals dating back to the First Dynasty of Babylon, 1830-1530 B.C. Over the next 1,500 years, however, Shala was gradually transformed into Virgo, and on a very late Mesopotamian tablet from the Seleucid period, she was depicted — alone — carrying in her left hand an ear or spike of grain. Several centuries later, Ptolemy depicted her in the heavens as a virgin holding a spike of grain in her left hand, which was marked in the heavens by the 1st-magnitude star Spica.

The association of the Lady of the Ear of Grain with Virgo, the young maiden, can be traced back to Near Eastern mythology. As noted, Shala was originally the goddess of ripe grain. Her husband, Ishkur, the Sumerian god of thunderstorms and lightning, was later identified with the Akkadian Adad, who evolved as the chief god of the Canaanites, Hadad. He later emerged in classical mythology as Zeus-Jupiter Pluvius, God, the Father, the Rain Maker. Shala-Ezinu, a relatively minor deity among the Sumerians, was thus elevated to the position of the wife of the king of the gods. As a late Babylonian psalm of the first millennium B.C. described her, she was "Shala, the great wife." And just as the Akkadian Ishtar, the primary goddess of Mesopotamia, absorbed the attributes of many of the lesser female deities, Shala gradually did the same. She was, in fact, eventually merged with Ishtar herself. As an indication that this actually occurred, Mesopotamian tablets and monuments reveal that the ear of grain was the symbol of both Shala and Ishtar. Moreover, one of the characteristics of Ishtar (Sumerian Inanna or Ninana) was that of a virgin or nubile young maiden who encouraged men to vie in combat for her hand in marriage. This, in turn, gave rise to Ishtar's (Inanna, Shala) attributes as goddess of battles and storms. She became a fitting mate for Ishkur (Adad, Baal), the god of thunder and lightning.

The association of virginity with the grain goddess was not limited just to the peoples of the ancient Near East but also to the Greeks and other peoples of the world. Ancient humans widely believed that if seeds were sown by a virgin — or a married woman who practiced abstinence during the planting season — the seed would be unspoiled and pure and would therefore produce a rich, bountiful crop.

To the peoples of the ancient Near East, "the Lady of the Grain Ear" was one of the constellations that ruled the sixth month, Ululu (Elul), when the sun appeared to enter the constellation Virgo. This appearance marked the end of summer and the beginning of the harvest — when the ears of grain were ripe, heavily laden with fruit, and ready for the reaper. During this time, the power or numen of the goddess Shala dominated the fields.

The Greeks and Romans linked several goddesses to this constellation. According to some historians, the constellation represents Dike (Justice), daughter of Zeus and Themis (Divine Law). Dike originally lives among human beings, but angered and disappointed by the injustice and inhumanity perpetrated by mankind, she deserts them to find instead a place in the heavens. Others historians suggest that the constellation depicts Demeter (Roman Ceres), the goddess of grain and agriculture.

See Reiner and Pingree, *Baby. Plan. Omens,* Part 2, p. 10; O'Neil, *Time and the Calendars*, pp. 57-58; Jacobsen, *Toward the Image of Tammuz,* pp. 28-30, 34; Langdon, *Babylonian Menologies,* p. 6; van der Waerden, "Babylonian Astronomy. II. The Thirty-Six Stars," pp. 14-15, 20; van der Waerden, "History of the Zodiac," pp. 219-20, 226; Contenau, *Everyday Life in Babylon and Assyria,* pp. 52, 246, 251-52, 257; Pritchard, *Anc. Near East. Texts,* p. 390; Pritchard, *Anc. Near East Picts.,* plates 703-4; Frymer-Krensky, "Adad," pp. 26-27; Jacobsen, "Mesopotamian Religions," pp. 458-59; van Buren, *Symbols of the Gods,* pp. 13-14, 35, 70-73, 190; Jacobsen, *Treasures of Darkness,* pp. 7, 135-43, 236; Frazer, *The Golden Bough,* Part V, *Spirits of the Corn and the Wild,* I, 113-16; van Buren, "Fish-Offerings in Ancient Mesopotamia," p. 119; Krupp, *Beyond the Blue Horizon,* p. 136; Hempel, "A Catalogue of Near Eastern Venus Deities;" Condos, *Star Myths,* pp. 205-7. For another constellation associated with the goddess Ishtar, see Pisces; for an account of Enki (Ea), see Aquarius.

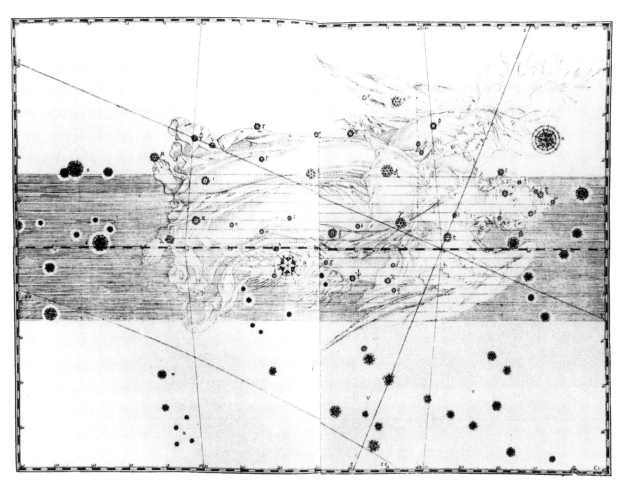

Figure 78. Bayer's Virgo. The bright 1st-magnitude star labeled R just above the spike of wheat is Beta Leo, Denebola, in Leo's tail. The letter S is the constellation Libra; T is Crater; V is Corvus; W is the tail of Hydra; and X is the head of Centaurus. The lines Y, Z, and ZZ are, respectively, the celestial equator, the equinoctial colure, and the Tropic of Capricorn. The shaded area is the zodiac.

Table 90. The lettered stars of Virgo

	MAGNITUDES			CATALOGUE NUMBERS				
Letter	Bayer	Lacaille	Visual	Flamsteed	Lacaille	Other	HR	HD
Alpha	1		0.98	67			5026	116658
Beta	3		3.61	5			4540	102870
Gamma	3		2.75c	29			4825/6	110379/80
Delta	3		3.38	43			4910	112300
Epsilon	3		2.83	47			4932	113226
Zeta	3		3.37	79			5107	118098
Eta	4		3.89	15			4689	107259
Theta	4		4.38	51			4963	114330
Iota	4		4.08	99			5338	124850
Kappa	4		4.19	98			5315	124294
Lambda	4		4.52	100			5359	125337
Mu	4		3.88	107			5487	129502
Nu	5		4.03	3			4517	102212
Xi+	5		4.85	2			4515	102124
Omicron	5		4.12	9			4608	104979
Pi	5		4.66	8			4589	104321
Rho	5		4.88	30			4828	110411
Sigma	5		4.80	60			5015	115521
Tau	5		4.26	93			5264	122408
Upsilon+	5		5.14	102			5366	125454
Phi	5		4.81	105			5409	126868
Chi	5		4.66	26			4813	110014
Psi	5		4.79	40			4902	112142
Omega	6		5.36	1			4483	101153
A^1+	6		5.32	4			4528	102510
A^2	6		5.58	6			4559	103484
b	6		5.37	7			4585	104181
c	6		4.96	16			4695	107328
d^1+	6		5.59	31			4829	110423
d^2	6		5.22	32			4847	110951
e	6		5.22	59			5011	115383
f	6		5.87	25			4799	109704
g+	6		5.55				4957	114113
h	6		5.21	76			5100	117818
i+	6		5.25	68			5064	116870
k	6		5.79	44			4921	112846
l^1+			6.09	72			5088	117436
l^2	6		4.69	74			5095	117675
m	6		5.01	82			5150	119149
n+	6							
o+	6		4.94	78			5105	118022
p	6		5.15	90			5232	121299
q	6		5.48	21			4781	109309
M+			5.53	1 Ser			5573	132132

The Lost, Missing, or Troublesome Stars of Virgo

Xi, 2, HR 4515, 4.85V, *SA, HRP, BS,* BV 2*, H 2*.

Bayer described Xi as a single star of the 5th magnitude, but Flamsteed designated his 2 as Xi-1 and a neighboring star, his 4, as Xi-2. See A.

Upsilon¹, 102, HR 5366, 5.14V, *SA* as Upsilon, *HRP* as Upsilon, *BS* as Upsilon, BF 1954*, H 157*, G 238 as Upsilon.
Upsilon², 103, HD 125817, 6.8V, BF 1964*, H 159*, G 241.

Although Bayer described Upsilon as a single 5th-magnitude star, Flamsteed designated these two neighboring stars, his 102 and 103, as Upsilon-1 and Upsilon-2. Argelander, unable to see the fainter star, 103, dropped its letter (*UN,* p. 103). Heis, though, saw both stars and restored Upsilon to 103. Gould also saw both stars, but he agreed with Argelander that 103 was too dim to merit a letter.

A¹, 4, HR 4528, 5.32V, *SA, HRP,* BV 3*, BF 1654*, H 4*, G 8*.
A², 6, HR 4559, 5.58V, *SA, HRP,* BV 5*, BF 1664*, H 10*, G 14*.

Bayer described A as a *binae* or pair of 6th-magnitude stars and labeled each with the letter A. In all probability, Flamsteed observed that the two stars synonymous with Bayer's pair were his 4 and 6, but he designated only his 6 as A. Noting that the more westerly of the pair, his 4, was much closer to Xi than to the other A, his 6, he designated it Xi-2. Bevis and Baily, though, decided to follow Bayer's designation. They lettered 4 and 6 as A¹ and A². See Xi; *BF*, p. 399. For Bayer's practice of labeling individual stars with the same letter, see A Ser.

b, 7, HR 4585, 5.37V, *SA, HRP.*

c, 16, HR 4695, 4.96V, *SA, HRP.*

d¹, 31, HR 4829, 5.59V, *SA, HRP.*
d², 32, HR 4847, 5.22V, *SA, HRP.*

Bayer described d as a couple of 6th-magnitude stars, and Flamsteed designated these two neighboring stars, his 31 and 32, as d-1 and d-2.

e, 59, HR 5011, 5.22V, *SA, HRP.*

f, 25, HR 4799, 5.87V, *SA, HRP.*

g, HR 4957, 5.55V, *SA, HRP,* BV 56*, H 70*, G 110*.

Bayer classified g as a 6th-magnitude star. Although Flamsteed assigned g to his 49 (HR 4955, 5.19V, BV 57, BF 1803 as g), Bevis felt it did not conform well to Bayer's chart. He switched g to this star, HR 4957, which more nearly reflected the position and magnitude of Bayer's g. Argelander, Heis, Gould, and most astronomers have accepted the switch (*UN,* p. 102; *Uran. Arg.,* p. 94). Baily also agreed that 49 was probably not Bayer's g, but he deferred to Flamsteed and included it as g in his revised edition of Flamsteed's catalogue.

h, 76, HR 5100, 5.21V, *SA, HRP.*

i, 68, HR 5064, 5.25V, *SA, HRP,* HF 54*, BV 75*, B 515*, BF 1847*.

Flamsteed's manuscripts regularly refer to this star as i, and it is labeled i in Halley's 1712 pirated edition of his catalogue, but Flamsteed's authorized "Catalogus Britannicus" of 1725 mistakenly designates it Iota, due probably to a copying or typographical error since the letters are very similar in appearance, i, ι. Bevis and Bode relettered it i.

j

Bayer did not use this letter.

k, 44, HR 4921, 5.79V, *SA, HRP.*

l¹, 72, HR 5088, 6.09V, *SA,* BV 69, BF 1854*, G 154.
l², 74, HR 5095, 4.69V, *SA, HRP* as l, BV 70 as l, BF 1856*, H 108 as l, G 157 as l.
l³, 80, HR 5111, 5.73V, BV 72, BF 1865, H 116, G 169.

    Bayer described l as a single 6th-magnitude star, but Flamsteed designated these three neighboring stars, his 72, 74, and 80, as l-1, l-2 and l-3. Bevis, though, labeled only the brightest one, 74, as l. Baily, on the other hand, felt that only 72 or 74 could be synonymous with Bayer's l. He disregarded 80 since it was somewhat apart from the other two. Argelander agreed with Bevis. Unable to see 72 with his naked eye, he believed only 74 was Bayer's l. In his opinion, 80 was too far away from 74 to be considered l (*UN,* p. 102). Gould saw all three stars, but he agreed with Bevis and Argelander because he felt 72 and 80 were too dim to be equivalent to l.

m, 82, HR 5150, 5.01V, *SA, HRP.*

n

    Bayer described n as a single star of the 6th magnitude. Flamsteed's 1725 catalogue lists his 13 as synonymous with Bayer's n, an obvious error since according to Bayer's chart, n was at least 25° due east of 13. Baily noted the error by dropping n from 13 and reassigning it to Flamsteed's 88 (HD 120235, 6.5V, BF 1888 as n), although he had some doubts whether that was the proper star. Argelander felt that n should be synonymous with HD 121481 (6.8V, H 133 as n, G 200) since it conformed better to Bayer's chart than 88 and since, despite its dimness, he claimed he was able to see it without telescopic assistance (*UN,* p. 103). Hoffleit has suggested n might be HD 121496 (6.85V, G 201, "Discordances," p. 65). Although Gould made no conjecture about n — in fact, it was not even mentioned in his catalogue — he noted that to his naked eye, HD 121481 and HD 121496 appeared as one image. Actually, these two stars, along with their close neighbor HD 121444, 7.6V, are bunched closely together and conform to Bayer's chart. In all probability, the combined light from these three stars is what Argelander saw. Is it possible that their combined light is also what Bayer saw?

    Bevis suggested another possibility. He felt that n (BV 96) was equivalent to Hevelius's HEV 25. He noted that both Hevelius and Flamsteed equated HEV 25 with Tycho's 19, although there was a difference of about ½° in longitude and ⅓° in latitude between the two stars. He also noted that the coordinates of Tycho's 19 were those that Bayer plotted for n on his chart of Virgo — 24½° Libra, 2½° north of the ecliptic. Consequently, Bevis surmised that n was synonymous with HEV 25. But HEV 25 is a problem star. There is no star at the coordinates listed for it by Hevelius. Baily has suggested it might be Piazzi's XIII, 287, which is equivalent to BAC 4680 and HD 122816 (6.6V). In his catalogue, Bevis described HEV 25 as Hevelius's 38 because he was using Flamsteed's edition of Hevelius's catalogue in which the stars are rearranged in order of right ascension. See also Baily, "Catalogues of Ptolemy *et. al.,"* n. 649, p. 164; n. 1427, p. 245.

    Bayer himself noted that n was synonymous with Ptolemy's 20. However, the latter star, Flamsteed's 86, is located below the ecliptic and n is north of the ecliptic (Toomer, *Ptolemy's Almagest,* p. 370). Moreover, as noted, n's coordinates are the same as Tycho's 19. Tycho described his 19 as "the star in the [Virgin's] left knee," the exact same description used by Ptolemy for his 20, but not only are their coordinates markedly different but Tycho's 19 seems to have vanished. Baily noted: "I cannot find any star that will accord with...[its] description" ("Catalogues of Ptolemy *et al.,"* n. 649, p. 164). The editor of Tycho's collected works, J. L. E. Dreyer, was also at a loss to identify 19 (Brahe, *Opera Omnia,* III, 349).

    Which then is Bayer's n? Is it the star or stars that conform best to Bayer's chart: HD 121444, HD 121481, or possibly HD 121496, or the combined light of all three? Is n HD 122816, which may be HEV 25 Vir, which may be synonymous with Tycho's 19, the coordinates of which Bayer plotted on his chart? Or is n, in reality, a nonexistent star, the coordinates of which arose largely through a series of errors on the part of Tycho and Bayer?

    In attempting to identify Bayer's stars, especially the fainter ones like n Vir, it should be borne in mind that it is not enough merely to compare Bayer's *Uranometria* with a modern atlas and then select those stars that seem best to conform to Bayer's charts. Like many stellar cartographers both before and after him, Bayer did not observe all the stars in his atlas. He based his charts partly on personal observations, especially of *informes* — stars outside the borders of the forty-eight Ptolemaic constellations — and partly on stellar positions already catalogued by others like Tycho Brahe, Keyzer and Houtman, and Johannes Schoener (see Swerdlow, "A Star Catalogue Used by Johannes Bayer") — and updated to 1600, the epoch of his *Uranometria.* Consequently, if Tycho's coordinates are in error, Bayer's atlas, in all probability, will likewise be in error as it is with i Leo and the stars in the legs of Ophiuchus. For the student interested in identifying Bayer's stars, then, it is important not only to compare star charts but also to understand what Bayer and his contemporaries intended.

o, 78, HR 5105, 4.94V, *SA, HRP,* BV 59*, H 113*, G 163*, the variable CW.

Bayer described o as a 6th-magnitude star. Flamsteed equated o with his 84 (HR 5159, 5.36V, BV 66, BF 1877 as o), and Baily agreed. Bevis, though, switched o to 78 since it was brighter and conformed better to Bayer's chart. Argelander (*UN,* p. 103), Heis, Gould, and most modern authorities have accepted Bevis's designation.

p, 90, HR 5232, 5.15V, *SA, HRP.*

q, 21, HR 4781, 5.48V, *SA, HRP.*

y, HR 5106, 5.91V, *SA,* G 165, ADS 8954.

Bayer did not letter this star. Gould noted that this star was variable and should be designated Y (*Uran. Arg.,* pp. 319-20), but he left it unlettered in the body of his catalogue. There is currently some question about this star's variability. It is part of the triple-star system Burnham 932, and Hoffleit suggested that component A or B might be variable (*BS,* p. 413). It appears as lower-case y not only in Tirion's *SA* and *Uranometria 2000.0* but also in Becvar's *Atlas Eclipticalis.* The confusion may stem from Gould's suggestion to label it Y. In any case, y should not be confused with the Mira-type variable Y Vir at $12^h 33.9^m$, -4°25'. The second edition of *SA* removes HR 5106's y.

M, 1 Ser in Virgo, HR 5573, 5.53V, *SA, BS* as M Ser, BAC 4931, H 17 in Libra, G 264.

Bayer did not letter this star M. Bode assigned M to a star about 5° north, HR 5584 (5.93V, B 731 as M, H 179), but it has somehow become confused with this star. Bode lettered stars from A to Q in Virgo, but none has survived in the literature except M.

VOLANS, VOL

Flying Fish

Keyzer and Houtman devised the constellation Volans (Figure 79). It first appeared in Houtman's catalogue of 1603 as De vlieghende Visch, Flying Fish. Bayer depicted it on his chart of the southern skies as Piscis Volans, and Lacaille lettered its stars from Alpha to Iota. Like many of the other constellations devised by Keyzer and Houtman, Volans was one of the exotic creatures that Europeans encountered as they traveled into strange and hitherto unexplored areas in the fifteenth and sixteenth centuries.

Table 91. The lettered stars of Volans

	Magnitudes			Catalogue Numbers				
Letter	Bayer	Lacaille	Visual	Flamsteed	Lacaille	Other	HR	HD
Alpha		5	4.00		829		3615	78045
Beta		5	3.77		768		3347	71878
Gamma^{1,2}+		5	3.62c		635		2735/6	55864/5
Delta		5	3.98		646		2803	57623
Epsilon		5	4.35		749		3223	68520
Zeta		6	3.95		708		3024	63295
Eta		5	5.29		769		3334	71576
Theta		6	5.20		797		3460	74405
Iota		6	5.40		592		2602	51557
Kappa^1+			5.37			G 25	3301	71046
Kappa^2			5.65			G 26	3302	71066

The Lost, Missing, or Troublesome Stars of Volans

Gamma^1, HR 2735, 5.67V, *SA, HRP, BS,* BR 1529*, G 8*.
Gamma^2, HR 2736, 3.78V, *SA, HRP, BS,* CA 635 as Gamma, BR 1530*, L 2746 as Gamma, BAC 2400 as Gamma, G 9*.

Lacaille described Gamma as a single star of the 5th magnitude. Later astronomers added the indices to identify the components of this double star, Dunlop 42, the combined magnitude of which is 3.62V.

Kappa^1, HR 3301, 5.37V, *SA, HRP, BS,* CA 762, BR 2018, L 3355, G 25*.
Kappa^2, HR 3302, 5.65V, *SA, HRP, BS,* CA 763, BR 2022, L 3357, G 26*.

Lacaille did not letter Kappa. Although he observed and catalogued these two adjacent stars, he left them unlettered. This is actually a triple star with a third component, C (BR 2032, GC 11431), the magnitude of which is 8.56V. Gould noted that the combined light of all three stars produced a magnitude of about 4.7 and designated the two major components Kappa^1 and Kappa^2 since he felt stars of their brightness merited letters.

Lettered Stars - Volans

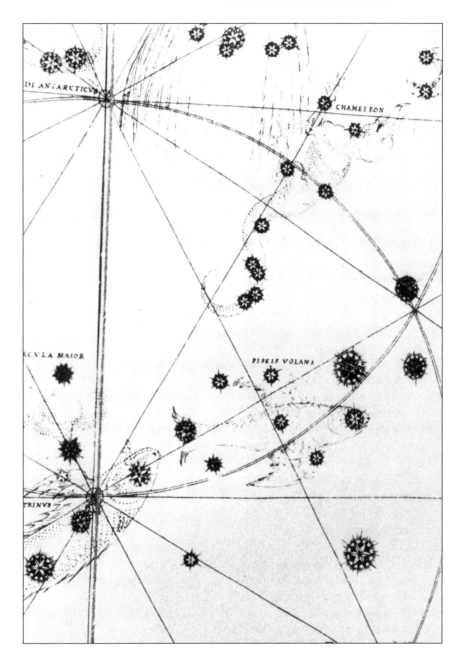

Figure 79. Bayer's Piscis Volans. This is an enlarged view of Piscis Volans from Bayer's Tabula XLIX. The chart depicts the south ecliptic pole in Dorado at the bottom left and the south celestial pole near Apus directly above it.

327

VULPECULA, VUL

Little Fox

Hevelius claimed that he devised the constellation Vulpecula (Figure 80) from twenty-seven new stars, including the Nova of 1672, that had never before been catalogued. He named his new constellation Vulpecula cum Ansere, the Little Fox with the Goose, and depicted it in his atlas as a fox with a goose dangling from its jaws. He placed it between Lyra and Aquila, next to another constellation that he also devised, the asterism Cerberus, the three-headed Hound of Hell, which Hercules holds in his left hand. Drawing on classical lore and astrology, Hevelius felt that this was an ideal spot to place Vulpecula. Since the fox was known for its astute, thievish, and voracious nature, it would be an appropriate neighbor for the nearby constellations Cerberus, Aquila the Eagle, and Lyra. Lyra was often depicted as a lyre clutched in the talons of an eagle or vulture. In fact, Bayer noted that the constellation was occasionally referred to as Vultur Cadens, the Preying Vulture.

Hevelius explained that a little tableau was being played out in the heavens. The Fox, he noted, had just stolen the Goose and was on its way to present it to Cerberus as a consolation for what was to come. The monstrous Cerberus, depicted by Hevelius with three serpentine heads, was preparing to breakfast on the Fox's offering. This was, however, to be his last meal since Hercules was about to strike him dead with his uplifted club.

Hevelius based this scene, to a certain extent, on the last and most difficult of the Twelve Labors of Hercules. According to legend, Hercules is ordered by Eurystheus, King of Mycena, to capture — not kill — Cerberus, the three-headed guardian of the gates of Hell, whose job it was not to keep trespassers out, but to keep inhabitants in.

The twelve seemingly impossible tasks of Hercules are:

1. to kill the Nemean lion, a huge creature that could not be killed with weapons devised by humans;
2. to kill the nine-headed Hydra that lived in the swamps of Lerna;
3. to capture alive the stag with golden horns that lived on Mount Cerynea;
4. to capture alive the Erymanthian boar, a huge creature that was terrorizing the countryside;
5. to clean in one day the stables of King Augeas where thousands of cattle had been kept for years on end;
6. to drive off the Stymphalian birds — creatures with iron beaks and claws whose feathers could be shot off like arrows — that were terrorizing the inhabitants of Stymphalus;
7. to capture and tame the wild, mad bull of Crete that was wreaking havoc on the island;
8. to capture the man-eating mares of King Diomedes of Thrace;
9. to steal the girdle of Hippolyte, Queen of the Amazons;
10. to rustle the cattle of Geryon — a monstrous giant with three bodies, three heads, and six hands and feet — who was assisted by his two-headed dog;
11. to steal the golden apples of the Hesperides that were guarded by the multi-headed dragon, Ladon; and
12. to capture Cerberus.

Flamsteed included Vulpecula in his atlas and catalogue but without Cerberus. He called the constellation the Little Fox and the Goose and credited Hevelius with having devised it. Baily found only one star brighter than magnitude 4½, which he lettered Alpha.

See Hevelius, *Prodromus Astronomiae,* p. 117; Hamilton, *Mythology,* pp. 231-34; Schwab, *Gods & Heroes,* pp. 161-75. See also Cancer and Draco for detailed accounts of Hercules' early career and second and eleventh labors.

Lettered Stars - Vulpecula

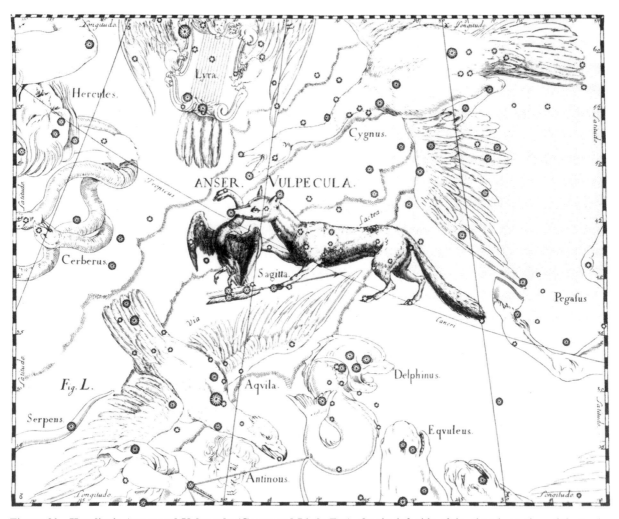

Figure 80. Hevelius's Anser and Vulpecula (Goose and Little Fox). On the left side of the chart is another of the twelve constellations devised by Hevelius, Cerberus, with its three serpentine heads. See Figure 42.

Table 92. The lettered stars of Vulpecula

	MAGNITUDES			CATALOGUE NUMBERS				
Letter	Bayer	Lacaille	Visual	Flamsteed	Lacaille	Other	HR	HD
Alpha			4.44	6		BAC 6674	7405	183439

The Lost, Missing, or Troublesome Stars of Vulpecula

q, 32, HR 8008, 5.01V, BV 33, BF 2855.
r, 31, HR 7995, 4.59V, BV 32, BF 2850.

    Hevelius devised Vulpecula from the stars in that sector of the heavens between Delphinus and Cygnus. Bayer included these two stars on his chart of Cygnus, but since he considered them *informes* outside the border of the constellation, he left them unlettered. Flamsteed considered them in Vulpecula and lettered them q and r, but Bevis and Baily removed their letters as superfluous (see i Aql). Flamsteed chose the letters q and r for these two stars because Bayer's last letter in neighboring Cygnus was P, the Nova of 1600. Flamsteed continued the lettering from Cygnus since 31 and 32 were very close to the Swan. Moreover, he was apparently unwilling to start a new series of letters for a new constellation like Vulpecula that had not been among the constellations in Bayer's *Uranometria*. For Bayer's use of upper-case Roman letters for miscellaneous astronomical objects, see P Cyg.

Part Two: Numbered Stars

ANDROMEDA, AND

Andromeda, Chained Lady

Andromeda (Figure 81) is a Ptolemaic Constellation that Flamsteed numbered from 1 to 66. Among the stellar objects he included in his catalogue was number 33, the Great Andromeda Galaxy, the nearest to our own Milky Way Galaxy.

Flamsteed's monumental "Catalogus Britannicus," which is included in the third volume of his *Historia Coelestis Britannica* of 1725, is arranged by constellation into three sections: ecliptical, southern, and northern. Within each constellation, the stars are listed in strict order of right ascension without numeration, unlike Edmund Halley's spurious or pirated edition of 1712 in which the stars are numbered. The pirated edition — Flamsteed called it the corrupted Catalogue — was based on Flamsteed's manuscript catalogue that had been loaned to Isaac Newton and the Royal Society with the understanding that it would not be published since Flamsteed felt it was only a preliminary working draft still in need of additional editing and correction. But Newton ignored Flamsteed's wishes and, as President of the Society, authorized Halley to edit and publish Flamsteed's draft manuscript, thus igniting a bitter feud between the first Astronomer Royal and Newton and Halley. The authorized edition of the catalogue was ultimately printed six years after Flamsteed's death. The pirated 1712 edition was published under the title *Historiae Coelestis Libri Duo*. It lists 2,682 stellar objects as compared to 2,935 in the authorized version. The 1725 volume also differs from the unauthorized edition in that it includes several more constellations. All in all, Flamsteed's "Catalogus Britannicus" contains a total of fifty-five constellations. Among these are all forty-eight Ptolemaic constellations except Ara and Corona Australis, which are too far south to be seen from Flamsteed's base at the Greenwich Observatory, just outside London. In addition, the "Catalogus Britannicus" includes Coma, which Ptolemy considered an asterism; Camelopardalis and Monoceros, two constellations probably devised by Petrus Plancius; and six newly created Hevelian constellations: Canes Venatici, Lacerta, Leo Minor, Lynx, Sextans, and Vulpecula. Halley's pirated edition lists only the forty-six Ptolemaic constellations that were visible from England, and Coma. For an old but detailed account of the feud and a look at many of the original documents, see Baily's *An Account of the Revd. John Flamsteed*; for a recent account of the feud, see the essays in Willmouth, *Flamsteed's Stars,* especially Alan Cook's "Edmund Halley and John Flamsteed at the Royal Observatory." See also, Wagman, "Who numbered Flamsteed's Stars?" and Adam Perkins's response.

Lost Stars

Figure 81. Flamsteed's Andromeda, Perseus, and Triangulum. Below Triangulum, is Hevelius's Little Triangle, which Flamsteed did not recognize as a separate constellation in his catalogue.

Flamsteed's charts display three different styles for determining stellar coordinates. Along the very top of the chart, in large Roman numerals, are hours of right ascension, which are based on time. Also along the top of the chart, but beneath the hours and minutes of right ascension, are arc degrees. These two methods, time and degrees, are based on equatorial coordinates. These two methods place the north celestial pole in the position indicated by Earth's axis, the spot marked approximately by the North Star, Polaris, in Ursa Minor. The apparent center of the heavens, the celestial equator, is marked by a line exactly 90° south, approximately where Orion's belt is located. All stellar atlases and catalogues today use hours of right ascension (r.a.) to determine celestial longitude.

The third style of determining stellar coordinates that Flamsteed used, the slanted lines on Flamsteed's chart, represents ecliptical longitude and latitude. This method measures stellar positions based on the ecliptic, the apparent path of the sun through the heavens. The ecliptic is the double line at the bottom-left corner of the chart marked by the sign of Taurus, ♉. See figs. 1 and 36.

The Andromeda Nebula, 33 Andromedae, is the small star to the right of Nu (ν) at right ascension (r.a.) 6½°, declination 50½° north polar distance (n.p.d.). Flamsteed measured declination starting with 0° at the celestial north pole whereas modern practice starts with 0° at the celestial equator.

Since Flamsteed's atlas is based on his "Catalogus Britannicus," 83 of the 84 nonexistent stars in his catalogue are plotted on his star charts. The one exception is 11 Vulpeculae, the Nova of 1670, whose coordinates Flamsteed copied — and updated for precession — from Hevelius's catalogue. Flamsteed did not observe the Nova, and it is not plotted on his atlas. On his chart of Andromeda, Perseus, and Triangulum, the star 19 Persei is nonexistent. It can be located using any one of the three styles of coordinates that Flamsteed employed on his charts. Using the right ascension (r.a.) time method, its coordinates would be 2^h34^m celestial longitude and 38°32' declination or celestial latitude north polar distance (n.p.d.); and using the right ascension degree

Table 93. The numbered stars of Andromeda

		MAGNITUDES		CATALOGUE NUMBERS	
Flamsteed Number	Letter	Flamsteed	Visual	HR	HD
1	Omicron	3½	3.62	8762	217675
2		6	5.10	8766	217782
3		6	4.65	8780	218031
4		6	5.33	8804	218452
5		6	5.70	8805	218470
6		6½	5.94	8825	218804
7		5½	4.52	8830	219080
8		6	4.85	8860	219734
9		6	6.02	8864	219815
10		6½	5.79	8876	219981
11		6	5.44	8874	219945
12		6	5.77	8885	220117
13		6	5.75	8913	220885
14		6	5.22	8930	221345
15		6	5.59	8947	221756
16	Lambda	4	3.82	8961	222107
17	Iota	4	4.29	8965	222173
18		6	5.30	8967	222304
19	Kappa	4	4.14	8976	222439
20	Psi	5½	4.95	9003	223047
21	Alpha	2	2.06	15	358
22		5	5.03	27	571
23		6	5.72	41	905
24	Theta	4½	4.61	63	1280
25	Sigma	5	4.52	68	1404
26		6	6.11	70	1438
27	Rho	5	5.18	82	1671
28		6	5.23	114	2628
29	Pi	4½	4.36	154	3369
30	Epsilon	4	4.37	163	3546

(table continued on next page)

---

method its coordinates would be 38°33' celestial longitude and 38°32' (n.p.d.) for its declination. Its ecliptic coordinates would be ♉ 23°53' ecliptic longitude and 34°15' ecliptic latitude. 19 Persei is the small, 6th-magnitude star slightly to the left and adjacent to Tau (τ), next to Perseus's right eye. Its ecliptic coordinates can be traced by noting the sign of Taurus, ♉, below the left ear of the Ram at the bottom of the chart. The dark slanting line to the right is 0° longitude in the sign of Taurus or 30° longitude from the First Point of Aries, ♈, the vernal equinox, which can be seen at the bottom right corner of the chart. 19 Persei is at longitude Taurus 23°53' or 53°53' from the First Point of Aries (30° plus 23°53') along the ecliptic and 34°15' latitude counting up from the ecliptic, the black and white line in the left corner of the chart.

All of Flamsteed's stars that were included in his "Catalogus Britannicus" were included in his atlas. The editors who prepared his atlas copied his stellar coordinates from his catalogue, which, in turn, was based on his manuscript catalogue and observation notes. Thus, if Flamsteed, for one reason or another, mistakenly copied a star's coordinates into his observation notes and from thence into his catalogue, the editors of his atlas plotted the erroneous coordinates on his star charts. They did not check each star in his catalogue against its position in the heavens.

Table 93. The numbered stars of Andromeda (continued)

		Magnitudes		Catalogue Numbers	
Flamsteed Number	Letter	Flamsteed	Visual	HR	HD
31	Delta	3	3.27	165	3627
32		6	5.33	175	3817
33+		Neb		182	
34	Zeta	4	4.06	215	4502
35	Nu	4	4.53	226	4727
36		6	5.47	258	5286
37	Mu	3¾	3.87	269	5448
38	Eta	4½	4.42	271	5516
39		6	5.98	290	6116
40+	Sigma Psc	6	5.50	291	6118
41		5	5.03	324	6658
42	Phi	5	4.25	335	6811
43	Beta	2	2.06	337	6860
44		6	5.65	340	6920
45		5½	5.81	348	7019
46	Xi	4½	4.88	390	8207
47		6	5.58	395	8374
48	Omega	5	4.83	417	8799
49	A	5	5.27	430	9057
50+		5¾	4.09	458	9826
51+	Upsilon Per	5	3.57	464	9927
52	Chi	6	4.98	469	10072
53	Tau	5	4.94	477	10205
54+	Phi Per	4	4.07	496	10516
55+		Neb	5.40	543	11428
56		6	5.67	557	11749
57	Gamma[1,2]	2½	2.20c	603/4	12533/4
58		6	4.82	620	13041
59		6	5.58c	628/9	13294/5
60b		6	4.83	643	13520
61+		6			
62c		6	5.30	670	14212
63		6	5.59	682	14392
64		6	5.19	694	14770
65		5	4.71	699	14872
66		6½	6.12	709	15138

The Lost, Missing, or Troublesome Stars of Andromeda

33, HR 182, *HRP, BS,* BF 58*, M 31, the Andromeda Nebula.

Besides the Andromeda Nebula, Flamsteed cited ten nebulous objects in his catalogue: 55 And, although not nebulous, is a double star (h 1094) and perhaps Flamsteed mistook its double image in his telescope for a patch of nebulosity; 39 and 41 Cnc, in M 44, Praesepe, the Beehive Cluster; 15 Com, in the open cluster Melotte 111; 39 Ori, in the bright defuse nebula Sharpless 2-264; 4 and 5 Sgr, not in any particular nebulosity, but in one of the richest areas of the Milky Way; 7 Sgr, in M 8, the Lagoon Nebula; and 32 and 35 Sgr, a closely located pair in an especially rich area of the Milky Way next to the globular cluster NGC 6717. Flamsteed also mentioned the Pleiades and Hyades (Suculae), but he did not describe them as *nebulosa,* as he did with the others.

40, in Pisces, HR 291, 5.50V, *HRP,* BF 99 and Table A, p. 645.

This star is the same as 69, Sigma Psc. It is one of Flamsteed's 22 duplicate stars. Flamsteed's authorized catalogue of 1725 contains a total of 2,935 stellar listings, but this includes twenty-two duplicates, like 40 And/69 Psc, and eighty-four stars that cannot now be located. In addition, Flamsteed observed 458 stars that for some inexplicable reason were not included in his 1725 catalogue. See Baily, "A Catalogue of 560 Stars Observed by Flamsteed but not Inserted in his British Catalogue"; *BF*, pp. 392-94.

50, HR 458, 4.09V, *SA, HRP, BS,* BF 184*, BAC 480*.

For a discussion of whether this star is Bayer's Upsilon, see Upsilon And in Part One.

51, Upsilon Per in Andromeda, HR 464, 3.57V, *SA, HRP, BS.*

For an account of whether this star belongs in Andromeda or Perseus, see Upsilon Per in Part One.

54, in Perseus as Phi Per, HR 496, 4.07V, *HRP,* BF 201*.

For an account of whether this star belongs in Andromeda or Perseus, see Phi Per in Part One.

55, HR 543, 5.40V, *SA, HRP, BS,* BF 217*.

Flamsteed's catalogue, as well as Baily's revised edition of Flamsteed's catalogue, describes 55 as nebulous. See 33 for a possible explanation.

61, BV 66*, B 262*, BF 278*.

Considerable confusion continues to surround this star. Baily noted that Flamsteed observed it on November 19, 1693 and reduced it to right ascension 29°31' and declination +47°22'. But Baily claimed that he could find no star at that position in the heavens. He felt Flamsteed mistakenly recorded the position of this star as south of the zenith instead of north. Based on this assumption, Baily surmised that the star Flamsteed actually observed was Piazzi's II, 35, which he felt was synonymous with his BAC 700, which is HD 14134, 6.55V, in the Double Cluster of Perseus, about 10° north of where Flamsteed originally placed it. Baily admitted that he had no authority to make such a radical switch in declination and advised his readers that William Herschel suggested that 61 was actually just where Flamsteed stated. Bevis and Bode agreed with Herschel. They believed that the star Flamsteed intended for 61 was indeed where his catalogue said it was: about 1½° north of 62. They equated 61 with BAC 705, which is HR 673, 6.37V (BV 66 and B 262), although there is about a 10' difference in declination between BAC 705 and Flamsteed's 61. To add to the confusion, Bevis's catalogue mistakenly lists 61 as Bayer's c although Bayer's c is actually 62. Because of the various problems involved with identifying 61, most astronomers have not accepted either Baily's supposition or that of Bevis and Bode. As a matter of fact, both Argelander and Heis omitted 61 from their catalogues.

Aquarius, Aqr

Water Bearer

Aquarius (Figure 82) is a Ptolemaic constellation that Flamsteed numbered from 1 to 108.

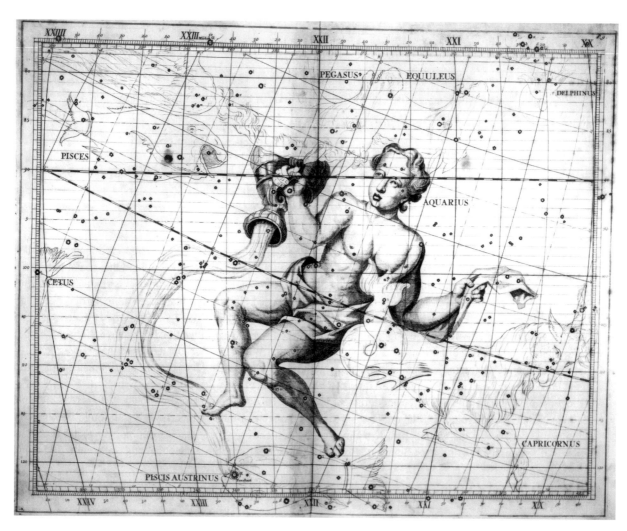

Figure 82. Flamsteed's Aquarius, Capricornus, and Piscis Austrinus. The head of the Sea Goat can be seen in Figure 102. The double straight line across the chart at 90° is the celestial equator. The curved double line below it is the ecliptic. The First Point of Aries, ♈, the vernal equinox, is where the two lines meet. The star 80 Aquarii at 340°38' r.a., 96°23' n.p.d. is nonexistent. See Figure 81 for an explanation of how nonexistent stars showed up on Flamsteed's charts.

Table 94. The numbered stars of Aquarius

		MAGNITUDES		CATALOGUE NUMBERS	
Flamsteed Number	Letter	Flamsteed	Visual	HR	HD
1		6	5.16	7897	196758
2	Epsilon	4¾	3.77	7950	198001
3	k	5	4.42	7951	198026
4		6	5.99	7982	198571
5		6	5.55	7985	198667
6	Mu	4½	4.73	7990	198743
7		6	5.51	8015	199345
8+		6½	6.6		199828
9+		6	6.6		200004
10+		6	6.6		199944
11		6	6.21	8041	199960
12		6	5.54c	8058/9	200496/7
13	Nu	5	4.51	8093	201381
14+		6	6.7		202466
15		6	5.82	8141	202753
16		6	5.87	8160	203222
17		6	5.99	8175	203525
18		6	5.49	8187	203705
19		6	5.70	8195	203875
20		6	6.36	8192	203843
21		6	5.49	8199	203926
22	Beta	3	2.91	8232	204867
23	Xi	6	4.69	8264	205767
24+		6	6.63		206058
25	d	6	5.10	8277	206067
26		6	5.67	8287	206445
27+		6	5.64	8328	207203
28		6	5.58	8390	209128
29		6	6.37	8396	209278
30		6	5.54	8401	209396
31	Omicron	5	4.69	8402	209409
32		6	5.30	8410	209625
33	Iota	4	4.27	8418	209819
34	Alpha	3	2.96	8414	209750
35		5¾	5.81	8439	210191
36+		6	7.00		210269
37+		6	6.70		210422
38	e	6	5.46	8452	210424
39		6	6.03	8462	210705
40+		7½	7.0		210845
41		6	5.32	8480	210960
42		7	5.34	8496	211361
43	Theta	4	4.16	8499	211391
44		6	5.75	8504	211434
45		6	5.95	8508	211676

(table continued on next page)

Table 94. The numbered stars of Aquarius (continued)

		MAGNITUDES		CATALOGUE NUMBERS	
Flamsteed Number	Letter	Flamsteed	Visual	HR	HD
46	Rho	5½	5.37	8512	211838
47		5½	5.13	8516	212010
48	Gamma	3	3.84	8518	212061
49		5	5.53	8529	212271
50		6	5.76	8534	212430
51		6	5.78	8533	212404
52	Pi	5	4.66	8539	212571
53	f	6	5.56c	8544/5	212697/8
54+		6	7.0		212741
55	Zeta[1,2]	4	3.65c	8558/9	210051/2
56		6	6.37	8567	213236
57	Sigma	5	4.82	8573	213320
58		6	6.38	8583	213464
59	Upsilon	5	5.20	8592	213845
60		6	5.89	8590	213789
61+		6	6.74		214028
62	Eta	4	4.02	8597	213998
63	Kappa	5	5.03	8610	214376
64+		6	7.16		214572
65+		6	7.1		215097
66	g[1]	6	4.69	8649	215167
67		6	6.41	8647	215143
68	g[2]	6	5.26	8670	215721
69	Tau[1]	5	5.66	8673	215766
70		6	6.19	8676	215874
71	Tau[2]	5¾	4.01	8679	216032
72+		6	7.2		216182
73	Lambda	4	3.74	8698	216386
74		6	5.80	8704	216494
75+		7	7.1		216567
76	Delta	3	3.27	8709	216627
77		6	5.56	8711	216640
78		6	6.19	8710	216637
79+	Alpha PsA	1¾	1.16	8728	216956
80+		7			
81		7	6.21	8757	217531
82		7	6.15	8763	217701
83	h	6	5.43	8782	218060
84+		7	7.5		218081
85+		6	6.9		218173
86	c[1]	6	4.47	8789	218240
87+		6	7.4		218331
88	c[2]	4	3.66	8812	218594
89	c[3]	5½	4.69	8817	218640

(table continued on next page)

Table 94. The numbered stars of Aquarius (continued)

Flamsteed Number	Letter	Magnitudes		Catalogue Numbers	
		Flamsteed	Visual	HR	HD
90	Phi	5	4.22	8834	219215
91	Psi^1	5	4.21	8841	219449
92	Chi	6	5.06	8850	219576
93	Psi^2	5	4.39	8858	219688
94		6	5.08	8866	219834
95	Psi^3	5	4.98	8865	219832
96		6½	5.55	8868	219877
97		6	5.20	8890	220278
98	b^1	5	3.97	8892	220321
99	b^2	5	4.39	8906	220704
100		5	6.29	8932	221357
101	b^3	5	4.71	8939	221565
102	Omega^1	5	5.00	8968	222345
103	A^1	5	5.34	8980	222547
104+	A^2		4.82	8982	222574
105	Omega^2	5	4.49	8988	222661
106	i^1	5	5.24	8998	222847
107	i^2	6	5.29	9002	223024
108	i^3	6	5.18	9031	223640

The Lost, Missing, or Troublesome Stars of Aquarius

8, HD 199828, 6.6V, *SA*, GC 29298*.

9, HD 200004, 6.6V, *SA*, GC 29337*.

10, HD 199944, 6.6V, *SA*, GC 29316*.

14, HD 202466, 6.7V, *SA*, GC 29742*.

24, HD 206058, 6.63V, *SA, BS Suppl.*, GC 30314*.

27, in Pegasus, HR 8328, 5.64V, *SA, HRP,* BF 2979* and Table A, p. 645.
   This star is the same as 11 Peg. It is one of Flamsteed's twenty-two duplicate stars.

36, HD 210269, 7.00V, *SA*, GC 30989*.

37, HD 210422, 6.70V, *SA*, GC 31017*.

40, HD 210845, 7.0V, *SA*, GC 31084*.

54, HD 212741, 7.0V, *SA*, GC 31353*.

61, HD 214028, 6.74V, *SA*, GC 31540*.

64, HD 214572, 7.16V, *SA,* GC 31614*.

65, HD 215097, 7.1V, *SA,* GC 31693*.

72, HD 216182, 7.2V, BF 3125*, BAC 7959*, B 237*, P XXII, 230*, CAP 4580*, T 10527*.

This star's number has been omitted from most modern atlases and catalogues because there is no star at the coordinates listed in Flamsteed's catalogue. In examining Flamsteed's manuscripts, Baily discovered that the Astronomer Royal made an error of 3° in reducing 72's declination: instead of -8°55'45", he wrote -5°55'45". Baily corrected the mistake in his revised edition of Flamsteed's catalogue. Actually, Bode was among the first to record the error in his *Vorstellung Der Gestirne* of 1782. The star is properly identified in the second edition of *SA*.

75, HD 216567, 7.1V, *SA,* GC 31931*.

79, in Piscis Austrinus, HR 8728, 1.16V, *HRP,* BF 3142 and Table A, p. 645.

This star is the same as 24 PsA (Fomalhaut). It is one of Flamsteed's twenty-two duplicate stars. For a discussion of Flamsteed's reason for placing Fomalhaut in both constellations, see Alpha PsA in Part One.

80, BF 3150* and Table C, p. 646.

Flamsteed observed this star on August 12, 1706, but Baily claimed that because of an error either in observation or reduction, he could not locate it. Peters, however, in "Flamsteed's Stars," p. 82, stated that Baily had erred and that the star had been observed by, among others, Lalande (*Hist. Cel.,* p. 188 and LL 45022) and Piazzi (XXII, 279). Peters claimed this star should be equated with BAC 8017 and H 99 and should be reassigned its Flamsteed number. He asserted that Flamsteed erred 1^m in noting 80's transit across his lens and suggested that if 1^m were to be added to this star's right ascension, it would be synonymous with the star cited above. If Peters's views are accepted, HR 8759, 5.94V, would become 80 Aqr. It should be noted that more than one hundred years have elapsed since Peters made his suggestion and, thus far, astronomers have preferred to follow Baily's assertion that 80 Aqr cannot be located.

84, HD 218081, 7.5V, *SA,* GC 32160*.

85, HD 218173, 6.9V, *SA,* GC 32175*.

87, HD 218331, 7.4V, *SA,* GC 32198*.

104, A^2, HR 8982, 4.82V, *SA, HRP, BS,* HF 92, BF 3245*.

Although Flamsteed's 1725 "Catalogus Britannicus" omits any magnitude for this star, Halley's 1712 edition lists its magnitude as 5.

Aquila, Aql

Eagle

Aquila (figures 83 and 102) is a Ptolemaic constellation that Flamsteed numbered from 1 to 71. He called the constellation Aquila with Antinous and noted that it contained several stars from the newly devised Hevelian constellation Scutum. Although Scutum was depicted on his atlas (Figure 107), Flamsteed did not consider it a separate constellation. Gould separated Scutum from Aquila and lettered its brighter stars.

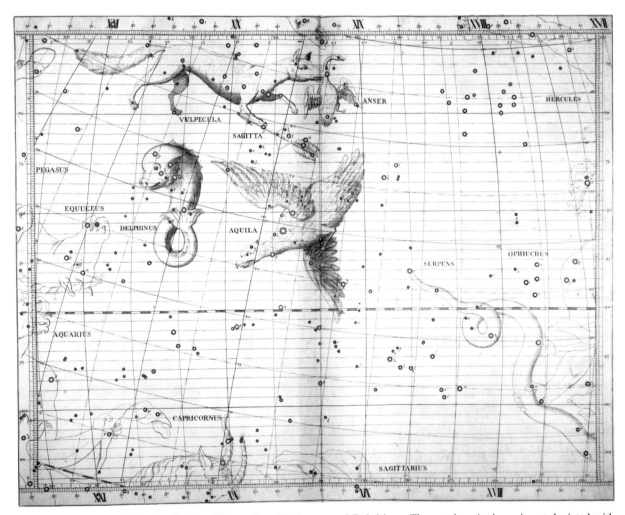

Figure 83. Flamsteed's Aquila, Sagitta, Vulpecula with Anser, and Delphinus. The asterism Antinous is not depicted with the Eagle although Flamsteed mentioned it in his catalogue. It is shown, though, in Figure 102. The following stars are nonexistent: **33 Aql,** 288° 0' r.a., 94°20' n.p.d.; **34 Aql,** 288°19' r.a., 93°53' n.p.d.; **40 Aql,** 290° 5' r.a., 93°36' n.p.d.; **43 Aql,** 290°32' r.a., 92°10' n.p.d.; **11 Vul,** ♒ 2°42' long., +47°28' lat. (ecliptical). The coordinates for 11 Vul were copied from Hevelius's catalogue. See Figure 81.

Lost Stars

Table 95. The numbered stars of Aquila

		MAGNITUDES		CATALOGUE NUMBERS	
Flamsteed Number	**Letter**	**Flamsteed**	**Visual**	**HR**	**HD**
1+	Alpha Sct	4	3.85	6973	171443
2+	Delta Sct	5	4.72	7020	172748
3+	Epsilon Sct	5	4.90	7032	173009
4		5	5.02	7040	173370
5		6	5.90	7059	173654
6+	Beta Sct	4	4.22	7063	173764
7+		6	6.9		174532
8		6	6.10	7101	174589
9+	Eta Sct	4¾	4.83	7149	175751
10		6	5.89	7167	176232
11		6	5.23	7172	176303
12	i	4	4.02	7193	176678
13	Epsilon	3½	4.02	7176	176411
14	g	6	5.42	7209	176984
15	h	6	5.42	7225	177463
16	Lambda	3	3.44	7236	177756
17	Zeta	3	2.99	7235	177724
18		6	5.09	7248	178125
19		6	5.22	7266	178596
20		5½	5.34	7279	179406
21		5	5.15	7287	179761
22		6	5.59	7303	180482
23		7	5.10	7319	180972
24		7	6.41	7321	181053
25	Omega[1]	6	5.28	7315	180868
26	f	6	5.01	7333	181391
27	d	6	5.49	7336	181440
28	A	6	5.53	7331	181333
29	Omega[2]	7	6.02	7332	181383
30	Delta	3	3.36	7377	182640
31	b	6	5.16	7373	182572
32	Nu	5	4.66	7387	182835
33+		6			
34+		6			
35	c	6	5.80	7400	183324
36	e	6	5.03	7414	183630
37		6	5.12	7430	184492
38	Mu	4	4.45	7429	184406
39	Kappa	3½	4.95	7446	184915
40+		6			
41	Iota	3½	4.36	7447	184930
42		6	5.46	7460	185124
43+		6			
44	Sigma	5	5.17	7474	185507
45		6	5.67	7480	185762

(table continued on next page)

344

Table 95. The numbered stars of Aquila *(continued)*

		MAGNITUDES		CATALOGUE NUMBERS	
Flamsteed Number	Letter	Flamsteed	Visual	HR	HD
46		6	6.34	7493	186122
47	Chi	6	5.27	7497	186203
48	Psi	6	6.26	7511	186547
49	Upsilon	6	5.91	7519	186689
50	Gamma	3	2.72	7525	186791
51		5	5.39	7553	187532
52	Pi	6	5.72	7544	187259
53	Alpha	1½	0.77	7557	187642
54	Omicron	5¾	5.11	7560	187691
55	Eta	3½	3.90	7570	187929
56		5	5.79	7584	188154
57		6	5.34c	7593/4	188293/4
58		6	5.61	7596	188350
59	Xi	5	4.71	7595	188310
60	Beta	3½	3.71	7602	188512
61	Phi	6	5.28	7610	188728
62		6	5.68	7667	190299
63	Tau	6	5.52	7669	190327
64		6	5.99	7690	191067
65	Theta	3	3.23	7710	191692
66		5½	5.47	7720	192107
67	Rho	5	4.95	7724	192425
68		6	6.13	7821	194939
69		5	4.91	7831	195135
70		5	4.89	7873	196321
71		4	4.32	7884	196574

The Lost, Missing, or Troublesome Stars of Aquila

1, in Scutum as Alpha Sct, HR 6973, 3.85V, *SA, HRP,* GC 25385*.

2, in Scutum as Delta Sct, HR 7020, 4.72V, *SA, HRP,* GC 25580*.

3, in Scutum as Epsilon Sct, HR 7032, 4.90V, *SA, HRP,* GC 25610*.

6, in Scutum as Beta Sct, HR 7063, 4.22V, *SA, HRP,* GC 25730*.

Flamsteed did not recognize Scutum as a separate constellation. He included its stars with Aquila. See Scutum in Part One.

7, HD 174532, 6.9V, *SA,* GC 25855*.

9, in Scutum as Eta Sct, HR 7149, 4.83V, *SA, HRP,* GC 26013*.

See 1, 2, 3, and 6 above.

33
34
40
43

These stars do not exist. In examining Flamsteed's manuscripts, Baily discovered that the Astronomer Royal had repeatedly made the same computation error of $1^h 10^s$ in determining the times of transit of 68, 69, 70, and 71 Aql on August 20, 1690. This resulted in the generation of new sets of coordinates that led him to believe he had observed new stars: 33, 34, 40, and 43. Flamsteed reobserved 68, 69, 70, and 71 and obtained their correct coordinates, but the coordinates of 33, 34, 40, and 43 were never rechecked or compared with their supposed positions in the heavens. See BF 2783/4, BF 2800, BF 2805 and Table B, p. 645. For a detailed account of Flamsteed's observational technique, his instruments, and his method of determining stellar coordinates, see *BF*, pp. 370-77.

Aries, Ari

Ram

Aries (Figure 84) is a Ptolemaic constellation that Flamsteed numbered from 1 to 66.

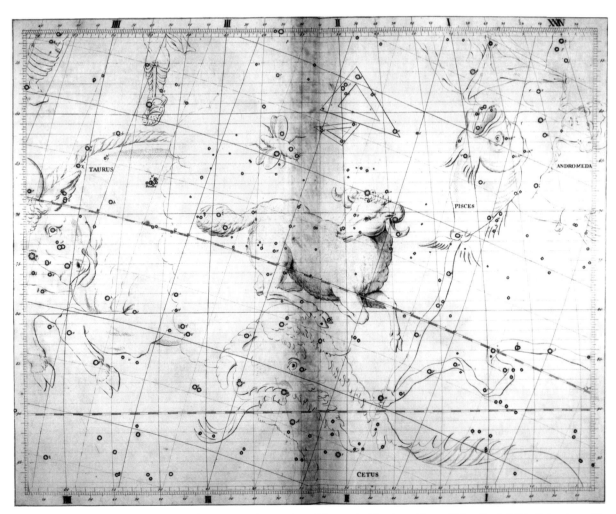

Figure 84. Flamsteed's Aries. Above the Ram are the asterisms Musca, the Northern Fly, and Hevelius's Little Triangle, neither of which is included in Flamsteed's "Catalogus Britannicus." Although ignored by Flamsteed, Musca has been treated as a separate constellation as late as 1914 in William Tyler Olcott's *A Field Book of the Stars,* 2nd edition. The following are nonexistent stars in Aries: **18 Ari,** 28°59' r.a., 71°33' n.p.d.; **28 Ari,** 33°33' r.a., 71°31' n.p.d.

Lost Stars

Table 96. The numbered stars of Aries

		MAGNITUDES		CATALOGUE NUMBERS	
Flamsteed Number	**Letter**	**Flamsteed**	**Visual**	**HR**	**HD**
1+		6¾	5.86	530	11154
2+		6¾	5.24	493	10476
3+		6	6.55	515	10845
4		6¾	5.84	522	10982
5	Gamma[1,2]	4	3.88c	545/6	11503/2
6	Beta	3	2.64	553	11636
7+		tel	5.74	559	11763
8	Iota	6	5.10	563	11909
9	Lambda	5	4.79	569	11973
10		6½	5.63	605	12558
11		5	6.15	615	12885
12	Kappa	5¾	5.03	613	12869
13	Alpha	2	2.00	617	12929
14		6	4.98	623	13174
15		6	5.70	631	13325
16		8	6.02	633	13363
17	Eta	6	5.27	646	13555
18+		7			
19		6½	5.71	648	13596
20		6	5.79	656	13871
21		7	5.58	657	13872
22	Theta	5¾	5.62	669	14191
23+		7	6.84		14305
24	Xi	6	5.47	702	14951
25+		7	6.49		15228
26		6½	6.15	729	15550
27		6½	6.23	731	15596
28+		6			
29		6½	6.04	741	15814
30		7	6.11c	764/5	16232/46
31		5½	5.68	763	16234
32	Nu	6	5.30	773	16432
33		5	5.30	782	16628
34	Mu	6	5.69	793	16811
35		4	4.66	801	16908
36		7	6.46	808	17017
37	Omicron	6	5.77	809	17036
38		7	5.18	812	17093
39		4	4.51	824	17361
40		6	5.82	828	17459
41	c	3	3.63	838	17573
42	Pi	6	5.22	836	17543
43	Sigma	6	5.49	847	17769
44+	Rho[1]	6	6.9		18091
45	Rho[2]	6½	5.91	867	18191

(table continued on next page)

Table 96. The numbered stars of Aries *(continued)*

		MAGNITUDES		CATALOGUE NUMBERS	
Flamsteed Number	Letter	Flamsteed	Visual	HR	HD
46	Rho³	6½	5.63	869	18256
47		6½	5.80	878	18404
48	Epsilon	5	4.63c	887/8	18519/20
49		7	5.90	905	18769
50+		7	7.0		18654
51+		7	6.72		18803
52		6	5.36c	927/8	19134/5
53		7	6.11	938	19374
54		6½	6.27	940	19460
55		7	5.72	944	19548
56		6½	5.79	954	19832
57	Delta	4	4.35	951	19787
58	Zeta	5	4.89	972	20150
59		7	5.90	995	20618
60		7	6.12	1000	20663
61	Tau¹	7	5.28	1005	20756
62		6	5.52	1012	20825
63	Tau²	6	5.09	1015	20893
64		6	5.50	1022	21017
65		7	6.08	1027	21050
66		7	6.03	1048	21467

The Lost, Missing, or Troublesome Stars of Aries

1, HR 530, 5.86V, *SA, HRP, BS,* BF 214*.

This star, 1 Ari, is not the lead star in the constellation. When reducing it to 1690, the epoch of his catalogue, Flamsteed erred 10^m in calculating its right ascension. Consequently, 2, 3, and 4 Ari precede it in right ascension.

2, in Pisces, HR 493, 5.24V, *HRP,* BF 204* and Table A, p. 645.

This star is the same as 107 Psc. It is one of Flamsteed's twenty-two duplicate stars.

3, in Pisces, HR 515, 6.55V, *SA, HRP,* GC 2156*, the variable VY Psc.

7, HR 559, 5.74V, *SA, HRP, BS,* BF 229*.

Both Halley's 1712 edition of Flamsteed's catalogue and the authorized edition of 1725 list this star's magnitude as "tel," *telescopica* or telescopic. But Baily noted that in Flamsteed's manuscripts, its magnitude was listed differently on three separate occasions as 8, 7, and 6.

18, BF 271*, *BAC's* errata, item 2.

This star does not exist. Flamsteed observed it on October 25, 1695, but it cannot now be located. In his manuscripts, Flamsteed, himself marked it as doubtful. Baily suggested that if 18's declination were to be reduced by 6', it would correspond to Piazzi's II, 12, which is synonymous with Taylor's 716 and HD 13566, 8.0V.

23, HD 14305, 6.84V, *SA,* GC 2790*.

25, in Cetus, HD 15228, 6.49V, *SA, BS Suppl.,* GC 2946*.

28, BF 314* and Table C, p. 646.

This star does not exist. Flamsteed observed it on December 10, 1692 at $8^h19^m58^s$, but it cannot now be located. Baily suggested there was an error in reducing the time of its transit. He supposed that if it crossed Flamsteed's lens 1^m earlier, at $8^h18^m58^s$, it would be the same as 26 Ari. Peters, in "Flamsteed's Stars," pp. 70-71, also identified it as 26 Ari.

44, Rho¹, HD 18091, 6.9V, *SA,* GC 3492*.

For a discussion of Bayer's Rho and the indices assigned to it, see Rho Ari in Part One.

50, HD 18654, 7.0V, BF 380*, BAC 933*, B 139*, T 1000*.

This star is missing from most modern atlases and catalogues because there is no star in the sky at the coordinates listed in Flamsteed's catalogue. Flamsteed erred 1^m in reducing this star to 1690, the epoch of his catalogue. In his revised edition of Flamsteed's catalogue, Baily corrected its right ascension by 15', the equivalent of 1^m in time, from 41°2'0" to 40°47'0". The star is properly identified in the second edition of *SA*.

Although Flamsteed was one of the first astronomers to use clock time in recording his observations, he arranged the right ascension of the stars in his catalogue by degrees, not hours. It should be noted:

    1 degree (°) equals 4 minutes (m) of sidereal time.
    15 degrees (°) equals 1 hour (h) of sidereal time.
    1 arc minute (') equals 4 seconds (s) of sidereal time.
    15 arc seconds (") equals 1 second (s) of sidereal time.

In modern astronomy, at least since the middle of the nineteenth century, right ascension, equivalent to longitude on Earth, is usually described in hours, minutes, and seconds of sidereal time along the celestial equator starting at the point of the vernal equinox.

51, HD 18803, 6.72V, *SA, BS Suppl.,* GC 3640*.

Auriga, Aur

Charioteer

Auriga (Figure 85) is a Ptolemaic constellation that Flamsteed numbered from 1 to 66.

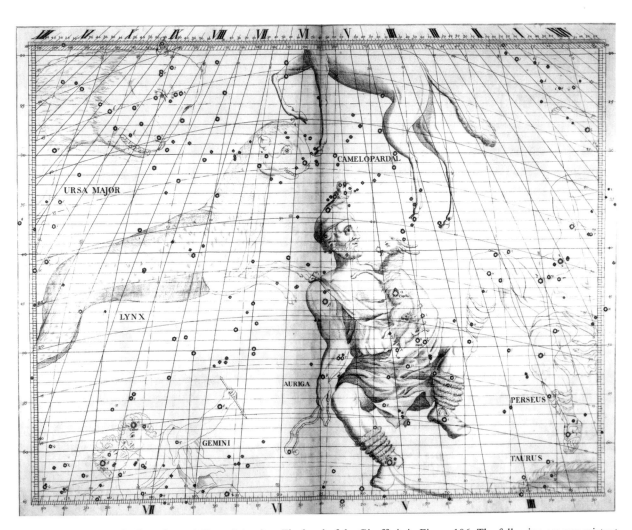

Figure 85. Flamsteed's Camelopardalis and Auriga. The head of the Giraffe is in Figure 106. The following are nonexistent stars in Camelopardalis: **13 Cam,** 70°39' r.a., 38°1' n.p.d.; **27 Cam,** [80°11'] r.a., 33°7' n.p.d.

Table 97. The numbered stars of Aurgia

		MAGNITUDES		CATALOGUE NUMBERS	
Flamsteed Number	Letter	Flamsteed	Visual	HR	HD
1+		5	4.88	1533	30504
2		5½	4.78	1551	30834
3	Iota	4	2.69	1577	31398
4	Omega	5	4.94	1592	31647
5		6	5.95	1599	31761
6		6	6.58	1602	31780
7	Epsilon	4	2.99	1605	31964
8	Zeta	4	3.75	1612	32068
9		5¾	5.00	1637	32537
10	Eta	4	3.17	1641	32630
11	Mu	5	4.86	1689	33641
12+		6	6.88		33988
13	Alpha	1	0.08	1708	34029
14		5	5.02	1706	33959
15	Lambda	5	4.71	1729	34411
16		6	4.54	1726	34334
17		6¾	6.14	1728	34364
18		8	6.49	1734	34499
19		6	5.03	1740	34578
20	Rho	6	5.23	1749	34759
21	Sigma	5½	4.99	1773	35186
22		6	6.46	1768	35076
23+	Gamma	2	1.65	1791	35497
24	Phi	5½	5.07	1805	35620
25	Chi	5½	4.76	1843	36371
26		6	5.40	1914	37269
27	Omicron	6	5.47	1971	38104
28+		7	6.80		38604
29	Tau	5	4.52	1995	38656
30	Xi	6	4.99	2029	39283
31	Upsilon	6	4.74	2011	38944
32	Nu	5	3.97	2012	39003
33	Delta	4	3.72	2077	40035
34	Beta	2	1.90	2088	40183
35	Pi	6	4.26	2091	40239
36		6	5.73	2101	40394
37	Theta	4	2.62	2095	40312
38		6½	6.10	2119	40801
39		6½	5.87	2132	41074
40		6	5.36	2143	41357
41		6	5.64c	2175/6	42126/7
42		6	6.52	2228	43244
43		6	6.38	2239	43380
44	Kappa	4½	4.35	2219	43039
45		6	5.36	2264	43905

(table continued on next page)

Table 97. The numbered stars of Aurgia *(continued)*

Flamsteed Number	Letter	Magnitudes		Catalogue Numbers	
		Flamsteed	Visual	HR	HD
46	Psi¹	5	4.91	2289	44537
47		6	5.90	2338	45466
48		6	5.55	2332	45412
49		5½	5.27	2398	46553
50	Psi²	5½	4.79	2427	47174
51		5½	5.69	2419	47070
52	Psi³	5	5.20	2420	47100
53		6	5.79	2425	47152
54		6	6.03	2438	47395
55	Psi⁴	5	5.02	2459	47914
56+	Psi⁵		5.25	2483	48682
57+	Psi⁶		5.22	2487	48781
58	Psi⁷	4½	5.02	2516	49520
59		6	6.12	2539	50018
60	Psi⁸	6	6.30	2541	50037
61	Psi⁸	6	6.48	2547	50204
62		6½	6.00	2600	51440
63		4½	4.90	2696	54716
64		5	5.75	2753	56221
65		5	5.13	2793	57264
66		5	5.19	2805	57669

The Lost, Missing, or Troublesome Stars of Auriga

1, in Perseus, HR 1533, 4.88V, *SA, HRP,* GC 5868*.

12, HD 33988, 6.88V, *SA, BS Suppl.,* GC 6424*.

23, Gamma, in Taurus, HR 1791, 1.65V, *HRP, BS,* BF 704 and Table A, p. 645, El Nath.

This star is the same as 112, Beta Tau. It is one of Flamsteed's twenty-two duplicate stars. Of these twenty-two, two were intentionally duplicated — this star and Fomalhaut (79 Aqr and 24 PsA). Flamsteed duplicated these two because Ptolemy had duplicated them in his catalogue. Actually, Ptolemy duplicated three stars: these two and Nu Boo and Psi Her. Flamsteed, however, included only Nu Boo (52/3 Boo) in his catalogue.

28, HD 38604, 6.80V, *SA,* GC 7265*.

56, Psi⁵, HR 2483, 5.25V, *SA, HRP, BS,* HF 58.
57, Psi⁶, HR 2487, 5.22V, *SA, HRP, BS,* HF 57*.

The magnitudes for 56 and 57 are not included in Flamsteed's 1725 "Catalogus Britannicus," but they are listed in Halley's 1712 edition as 5 and 6 respectively.

Bootes, **Boo**

Herdsman, Bear Keeper, Bear Driver

Bootes (Figure 86) is a Ptolemaic constellation that Flamsteed numbered from 1 to 54.

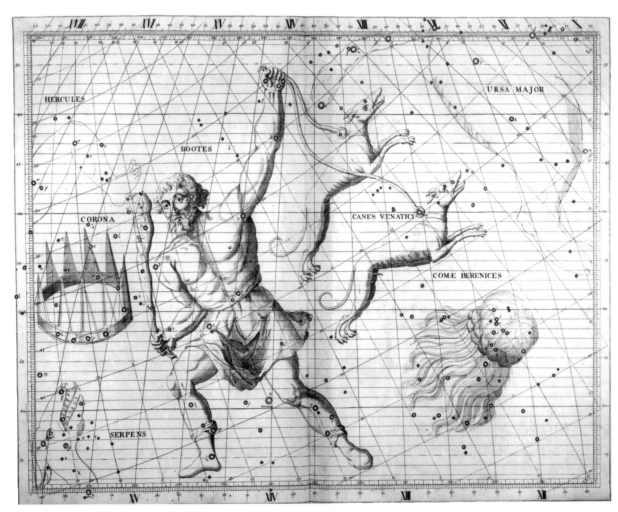

Figure 86. Flamsteed's Coma Berenices, Bootes, and Canes Venatici. The following are nonexistent stars: **19 Com,** 183°29' r.a., 61°10' n.p.d.; **34 Com,** 189°30' r.a., 61° 4' n.p.d.; **22 CVn,** 196° r.a., 36°47' n.p.d.

Table 98. The numbered stars of Bootes

		MAGNITUDES		CATALOGUE NUMBERS	
Flamsteed Number	Letter	Flamsteed	Visual	HR	HD
1		6	5.75	5144	119055
2		6	5.62	5149	119126
3		6	5.95	5182	120064
4	Tau	4	4.50	5185	120136
5	Upsilon	4	4.07	5200	120477
6	e	5½	4.91	5201	120539
7		7	5.70	5225	121107
8	Eta	3	2.68	5235	121370
9		5	5.01	5247	121710
10		7	5.76	5255	121996
11		6¾	6.23	5263	122405
12	d	5	4.83	5304	123999
13		6	5.25	5300	123782
14		6	5.54	5323	124570
15		6	5.29	5330	124679
16	Alpha	1	-0.04	5340	124897
17	Kappa[1,2]	4	4.44c	5328/9	124674/5
18		6	5.41	5365	125451
19	Lambda	4	4.18	5351	125162
20		5	4.86	5370	125560
21	Iota	4	4.75	5350	125161
22	f	5	5.39	5405	126661
23	Theta	4	4.05	5404	126660
24	g	6½	5.59	5420	127243
25	Rho	4	3.58	5429	127665
26		7	5.92	5434	127739
27	Gamma	3	3.03	5435	127762
28	Sigma	5	4.46	5447	128167
29	Pi[1,2]	3¾	4.53c	5475/6	129174/5
30	Zeta	3	3.78c	5477/8	129246/7
31		5	4.86	5480	129312
32		6	5.56	5481	129336
33		6	5.39	5468	129002
34		6	4.81	5490	129712
35	Omicron	4½	4.60	5502	129972
36	Epsilon	3	2.37c	5505/6	129988/9
37	Xi	4	4.55	5544	131156
38	h	6	5.74	5533	130945
39		6	5.69	5538	131041
40		6½	5.64	5588	132772
41	Omega	5	4.81	5600	133124
42	Beta	3	3.50	5602	133208
43	Psi	5	4.54	5616	133582
44	i	6	4.76	5618	133640
45	c	5	4.93	5634	134083

(table continued on next page)

Table 98. The numbered stars of Bootes *(continued)*

		MAGNITUDES		CATALOGUE NUMBERS	
Flamsteed Number	Letter	Flamsteed	Visual	HR	HD
46	b	6	5.67	5638	134320
47+	k		5.57	5627	133962
48	Chi	5	5.26	5676	135502
49	Delta	3	3.47	5681	135722
50		5	5.37	5718	136849
51	Mu^{1,2}	4	4.33c	5733/4	137391/2
52+	Nu^1	6	5.02	5763	138481
53+	Nu^2	6	5.02	5774	138629
54	Phi	6	5.24	5823	139641

The Lost, Missing, or Troublesome Stars of Bootes

47, k, HR 5627, 5.57V, *SA, HRP, BS,* HF 47*.

Flamsteed's "Catalogus Britannicus" of 1725 omits this star's magnitude, but it is listed as magnitude 7 in Halley's 1712 pirated edition of his catalogue.

52, Nu^1, HR 5763, 5.02V, *SA, HRP, BS.*
53, Nu^2, HR 5774, 5.02V, *SA, HRP, BS.*

Nu Boo is the same as Psi Her. Flamsteed assigned Nu-1 and Nu-2 to his 52 and 53 Boo. See Nu Boo and Psi Her in Part One.

Camelopardalis, Cam

Giraffe

Probably devised by Petrus Plancius, the constellation Camelopardalis (Figure 85) was included in the catalogues of Hevelius and Flamsteed. The latter, unaware of Plancius's work, asserted that it was a Hevelian constellation (*Constellatio Heveliana*). If Flamsteed had carefully read Hevelius's *Prodromus Astronomiae,* he would have noticed that it was not among the twelve new constellations that Hevelius took credit for devising (see Canes Venatici and Camelopardalis in Part One). Flamsteed numbered the stars of Camelopardalis's from 1 to 58.

Table 99. The numbered stars of Camelopardalis

		MAGNITUDES		CATALOGUE NUMBERS	
Flamsteed Number	Letter	Flamsteed	Visual	HR	HD
1		6	5.77	1417	28446
2		5	5.35	1466	29316
3		6	5.05	1467	29317
4		6	5.34	1511	30121
5		6	5.52	1555	30958
6+		6	6.9		31189
7		5	4.47	1568	31278
8		7	6.08	1588	31579
9	Alpha	4½	4.29	1542	30614
10	Beta	4¾	4.03	1603	31910
11		5	5.08	1622	32343
12		6	6.25	1623	32357
13+		4½			
14		5	6.50	1678	33296
15		6	6.13	1719	34233
16		6	5.23	1751	34787
17		6	5.42	1802	35583
18		6	6.43	1828	36066
19		6	6.15	1857	36570
20+		7	7.45		36770
21+		6½	6.7		37136
22+		7½	7.05		37070
23		6	6.15	1943	37638
24		6	6.05	1941	37601
25+		7½	6.8		37735

(table continued on next page)

Table 99. The numbered stars of Camelopardalis *(continued)*

Flamsteed Number	Letter	Magnitudes		Catalogue Numbers	
		Flamsteed	Visual	HR	HD
26		5½	5.94	1969	38091
27+		5½			
28+		6½	6.8		38129
29		5½	6.54	1992	38618
30		6	6.14	2006	38831
31		5	5.20	2027	39220
32+	Xi Aur	5	4.99	2029	39283
33+		7	7.1		39724
34+		6	6.44	2079	40062
35+		5½	6.45	2123	40873
36		6	5.32	2165	41927
37		5½	5.36	2152	41597
38		7	7.1		41783
39+		6½	6.7		41782
40		6½	5.35	2201	42633
41+		7	6.6		47005
42		4½	5.14	2490	48879
43		4½	5.12	2511	49340
44+		6	7.4		55944
45+		7	7.6		56099
46+		7	6.9		56243
47		6	6.35	2772	56820
48+		6	7.1		60844
49		5	6.49	2977	62140
50+		6	5.27	2969	61931
51		5	5.92	2975	62066
52+		5	6.72	3077	64347
53		6	6.01	3109	65339
54+		6	6.49	3119	65626
55+		5	5.32	3182	67447
56+		6	6.45	3221	68457
57+		5	5.71	3245	69148
58+		5	5.89	3254	69548

The Lost, Missing, or Troublesome Stars of Camelopardalis

6, HD 31189, 6.9V, *SA,* GC 6002*.

13

This star does not exist. Flamsteed observed it on January 20, 1696, but no star exists at its coordinates. Baily believed that Flamsteed observed this star south of the zenith, not north as he recorded it. If this assumption is correct, the star Flamsteed observed was 9 Aur. See BF 644 and Table B, p. 645.

Numbered Stars - Camelopardalis

20, HD 36770, 7.45V, *SA,* BF 720*, BAC 1735*.

This star's number is often omitted from atlases and catalogues because no star exits at the position listed in Flamsteed's catalogue. He observed this star on January 26, 1696, but while reducing its coordinates to 1690, he erred in computing its declination; he confused algebraic signs. Baily corrected the error in his revised edition of Flamsteed's catalogue.

21, HD 37136, 6.7V, *SA,* GC 7015*.

22, HD 37070, 7.05V, *SA, BS Suppl.,* GC 6990*.

Boss's *General Catalogue* mistakenly lists two stars as 22 Cam: this star, which is Flamsteed's 22 Cam, and GC 8020, HD 42818, 4.80V.

25, in Auriga, HD 37735, 6.8V, *SA,* GC 7133*.

27, BF 756* and Table C, p. 646.

This star does not exist. Flamsteed observed it on January 22, 1696, but he made a 30' error in its declination when entering it in his catalogue. He also omitted its right ascension. Although Baily corrected the mistakes, he noted that there is no star at its position. Argelander suggested that Flamsteed might have inadvertently reobserved 24 Cam. Peters, on the other hand, felt he might have observed 28 Cam or possibly one of three 9th-magnitude stars: BD+56 1060, BD+56 1063, or BD+56 1064. See Peters, "Flamsteed's Stars," p. 74.

28, HD 38129, 6.8V, *SA,* GC 7201*.

32, in Auriga, HR 2029, 4.99V, *HRP, BS,* BF 778 and Table A, p. 645.

This star is the same as 30, Xi Aur. It is one of Flamsteed's twenty-two duplicate stars.

33, in Auriga, HD 39724, 7.1V, BF 790*, BAC 1872*, B 98*.

Although Flamsteed included this star in Camelopardalis, Baily, thought that it more properly belonged in Auriga and switched it there.

34, in Auriga, HR 2079, 6.44V, *HRP,* BF 797*, BAC 1887.

Flamsteed observed this star on January 22, 1696. In reducing its position to epoch 1690 for his catalogue, he erred in computing its declination; he used the wrong algebraic sign. Baily made the necessary correction in his revised edition of Flamsteed's catalogue and switched it to Auriga, where he felt it belonged.

35, in Auriga, HR 2123, 6.45V, *SA, HRP,* GC 7663*.

38, HD 41783, 7.1V, *SA,* BF 829*, BAC 1949*.

39, HD 41782, 6.7V, *SA,* BF 827*, BAC 1950*.

41, HD 47005, 6.6V, *SA,* GC 8666*.

44, in Lynx, HD 55944, 7.4V, *SA,* BF 1000*, BAC 2365*.

45, in Lynx, HD 56099, 7.6V, *SA,* GC 9689*.

46, in Lynx, HD 56243, 6.9V, *SA,* GC 9711*.

48, in Lynx, HD 60844, 7.1V, *SA,* GC 10252*.

50, in Lynx, HR 2969, 5.27V, *HRP,* BF 1065*, BAC 2532*.

Noting that Flamsteed erroneously included this star in Camelopardalis, Baily switched it to Lynx. Unlike other stars in this constellation that hover near the Lynx-Camelopardalis border, this one is clearly within Lynx — almost 10° inside its border.

52, in Lynx, HR 3077, 6.72V, *SA, HRP,* GC 10700*.

54, in Lynx, HR 3119, 6.49V, *SA, HRP, BS,* GC 10864*.
55, in Ursa Major, HR 3182, 5.32V, *SA, HRP, BS,* GC 11100*.
56, in Ursa Major, HR 3221, 6.45V, *SA, HRP, BS,* GC 11199*.

BS notes that these three stars are Flamsteed's 54, 55, and 56 Cam.

57, in Ursa Major, HR 3245, 5.71V, *SA, HRP, BS,* GC 11291*.

58, in Lynx, HR 3254, 5.89V, *HRP,* BF 1151* and Table A, p. 645.

Flamsteed provided only the declination of this star in his catalogue. Baily asserted that this star was the same as 30 Lyn, one of Flamsteed's twenty-two duplicate stars.

CANCER, CNC

Crab

Cancer (Figure 87) is a Ptolemaic constellation that Flamsteed numbered from 1 to 83.

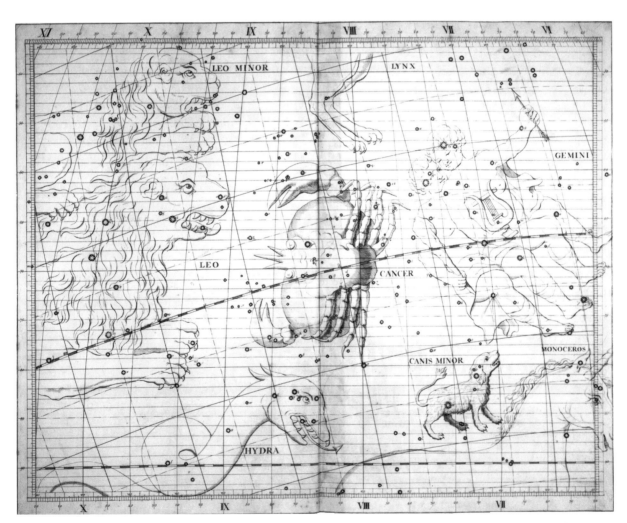

Figure 87. Flamsteed's Cancer. The following are nonexistent stars in Cancer: **26 Cnc,** 122°10' r.a., 61° 9' n.p.d.; **56 Cnc,** 128°48' r.a., 60°32' n.p.d.; **73 Cnc,** 132°27' r.a., 73°30' n.p.d.; **74 Cnc,** 132°32' r.a., 74°15' n.p.d.

Table 100. The numbered stars of Cancer

		MAGNITUDES		CATALOGUE NUMBERS	
Flamsteed Number	Letter	Flamsteed	Visual	HR	HD
1		6	5.78	3095	64960
2	Omega[1]	6	5.83	3124	65714
3		6	5.55	3128	65759
4	Omega[2]	6	6.31	3132	65856
5		6	5.99	3134	65873
6+	Chi Gem	5	4.94	3149	66216
7+		8	6.84		66347
8		6	5.12	3163	66664
9	Mu[1]	7	5.99	3169	66875
10	Mu[2]	5	5.30	3176	67228
11+		6	6.88		67402
12		6	6.27	3184	67483
13+		6½	6.41		67690
14	Psi	4	5.73	3191	67767
15+	Psi Gem	5	5.64	3215	68351
16	Zeta[1,2]	5½	4.67c	3208/9/10	68257/55/56
17	Beta	3¾	3.52	3249	69267
18	Chi	6	5.14	3262	69897
19	Lambda	6	5.98	3268	70011
20	d[1]	6	5.84	3284	70569
21		6	6.08	3290	70734
22	Phi[1]	6½	5.57	3304	71093
23	Phi[2]	6	5.56c	3310/1	71150/1
24		6	6.58c	3312/3	71152/3
25	d[2]	6	6.14	3299	71030
26+		6			
27		6	5.50	3319	71250
28		6½	6.10	3329	71496
29		6½	5.95	3333	71555
30	Upsilon[1]	6	5.75	3355	72041
31	Theta	5¾	5.35	3357	72094
32	Upsilon[2]	7½	6.36	3369	72324
33	Eta	6½	5.33	3366	72292
34		6	6.46	3372	72359
35		7	6.58	3387	72779
36	c	6	5.88	3406	73143
37		6	6.53	3412	73316
38+		8	6.66		73575
39		6	6.39	3427	73665
40+		6	6.61		73666
41	Epsilon	7	6.30	3429	73731
42+		7½	6.85		73785
43	Gamma	4	4.66	3449	74198

(table continued on next page)

Table 100. The numbered stars of Cancer *(continued)*

		MAGNITUDES		CATALOGUE NUMBERS	
Flamsteed Number	Letter	**Flamsteed**	**Visual**	**HR**	**HD**
44+		6	7.9		74200
45	A^1	6	5.62	3450	74228
46		6	6.13	3464	74485
47	Delta	4	3.94	3461	74442
48	Iota	5	4.09c	3474/5	74738/9
49	b	6	5.66	3465	74521
50	A^2	6	5.87	3481	74873
51	Sigma^1	6	5.66	3519	75698
52+		6	7.2		75558
53		6	6.23	3521	75716
54		7	6.38	3510	75528
55	Rho^1	6	5.95	3522	75732
56+		6			
57		5½	5.39	3532	75959
58	Rho^2	6	5.22	3540	76219
59	Sigma^2	5½	5.45	3555	76398
60		4½	5.41	3550	76351
61		6	6.29	3563	76572
62	Omicron^1	6	5.20	3561	76543
63	Omicron^2	6	5.67	3565	76582
64	Sigma^3	6	5.20	3575	76813
65	Alpha	4	4.25	3572	76756
66		6	5.82	3587	77104
67		6½	6.07	3589	77190
68+		6	7.1		77230
69	Nu	6	5.45	3595	77350
70		6½	6.38	3601	77557
71+		7	8.0		77892
72	Tau	6½	5.43	3621	78235
73+		6			
74+		6			
75		6½	5.98	3626	78418
76	Kappa	4½	5.24	3623	78316
77	Xi	5½	5.14	3627	78515
78+		7½	7.19		78479
79		8	6.01	3640	78715
80+		7	6.89		79009
81	Pi^1	7	6.51	3650	79096
82	Pi^2	6	5.34	3669	79554
83+		6	6.60		80218

The Lost, Missing, or Troublesome Stars of Cancer

6, In Gemini as Chi Gem, HR 3149, 4.94V, *HRP, BS,* BF 1120*, BAC 2672*.

Bayer designated this star as Chi Gem, but Flamsteed switched it to Cancer with the notation that it was Bayer's Chi Gem. When fixed boundaries for the constellations were established in 1930, the IAU agreed that this star belongs in Gemini, as Bayer had originally proposed.

7, HD 66347, 6.84V, *SA,* GC 10928*.

11, HD 67402, 6.88V, *SA,* GC 11043*.

13, HD 67690, 6.41V, *SA,* GC 11082*.

15, HR 3215, 5.64V, *SA, HRP, BS.*

Bayer designated this star Psi Gem. See Psi Gem in Part One.

26

This star does not exist. Flamsteed made a computation error of 1^m in one of his observations of 22 Cnc. This produced a new set of coordinates that led him to believe he was observing a new star, 26 Cnc. See BF 1173 and Table B, p. 645.

38, HD 73575, 6.66V, *SA, BS Suppl.,* GC 11874*.

40, HD 73666, 6.61V, *SA, BS Suppl.,* GC 11889*.

42, HD 73785, 6.85V, *SA, BS Suppl.,* GC 11916*.

44, HD 74200, 7.9V, BF 1223*, BAC 2938*, B 114*, P VIII, 143*, T 3723*.

52, HD 75558, 7.2V, *SA,* GC 12215*.

56

This star does not exist. Flamsteed made a 1^m error in reducing one of his observations of 55 Cnc. This produced a new set of coordinates that led him to believe that he was observing a new star, 56 Cnc. See BF 1250 and Table B, p. 645.

68, HD 77230, 7.1V, *SA,* GC 12475*.

71, HD 77892, 8.0V, *SA,* GC 12556*.

73

This star does not exist. In one of his observations of 62 Cnc, Flamsteed made a 10^m error in right ascension in reducing it to 1690. This produced a new set of coordinates that led him to believe he was observing a new star, 73 Cnc. See BF 1261 and Table B, p. 645.

74

This star does not exist. In one of his observations of 63 Cnc, Flamsteed made two errors of reduction: one of 10^m in determining right ascension and another of 1° in determining declination. This generated a new set of coordinates that led him to believe he was observing a new star, 74 Cnc. See BF 1262 and Table B, p. 645.

78, HD 78479, 7.19V, *SA,* GC 12625*.

80, HD 79009, 6.89V, *SA, BS Suppl.,* GC 12685*.

83, HD 80218, 6.60V, *SA,* GC 12841*.

CANES VENATICI, CVn

Hunting Dogs

Hevelius devised the constellation Canes Venatici (Figure 86). Flamsteed acknowledged that it was a Hevelian constellation and included it in both his catalogue and atlas. He numbered its stars from 1 to 25.

Table 101. The numbered stars of Canes Venatici

		MAGNITUDES		CATALOGUE NUMBERS	
Flamsteed Number	Letter	Flamsteed	Visual	HR	HD
1+		6	6.16	4654	106478
2		5	5.66	4666	106690
3		6	5.29	4690	107274
4		6	6.06	4715	107904
5		6	4.80	4716	107950
6		5	5.02	4728	108225
7		7	6.21	4761	108845
8	Beta	4½	4.26	4785	109358
9		6½	6.37	4811	109980
10		6	5.95	4845	110897
11		6	6.27	4866	111421
12	Alpha[1,2]	2½	2.80c	4914/5	112412/3
13+		4½	4.90	4924	112989
14		5	5.25	4943	113797
15		5¾	6.28	4967	114376
16+		6	7.17		114427
17		6	5.91	4971	114447
18+		6	7.16		114674
19		7	5.79	5004	115271
20		6	4.73	5017	115604
21		6	5.15	5023	115735
22+		6			
23		7	5.60	5032	116010
24		5½	4.70	5112	118232
25		5	4.82	5127	118623

The Lost, Missing, or Troublesome Stars of Canes Venatici

1, in Ursa Major, HR 4654, 6.16V, *SA, HRP,* GC 16721*.

Although Hoffleit did not identify this star with its Flamsteed number in the *BS,* it is properly identified in her "Additions and Corrections," p. 190.

13, in Coma, HR 4924, 4.90V, *HRP,* BF 1793 and Table A, p. 645 where 37 Com is inadvertently listed as 31 Com.

This star is the same as 37 Com. It is one of Flamsteed's twenty-two duplicate stars.

16, HD 114427, 7.17V, *SA,* GC 17834*.

18, HD 114674, 7.16V, *SA,* GC 17868*.

22, BF 1838* and Table D, p. 646, BAC, p. 77.

This star does not exist. Its arc minutes and seconds of right ascension are missing from Flamsteed's catalogue. Acknowledging that there was no record that Flamsteed ever observed this star, Baily was at a loss to explain how it ever found its way into his catalogue. All in all, there are eleven of these nonobserved and nonexistent objects in the "Catalogus Britannicus."

Canis Major, CMa

Greater, Larger, or Big Dog

Canis Major (Figure 88) is a Ptolemaic constellation that Flamsteed numbered from 1 to 31.

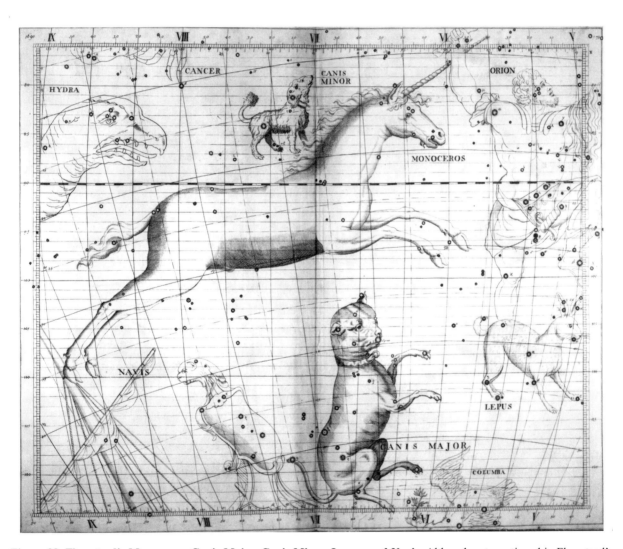

Figure 88. Flamsteed's Monoceros, Canis Major, Canis Minor, Lepus, and Navis. Although not mentioned in Flamsteed's catalogue, Columba, Noah's Dove, is depicted in his atlas. Also depicted is the upper portion of Navis, the Ship, showing the stars described in Flamsteed's "Catalogus Britannicus." This section of the Ship is currently called Puppis. The following are nonexistent stars: **12 CMi,** 113°48' r.a., 76°41' n.p.d.; **17 Pup,** 118°49' r.a., 102°22' n.p.d.

Lost Stars

Table 102. The numbered stars of Canis Major

		MAGNITUDES		CATALOGUE NUMBERS	
Flamsteed Number	**Letter**	**Flamsteed**	**Visual**	**HR**	**HD**
1	Zeta	3	3.02	2282	44402
2	Beta	2	1.98	2294	44743
3+	Delta Col	4	3.85	2296	44762
4	Xi^1	5	4.33	2387	46328
5	Xi^2	5	4.54	2414	46933
6	Nu^1	5	5.70	2423	47138
7	Nu^2	5	3.95	2429	47205
8	Nu^3	5	4.43	2443	47442
9	Alpha	1	-1.46	2491	48915
10		6	5.20	2492	48917
11		5	5.29	2504	49229
12		6	6.08	2509	49333
13	Kappa	5	3.96	2538	50013
14	Theta	5	4.07	2574	50778
15		6	4.83	2571	50707
16	Omicron^1	5	3.87	2580	50877
17		6	5.74	2588	51055
18	Mu	4	5.00	2593	51250
19	Pi	6	4.68	2590	51199
20	Iota	4	4.37	2596	51309
21	Epsilon	2¾	1.50	2618	52089
22	Sigma	4	3.47	2646	52877
23	Gamma	3	4.12	2657	53244
24	Omicron^2	4¾	3.02	2653	53138
25	Delta	2½	1.84	2693	54605
26		7	5.92	2718	55522
27		7	4.66	2745	56014
28	Omega	3½	3.85	2749	56139
29		5	4.98	2781	57060
30	Tau	5	4.40	2782	57061
31	Eta	2¾	2.45	2827	58350

The Lost, Missing, or Troublesome Stars of Canis Major

3, in Columba as Delta Col, HR 2296, 3.85V, *SA, HRP*

Although Flamsteed included Columba in his atlas, he did not recognize it as a separate constellation in his catalogue. Lacaille later placed several of Canis Major's southern stars, including this one, Flamsteed's 3, within the border of Columba. He designated it Delta Col.

Canis Minor, CMi

Lesser, Smaller, or Little Dog

Canis Minor (Figure 88) is a Ptolemaic constellation that Flamsteed numbered from 1 to 14.

Table 103. The numbered stars of Canis Minor

Flamsteed Number	Letter	Magnitudes		Catalogue Numbers	
		Flamsteed	Visual	HR	HD
1		6¾	5.30	2820	58187
2	Epsilon	6	4.99	2828	58367
3	Beta	3	2.90	2845	58715
4	Gamma	6	4.32	2854	58972
5	Eta	6	5.25	2851	58923
6		6	4.54	2864	59294
7	Delta[1]	6	5.25	2880	59881
8	Delta[2]	6	5.59	2887	60111
9	Delta[3]	6	5.81	2901	60357
10	Alpha	1½	0.38	2943	61421
11		6	5.30	3008	62832
12+		5½			
13	Zeta	5	5.14	3059	63975
14		6	5.29	3110	65345

The Lost, Missing, or Troublesome Stars of Canis Minor

12, BF 1098* and Table D, p. 646.

This star does not exist. There is no star at its coordinates in the heavens and there is no record that Flamsteed ever observed it. It is one of eleven such nonobserved and nonexistent objects in Flamsteed's catalogue.

CAPRICORNUS, CAP

Goat, Sea Goat

Capricornus (figures 82 and 102) is a Ptolemaic constellation that Flamsteed numbered from 1 to 51.

Table 104. The numbered stars of Capricornus

		MAGNITUDES		CATALOGUE NUMBERS	
Flamsteed Number	**Letter**	**Flamsteed**	**Visual**	**HR**	**HD**
1	Xi¹	6	6.34	7712	191753
2	Xi²	6	5.85	7715	191862
3		6	6.32	7738	192666
4		6	5.87	7748	192879
5	Alpha¹	4	4.24	7747	192876
6	Alpha²	3	3.57	7754	192947
7+	Sigma	Obs	5.28	7761	193150
8	Nu	6	4.76	7773	193432
9	Beta	3	3.08	7776	193495
10+	Pi	Obs	5.25	7814	194636
11	Rho	6	4.78	7822	194943
12+	Omicron	Obs	5.58c	7829/30	195093/4
13+		6	6.76		196348
14	Tau	6	5.22	7889	196662
15	Upsilon	6	5.10	7900	196777
16	Psi	5	4.14	7936	197692
17		6	5.93	7937	197725
18	Omega	6	4.11	7980	198542
19		6½	5.78	8000	199012
20		6½	6.25	8033	199728
21+		6	6.5		199947
22	Eta	5	4.84	8060	200499
23	Theta	5	4.07	8075	200761
24	A	6	4.50	8080	200914
25	Chi	6	6.02	8087	201184
26+		6	6.8		201301
27		6	6.25	8091	201352
28	Phi	6	5.24	8127	202320

(table continued on next page)

Table 104. The numbered stars of Capricornus (continued)

Flamsteed Number	Letter	Magnitudes Flamsteed	Visual	Catalogue Numbers HR	HD
29		6	5.28	8128	202369
30		6	5.43	8137	202671
31		7	7.05	8139	202723
32	Iota	5	4.28	8167	203387
33		6	5.41	8183	203638
34	Zeta	5	3.74	8204	204075
35		6	5.78	8207	204139
36	b	6	4.51	8213	204381
37		6	5.69	8245	205289
38+		6	7.1		205306
39	Epsilon	4	4.68	8260	205637
40	Gamma	4	3.68	8278	206088
41		6	5.24	8285	206356
42		6	5.18	8283	206301
43	Kappa	5	4.73	8288	206453
44		6	5.88	8295	206561
45		6	5.99	8302	206677
46	c^1	6	5.09	8311	206834
47	c^2	6	6.00	8318	207005
48	Lambda	5	5.58	8319	207052
49	Delta	3	2.87	8322	207098
50+		6	7.00		207061
51	Mu	5	5.08	8351	207958

The Lost, Missing, or Troublesome Stars of Capricornus

7, Sigma, HR 7761, 5.28V, *SA, HRP, BS,* HF 7*.
10, Pi, HR 7814, 5.25V, *SA, HRP, BS,* HF 10*.
12, Omicron, HR 7829/30, 6.74V, 5.94V, comb. mag. 5.58V, *SA, BS,* HF 12*.

In Flamsteed's "Catalogus Britannicus," his authorized catalogue of 1725, the magnitude for each of these three stars is noted as "Obs," *obscura,* obscure or unknown. However, in Halley's pirated edition of 1712, their magnitudes are given as 6, 6, and 5½, respectively.

HR 7829/30 is the double-star ADS 13902. Baily, in the *BAC*, assigned 12, Omicron, only to HR 7830 (BAC 7054) and left the dimmer star, HR 7829 (BAC 7053), unlettered and unnumbered. *HRP* assigns Omicron to both but assigns 12 only to the brighter star, HR 7830. *BS* lists both stars as 12, Omicron.

13, HD 196348, 6.76V, *SA, BS Suppl.,* GC 28694*.

21, HD 199947, 6.5V, *SA,* GC 29322*.

26, HD 201301, 6.8V, *SA,* GC 29559*.

38, HD 205306, 7.1V, *SA,* GC 30208*.

50, HD 207061, 7.00V, *SA,* GC 30484*.

CASSIOPEIA, CAS

Cassiopeia, Queen, Lady in the Chair

Cassiopeia (Figure 89) is a Ptolemaic constellation that Flamsteed numbered from 1 to 55.

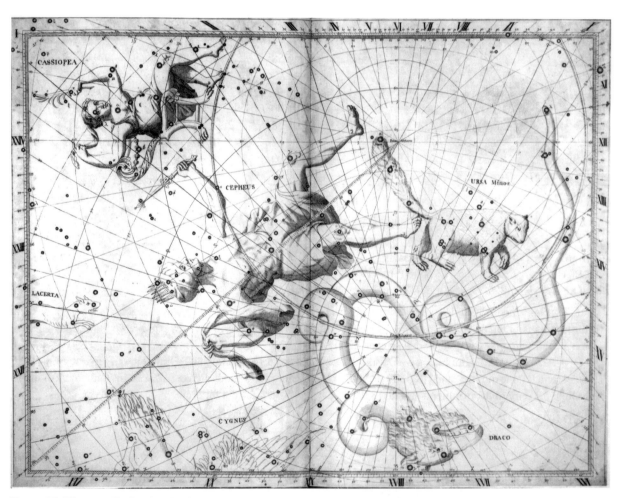

Figure 89. Flamsteed's Cassiopeia, Cepheus, Ursa Minor, and Draco. This chart depicts both the celestial north pole, Polus Arcticus, and the ecliptic north pole, Polus Eclipticus. The following are nonexistent stars: **3 Cas,** 347°37' r.a., 33° 2' n.p.d.; **29 Cas,** 11°29' r.a., 26°38' n.p.d.; **41 Cas,** 19°56' r.a., 18°58' n.p.d.; **56 Dra,** 287°53' r.a., 13°59' n.p.d.; **62 Dra,** 296°48' r.a., 17°42' n.p.d.; **70 Dra,** 302°39' r.a., 26° 0' n.p.d.

Table 105. The numbered stars of Cassiopeia

		Magnitudes		Catalogue Numbers	
Flamsteed Number	Letter	Flamsteed	Visual	HR	HD
1		6	4.85	8797	218376
2		7	5.70	8822	218753
3+		6			
4		5	4.98	8904	220652
5	Tau	5	4.87	9008	223165
6		6	5.43	9018	223385
7+	Rho	6	4.54	9045	224014
8	Sigma	6	4.88	9071	224572
9		6	5.88	9100	225180
10		6	5.59	7	144
11	Beta	2¾	2.27	21	432
12		6	5.40	93	2011
13		6	6.18	121	2729
14	Lambda	5	4.73	123	2772
15	Kappa	4	4.16	130	2905
16		6	6.48	137	3038
17	Zeta	4	3.66	153	3360
18	Alpha	3	2.23	168	3712
19	Xi	6	4.80	179	3901
20	Pi	6	4.94	184	4058
21		6	5.66	192	4161
22	Omicron	6	4.54	193	4180
23		6	5.41	208	4382
24	Eta	4	3.44	219	4614
25	Nu	5	4.89	223	4636
26	Upsilon[1]	7	4.83	253	5234
27	Gamma	3	2.47	264	5394
28	Upsilon[2]	6	4.63	265	5395
29+		6			
30	Mu	5	5.17	321	6582
31		6	5.29	336	6829
32		6	5.57	345	6972
33	Theta	4	4.33	343	6961
34	Phi	6	4.98	382	7927
35		7	6.34	384	8003
36	Psi	5½	4.74	399	8491
37	Delta	3	2.68	403	8538
38		6	5.81	427	9021
39	Chi	6	4.71	442	9408
40		6	5.28	456	9774
41+		6			
42		6	5.18	480	10250
43		6	5.59	478	10221
44		6	5.78	491	10425
45	Epsilon	3	3.38	542	11415

(table continued on next page)

Lost Stars

Table 105. The numbered stars of Cassiopeia (continued)

Flamsteed Number	Letter	Magnitudes		Catalogue Numbers	
		Flamsteed	Visual	HR	HD
46	Omega	6	4.99	548	11529
47		5	5.27	581	12230
48	A	5	4.54	575	12111
49		6	5.22	592	12339
50		4½	3.98	580	12216
51+		6	7.5		12441
52		7	6.00	586	12279
53		7	5.58	589	12301
54+		6	6.55		12800
55		6	6.05	640	13474

The Lost, Missing, or Troublesome Stars of Cassiopeia

3, BF 3213* and Table C, p. 645.

Baily believed that Flamsteed observed this star but that he erred mathematically in determining its position. Baily could find no star at the coordinates listed in Flamsteed's catalogue. Some astronomers have recently suggested, however, that Flamsteed's coordinates may indeed have been accurate and that he may have observed a 6th-magnitude remnant of the Supernova of 1653, which is located, according to radio telescope data, at the approximate position of 3 Cas. Objects detected with radio telescopes are usually identified with upper-case Roman letters and this star remnant is known as Cassiopeia A. For a summary of the debate regarding the existence of 3 Cas, see *Sky & Telescope,* April 1976, pp. 295-96.

7, Rho, HR 9045, 4.54V, *SA, HRP.*

BS listed this star as 71 Eta, but it was corrected in the "Additions and Corrections," p. 165.

29

This star does not exist. Baily suggested that Flamsteed probably confused 29 with 32, which he observed twice. He believed that at Flamsteed's first observation of 32 he erred 6^m or 90' in recording its transit. Such an error would account for the difference in right ascension, 1°29' or 89', between 29 and 32. Their declination is exactly the same. See BF 120 and Table B, p. 645.

41

This star does not exist. Flamsteed confused 41 with 42, which he observed twice. On his first observation of 42, he erred 2° in copying its declination. On his second observation, he recorded the declination correctly but this produced a different set of coordinates that led him to believe he had observed two different stars. See BF 189 and Table B, p. 645.

51, HD 12441, 7.5V, *SA,* GC 2493*.

54, HD 12800, 6.55V, *SA,* GC 2544*.

CENTAURUS, CEN

Centaur

Centaurus (Figure 99) is a Ptolemaic constellation that Flamsteed numbered from 1 to 5. All of these stars are listed in the *BS* and there are no missing or problem stars among them. Although a large constellation, the Centaur is mostly below the horizon in England, and Flamsteed was able to observe only a limited number of stars in the Centaur's upper torso. See Lupus, below.

Table 106. The numbered stars of Centaurus

Flamsteed Number	Letter	Magnitudes		Catalogue Numbers	
		Flamsteed	Visual	HR	HD
1	i	4½	4.23	5168	119756
2	g	4½	4.19	5192	120323
3	k	4½	4.32c	5210/1	120709/10
4	h	4½	4.73	5221	120955
5	Theta	2½	2.06	5288	123139

CEPHEUS, CEP

Cepheus, King

Cepheus (Figure 89) is a Ptolemaic constellation that Flamsteed numbered from 1 to 35.

Table 107. The numbered stars of Cepheus

Flamsteed Number	Letter	Magnitudes		Catalogue Numbers	
		Flamsteed	Visual	HR	HD
1	Kappa	5	4.39	7750	192907
2	Theta	5	4.22	7850	195725
3	Eta	4	3.43	7957	198149
4		6	5.60	7945	197950
5	Alpha	3	2.44	8162	203280
6		6	5.18	8171	203467
7		6	5.44	8227	204770
8	Beta	3	3.23	8238	205021
9		6	4.73	8279	206165
10	Nu	5	4.29	8334	207260
11		5	4.56	8317	206952
12		7	5.52	8339	207528
13		6	5.80	8371	208501
14		6	5.56	8406	209481
15+		6¾	6.70		209744
16		5½	5.03	8400	209369
17	Xi	5	4.29	8417	209790
18+			5.29	8416	209772
19		6	5.11	8428	209975
20		6	5.27	8426	209960
21	Zeta	4½	3.35	8465	210745
22	Lambda	6	5.04	8469	210839
23	Epsilon	4	4.19	8494	211336
24		5½	4.79	8468	210807
25		7	5.75	8511	211833
26		6	5.46	8561	213087
27	Delta	4½	3.75	8571	213306
28	Rho[1]	6	5.83	8578	213403
29	Rho[2]	6	5.52	8591	213798

(table continued on next page)

Table 107. The numbered stars of Cepheus *(continued)*

Flamsteed Number	Letter	Magnitudes		Catalogue Numbers	
		Flamsteed	Visual	HR	HD
30		6	5.19	8627	214734
31		6	5.08	8615	214470
32	Iota	4	3.52	8694	216228
33	Pi	5	4.41	8819	218658
34	Omicron	5	4.75	8872	219916
35	Gamma	3	3.21	8974	222404

The Lost, Missing, or Troublesome Stars of Cepheus

15, HD 209744, 6.70V, *SA, BS Suppl.,* GC 30874*.

18, HR 8416, 5.29V, *SA, HRP, BS,* BF 3031*.

In Flamsteed's catalogue, only the declination of this star is given. See Baily's note to BF 3031.

Cetus, Cet

Sea Monster, Whale

Cetus (Figure 90) is a Ptolemaic constellation that Flamsteed numbered from 1 to 97.

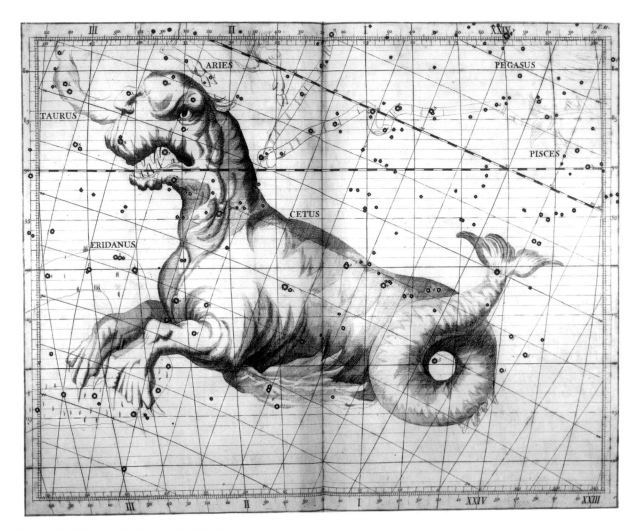

Figure 90. Flamsteed's Cetus. The following are nonexistent stars in Cetus: **24 Cet,** 11°18' r.a., 86° 1' n.p.d.; **74 Cet,** 33°14' r.a., 113°15' n.p.d.

Table 108. The numbered stars of Cetus

Flamsteed Number	Letter	Magnitudes		Catalogue Numbers	
		Flamsteed	Visual	HR	HD
1		6	6.26	9065	224481
2		4½	4.55	9098	225132
3		6	4.94	9103	225212
4+		6	6.43	11	315
5+		6	6.07	14	352
6		5	4.89	33	693
7		5	4.44	48	1038
8	Iota	3	3.56	74	1522
9		6	6.39	88	1835
10		6	6.19	101	2273
11+		6	7.48		2629
12		6	5.72	117	2637
13		6	5.20	142	3196
14		6	5.93	143	3229
15+		6	6.68		3512
16	Beta	3	2.04	188	4128
17	Phi[1]	5	4.76	194	4188
18		6	6.15	203	4307
19	Phi[2]	5	5.19	235	4813
20		6	4.77	248	5112
21		6	6.16	255	5268
22	Phi[3]	5	5.31	267	5437
23	Phi[4]	5	5.61	279	5722
24+		6			
25		6	5.43	296	6203
26		6	6.04	301	6288
27		6	6.12	315	6482
28		6	5.58	317	6530
29+		6	6.46		6734
30		6	5.82	329	6706
31	Eta	3	3.45	334	6805
32		6	6.40	346	6976
33		6	5.95	347	7014
34		6	5.94	353	7147
35+		6	6.8		7218
36+		6	6.60		7268
37		5	5.13	366	7439
38		6	5.70	368	7476
39		6	5.41	373	7672
40+		6	6.52		7727
41+		6	6.9		7812
42		6	5.87	385	8036
43		6	6.49	393	8335
44		6	6.21	401	8511
45	Theta	3	3.60	402	8512

(table continued on next page)

Table 108. The numbered stars of Cetus (continued)

Flamsteed Number	Letter	Magnitudes		Catalogue Numbers	
		Flamsteed	Visual	HR	HD
46		5	4.90	412	8705
47		6	5.66	421	8829
48		6	5.12	433	9132
49		6	5.63	451	9672
50		6	5.42	459	9856
51+	Nu Psc	6	4.44	489	10380
52	Tau	3½	3.50	509	10700
53	Chi	5	4.67	531	11171
54+		6	5.94	534	11257
55	Zeta	3	3.73	539	11353
56		4	4.85	565	11930
57		5	5.41	583	12255
58+		6	6.6		12020
59	Upsilon	4½	4.00	585	12274
60		6	5.43	607	12573
61		7	5.93	610	12641
62+		6½	7.07		13228
63		6	5.93	639	13468
64		6	5.63	635	13421
65	Xi^1	6	4.37	649	13611
66		6	5.51	650	13612
67		6	5.51	666	14129
68	Omicron	2¾	3.04	681	14386
69		6	5.28	689	14652
70		6	5.42	691	14690
71		6	6.33	704	15004
72	Rho	4	4.89	708	15130
73	Xi^2	4¾	4.28	718	15318
74+		6			
75		5½	5.35	739	15779
76	Sigma	4	4.75	740	15798
77		6	5.75	752	16074
78	Nu	4½	4.86	754	16161
79+		6	6.78		16141
80		6	5.53	759	16212
81		6	5.65	771	16400
82	Delta	3	4.07	779	16582
83	Epsilon	3	4.84	781	16620
84		6	5.71	790	16765
85+		6	6.30	797	16861
86	Gamma	3	3.47	804	16970
87	Mu	4	4.27	813	17094
88+		6	5.18	812	17093
89	Pi	3¾	4.25	811	17081

(table continued on next page)

Table 108. The numbered stars of Cetus *(continued)*

		MAGNITUDES		CATALOGUE NUMBERS	
Flamsteed Number	Letter	Flamsteed	Visual	HR	HD
90+	Tau¹ Eri	8	4.47	818	17206
91	Lambda	4	4.70	896	18604
92	Alpha	2	2.53	911	18884
93		6	5.61	910	18883
94		6	5.06	962	19994
95		6	5.38	992	20559
96	Kappa¹	5	4.83	996	20630
97	Kappa²	4	5.69	1007	20791

The Lost, Missing, or Troublesome Stars of Cetus

4, in Pisces, HR 11, 6.43V, *SA, HRP, BS,* GC 114*.

5, in Pisces, HR 14, 6.07V, *SA, HRP, BS,* GC 124*.

11, HD 2629, 7.48V, *SA,* GC 581*.

15, HD 3512, 6.68V, *SA, BS Suppl.,* GC 752*.

24

This star does not exist. In one of his observations of 73 Psc, Flamsteed erred by $3^m 40^s$ in reducing it to 1690, the epoch of his catalogue. This generated a new set of coordinates which led him to believe he had observed a new star, 24 Cet. See BF 107 and Table B, p. 645.

29, HD 6734, 6.46V, *SA, BS Suppl.,* GC 1374*.

35, HD 7218, 6.8V, *SA,* GC 1454*.

36, HD 7268, 6.60V, *SA,* GC 1463*.

40, HD 7727, 6.52V, *SA, BS Suppl.,* GC 1542*.

41, HD 7812, 6.9V, *SA,* GC 1564*.

51, in Pisces, HR 489, 4.44V, *HRP,* BF 202 and Table A, p. 645.

This star is the same as 106, Nu Psc. It is one of Flamsteed's twenty-two duplicate stars.

54, in Aries, HR 534, 5.94V, *SA, HRP, BS,* GC 2229*.

58, HD 12020, 6.6V, *SA,* GC 2375*.

62, HD 13228, 7.07V, *SA,* GC 2574*.

74

This star does not exist. Flamsteed made three observations of 48 Cet, but, when reducing one of these observations, he made several mathematical errors that resulted in new coordinates. As a result, he believed he had observed an entirely new star, 74 Cet. See BF 179 and Table B, p. 645.

79, HD 16141, 6.78V, *SA, BS Suppl.,* GC 3106*.

85, in Aries, HR 797, 6.30V, *SA, HRP,* GC 3260*.

88, in Aries, HR 812, 5.18V, *HRP,* BF 346 and Table A, p. 645, the variable UV Ari.

This star is the same as 38 Ari. It is one of Flamsteed's twenty-two duplicate stars.

90, in Eridanus, HR 818, 4.47V, *HRP,* HF 71, BF 352 and Table A, p. 645.

This star is the same as 1, Tau¹ Eri. It is one of Flamsteed's twenty-two duplicate stars. Both in Halley's 1712 edition of Flamsteed's catalogue and in the authorized edition of 1725, this star's magnitude is listed as 8, due probably to a copying error. In his revised edition of Flamsteed's catalogue, Baily listed this star's magnitude as 4, the magnitude Flamsteed himself assigned to 1 Eri.

Coma Berenices, Com

Berenice's Hair

Although Coma Berenices (Figure 86) was devised in classical times and mentioned in Ptolemy's catalogue, Tycho was the first astronomer to consider it a separate, independent constellation. Hevelius and Flamsteed included it in their catalogues, and the latter numbered its stars from 1 to 43.

Table 109. The numbered stars of Coma Berenices

		MAGNITUDES		CATALOGUE NUMBERS	
Flamsteed Number	Letter	Flamsteed	Visual	HR	HD
1+		7	6.58		104452
2		6	5.87	4602	104827
3		6	6.39	4632	105778
4		6	5.66	4640	105981
5		6	5.57	4643	106057
6		5	5.10	4663	106661
7		4½	4.95	4667	106714
8		7	6.27	4685	107168
9		6	6.33	4688	107213
10+		6	6.67		107276
11		4½	4.74	4697	107383
12		5	4.81	4707	107700
13		4½	5.18	4717	107966
14		4½	4.95	4733	108283
15	Gamma	4½	4.36	4737	108381
16		4½	5.00	4738	108382
17		4½	5.29	4752	108662
18		5	5.48	4753	108722
19+		6			
20		6	5.69	4756	108765
21		5	5.46	4766	108945
22		7	6.29	4780	109307
23		4	4.81	4789	109485
24		5	4.94c	4791/2	109510/1
25		6	5.68	4801	109742
26		5	5.46	4815	110024

(table continued on next page)

Table 109. The numbered stars of Coma Berenices (continued)

Flamsteed Number	Letter	Magnitudes		Catalogue Numbers	
		Flamsteed	Visual	HR	HD
27		5	5.12	4851	111067
28		6	6.56	4861	111308
29+		6	5.70	4865	111397
30		6	5.78	4869	111469
31		5½	4.94	4883	111812
32		7	6.32	4884	111862
33+		7	6.92		111892
34+		5			
35		4½	4.90	4894	112033
36		5	4.78	4920	112769
37		5½	4.90	4924	112989
38		6	5.96	4929	113095
39		5	5.99	4946	113848
40		6	5.60	4949	113866
41		4¾	4.80	4954	113996
42	Alpha	4½	4.32c	4968/9	114378/9
43	Beta	5½	4.26	4983	114710

The Lost, Missing, or Troublesome Stars of Coma Berenices

1, HD 104452, 6.58V, *SA,* GC 16442*.

10, HD 107276, 6.67V, *SA, BS Suppl.,* BF 1702*, BAC 4147*, H 13.

19

This star does not exist. Flamsteed observed 18 Com on several occasions, but while observing it on April 8, 1692, he erred 3° in copying its zenith distance. This produced a new but erroneous set of coordinates that led him to believe he had seen a new star, 19 Com. See BF 1729 and Table B, p. 645.

29, HR 4865, 5.70V, *SA, HRP, BS,* BF 1770 and Table A, p. 645.

This star is the same as 36 Vir. It is one of Flamsteed's twenty-two duplicate stars.

33, HD 111892, 6.92V, *SA,* GC 17466*.

34

This star does not exist. Flamsteed observed 35 Com on several occasions, but while reducing his observations of March 9, 1692, he erred 6° in determining its declination. This generated a new set of coordinates that led him to believe he had seen a new star, 34 Com. See BF 1782 and Table B, p. 645.

Corona Borealis, CrB

Northern Crown

Corona Borealis (Figure 91) is a Ptolemaic constellation that Flamsteed numbered from 1 to 21. All its numbered stars are listed in the *BS,* and there are no missing or problem numbered stars in the constellation.

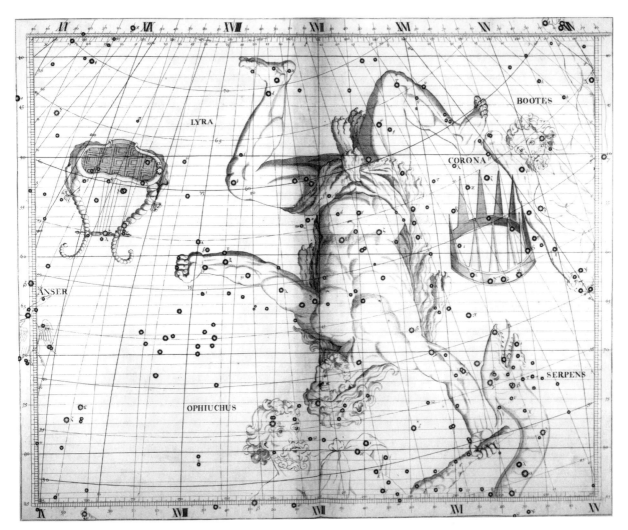

Figure 91. Flamsteed's Hercules, Corona Borealis, and Lyra. The following are nonexistent stars in Hercules: **55 Her,** 250°29' r.a., 71° 4' n.p.d.; **71 Her,** 257° 6' r.a., 65°10' n.p.d.; **80 Her,** 261°32' r.a., 42°25' n.p.d.; **81 Her,** 261°34' r.a., 42°26' n.p.d.

Table 110. The numbered stars of Corona Borealis

		MAGNITUDES		CATALOGUE NUMBERS	
Flamsteed Number	**Letter**	**Flamsteed**	**Visual**	**HR**	**HD**
1	Omicron	6	5.51	5709	136512
2	Eta	5	4.98c	5727/8	137107/8
3	Beta	4	3.68	5747	137909
4	Theta	4½	4.14	5778	138749
5	Alpha	2½	2.23	5793	139006
6	Mu	5	5.11	5800	139153
7	Zeta[1,2]	4	4.69c	5833/4	139891/2
8	Gamma	4	3.84	5849	140436
9	Pi	5	5.56	5855	140716
10	Delta	4	4.63	5889	141714
11	Kappa	5	4.82	5901	142091
12	Lambda	5	5.45	5936	142908
13	Epsilon	4½	4.15	5947	143107
14	Iota	5½	4.99	5971	143807
15	Rho	6	5.41	5968	143761
16	Tau	6	4.76	6018	145328
17	Sigma	6	5.22c	6063/4	146361/2
18	Upsilon	6	5.78	6074	146738
19	Xi	5	4.85	6103	147677
20	Nu[1]	5	5.20	6107	147749
21	Nu[2]	5	5.39	6108	147767

Corvus, Crv

Raven, Crow

Corvus (figures 92, 92a, and 105) is a Ptolemaic constellation that Flamsteed numbered from 1 to 9. All its numbered stars are listed in the *BS,* and there are no missing or problem numbered stars in the constellation.

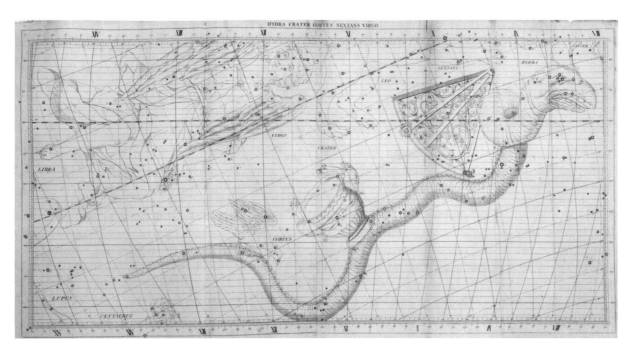

Figure 92. Flamsteed's Hydra, Crater, Corvus, Sextans, and Virgo. The following are nonexistent stars: **8 Hya,** 126°49' r.a., 96° 3' n.p.d.; **36 Hya,** 141°10' r.a., 98° 3' n.p.d.; **28 Sex,** 152°52' r.a., 91°10' n.p.d.

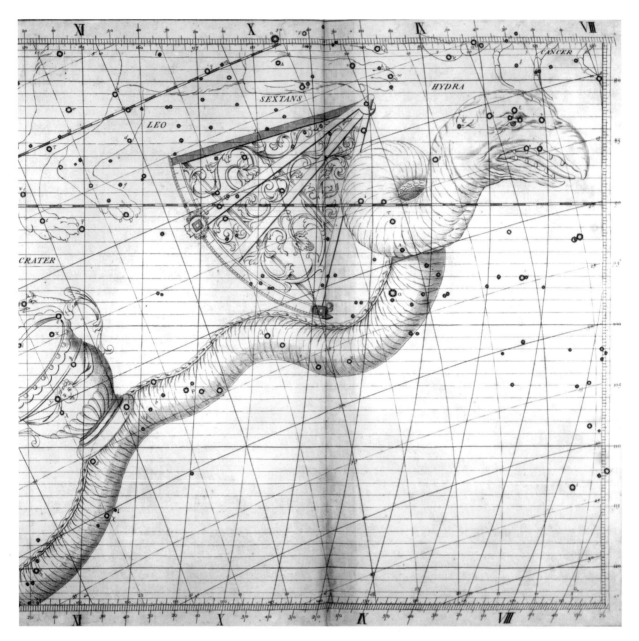

Figure 92a. Flamsteed's Hydra, Crater, Corvus, Sextans, and Virgo from a defective copy of the *Atlas Coelestis* at the United States Naval Observatory. In examining Flamsteed's atlas, several scholars (see, for example, Warner's *Sky Explored,* p. 82) have noted that this chart, plate 14, is incomplete, showing only Hydra's head and omitting her tail. But this is incorrect. Flamsteed's chart of Hydra is indeed complete. When preparing this chart for publication, the printers observed that it would not fit on one page. Hydra is the longest constellation in the heavens, extending from head to tail over 7 hours of right ascension or 30% of the celestial globe. Consequently, it was decided to attach an additional sheet of paper to the left side of the page in order to accommodate the Water Snake's hind quarters. Unfortunately, this additional sheet, for one reason or another, is missing from several copies of the atlas, leaving the erroneous impression that the capstone of Flamsteed's life work, his *Atlas Coelestis,* is somehow incomplete. There are other less significant differences among the various copies of the *Atlas Coelestis;* for example, the positioning by hand of the plate numbers — on some copies in the upper right corner, on other copies on the back of the previous page.

Table 111. The numbered stars of Corvus

Flamsteed Number	Letter	Magnitudes		Catalogue Numbers	
		Flamsteed	Visual	HR	HD
1	Alpha	4	4.02	4623	105452
2	Epsilon	4	3.00	4630	105707
3		6	5.46	4635	105850
4	Gamma	3	2.59	4662	106625
5	Zeta	5	5.21	4696	107348
6		6	5.68	4711	107815
7	Delta	3	2.95	4757	108767
8	Eta	5	4.31	4775	109085
9	Beta	3	2.65	4786	109379

CRATER, CRT

Cup, Bowl

Crater (figures 92 and 105) is a Ptolemaic constellation that Flamsteed numbered from 1 to 31. Flamsteed called this constellation Hydra and Crater and included within it not only the stars in Crater but those in Hydra that are directly below Crater. Most astronomers, however, have refused to recognize Hydra and Crater as a new constellation and they continue to keep the stars of each constellation separate and distinct. See Hydra.

Table 112. The numbered stars of Crater

		MAGNITUDES		CATALOGUE NUMBERS	
Flamsteed Number	Letter	Flamsteed	Visual	HR	HD
1+	Phi² Hya	6	6.03	4156	91880
2+	Phi³ Hya	5	4.91	4171	92214
3+	b¹ Hya	6	5.42	4214	93397
4+	Nu Hya	4	3.11	4232	93813
5+	b² Hya	6	6.6		94046
6+	b³ Hya	6	5.24	4251	94388
7	Alpha	4	4.08	4287	95272
8+		6	6.23	4302	95678
9+	Chi¹ Hya	5	4.94	4314	96202
10+		6	5.44	4334	96819
11	Beta	3½	4.48	4343	97277
12	Delta	4	3.56	4382	98430
13	Lambda	5½	5.09	4395	98991
14	Epsilon	4	4.83	4402	99167
15	Gamma	4	4.08	4405	99211
16	Kappa	5	5.94	4416	99564
17+	N Hya	6	5.00c	4443/4	100286/7
18+		6	5.04	4449	100393
19+	Xi Hya	4	3.54	4450	100407
20+		6	5.98	4458	100623
21	Theta	4	4.70	4468	100889
22+		7	5.74	4469	100893
23+		6	6.29	4473	100953
24	Iota	5	5.48	4488	101198
25+	Omicron Hya	5	4.70	4494	101431

(table continued on next page)

Table 112. The numbered stars of Crater (continued)

Flamsteed Number	Letter	Magnitudes Flamsteed	Visual	Catalogue Numbers HR	HD
26+		6	5.22	4503	101666
27	Zeta	4	4.73	4514	102070
28+	Beta Hya	4	4.28	4552	103192
29+		6	5.93	4565	103596
30	Eta	4	5.18	4567	103632
31+		5½	5.26	4590	104337

The Lost, Missing, or Troublesome Stars of Crater

1, in Hydra as Phi² Hya, HR 4156, 6.03V, *HRP,* BF 1513*.

2, in Hydra as Phi³ Hya, HR 4171, 4.91V, *HRP,* BF 1517*.

3, in Hydra as b¹ Hya, HR 4214, 5.42V, *HRP,* BF 1537*.

4, in Hydra as Nu Hya, HR 4232, 3.11V, *HRP,* BF 1542*.

5, in Hydra as b² Hya, HD 94046, 6.6V, BF 1547*, BAC 3722, P X, 176*, T 4884, H 113 in Hydra.

This star's Flamsteed number is often omitted from atlases and catalogues because no star exists at the coordinates listed in Flamsteed's catalogue. On examining Flamsteed's manuscripts, Baily discovered that Flamsteed had erred 9'15" in calculating this star's declination. Baily corrected the error in his revised edition of Flamsteed's catalogue.

6, in Hydra as b³ Hya, HR 4251, 5.24V, *HRP,* BF 1553*.

8, in Hydra, HR 4302, 6.23V, *HRP,* BF 1570*.

9, in Hydra as Chi¹ Hya, HR 4314, 4.94V, *HRP,* BF 1576*.

10, in Hydra, HR 4334, 5.44V, *HRP,* BF 1582*.

17, in Hydra as N Hya, HR 4443/4, 5.76V, 5.64V, *SA, HRP, BS.*

18, in Hydra, HR 4449, 5.04, *HRP,* BF 1626*.

19, in Hydra as Xi Hya, HR 4450, 3.54V, *HRP,* BF 1627*.

20, in Hydra, HR 4458, 5.98V, *HRP,* BF 1630*.

22, in Hydra, HR 4469, 5.74V, *HRP,* BF 1633*.

23, in Hydra, HR 4473, 6.29V, *HRP,* BF 1635*.

25, in Hydra as Omicron Hya, HR 4494, 4.70V, *HRP,* BF 1643*.

26, in Hydra, HR 4503, 5.22V, *HRP,* BF 1646*.

28, in Hydra as Beta Hya, HR 4552, 4.28V, *HRP,* BF 1660*.

29, in Hydra, HR 4565, 5.93V, *HRP,* BF 1667*.

31, in Corvus, HR 4590, 5.26V, *SA, HRP, BS,* BF 1671*, GC 16423*.

Cygnus, Cyg

Swan

Cygnus (Figure 93) is a Ptolemaic constellation that Flamsteed numbered from 1 to 81.

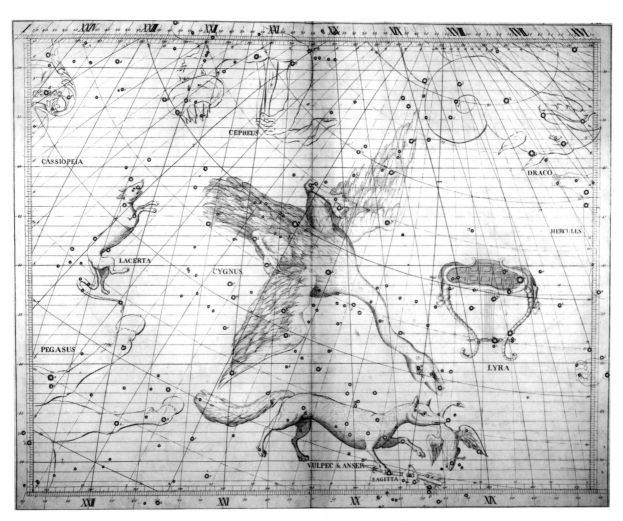

Figure 93. Flamsteed's Lyra, Cygnus, Lacerta, Vulpecula with Anser, and Sagitta. The following are nonexistent stars in Cygnus: **5 Cyg**, [289°] r.a., 60°52' n.p.d.; **38 Cyg**, [303°] r.a., 41°27' n.p.d.

Table 113. The numbered stars of Cygnus

		MAGNITUDES		CATALOGUE NUMBERS	
Flamsteed Number	Letter	Flamsteed	Visual	HR	HD
1	Kappa	4	3.77	7328	181276
2		5	4.97	7372	182568
3+		6	6.19	7386	182807
4		6	5.15	7395	183056
5+					
6	Beta[1,2]	3½	3.10c	7417/8	183912/14
7	Iota[1]	6	5.75	7408	183534
8		6	4.74	7426	184171
9+			5.38	7441	184759
10	Iota[2]	6	3.79	7420	184006
11		6	6.05	7457	185037
12	Phi	5	4.69	7478	185734
13	Theta	4	4.48	7469	185395
14		6	5.40	7483	185872
15		6	4.89	7517	186675
16	c	6	5.56c	7503/4	186408/27
17		5	4.99	7534	187013
18	Delta	3½	2.87	2528	186882
19		6	5.12	7566	187849
20	d	5½	5.03	7576	188056
21	Eta	6	3.89	7615	188947
22		6	4.94	7613	188892
23		6	5.14	7608	188665
24	Psi	5	4.92	7619	189037
25		6	5.19	7647	189687
26	e	6	5.05	7660	190147
27	b[1]	5	5.36	7689	191026
28	b[2]	5	4.93	7708	191610
29	b[3]	6	4.97	7736	192640
30	Omicron[1]	4	4.83	7730	192514
31	Omicron[1]	5	3.79	7735	192577
32	Omicron[2]	5½	3.98	7751	192909
33+			4.30	7740	192696
34	P	6	4.81	7763	193237
35		6	5.17	7770	193370
36		6	5.58	7769	193369
37	Gamma	3	2.20	7796	194093
38+					
39		6	4.43	7806	194317
40		6	5.62	7826	195050
41		4	4.01	7834	195295
42		6	5.88	7835	195324
43		5	5.69	7828	195068
44		6	6.19	7847	195593

(table continued on next page)

Table 113. The numbered stars of Cygnus *(continued)*

		Magnitudes		Catalogue Numbers	
Flamsteed Number	Letter	Flamsteed	Visual	HR	HD
45	Omega[1]	5	4.95	7844	195556
46	Omega[2]	5	5.44	7851	195774
47		6	4.61	7866	196093
48		6	6.32	7885	196606
49		6	5.51	7921	197177
50	Alpha	2	1.25	7924	197345
51		6	5.39	7929	197511
52		6	4.22	7942	197912
53	Epsilon	3	2.46	7949	197989
54	Lambda	4	4.53	7963	198183
55		6	4.84	7977	198478
56		6	5.04	7984	198639
57		6	4.78	8001	199081
58	Nu	4	3.94	8028	199629
59	f[1]	5½	4.74	8047	200120
60		6	5.37	8053	200310
61		6	4.81c	8085/6	201191/2
62	Xi	4	3.72	8079	200905
63	f[2]	6	4.55	8089	201251
64	Zeta	3	3.20	8115	202109
65	Tau	4	3.72	8130	202444
66	Upsilon	5	4.43	8146	202904
67	Sigma	4	4.23	8143	202850
68	A	6	5.00	8154	203064
69		6	5.94	8209	204172
70		6	5.31	8215	204403
71	g	6	5.24	8228	204771
72		6	4.90	8255	205512
73	Rho	4	4.02	8252	205435
74		6	5.01	8266	205835
75		6	5.11	8284	206330
76		6	6.11	8291	206538
77		6	5.69	8300	206644
78	Mu[1,2]	3½	4.50c	8309/10	206826/7
79		6	5.65	8307	206774
80	Pi[1]	4	4.67	8301	206672
81	Pi[2]	5	4.23	8335	207330

The Lost, Missing, or Troublesome Stars of Cygnus

3, in Vulpecula, HR 7386, 6.19V, *HRP,* BF 2629*.

5

This star does not exist. Flamsteed observed 2 Cyg on four separate occasions, but on September 5, 1690, while copying the results of his night's work, he made a 6' error in this star's declination. This produced a new but erroneous set of coordinates that led him to believe he had observed a new star, 5 Cyg, which was listed without its right ascension in his catalogue. See BF 2626 and Table B, p. 645.

9, HR 7441, 5.38V, *SA, HRP, BS,* BF 2658*.

Only this star's declination is included in Flamsteed's "Catalogus Britannicus" of 1725, but Baily was able to reconstruct 9's coordinates and its magnitude, 6, from Flamsteed's observational notes.

33, HR 7740, 4.30V, *SA, HRP, BS,* BF 2769*.

As with 9, only this star's declination is included in Flamsteed's catalogue. Once again, Baily was able to reconstruct 33's coordinates and its magnitude, 5, from Flamsteed's manuscripts.

38

This star does not exist. Flamsteed observed 43 Cyg twice, September 24 and 25, 1690, but he erred in determining the declination of his second observation. This generated a new set of coordinates that led him to believe he had seen a new star, 38 Cyg, which was listed without its right ascension in his catalogue. See BF 2792 and Table B, p. 645.

DELPHINUS, DEL

Dolphin

Delphinus (Figure 83) is a Ptolemaic constellation that Flamsteed numbered from 1 to 18. All its numbered stars are listed in the *BS*, and there are no missing or problem numbered stars in the constellation.

Table 114. The numbered stars of Delphinus

		MAGNITUDES		CATALOGUE NUMBERS	
Flamsteed Number	Letter	Flamsteed	Visual	HR	HD
1		6	6.08	7836	195325
2	Epsilon	3	4.03	7852	195810
3	Eta	6	5.38	7858	195943
4	Zeta	5	4.68	7871	196180
5	Iota	6	5.43	7883	196544
6	Beta	3	3.63	7882	196524
7	Kappa	6	5.05	7896	196755
8	Theta	6	5.72	7892	196725
9	Alpha	3	3.77	7906	196867
10		6	5.99	7918	197121
11	Delta	3½	4.43	7928	197461
12	Gamma[1,2]	3	4.12c	7947/8	197963/4
13		5	5.58	7953	198069
14		6	6.33	7974	198391
15		6	5.98	7973	198390
16		6	5.58	8012	199254
17		6	5.17	8011	199253
18		6	5.48	8030	199665

Draco, Dra

Dragon

Draco (Figure 89) is a Ptolemaic constellation that Flamsteed numbered from 1 to 80.

Table 115. The numbered stars of Draco

		Magnitudes		Catalogue Numbers	
Flamsteed Number	Letter	Flamsteed	Visual	HR	HD
1	Lambda	3½	3.84	4434	100029
2		6	5.20	4461	100696
3		6	5.30	4504	101673
4		6	4.95	4765	108907
5	Kappa	3	3.87	4787	109387
6		6	4.94	4795	109551
7		6	5.43	4863	111335
8		6	5.24	4916	112429
9		6	5.32	4928	113092
10	i	5	4.65	5226	121130
11	Alpha	2	3.65	5291	123299
12	Iota	3	3.29	5744	137759
13	Theta	3	4.01	5986	144284
14	Eta	3	2.74	6132	148387
15	A	4	5.00	6161	149212
16		5	5.53	6184	150100
17		5	5.08c	6185/6	150117/8
18	g	5	4.83	6223	151101
19	h	5	4.89	6315	153597
20		6	6.42	6319	153697
21	Mu	4¾	4.92c	6369/70	154905/6
22	Zeta	2	3.17	6396	155763
23	Beta	2½	2.79	6536	159181
24	Nu^1	4	4.88	6554	159541
25	Nu^2	4	4.87	6555	159560
26		6	5.23	6573	160269
27	f	5	5.05	6566	159966
28	Omega	4	4.80	6596	160922
29+		6	6.55		160538

(table continued on next page)

Table 115. The numbered stars of Draco *(continued)*

		MAGNITUDES		CATALOGUE NUMBERS	
Flamsteed Number	Letter	Flamsteed	Visual	HR	HD
30		6	5.02	6656	162579
31+	Psi[1]	7	4.58c	6636/7	162003/4
32	Xi	3	3.75	6688	163588
33	Gamma	2	2.23	6705	164058
34	Psi[2]	4½	5.48	6725	164613
35		6	5.04	6701	163989
36		6	5.03	6850	168151
37		6	5.95	6865	168653
38+		6	6.79		169027
39	b	5	4.98	6923	170073
40		5	6.04	6809	166865
41		5	5.68	6810	166866
42		6	4.82	6945	170693
43	Phi	5	4.22	6920	170000
44	Chi	4	3.57	6927	170153
45	d	5	4.77	6978	171635
46	c	5	5.04	7049	173524
47	Omicron	4	4.66	7125	175306
48		6	5.66	7175	176408
49		6	5.48	7218	177249
50		4½	5.35	7124	175286
51		5½	5.38	7251	178207
52	Upsilon	4½	4.82	7180	176524
53		5	5.12	7295	180006
54		5	4.99	7309	180610
55		6	6.16	7290	179933
56+		6			
57	Delta	3½	3.07	7310	180711
58	Pi	4	4.59	7371	182564
59		6	5.13	7312	180777
60	Tau	4½	4.45	7352	181984
61	Sigma	4½	4.68	7462	185144
62+		6			
63	Epsilon	5½	3.83	7582	188119
64	e	5½	5.27	7676	190544
65		6½	6.57	7682	190713
66		6	5.39	7701	191277
67	Rho	5	4.51	7685	190940
68		6	5.75	7727	192455
69		6	6.20	7686	190960
70+		6			
71		6	5.72	7792	193964
72+		6	8.5		194666
73		5½	5.20	7879	196502
74		6	5.96	7908	196925

(table continued on next page)

Table 115. The numbered stars of Draco *(continued)*

Flamsteed Number	Letter	Magnitudes		Catalogue Numbers	
		Flamsteed	Visual	HR	HD
75		6	5.46	7901	196787
76		5	5.68	8002	199095
77+		5	5.91	8112	201908
78+		5	5.17	8324	207130
79+		7	6.64		208509
80+		6	6.37	8473	210873

The Lost, Missing, or Troublesome Stars of Draco

29, HD 160538, 6.55V, *SA, BS Suppl.*, GC 23865*.

31, Psi¹, HR 6636/7, 4.58V, 5.79V, *SA, HRP, BS,* BF 2458*, ADS 10759.

This is a multi-star system. HR 6636 is component A and HR 6637 is component B. There are, in addition, two smaller components, C and D, with magnitudes of 11.4V and 12.9V, respectively. Flamsteed's catalogue lists 31 as a 7th-magnitude star, but on the three occasions Flamsteed observed it, he noted its magnitude as 4½, 3½, and 5. Did he perhaps split the pair and observe its two major components?

38, HD 169027, 6.79V, *SA, BS Suppl.*, GC 25008*.

56

This star does not exist. Flamsteed made several observations of 59 Dra, but during one of his observations he erred in recording its time of transit by more than 7^m. This resulted in a new set of coordinates that led him to believe he had observed a new star, 56 Dra. See BF 2648 and Table B, p. 645.

62

This star does not exist. Flamsteed observed 31 Dra four times, but on one occasion he erred 2^h in the process of reducing its coordinates to 1690. This produced a new set of coordinates that led him to believe he had observed a new star, 62 Dra. See BF 2458 and Table B, p. 645.

70

This star does not exist. Flamsteed observed 37 Cyg ten times, but on July 1, 1703, he mistakenly recorded its sighting as north of the zenith instead of south. This generated a new but erroneous set of coordinates that led him to believe he had observed a new star, 70 Dra. See BF 2778 and Table B, p. 645.

72, HD 194666, 8.5V, BF 2793*, BAC 7051*, B 276*, P XX, 162*, GC 28399.

This is the dimmest star in Flamsteed's catalogue. Piazzi recorded its magnitude as 9.1 and noted that it is barely visible. It is not in *SA,* but it is shown, without its Flamsteed number and about 25' due east of 71, in Tirion's *Uranometria 2000.0.*

77, in Cepheus, HR 8112, 5.91V, *SA, HRP, BS,* GC 29563*.

78, in Cepheus, HR 8324, 5.17V, *SA, HRP, BS,* GC 30452*.

79, in Cepheus, HD 208509, 6.64V, *SA,* GC 30669*.

80, in Cepheus, HR 8473, 6.37V, *HRP,* BF 3052*.

Equuleus, Equ

Little Horse

Equuleus (Figure 94) is a Ptolemaic constellation that Flamsteed numbered from 1 to 10.

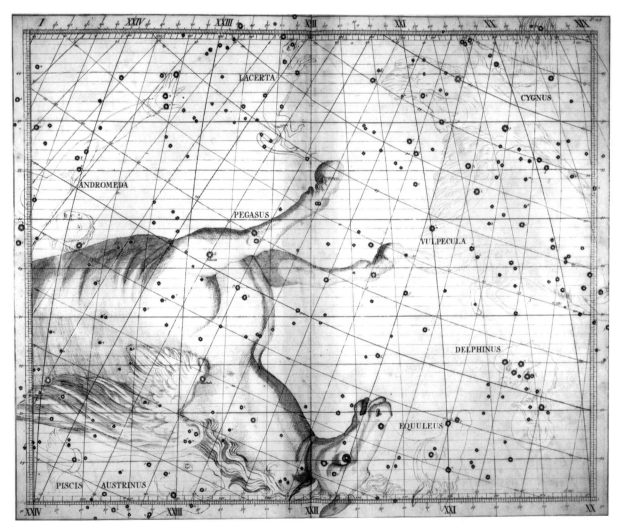

Figure 94. Flamsteed's Pegasus and Equuleus. Below Pegasus is the more westerly of the two Fishes, erroneously marked on the chart as Piscis Austrinus, the Southern Fish.

Table 116. The numbered stars of Equuleus

Flamsteed Number	Letter	Magnitudes		Catalogue Numbers	
		Flamsteed	Visual	HR	HD
1	Epsilon	5	5.23	8034	199766
2+	Lambda	6	6.64		200256
3		6	5.61	8066	200644
4		6	5.94	8077	200790
5	Gamma	4	4.69	8097	201601
6		6	6.07	8098	201616
7	Delta	4	4.49	8123	202275
8	Alpha	4	3.92	8131	202447
9		6	5.82	8163	203291
10	Beta	4	5.16	8178	203562

The Lost, Missing, or Troublesome Stars of Equuleus

2, Lambda, HD 200256, 6.64V, *SA,* GC 29361*.

Eridanus, Eri

River Eridanus

Eridanus (Figure 95) is a Ptolemaic constellation that Flamsteed numbered from 1 to 69.

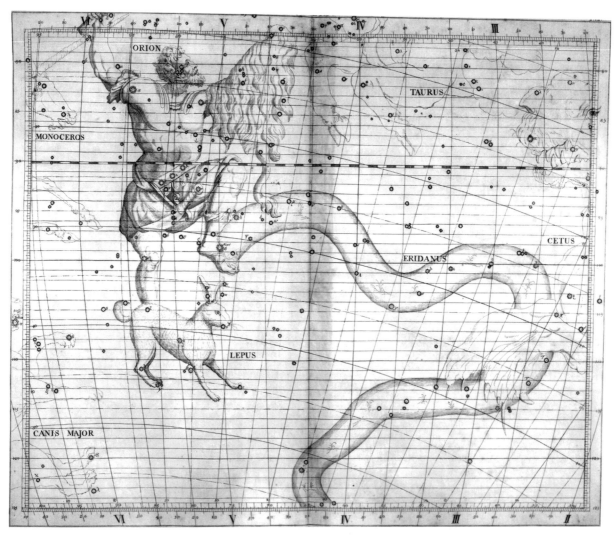

Figure 95. Flamsteed's Eridanus, Orion, and Lepus. The upper portion of Orion is in Figure 103; the rest of the River is in Figure 107. The following are nonexistent stars: **31 Eri,** 54°36' r.a., 96°14' n.p.d.; **12 Ori,** 72°39' r.a., 81° 7' n.p.d.; **26 Ori,** 77°10' r.a., 88°49' n.p.d.; **65 Ori,** 87° 8' r.a., 69°54' n.p.d.; **76 Ori,** 91°53' r.a., 84°37' n.p.d.

Table 117. The numbered stars of Eridanus

		MAGNITUDES		CATALOGUE NUMBERS	
Flamsteed Number	Letter	Flamsteed	Visual	HR	HD
1	Tau¹	4	4.47	818	17206
2	Tau²	4	4.75	850	17824
3	Eta	3	3.89	874	18322
4		6	5.45	883	18454
5		6	5.56	899	18633
6		6	5.84	889	18535
7		6	6.11	904	18760
8	Rho¹	6	5.75	907	18784
9	Rho²	5	5.32	917	18953
10	Rho³	4	5.26	925	19107
11	Tau³	3½	4.09	919	18978
12+	Alpha For	3	3.87	963	20010
13	Zeta	3	4.80	984	20320
14		6	6.14	988	20395
15		6	4.88	994	20610
16	Tau⁴	4	3.69	1003	20720
17	v	4½	4.73	1070	21790
18	Epsilon	3½	3.73	1084	22049
19	Tau⁵	4	4.27	1088	22203
20		5½	5.23	1100	22470
21		6	5.96	1111	22713
22		5½	5.53	1121	22920
23	Delta	3½	3.54	1136	23249
24		5	5.25	1146	23363
25		6	5.55	1150	23413
26	Pi	4	4.42	1162	23614
27	Tau⁶	4	4.23	1173	23754
28	Tau⁷	5¾	5.24	1181	23878
29+		6½	7.1		24371
30		5½	5.48	1202	24388
31+		5½			
32	w	4½	4.68c	1211/2	24554/5
33	Tau⁸	4½	4.65	1213	24587
34	Gamma	2	2.95	1231	25025
35		5	5.28	1244	25340
36	Tau⁹	4	4.66	1240	25267
37		6	5.44	1290	26409
38	Omicron¹	3½	4.04	1298	26574
39	A	5	4.87	1318	26846
40	Omicron²	5	4.43	1325	26965
41	Upsilon⁴	3¾	3.56	1347	27376
42	Xi	3¾	5.17	1383	27861
43	Upsilon³	5	3.96	1393	28028
44+		5½	5.55	1415	28375

(table continued on next page)

Table 117. The numbered stars of Eridanus (continued)

Flamsteed Number	Letter	Magnitudes		Catalogue Numbers	
		Flamsteed	Visual	HR	HD
45		5½	4.91	1437	28749
46		5	5.72	1449	29009
47		4	5.11	1451	29064
48	Nu	4	3.93	1463	29248
49+		5½	5.31	1469	29335
50	Upsilon¹	4	4.51	1453	29085
51	c	4	5.23	1474	29391
52	Upsilon²	3	3.82	1464	29291
53	l	3½	3.87	1481	29503
54		3½	4.32	1496	29755
55		6	5.96c	1505/6	30020/1
56		6	5.90	1508	30076
57	Mu	4	4.02	1520	30211
58		5½	5.51	1532	30495
59		6	5.77	1538	30606
60		6	5.03	1549	30814
61	Omega	5	4.39	1560	31109
62	b	6	5.51	1582	31512
63		6	5.38	1608	32008
64		6	4.79	1611	32045
65	Psi	5	4.81	1617	32249
66		6	5.12	1657	32964
67	Beta	3	2.79	1666	33111
68		6	5.12	1673	33256
69	Lambda	4	4.27	1679	33328

The Lost, Missing, or Troublesome Stars of Eridanus

12, in Fornax as Alpha For, HR 963, 3.87V, *HRP, BS,* BF 411*.

Lacaille shifted part of Eridanus to Fornax and lettered this star Alpha For. See Alpha For in Part One.

29, HD 24371, 7.1V, BF 485*, BAC 1209*, T 1327*, G 143*.

This star is properly identified in the second edition of *SA*.

31

This star does not exist. Flamsteed observed 30 Eri twice, but on his second observation he made two errors in reduction — 1ᵐ in right ascension and 5' in declination. This produced a new but erroneous set of coordinates that led him to believe he had seen a new star, 31 Eri. See BF 486 and Table B, p. 645.

44, in Taurus, HR 1415, 5.55V, *SA, HRP, BS,* GC 5441*.

49, in Taurus, HR 1469, 5.31V, *SA, HRP, BS,* GC 5627*.

Gemini, Gem

Twins

Gemini (Figure 96) is a Ptolemaic constellation that Flamsteed numbered from 1 to 85.

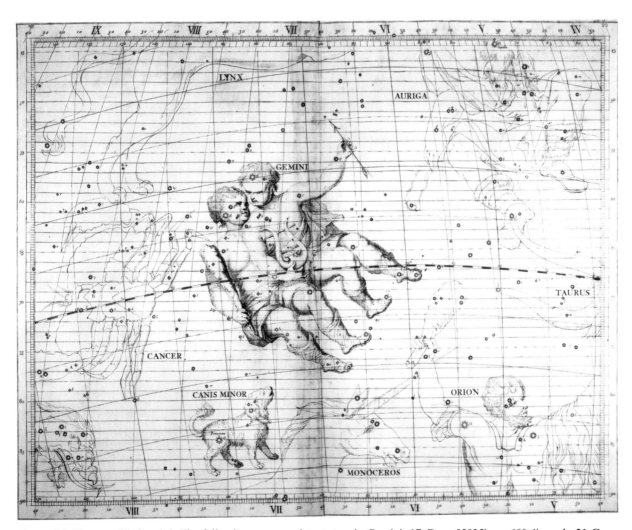

Figure 96. Flamsteed's Gemini. The following are nonexistent stars in Gemini: **17 Gem,** 92°35' r.a., 69° 4' n.p.d.; **21 Gem,** 93°49' r.a., 72° 2' n.p.d.; **29 Gem,** 96°23' r.a., 60°47' n.p.d.; **50 Gem,** 103°38' r.a., 74°22' n.p.d.; **72 Gem,** 110° 2' r.a., 61°18' n.p.d.; **73 Gem,** 110° 9' r.a., 61°31' n.p.d.

Table 118. The numbered stars of Gemini

		MAGNITUDES		CATALOGUE NUMBERS	
Flamsteed Number	**Letter**	**Flamsteed**	**Visual**	**HR**	**HD**
1		5	4.16	2134	41116
2+		8	6.72		41543
3		8	5.75	2173	42087
4+		7	6.82		42216
5		7	5.80	2185	42398
6		7	6.39	2197	42543
7	Eta	4½	3.28	2216	42995
8		7½	6.08	2230	43261
9		7	6.25	2240	43384
10+		8	6.5		43740
11+		8	6.92		43818
12+		8	6.95		43836
13	Mu	3	2.88	2286	44478
14+		7½	6.53		44974
15+		7	6.5		45352
16		7	6.22	2330	45394
17+		8			
18	Nu	4	4.15	2343	45542
19		7	6.40	2371	46031
20+		7½	6.28		46136
21+		6½			
22+		7	6.9		46516
23+		5	6.69		46781
24	Gamma	2½	1.93	2421	47105
25		7	6.42	2453	47731
26		5	5.21	2466	48097
27	Epsilon	3	2.98	2473	48329
28		6	5.44	2480	48450
29+		6½			
30		6	4.49	2478	48433
31	Xi	4¾	3.36	2484	48737
32		6	6.46	2489	48843
33		6½	5.85	2519	49606
34	Theta	4	3.60	2540	50019
35		6	5.65	2525	49738
36	d	6	5.27	2529	49908
37		6	5.73	2569	50692
38	e	6	4.65	2564	50635
39		6½	6.10	2601	51530
40		6	6.40	2605	51688
41		6	5.68	2615	52005
42	Omega	6	5.18	2630	52497
43	Zeta	3½	3.79	2650	52973

(table continued on next page)

Table 118. The numbered stars of Gemini *(continued)*

		MAGNITUDES		CATALOGUE NUMBERS	
Flamsteed Number	Letter	Flamsteed	Visual	HR	HD
44		6½	5.93	2659	53257
45		6	5.44	2684	54131
46	Tau	5	4.41	2697	54719
47		6	5.62	2700	54801
48		6	5.85	2706	55052
49+		7	7.07		55156
50+		6			
51		5¾	5.00	2717	55383
52		6¾	5.82	2725	55621
53		6½	5.71	2738	55870
54	Lambda	5	3.58	2763	56537
55	Delta	3	3.53	2777	56986
56		6½	5.10	2795	57423
57	A	5½	5.03	2808	57727
58		7½	6.02	2810	57744
59		6	5.76	2816	57927
60	Iota	4½	3.79	2821	58207
61		6	5.93	2837	58579
62	Rho	5	4.18	2852	58946
63		6	5.22	2846	58728
64	b¹	6	5.05	2857	59037
65	b²	6	5.01	2861	59148
66	Alpha	1	1.57c	2890/1	60178/9
67+		7½	6.6		60081
68		6	5.25	2886	60107
69	Upsilon	5	4.06	2905	60522
70		5	5.56	2924	60986
71	Omicron	5	4.90	2930	61110
72+		6			
73+		6½			
74	f	6	5.05	2938	61338
75	Sigma	5	4.28	2973	62044
76	c	6	5.31	2983	62285
77	Kappa	4½	3.57	2985	62345
78	Beta	2	1.14	2990	62509
79		7	6.33	2991	62510
80	Pi	5	5.14	3013	62898
81	g	6	4.88	3003	62721
82		6	6.18	3021	63208
83	Phi	5	4.97	3067	64145
84+		5	7.1		64092
85		6	5.35	3086	64648

The Lost, Missing, or Troublesome Stars of Gemini

2, HD 41543, 6.72V, *SA, BS Suppl.*, GC 7739*.

4, HD 42216, 6.82V, *SA, BS Suppl.*

10, HD 43740, 6.5V, *SA*, GC 8092*.

11, HD 43818, 6.92V, *SA, BS Suppl.*, GC 8104*.

12, HD 43836, 6.95V, *SA, BS Suppl.*

14, HD 44974, 6.53V, *SA*, GC 8296*.

15, HD 45352, 6.5V, *SA*, GC 8359*.

17

This star does not exist. Flamsteed observed 15 Gem on several occasions, but in reducing his observation of February 7, 1690, he made errors in right ascension and declination. These errors generated a new set of coordinates that led him to believe he had seen a new star, 17 Gem. See BF 895 and Table B, p. 645.

20, HD 46136, 6.28V, *SA, BS Suppl.*, ADS 5166.

See 21.

21, BF 913* and Table C, p. 646.

Flamsteed observed this star on February 19, 1696, but there is no star located at its coordinates. Baily noted that some astronomers believe that Flamsteed erred 1^m in recording its transit. If this did occur — he is quick to point out there is nothing in Flamsteed's manuscripts to suggest it did — he believed that it would mean that Flamsteed probably observed one of the components of 20 Gem, a double star, the coordinates of which are almost exactly the same as those of 21 Gem after the adjustment is made for the 1^m error. Peters also believed that Flamsteed erred 1^m, but he suggested that Flamsteed saw the combined light of the double star since he could not possibly have split the two stars, a relatively fixed pair only about 19" apart, with his primitive telescope. Consequently, Peters believed that 21 should be eliminated from Flamsteed's catalogue and that 20 was the star that he observed ("Flamsteed's Stars," p. 75). The pair in question, Struve 916, is made up of the following components:

20, HD 46136A, 6.28V, ADS 5166A, BF 912 as 20 Gem.
21, HD 46136B, 6.95, ADS 5166B, BF 913 as 21 Gem.

22, HD 46516, 6.9V, *SA*, GC 8544*.

23, HD 46781, 6.69V, *SA*, GC 8583*.

29

This star does not exist. A computation error in one of Flamsteed's observations of 28 Gem gave rise to a set of false coordinates that, in turn, led him to believe he had observed a new star, 29 Gem. See BF 944 and Table B, p. 645.

49, HD 55156, 7.07V, *SA, BS Suppl.*, GC 9534*.

50

This star does not exist. In one of his observations of 51 Gem, Flamsteed made errors in computation in both right ascension and declination. This produced new coordinates that led him to believe he had seen a new star, 50 Gem. See BF 1011 and Table B, p. 645.

67, HD 60081, 6.6V, *SA*, GC 10103*.

72
73

These two stars do not exist. Flamsteed observed 64 and 65 Gem on February 12, 1696, but erred in recording their times of transit. This generated new sets of coordinates that led him to believe he had seen two new stars, 72 and 73. See BF 1044 and 1046 and Table B, p. 645.

84, HD 64092, 7.1V, *SA,* GC 10642*.

Hercules, Her

Hercules, Kneeler

Hercules (Figure 91) is a Ptolemaic constellation that Flamsteed numbered from 1 to 113.

Table 119. The numbered stars of Hercules

		MAGNITUDES		CATALOGUE NUMBERS	
Flamsteed Number	**Letter**	**Flamsteed**	**Visual**	**HR**	**HD**
1	Chi	6	4.62	5914	142373
2		6	5.37	5932	142780
3+		5	5.83	5963	143553
4		6	5.75	5938	142926
5+	r	3	5.12	5966	143666
6	Upsilon	5	4.76	5982	144206
7	Kappa	5	5.02c	6008/9	145001/00
8		5½	6.14	6013	145122
9		6	5.48	6047	145892
10		5	5.70	6039	145713
11	Phi	6	4.26	6023	145389
12+		6	6.8		146169
13+		5½	7.4		146279
14+		7	6.67		145675
15+		6	7.4		146452
16		6	5.69	6065	146388
17+		6	6.56		146604
18+		7	7.1		146915
19+		6	6.68		147025
20	Gamma	3	3.75	6095	147547
21	o	6	5.85	6111	147869
22	Tau	4	3.89	6092	147394
23+		5	6.40	6110	147835
24	Omega	6	4.57	6117	148112
25		5	5.52	6123	148283
26+		6¾	6.9		148616
27	Beta	3	2.77	6148	148856
28	n	6	5.63	6158	149121

(table continued on next page)

Table 119. The numbered stars of Hercules *(continued)*

		MAGNITUDES		CATALOGUE NUMBERS	
Flamsteed Number	Letter	Flamsteed	Visual	HR	HD
29	h	4	4.84	6159	149161
30	g	5	5.04	6146	148783
31+		7	7.11		149141
32+		6	6.87		149420
33+		6	7.6		149805
34		6	6.45	6156	149081
35	Sigma	4	4.20	6168	149630
36	m¹	6	6.93	6194	150379
37	m²	6	5.77	6195	150378
38+		6	6.98		150525
39		5	5.92	6213	150682
40	Zeta	3	2.81	6212	150680
41+		6	6.58		151090
42		5	4.90	6200	150450
43	i	5½	5.15	6228	151217
44	Eta	3	3.53	6220	150997
45	l	5	5.24	6234	151525
46+		7	7.20		151237
47	k	5	5.49	6250	151956
48+		6	6.58		151937
49		6	6.52	6268	152308
50		5	5.72	6258	152173
51		5	5.04	6270	152326
52		5½	4.82	6254	152107
53		5	5.32	6279	152598
54		5	5.35	6293	152879
55+		5			
56		6	6.08	6292	152863
57		6	6.65	6305	153287
58	Epsilon	3	3.92	6324	153808
59	d	6	5.25	6332	154029
60		6	4.91	6355	154494
61		6	6.69	6346	154356
62+		6	6.8		155104
63		6	6.19	6391	155514
64	Alpha	3	3.08c	6406/7	156014/5
65	Delta	4	3.14	6410	156164
66+	e Oph	6	5.03	6433	156681
67	Pi	3½	3.16	6418	156283
68	u	5	4.82	6431	156633
69	e	4½	4.65	6436	156729
70		4	5.12	6457	157198
71+		5			
72	w	6	5.39	6458	157214
73		6	5.71	6480	157728

(table continued on next page)

Lost Stars

Table 119. The numbered stars of Hercules *(continued)*

		MAGNITUDES		CATALOGUE NUMBERS	
Flamsteed Number	Letter	Flamsteed	Visual	HR	HD
74		6	5.59	6464	157325
75	Rho	4	4.17c	6484/5	157778/9
76	Lambda	4½	4.41	6526	158899
77	x	6	5.80	6509	158414
78		6½	5.62	6533	159139
79		6	5.62	6571	160181
80+		4			
81+		4			
82	y	6	5.37	6574	160290
83		7	5.52	6602	161074
84		7	5.71	6608	161239
85	Iota	4	3.80	6588	160762
86	Mu	4	3.42	6623	161797
87		6	5.12	6644	162211
88	z	6	6.68	6664	162732
89		6	5.46	6685	163506
90	f	6	5.16	6677	163217
91	Theta	4	3.86	6695	163770
92	Xi	4	3.70	6703	163993
93		5	4.67	6713	164349
94	Nu	5	4.41	6707	164136
95		4	4.42c	6729/30	164668/9
96		5	5.28	6738	164852
97		5½	6.21	6741	164900
98		5	5.06	6765	165625
99	b	5	5.04	6775	165908
100		6	5.21c	6781/2	166045/6
101		5	5.10	6794	166230
102		4½	4.36	6787	166182
103	Omicron	4	3.83	6779	166014
104	A	4½	4.97	6815	167006
105		5	5.27	6860	168532
106		5½	4.95	6868	168720
107	t	6	5.12	6877	168914
108		6	5.63	6876	168913
109		4	3.84	6895	169414
110		4½	4.19	7061	173667
111		4	4.36	7069	173880
112		5	5.48	7113	174933
113		5	4.59	7133	175492

The Lost, Missing, or Troublesome Stars of Hercules

3, in Serpens, HR 5963, 5.83V, *HRP,* BF 2190*.

This star's Flamsteed number is often omitted from atlases and catalogues because no star exists at the position listed in the "Catalogus Britannicus." In examining Flamsteed's manuscripts, Baily discovered that Flamsteed had made several errors in right ascension and declination while computing 3's coordinates. Baily corrected these errors in his revised edition of Flamsteed's catalogue.

5, r, HR 5966, 5.12V, *SA, HRP, BS,* HF 7, BF 2193*.

In Halley's 1712 edition of Flamsteed's catalogue, this star's magnitude is listed as 5 and in Flamsteed's manuscripts it is referred to as magnitude 6, but in his authorized catalogue of 1725 its magnitude is mistakenly given as 3.

12, HD 146169, 6.8V, *SA,* GC 21854*.

13, HD 146279, 7.4V, BF 2229*, BAC 5422*, P XVI, 30*, T 7533*.

This star is properly identified in the second edition of *SA*. See also p Her in Part One.

14, HD 145675, 6.67V, *SA, BS Suppl.,* GC 21761*.

15, HD 146452, 7.4V, *SA,* GC 21875*.

17, HD 146604, 6.56V, *SA,* GC 21887*.

18, HD 146915, 7.1V, BF 2240*, B 42*, P XVI, 51*, T 7561*.

The star is properly identified in the second edition of *SA*.

19, HD 147025, 6.68V, *SA,* GC 21937*.

23, in Corona Borealis, HR 6110, 6.40V, *SA, HRP, BS,* GC 22040*.

26, HD 148616, 6.9V, *SA,* BF 2266*.

31, HD 149141, 7.11V, *SA,* GC 22235*.

32, HD 149420, 6.87V, *SA,* GC 22276*.

33, HD 149805, 7.6V, BF 2283*, BAC 5553*, HF 33*, B 102*, P XVI, 129*, T 7677*, GC 22333*.

This star's Flamsteed number is often omitted from atlases and catalogues because, as Baily noted, its right ascension in Flamsteed's catalogue is about 2' in error. Baily pointed out that Flamsteed's original observations of this star are correct, as are the coordinates in Halley's pirated catalogue of 1712, but that somehow an error was made in the authorized catalogue of 1725. It is correctly identified in the second edition of *SA*.

This star, though labeled Y in *SA* and other atlases and catalogues, is no longer considered variable.

38, HD 150525, 6.98V, BF 2293*, BAC 5590*, P XVI, 156*, T 7715*.

The star is properly identified in the second edition of *SA*.

41, HD 151090, 6.58V, *SA, BS Suppl.,* GC 22536*.

46, HD 151237, 7.20V, *SA,* BF 2314*.

48, HD 151937, 6.58V, *SA, BS Suppl.,* GC 22650*.

55, BF 2335* and Table C, p. 646.

This star does not exist. Baily examined Flamsteed's manuscripts and found an imperfect observation of this star that was bracketed to the star he had observed immediately before it, 54 Her. In the "Catalogus Britannicus" of 1725, the two stars are also bracketed together. Baily and Peters ("Flamsteed's Stars," p. 82) believed, therefore, that 55 was the same as 54.

62, HD 155104, 6.8V, *SA,* GC 23137*.

66, in Ophiuchus as e Oph, HR 6433, 5.03V, *HRP,* BF 2377*.

This star's Flamsteed number is often omitted from atlases and catalogues because no star exists at the coordinates listed in his "Catalogus Britannicus." In examining Flamsteed's manuscripts, Baily discovered Flamsteed had erred 4' in copying this star's declination and had mislabeled it Omega. Baily corrected these errors in his revised edition of Flamsteed's catalogue.

71

This star does not exist. In one of his observations of 70 Her, Flamsteed erred in copying its right ascension. This produced a new but erroneous set of coordinates that led him to believe he had seen a new star, 71 Her. See BF 2384 and Table B, p. 645.

80
81

These two stars do not exist. While observing 24 and 25 Dra on July 1, 1690, Flamsteed mistakenly recorded that he saw them south of the zenith instead of north. This produced two new but erroneous sets of coordinates that led him to believe he had observed two new stars, 80 and 81 Her. See BF 2416/7 and Table B, p.645.

Hydra, Hya

Female Water Snake, Large Water Snake

A Ptolemaic constellation, Hydra (figures 92 and 92a) is one of the largest in the heavens, stretching from 8^h right ascension to 15^h, thus reaching across almost one-third of the sky. The extraordinary length of Hydra convinced Flamsteed to split the constellation. He numbered the stars in the western part from 1 to 44 and those in the east from 45 to 60. Flamsteed merged the stars of the central part of the Water Snake with the stars of the constellation directly above it, Crater, the Cup, thus creating a new constellation he called Hydra and Crater. In essence, he split Hydra into three parts. The two extremities of the Water Snake, the eastern and western parts, with a total of sixty stars, kept the name Hydra. The third part, consisting of stars of Crater and the stars of the central part of Hydra, made up the new constellation, Hydra and Crater. Since this move flew into the face of almost two thousand years of astronomical tradition, most astronomers and the IAU have rejected Flamsteed's innovation and have continued to recognize Hydra and Crater as two separate, distinct constellations. See Crater.

Table 120. The numbered stars of Hydra

		MAGNITUDES		CATALOGUE NUMBERS	
Flamsteed Number	Letter	Flamsteed	Visual	HR	HD
1		4	5.61	3297	70958
2		4	5.59	3321	71297
3		6	5.72	3398	72968
4	Delta	4	4.16	3410	73262
5	Sigma	5	4.44	3418	73471
6		6	4.98	3431	73840
7	Eta	4	4.30	3454	74280
8+		6			
9		6	4.88	3441	74137
10		5	6.13	3469	74591
11	Epsilon	4	3.38	3482	74874
12	D	6	4.32	3484	74918
13	Rho	5	4.36	3492	75137
14		5½	5.31	3500	75333
15		6	5.54	3523	75737
16	Zeta	4	3.11	3547	76294
17		6	6.08c	3552/3	76369/70
18	Omega	6	4.97	3613	77996
19		6	5.60	3630	78556

(table continued on next page)

Table 120. The numbered stars of Hydra (continued)

		MAGNITUDES		CATALOGUE NUMBERS	
Flamsteed Number	Letter	Flamsteed	Visual	HR	HD
20		6	5.46	3641	78732
21		6	6.11	3655	79193
22	Theta	4	3.88	3665	79469
23		6	5.24	3681	79910
24		6	5.47	3683	79931
25+		6	7.07		80105
26		6	4.79	3706	80499
27	P	6	4.80	3709	80586
28		6	5.59	3738	81420
29		6	6.54	3744	81728
30	Alpha	2	1.98	3748	81797
31	Tau^1	5	4.60	3759	81997
32	Tau^2	5	4.57	3787	82446
33	A	6	5.56	3814	82870
34		6	6.40	3832	83373
35	Iota	4	3.91	3845	83618
36+					
37		6	6.31	3846	83650
38	Kappa	4½	5.06	3849	83754
39	Upsilon^1	5	4.12	3903	85444
40	Upsilon^2	5	4.60	3970	87504
41	Lambda	4	3.61	3994	88284
42	Mu	4	3.81	4094	90432
43+	Phi^1	5	7.6		91369
44		6	5.08	4145	91550
45	Psi	6	4.95	4958	114149
46	Gamma	3	3.00	5020	115659
47		6	5.15	5250	121847
48		6	5.77	5257	122066
49	Pi	4	3.27	5287	123123
50		6	5.08	5312	124206
51	k	5	4.77	5381	125932
52	l	5	4.97	5407	126769
53+		6	5.73	5484	129433
54	m	5½	4.94	5497	129926
55		6	5.63	5514	130158
56		6	5.24	5516	130259
57		7	5.77	5517	130274
58	E	5	4.41	5526	130694
59		6	5.65	5577	132219
60		6½	5.85	5591	132851

The Lost, Missing, or Troublesome Stars of Hydra

8

This star does not exist. Flamsteed made an error in computation in reducing one of his observations of 15 Hya. This produced a new set of coordinates that led him to believe he had observed a new star, 8 Hya. See BF 1254 and Table B, p. 645.

25, HD 80105, 7.6V, *SA, BS Suppl.,* GC 12822*.

36

This star does not exist. In one of his observations of 34 Hya, Flamsteed made an error in computing its right ascension. This generated a new set of coordinates that led him to believe he had seen a new star, 36 Hya. See BF 1364 and Table B, p. 645.

43, Phi¹, HD 91369, 7.7V, BF 1502*, BAC 3611*, P X, 104*, CAP 1910*, T 4709*.

See Phi Hya in Part One.

53, in Libra, HR 5484, 5.73V, *HRP,* BF 1997 and Table A, p. 645.

This star is the same as 4 Lib. It is one of Flamsteed's twenty-two duplicate stars.

Lacerta, Lac

Lizard

Hevelius devised the constellation Lacerta (Figure 93). Flamsteed acknowledged that it was a Hevelian constellation and included it in his atlas and catalogue, numbering its stars from 1 to 16. All Lacerta's numbered stars are listed in the *BS,* and there are no missing or problem numbered stars in the constellation.

While many astronomers in the past have devised new constellations of all kinds and shapes, with rare exception most have long since been discarded and forgotten. The Hevelian constellations, like Lacerta, owe their survival to the fact that Flamsteed included them in his "Catalogus Britannicus," which proved to be such an outstanding scientific work that it became the standard by which all later catalogues were judged.

The Hevelian constellations are not included in Halley's 1712 pirated edition of Flamsteed's catalogue, probably because the Astronomer Royal either did not yet have a copy of Hevelius's *Prodromus Astronomiae* of 1690 or, more likely, did not have an opportunity to analyze its contents. When Flamsteed's three-volume magnum opus was finally completed, it included a revised version of Hevelius's catalogue, with all of its 1,564 stars painstakingly rearranged by constellation in order of right ascension and keyed to Bayer's lettering system.

Table 121. The numbered stars of Lacerta

Flamsteed Number	Letter	Magnitudes		Catalogue Numbers	
		Flamsteed	Visual	HR	HD
1		5	4.13	8498	211388
2		5	4.57	8523	212120
3	Beta	4½	4.43	8538	212496
4		5	4.57	8541	212593
5		4½	4.36	8572	213310
6		5	4.51	8579	213420
7	Alpha	4	3.77	8585	213558
8		6	5.73	8603	214168
9		6	4.63	8613	214454
10		6	4.88	8622	214680
11		5	4.46	8632	214868
12		6	5.25	8640	214993
13		6	5.08	8656	215373
14		6	5.92	8690	216200
15		5	4.94	8699	216397
16		6	5.59	8725	216916

Leo, Leo

Lion

Leo (Figure 97) is a Ptolemaic constellation that Flamsteed numbered from 1 to 95.

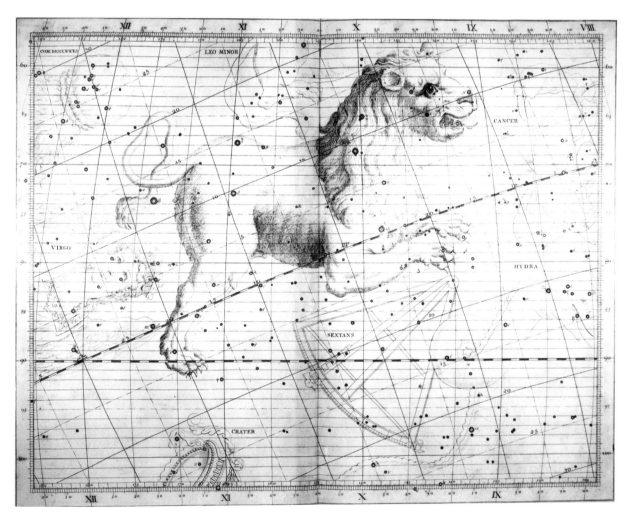

Figure 97. Flamsteed's Leo. The following are nonexistent stars in Leo: **25 Leo,** 145°15' r.a., 77°13' n.p.d.; **28 Leo,** 145°54' r.a., 80°14' n.p.d.; **38 Leo,** 150° 0' r.a., 68°45' n.p.d.

Table 122. The numbered stars of Leo

		MAGNITUDES		CATALOGUE NUMBERS	
Flamsteed Number	Letter	Flamsteed	Visual	HR	HD
1	Kappa	4	4.46	3731	81146
2	Omega	5	5.41	3754	81858
3		6	5.71	3755	81873
4	Lambda	4	4.31	3773	82308
5	Xi	4	4.97	3782	82395
6	h	6	5.07	3779	82381
7		6	6.36	3818	83023
8		6	5.69	3826	83189
9+		6	6.65		83273
10+		5	5.00	3827	83240
11+		6	6.64		83343
12+		7½	7.21		83469
13		6	6.24	3853	83821
14	Omicron	3½	3.52	3852	83808
15	f	6	5.64	3861	84107
16	Psi	6	5.35	3866	84194
17	Epsilon	3	2.98	3873	84441
18		6	5.63	3877	84561
19		7	6.45	3880	84722
20		6	6.09	3889	85040
21+		7	6.7		85259
22	g	6	5.32	3900	85376
23		6	6.46	3896	85268
24	Mu	3½	3.88	3905	85503
25+		6½			
26+		7	7.5		86359
27	Nu	4½	5.26	3937	86360
28+		7			
29	Pi	4	4.70	3950	86663
30	Eta	3½	3.52	3975	87737
31	A	5	4.37	3980	87837
32	Alpha	1	1.35	3982	87901
33+		6½	7.8		88233
34		7	6.44	3998	88355
35		6	5.97	4030	89010
36	Zeta	3	3.44	4031	89025
37		6	5.41	4035	89056
38+		6			
39		6	5.82	4039	89125
40		6	4.79	4054	89449
41	Gamma[1,2]	2	1.90c	4057/8	89484/5
42		6	6.12	4070	89774
43		6	6.07	4077	89962
44		5½	5.61	4088	90254

(table continued on next page)

Table 122. The numbered stars of Leo *(continued)*

		MAGNITUDES		CATALOGUE NUMBERS	
Flamsteed Number	Letter	Flamsteed	Visual	HR	HD
45		6	6.04	4101	90569
46	i	6	5.46	4127	91232
47	Rho	4	3.85	4133	91316
48		6	5.08	4146	91612
49		6	5.67	4148	91636
50+		6¾	6.62		92196
51	m	6	5.49	4208	93257
52	k	6	5.48	4209	93291
53	l	6	5.25	4227	93702
54		4½	4.32c	4259/60	94601/2
55		5¾	5.91	4265	94672
56		6½	5.81	4267	94705
57+		6	6.8		94738
58	d	5¾	4.84	4291	95345
59	c	5	4.99	4294	95382
60	b	5	4.42	4300	95608
61	p^2	5	4.74	4299	95578
62	p^3	6	5.95	4306	95849
63	Chi	4½	4.63	4310	96097
64		6	6.46	4322	96528
65	p^4	6	5.52	4319	96436
66+		6	6.8		96855
67		6	5.68	4332	96738
68	Delta	2½	2.56	4357	97603
69	p^5	5½	5.42	4356	97585
70	Theta	3	3.34	4359	97633
71+		6	7.03		98824
72		5	4.63	4362	97778
73	n	6	5.32	4365	97907
74	Phi	4	4.47	4368	98058
75		6	5.18	4371	98118
76		7	5.91	4381	98366
77	Sigma	4½	4.05	4386	98664
78+	Iota	4	3.94	4399	99028
79		5½	5.39	4400	99055
80		6	6.37	4410	99329
81		6	5.57	4408	99285
82+		7½	6.7		99305
83		8	6.50	4414	99491
84	Tau	4	4.95	4418	99648
85		6	5.74	4426	99902
86		6	5.52	4433	100006
87	e	4½	4.77	4432	99998
88		6	6.20	4437	100180

(table continued on next page)

Table 122. The numbered stars of Leo *(continued)*

Flamsteed Number	Letter	Magnitudes		Catalogue Numbers	
		Flamsteed	Visual	HR	HD
89		6	5.77	4455	100563
90		6	5.95	4456	100600
91	Upsilon	4	4.30	4471	100920
92		6	5.26	4495	101484
93		4	4.53	4527	102509
94	Beta	1½	2.14	4534	102647
95	o	6	5.53	4564	103578

The Lost, Missing, or Troublesome Stars of Leo

9, HD 83273, 6.65V, *SA, BS Suppl.,* GC 13291*.

10, HR 3827, 5.00V, *SA, BS,* BF 1357 and Table A, p. 645.

This star is the same as 1 Sex. It is one of Flamsteed's twenty-two duplicate stars.

11, HD 83343, 6.64V, *SA, BS Suppl.,* GC 13303*.

12, HD 83469, 7.21V, BF 1360*, BAC 3294*, B 44*, P IX, 136*, T 4230*.

The star is properly identified in the second edition of *SA*.

21, HD 85259, 6.7V, *SA,* GC 13548*.

25

This star does not exist. Baily examined Flamsteed's manuscripts, and discovered that the Astronomer Royal had observed BF 1310 — one of 458 stars that is not included in his printed catalogue — on at least three different occasions, but on one he erred 10° in reducing its right ascension to 1690, the epoch of his catalogue. This produced a new set of coordinates that led him to believe he had seen a new star, 25 Leo. BF 1310 is synonymous with BAC 3164 and HR 3689, 6.41V. See BF 1310 and Table B, p. 645.

26, HD 86359, 7.5V, *SA,* GC 13720*.

28

This star does not exist. In examining Flamsteed's manuscripts, Baily noticed that Flamsteed had erred in one of his observations of 11 Sex. Flamsteed apparently caught the mistake himself and erased the erroneous figures, but in such a way that they were mistaken for another set of figures. This generated a new set of coordinates that led him to believe he had seen a new star, 28 Leo. See BF 1411 and Table B, p. 645.

33, HD 88233, 7.8V, BF 1436*, BAC 3469*, B 115*, P IX, 256*, T 4498*.

The star is properly identified in the second edition of *SA*.

38

This star does not exist. In one of his observations of 37 Leo, Flamsteed erred 6° in computing its declination. He corrected the error in the margin of his manuscript but copied the erroneous figure into his computation book. This produced a new set of coordinates that led him to believe he had seen a new star, 38 Leo. See BF 1455 and Table B, p. 645.

50, HD 92196, 6.62V, BF 1515*, BAC 3643*, P X, 125*, T 4760*, H 79*, GC 14633*.

The star is properly identified in the second edition of *SA*.

57, HD 94738, 6.8V, *SA,* GC 15036*.

66, HD 96855, 6.8V, *SA,* GC 15336*.

71, HD 98824, 7.03V, *SA, BS Suppl.,* GC 15618*.

78, Iota, HR 4399, 3.94V, *SA, HRP.*

BS lists this star as Kappa, but it is corrected in the "Additions and Corrections," p. 164.

82, HD 99305, 6.7V, *SA,* GC 15681*.

Leo Minor, LMi

Lesser, Smaller, or Little Lion

Hevelius devised the constellation Leo Minor (Figure 98). Flamsteed acknowledged that it was a Hevelian constellation and included it in his atlas and catalogue, numbering its stars from 1 to 53.

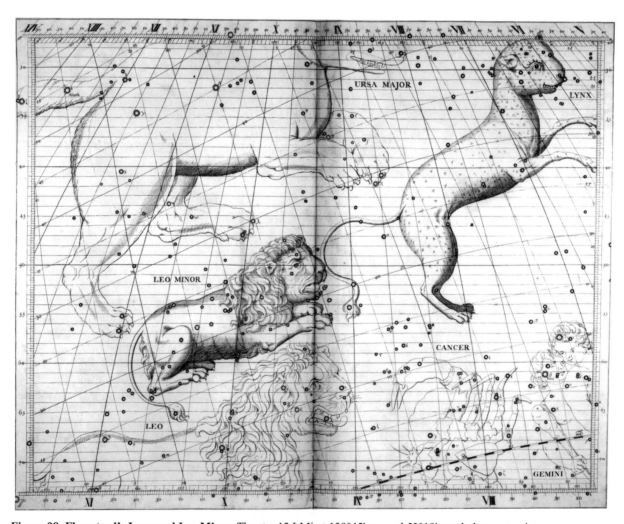

Figure 98. Flamsteed's Lynx and Leo Minor. The star 12 LMi at 139°45' r.a. and 53°19' n.p.d. does not exist.

Table 123. The numbered stars of Leo Minor

		MAGNITUDES		CATALOGUE NUMBERS	
Flamsteed Number	Letter	Flamsteed	Visual	HR	HD
1+		7	5.94	3409	73192
2+		6	6.8		73427
3+		6	6.8		73508
4+		7	6.10	3423	73596
5+		7	6.25	3499	75332
6+		6	6.41	3723	80956
7		6	5.85	3764	82087
8		5	5.37	3769	82198
9		6	6.18	3791	82522
10		4½	4.55	3800	82635
11		6	5.41	3815	82885
12+		5			
13		6	6.14	3857	83951
14+		6	6.82		84453
15+		6	5.09	3881	84737
16+		6	6.65		85029
17+		6	6.73		85373
18+		6	6.6		86012
19		5½	5.14	3928	86146
20		6	5.36	3951	86728
21		5	4.48	3974	87696
22		6½	6.46	4014	88786
23		5½	5.35	4024	88960
24		6	6.49	4027	88986
25+		6	6.79		89572
26+		6	7.1		89892
27		6	5.72	4075	89904
28		6	5.50	4081	90040
29+		6	6.52		90250
30		4¾	4.74	4090	90277
31	Beta	5	4.21	4100	90537
32		6	5.77	4113	90840
33		4½	5.90	4124	91130
34		4½	5.58	4137	91365
35		5½	6.28	4150	91752
36+		6	6.6		92000
37+		3	4.71	4166	92125
38		6	5.85	4168	92168
39+		6	6.9		92371
40		6	5.51	4189	92769
41		5	5.08	4192	92825
42		4½	5.24	4203	93152
43		6	6.15	4223	93636
44		6	6.04	4230	93765

(table continued on next page)

Lost Stars

Table 123. The numbered stars of Leo Minor *(continued)*

		MAGNITUDES		CATALOGUE NUMBERS	
Flamsteed Number	Letter	Flamsteed	Visual	HR	HD
45+		6	7.4		94218
46	o	4½	3.83	4247	94264
47+		6	5.72	4256	94497
48		6	6.20	4254	94480
49+		6	7.3		94671
50		6	6.35	4270	94747
51+		6	7.62		96094
52+		5½	6.87		96418
53+		5½	5.68	4332	96738

The Lost, Missing, or Troublesome Stars of Leo Minor

1, in Cancer, HR 3409, 5.94V, *HRP,* BF 1200*, BAC 2896.

2, in Cancer, HD 73427, 6.8V, BF 1204*, BAC 2905, B 2*, P VIII, 113*, H 45 in Cancer.
3, in Cancer, HD 73508, 6.8V, BF 1206*, BAC 2908, B 3*, P VIII, 117*, T(P) 3700*, H 45 in Cancer.
4, in Cancer, HR 3423, 6.10V, *HRP,* BF 1209*, BAC 2912, H 46 in Cancer.
5, in Lynx, HR 3499, 6.25V, *HRP,* BF 1242*, BAC 2984, H 55 in Cancer.

The Flamsteed numbers for these four stars are often omitted from atlases and catalogues because no stars exist at the coordinates listed in Flamsteed's catalogue. Baily discovered that Flamsteed observed these stars on February 15, 1704, and made an error of 5' in computing the right ascension of all four stars. Baily alerted astronomers to these errors in his revised edition of Flamsteed's catalogue (see BF 1204). Several years later, while editing the *BAC,* Baily switched these stars to Lynx. Heis, on the other hand, switched all four to Cancer. Moreover, he felt the combined light of 2 and 3 appeared as a single image to his naked eye. The four stars have been permanently assigned to the constellations as in the heading above.

6, in Leo, HR 3723, 6.41V, *HRP,* BF 1322*, BAC 3194*.

12

This star does not exist. Flamsteed made several observations of 42 Lyn, but during one observation he erred several degrees in computing its declination. This produced a new but erroneous set of coordinates that led him to believe he had seen a new star, 12 LMi. See BF 1352 and Table B, p. 645.

14, in Ursa Major, HD 84453, 6.82V, *SA, BS Suppl.,* GC 13451*.

15, in Ursa Major, HR 3881, 5.09V, *SA, HRP,* GC 13497*.

16, HD 85029, 6.65V, *SA,* GC 13533*.

17, HD 85373, 6.73V, *SA,* GC 13573*.

18, in Leo, HD 86012, 6.6V, *SA,* GC 13673*.

25, in Ursa Major, HD 89572, 6.79V, BF 1459*, BAC 3525, B 72*, P X, 40*.

26, HD 89892, 7.1V, *SA,* BF 1466*, BAC 3539*.

29, HD 90250, 6.52V, *SA, BS Suppl.,* GC 14312*.

36, HD 92000, 6.6V, *SA,* H 28*, GC 14609*.

37, HR 4166, 4.71V, *SA, HRP, BS,* HF 46 in Leo, BF 1512*.

In Halley's edition of Flamsteed's catalogue, this star's magnitude is listed as 3, as it is in the authorized edition of 1725. In examining Flamsteed's manuscripts, Baily discovered that the Astronomer Royal observed 37 LMi on three different occasions and recorded three different magnitudes: 4, 4½, and 6.

39, HD 92371, 6.9V, *SA,* GC 14668*.

45, HD 94218, 7.4V, BF 1545*, BAC 3727*, B 121*, P X, 180*, T 4888*.

The star is properly identified in the second edition of *SA*.

47, in Ursa Major, HR 4256, 5.72V, *SA, HRP, BS,* GC 15006*.

49, in Leo, HD 94671, 7.3V, BF 1555*, BAC 3748*, B 130*, P X, 192*, T 4918*.

51, HD 96094, 7.62V, *SA,* GC 15239*.

52, in Leo, HD 96418, 6.87V, *SA, BS Suppl.,* GC 15284*.

53, in Leo, HR 4332, 5.68V, *HRP,* BF 1580 and Table A, p. 645.

This star is the same as 67 Leo. It is one of Flamsteed's twenty-two duplicate stars.

Lepus, Lep

Hare

Lepus (Figure 95) is a Ptolemaic constellation that Flamsteed numbered from 1 to 19.

Table 124. The numbered stars of Lepus

Flamsteed Number	Letter	Magnitudes		Catalogue Numbers	
		Flamsteed	Visual	HR	HD
1+		9	5.75	1634	32503
2	Epsilon	4	3.19	1654	32887
3	Iota	5	4.45	1696	33802
4	Kappa	5	4.36	1705	33949
5	Mu	4	3.31	1702	33904
6	Lambda	4½	4.29	1756	34816
7	Nu	5½	5.30	1757	34863
8		6	5.25	1783	35337
9	Beta	3	2.84	1829	36079
10		6	5.55	1849	36473
11	Alpha	3	2.58	1865	36673
12		6	5.87	1968	38090
13	Gamma	3½	3.60	1983	38393
14	Zeta	4	3.55	1998	38678
15	Delta	3¾	3.81	2035	39364
16	Eta	4	3.71	2085	40136
17		6	4.93	2148	41511
18	Theta	4	4.67	2155	41695
19		6	5.31	2168	42042

The Lost, Missing, or Troublesome Stars of Lepus

1, HR 1634, 5.75V, *SA, HRP, BS,* HF 1*, BF 658*.

The magnitude of 9 listed for this star in the "Catalogus Britannicus" is probably a typographical error since both Halley's 1712 edition of Flamsteed's catalogue and Flamsteed's manuscript catalogue list it as 6.

Libra, Lib

Scales, Balance

Libra (Figure 99) is a Ptolemaic constellation that Flamsteed numbered from 1 to 51.

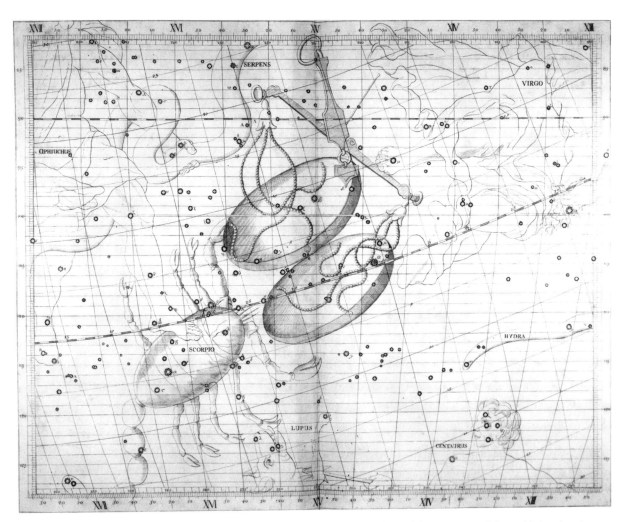

Figure 99. Flamsteed's Libra and Scorpius. Also depicted are the upper portions of the Centaur and the Wolf showing the stars included in Flamsteed's "Catalogus Britannicus." The rest of these two constellations are in Figure 107. The star 1 Lib at 209°47' r.a. and 115°47' n.p.d. does not exist.

Table 125. The numbered stars of Libra

Flamsteed Number	Letter	Magnitudes		Catalogue Numbers	
		Flamsteed	Visual	HR	HD
1+		5½			
2		7	6.21	5383	126035
3+		6	8.09		128756
4		6	5.73	5484	129433
5		6	6.33	5503	129978
6+	e Hya	5	4.41	5526	130694
7	Mu	5	5.31	5523	130559
8	Alpha¹	6	5.15	5530	130819
9	Alpha²	2	2.75	5531	130841
10+		6	6.82		131027
11		6	4.94	5535	130952
12		6	5.30	5548	131430
13	Xi¹	6	5.80	5554	131530
14+		6	7.2		131992
15	Xi²	6	5.46	5564	131918
16		5½	4.49	5570	132052
17		7	6.60	5578	132230
18		6	5.87	5582	132345
19	Delta	4½	4.92	5586	132742
20	Sigma	3	3.29	5603	133216
21	Nu	5	5.20	5622	133774
22+		6	6.39		133800
23		7	6.45	5657	134987
24	Iota¹	3¾	4.54	5652	134759
25	Iota²	6	6.08	5656	134967
26		6	6.17	5662	135230
27	Beta	2	2.61	5685	135742
28		6	6.17	5701	136366
29	Omicron	7	6.30	5703	136407
30+		6	6.47		136801
31	Epsilon	4	4.94	5723	137052
32	Zeta¹	6	5.64	5743	137744
33+	Zeta²	7	6.69		137949
34	Zeta³	6	5.82	5750	138137
35	Zeta⁴	4	5.50	5764	138485
36		6	5.15	5775	138688
37		6	4.62	5777	138716
38	Gamma	3½	3.91	5787	138905
39	Upsilon	4	3.58	5814	139446
40	Tau	4	3.66	5812	139365
41		6	5.38	5794	139063
42		6	4.96	5824	139663
43	Kappa	4	4.74	5838	139997
44	Eta	4	5.41	5848	140417
45	Lambda	4	5.03	5902	142096

(table continued on next page)

Table 125. The numbered stars of Libra (continued)

Flamsteed Number	Letter	MAGNITUDES		CATALOGUE NUMBERS	
		Flamsteed	Visual	HR	HD
46	Theta	4	4.15	5908	142198
47		6	5.94	5915	142378
48		4	4.88	5941	142983
49		6	5.47	5954	143333
50		6	5.55	5959	143459
51+	Xi Sco	4½	4.16c	5977/8	144069/70

The Lost, Missing, or Troublesome Stars of Libra

1

This star does not exist. In one of his observations of 50 Hya, Flamsteed made an error of 1° in right ascension. This produced a new set of coordinates that led him to believe he had seen a new star, 1 Lib. See BF 1935 and Table B, p. 645.

3, HD 128756, 8.09V, *SA,* GC 19738*.

6, in Hydra as E Hya, HR 5526, 4.41V, *HRP,* BF 2019 and Table A, p. 645.

This star is the same as 58 Hya. It is one of Flamsteed's twenty-two duplicate stars.

10, HD 131027, 6.82V, *SA,* GC 19990*.

14, in Hydra, HD 131992, 7.2V, BF 2035*, BAC 4925*, B 42*, P XIV, 213*, CAP 2655*, T 6947*.

22, HD 133800, 6.39V, *SA, BS Suppl.,* GC 20320*.

30, HD 136801, 6.47V, *SA, BS Suppl.,* GC 20683*.

33, Zeta^2, HD 137949, 6.69V, *SA,* GC 20814*.

See Zeta Lib in Part One for an account of whether this star should be labeled Zeta^2.

51, in Scorpius as Xi Sco, HR 5977/8, 5.07V, 4.77V, comb. mag. 4.16V, *HRP,* BF 2192*.

For an account of how this star has been switched back and forth between Libra and Scorpius, see Xi Sco in Part One.

LUPUS, LUP

Wolf

Lupus (Figure 99) is a Ptolemaic constellation that Flamsteed numbered from 1 to 5. Since its stars are mostly below the horizon in England, Flamsteed could observe only the very northern part of the constellation. In Halley's 1712 edition of Flamsteed's catalogue, Flamsteed merged Lupus and Centaurus into a single constellation, Centaurus cum Lupo (Centaur with the Wolf), with a total of thirteen stars. In his "Catalogus Britannicus" of 1725, the two are listed separately with five stars in each constellation. The three stars in the 1712 catalogue that are omitted from the authorized catalogue are:

HF 1, Cen & Lup, r Cen, HR 5006, 5.10V, BF 1825, CA 1139, BR 4386, BAC 4437.

HF 9, Cen & Lup, ϕ^1 Lup, HR 5705, 3.56V, BF 2081, CA 1286, BR 5293, BAC 5054.

HF 10, Cen & Lup, ϕ^2 Lup, HR 5712, 4.54V, BF 2084, CA 1287, BR 5299, BAC 5060.

Baily noted that Flamsteed had observed these stars but he was unable to explain why they were omitted from his definitive catalogue. Referring to 1 Cen & Lup, Baily wrote: "I know not why it was afterwards rejected (BF 1825)."

All the numbered stars in Lupus are listed in the *BS,* and there are no missing or problem numbered stars in the constellation.

Table 126. The numbered stars of Lupus

Flamsteed Number	Letter	Magnitudes		Catalogue Numbers	
		Flamsteed	Visual	HR	HD
1	i	5	4.91	5660	135153
2	f	5½	4.34	5686	135758
3	Psi¹	5½	4.67	5820	139521
4	Psi²	5½	4.75	5839	140008
5	Chi	5	3.95	5883	141556

Lynx, Lyn

Lynx or Tiger

Hevelius devised the constellation Lynx (Figure 98). Flamsteed acknowledged that it was a Hevelian constellation and included its stars in his atlas and catalogue, numbering them from 1 to 44.

Table 127. The numbered stars of Lynx

		MAGNITUDES		CATALOGUE NUMBERS	
Flamsteed Number	Letter	Flamsteed	Visual	HR	HD
1		5½	4.98	2215	42973
2		4	4.48	2238	43378
3+		6	7.2		43749
4		6	5.94	2257	43812
5		6	5.21	2293	44708
6		6½	5.88	2331	45410
7		6½	6.45	2376	46101
8		6½	5.94	2394	46480
9+		7	6.56		46318
10+		6½	7.0		46635
11		6	5.85	2402	46590
12+		7	4.87	2470	48250
13		6	5.35	2477	48432
14		5	5.33	2520	49618
15		5	4.35	2560	50522
16	Psi[10] Aur	6	4.90	2585	50973
17+		7	6.7		53633
18		6	5.20	2715	55280
19		5	5.22c	2783/4	57102/3
20+		6	6.86		57066
21		5	4.64	2818	58142
22		6	5.36	2849	58855
23		7	6.06	2929	61106
24		5	4.99	2946	61497
25		6	6.25	3065	64106
26		5	5.45	3066	64144
27		5	4.84	3173	67006

(table continued on next page)

Table 127. The numbered stars of Lynx *(continued)*

		MAGNITUDES		CATALOGUE NUMBERS	
Flamsteed Number	Letter	Flamsteed	Visual	HR	HD
28		7	6.26	3167	66824
29		5	6.55	3235	68930
30+		6	5.89	3254	69548
31		5	4.25	3275	70272
32		7	6.24	3365	72291
33		6	5.78	3377	72524
34		6	5.37	3422	73593
35		7	5.15	3508	75506
36		5½	5.32	3652	79158
37+		6	6.13	3697	80290
38		4	3.82	3690	80081
39+		6	6.9		80608
40	Alpha	4	3.13	3705	80493
41+		6	5.41	3743	81688
42		6	5.25	3829	83287
43		6¾	5.62	3851	83805
44+		5½	5.16	3870	84335

The Lost, Missing, or Troublesome Stars of Lynx

3, HD 43749, 7.2V, *SA,* GC 8140*.

9, HD 43618, 6.56V, *SA,* GC 8549*.

10, HD 46635, 7.0V, *SA,* GC 8606*.

12, HR 2470, 4.87V, *SA, HRP, BS,* HF 40 in Ursa Major, BF 927*.

This star's magnitude is listed as 7 in Flamsteed's 1725 catalogue. This is an obvious error since it is listed as 5½ in Halley's 1712 edition of his catalogue and as either 5 or 6 in his observational notes.

17, HD 53633, 6.7V, *SA,* GC 9411*.

20, HD 57066, 6.86V, *SA, BS Suppl.*

30, HR 3254, 5.89V, *SA, HRP, BS,* BF 1151 and Table A, p. 645.

This star is the same as 58 Cam. It is one of Flamsteed's twenty-two duplicate stars.

37, in Ursa Major, HR 3697, 6.13V, *SA, HRP,* GC 12865*.

39, in Ursa Major, HD 80608, 6.9V, *SA,* GC 12911*.

41, in Ursa Major, HR 3743, 5.41V, *SA, HRP, BS,* GC 13051*.

44, in Ursa Major, HR 3870, 5.16V, *HRP,* BF 1370*, BAC 3324*, B 163*.

This star's Flamsteed number is often omitted from atlases and catalogues because it is about 15° outside the border of Lynx. Bode was among the first to switch it to Ursa Major.

Lyra, Lyr

Lyre, Harp

Lyra (figures 91 and 93) is a Ptolemaic constellation that Flamsteed numbered from 1 to 21. All its numbered stars are listed in the *BS,* and there are no missing or problem numbered stars in the constellation.

Table 128. The numbered stars of Lyra

Flamsteed Number	Letter	Magnitudes		Catalogue Numbers	
		Flamsteed	Visual	HR	HD
1	Kappa	5	4.33	6872	168775
2	Mu	6	5.12	6903	169702
3	Alpha	1	0.03	7001	172167
4	Epsilon^1	5	4.66c	7051/2	173582/3
5	Epsilon^2	6	4.59c	7053/4	173607/8
6	Zeta^1	5	4.36	7056	173648
7	Zeta^2	5	5.73	7057	173649
8	Nu^1	6	5.91	7100	174585
9	Nu^2	6	5.25	7102	174602
10	Beta	3	3.45	7106	174638
11	Delta^1	4½	5.58	7131	175426
12	Delta^2	4	4.30	7139	175588
13		6	4.04	7157	175865
14	Gamma	3	3.24	7178	176437
15	Lambda	6	4.93	7192	176670
16		6	5.01	7215	177196
17		6	5.23	7261	178449
18	Iota	5	5.28	7262	178475
19		6	5.98	7283	179527
20	Eta	6	4.39	7298	180163
21	Theta	6	4.36	7314	180809

Monoceros, Mon

Unicorn

The constellation Monoceros (Figure 88) was devised by Petrus Plancius. Hevelius and Flamsteed included Monoceros in their catalogues, but the latter, unaware of Plancius's work, called it a Hevelian constellation even though it was not included among the twelve constellations that Hevelius claimed to have devised. Flamsteed numbered its stars from 1 to 31.

Table 129. The numbered stars of Monoceros

		MAGNITUDES		CATALOGUE NUMBERS	
Flamsteed Number	Letter	Flamsteed	Visual	HR	HD
1		6	6.12	2107	40535
2		6	5.03	2108	40536
3		6	4.95	2128	40967
4+		6	6.8		42116
5	Gamma	4½	3.98	2227	43232
6		6½	6.75	2255	43760
7		6	5.27	2273	44112
8	Epsilon	4	4.30c	2298/9	44769/70
9+		5	6.50		45418
10		6	5.06	2344	45546
11	Beta	5	3.74c	2356/7/8	45725/6/7
12		5	5.84	2382	46241
13		4	4.50	2385	46300
14		5½	6.45	2404	46642
15		4	4.66	2456	47839
16		6	5.93	2494	48977
17		5	4.77	2503	49161
18		4	4.47	2506	49293
19		5	4.99	2648	52918
20		6	4.92	2701	54810
21		5	5.45	2707	55057
22	Delta	4½	4.15	2714	55185
23+		6½	6.7		55533
24		6	6.41	2744	56003
25		6	5.13	2927	61064

(table continued on next page)

Table 129. The numbered stars of Monoceros *(continued)*

		MAGNITUDES		CATALOGUE NUMBERS	
Flamsteed Number	Letter	Flamsteed	Visual	HR	HD
26	Alpha	4½	3.93	2970	61935
27		5	4.93	3122	65695
28		5	4.68	3141	65953
29	Zeta	6	4.34	3188	67594
30+	C Hya	6	3.90	3314	71155
31+	F Hya	4	4.62	3459	74395

The Lost, Missing, or Troublesome Stars of Monoceros

4, in Lepus, HD 42116, 6.8V, *SA,* BF 856*.

9, HD 45418, 6.50V, *SA, BS Suppl.,* GC 8351*.

23, HD 55533, 6.7V, BF 1018*, BAC 2366*, B 140*, P VII, 24*, CAP 1177*, T 2886*.

The star is properly identified in the second edition of *SA*.

30, in Hydra as C Hya, HR 3314, 3.90V, *SA, HRP,* HF 1 in Hydra, GC 11499*.

Flamsteed's catalogue lists this star's magnitude as 6, an obvious error. Halley's edition of Flamsteed's catalogue lists it as 3¾. Baily included this star in Hydra. See C Hydra in Part One.

31, in Hydra as F Hya, HR 3459, 4.62V, *SA, HRP,* GC 12006*.

Baily included this star in Hydra. See F Hydra in Part One.

Ophiuchus, Oph

Serpent Bearer

Ophiuchus (Figure 100) is a Ptolemaic constellation that Flamsteed numbered from 1 to 74.

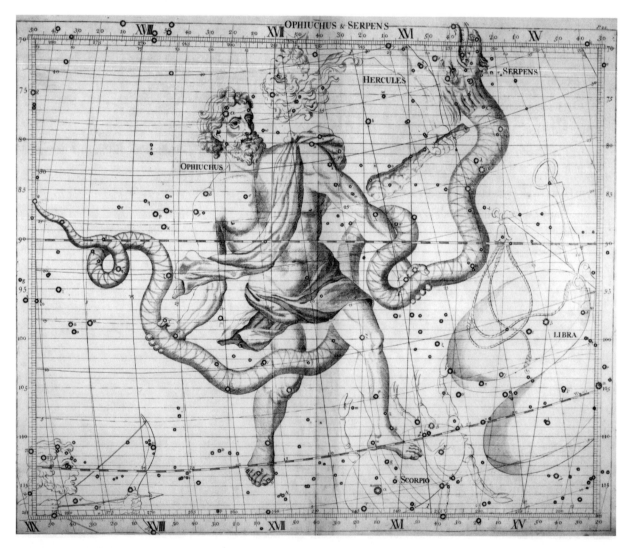

Figure 100. Flamsteed's Ophiuchus and Serpens. The following stars do not exist: **6 Oph**, 242°16' r.a., 84°10' n.p.d.; **46 Oph**, 257°14' r.a., 75°26' n.p.d.; **48 Oph**, 257°45' r.a., 76°49' n.p.d.; **59 Oph**, 262° 1' r.a., 111°41' n.p.d.; **65 Oph**, 265°34' r.a., 107°56' n.p.d.; **33 Ser**, 233°25' r.a., 72°27' n.p.d.; **42 Ser**, 236°33' r.a., 98° 9' n.p.d.; **52 Ser**, 250°56' r.a., 102°30' n.p.d.; **54 Ser**, 256° 5' r.a., 102°40' n.p.d.

Table 130. The numbered stars of Ophiuchus

		MAGNITUDES		CATALOGUE NUMBERS	
Flamsteed Number	Letter	Flamsteed	Visual	HR	HD
1	Delta	3	2.74	6056	146051
2	Epsilon	3½	3.24	6075	146791
3	Upsilon	5	4.63	6129	148367
4	Psi	5	4.50	6104	147700
5	Rho	5	4.63c	6112/3	147933/4
6+		6			
7	Chi	6	4.42	6118	148184
8	Phi	4	4.28	6147	148786
9	Omega	5	4.45	6153	148898
10	Lambda	4	3.82	6149	148857
11+	n Her	6	5.63	6158	149121
12		6	5.75	6171	149661
13	Zeta	3	2.56	6175	149757
14		6	5.74	6205	150557
15+		6	6.8		150937
16		6	6.03	6224	151133
17+	i Her	6	5.15	6228	151217
18+		6½	7.1		151659
19		6	6.10	6232	151431
20		5½	4.65	6243	151769
21		6	5.51	6255	152127
22+		7	7.0		152534
23		6	5.25	6280	152601
24		7	5.58	6291	152849
25	Iota	4	4.38	6281	152614
26		6	5.75	6310	153363
27	Kappa	4	3.20	6299	153210
28+		6	6.7		154021
29		6	6.26	6321	153727
30		6	4.82	6318	153687
31+		6	6.60		154132
32+		6	4.98	6337	154143
33+		6	5.93	6341	154228
34+		6	6.08	6342	154278
35	Eta	3	2.43	6378	155125
36	A	5¾	4.31c	6401/2	155885/6
37		6	5.33	6393	155644
38+		6½	6.81		156252
39	Omicron	6	5.15c	6424/5	156349/50
40	Xi	4	4.39	6445	156897
41		6	4.73	6415	156266
42	Theta	3¾	3.27	6453	157056
43		4½	5.35	6459	157236
44	b	4¾	4.17	6486	157792

(table continued on next page)

Table 130. The numbered stars of Ophiuchus *(continued)*

		MAGNITUDES		CATALOGUE NUMBERS	
Flamsteed Number	Letter	Flamsteed	Visual	HR	HD
45	d	6	4.29	6492	157919
46+		6			
47+		6	6.21	6496	157968
48+		6			
49	Sigma	5	4.34	6498	157999
50+		7	7.4		158527
51	c	6	4.81	6519	158643
52		6	6.57	6545	159376
53	f	6	5.81	6548	159480
54+		6	6.6		159466
55	Alpha	2	2.08	6556	159561
56+		6	7.3		159610
57	Mu	4	4.62	6567	159975
58		6	4.87	6595	160915
59+		6			
60	Beta	3	2.77	6603	161096
61		6	6.17	6609	161270
62	Gamma	3	3.75	6629	161868
63+		5	6.20	6672	162978
64	Nu	4	3.34	6698	163917
65+		6			
66		4½	4.64	6712	164284
67		4	3.97	6714	164353
68		4	4.45	6723	164577
69	Tau	5	4.78c	6733/4	164764/5
70		4	4.03	6752	165341
71		6	4.64	6770	165760
72+		6	3.73	6771	165777
73		6	5.73	6795	166233
74		6	4.86	6866	168656

The Lost, Missing, or Troublesome Stars of Ophiuchus

6

This star does not exist. Flamsteed made an error in copying the declination of 21 Her; he mistakenly copied 9 Her's declination. This produced a new set of coordinates that led him to believe he had seen a new star, which he then numbered 6 Oph. See BF 2247 and Table B, p. 645.

11, in Hercules as n Her, HR 6158, 5.63V, *HRP,* BF 2271 and Table A, p. 645.

This star is the same as 28 Her. It is one of Flamsteed's twenty-two duplicate stars.

15, HD 150937, 6.8V, *SA,* GC 22527*.

17, in Hercules as i Her, HR 6228, 5.15V, *HRP,* BF 2308 and Table A, p. 645.

This star is the same as 43 Her. It is one of Flamsteed's twenty-two duplicate stars.

18, HD 151659, 7.1V, *SA,* GC 22632*.

22, HD 152534, 7.0V, *SA,* GC 22778*.

28, HD 154021, 6.7V, *SA,* GC 23006*.

31, HD 154132, 6.60V, *SA,* GC 23024*.

32, in Hercules, HR 6337, 4.98V, *HRP,* BF 2348*.

This star's Flamsteed number is often omitted from atlases and catalogues because there is no star at the coordinates listed in Flamsteed's catalogue. Flamsteed erred 40^s in noting 32's time of transit across his lens on May 5, 1692. This created an error of 10' in right ascension, which Baily later corrected in his revised edition of Flamsteed's catalogue by increasing Flamsteed's original figure of 252°5'15" to 252°15'15".

33, in Hercules, HR 6341, 5.93V, *HRP,* BF 2349*.

34, in Hercules, HR 6342, 6.08V, *HRP,* BF 2350*.

38, HD 156252, 6.81V, BF 2366 and Table A, p. 645, BAC 5822*.

This star is the same as 31 Sco in Oph. Tirion's *SA* and his *Uranometria 2000.0* erroneously assign 38 to the variable U Oph, HR 6414. *HRP* makes the same mistake and it is repeated in later editions of the *BS.* It is finally corrected in the fourth edition. The second edition of Tirion's *SA* also corrects it. See Hoffleit, "Discordances," p. 55.

46

This star does not exist. After observing 32 Oph on April 12, 1703, Flamsteed erred 5° in reducing its right ascension to the epoch of his catalogue, 1690. This led him to compute a new set of coordinates and to believe he had seen a new star, 46 Oph. See BF 2348 and Table B, p. 645.

47, in Serpens, HR 6496, 6.21V, *HRP,* BF 2387*.

Tirion's *SA* and his *Uranometria 2000.0* erroneously assign 47 Oph to HR 6493, 4.54V, about 8° due north of HR 6496. A typographical error in the notes section of the *BS* describes this star as 470. The second edition of *SA* removes 47 from HR 6493 but fails to identify HR 6496 as 47.

48

This star does not exist. In one of his observations of 60 Her, Flamsteed erred 5° in determining its right ascension. This led him to compute a new set of coordinates and to believe he had seen a new star, 48 Oph. See BF 2353 and Table B, p. 645.

50, HD 158527, 7.4V, BF 2394*, BAC 5904*, B 216*, P XVII, 113*, CAP 3310*, T 8066*.

The star is properly identified in the second edition of *SA*.

54, HD 159466, 6.6V, *SA,* GC 23819*.

56, HD 159610, 7.3V, BF 2410*, BAC 5942*, B 245*, P XVII, 154*, T 8113*.

The star is properly identified in the second edition of *SA*.

59

This star does not exist. In one of his observations of 3 Sgr, Flamsteed erred 6° in calculating its declination. This produced a new set of coordinates that led him to believe he had seen a new star, 59 Oph. See BF 2418 and Table B, p. 645.

63, in Sagittarius, HR 6672, 6.20V, *SA, HRP,* GC 24347*.

65, BF 2441* and Table C, p. 646.

There is some question about the existence of this star. Baily noted that Flamsteed observed it on May 6, 1691, and properly reduced it to 1690, but there is no star in the sky at the coordinates Flamsteed listed. Peters ("Flamsteed's Stars," p. 82) suggested that Flamsteed made errors both in right ascension and declination when he observed this star. Peters proposed that if these errors were to be corrected, the star Flamsteed observed would be 6 Sgr.

72, HR 6771, 3.73V, *SA, HRP, BS,* HF 67, BF 2468*.

Flamsteed's catalogue of 1725 mistakenly lists this star's magnitude as 6, possibly a copying or typographical error since his observational notes describe it once as magnitude 5 and twice as magnitude 4. Halley's 1712 edition of his catalogue lists it correctly as 4.

ORION, ORI

Orion, Great Hunter

Orion (figures 95 and 103) is a Ptolemaic constellation that Flamsteed numbered from 1 to 78.

Table 131. The numbered stars of Orion

		MAGNITUDES		CATALOGUE NUMBERS	
Flamsteed Number	Letter	Flamsteed	Visual	HR	HD
1	Pi³	4	3.19	1543	30652
2	Pi²	4	4.36	1544	30739
3	Pi⁴	4	3.69	1552	30836
4	Omicron¹	4½	4.74	1556	30959
5		6	5.33	1562	31139
6	g	6	5.19	1569	31283
7	Pi¹	6	4.65	1570	31295
8	Pi⁵	4	3.72	1567	31237
9	Omicron²	4½	4.07	1580	31421
10	Pi⁶	4½	4.47	1601	31767
11		5	4.68	1638	32549
12+		6			
13		6	6.17	1662	33021
14	i	5	5.34	1664	33054
15		5	4.82	1676	33276
16	h	6	5.43	1672	33254
17	Rho	4½	4.46	1698	33856
18		5¾	5.50	1718	34203
19	Beta	1	0.12	1713	34085
20	Tau	4	3.60	1735	34503
21		6	5.34	1746	34658
22	o	5	4.73	1765	35039
23	m	6	5.00	1770	35149
24	Gamma	2	1.64	1790	35468
25	Psi¹	5	4.95	1789	35439
26+		6			
27	p	6	5.08	1787	35410
28	Eta	3	3.36	1788	35411

(table continued on next page)

Table 131. The numbered stars of Orion (continued)

Flamsteed Number	Letter	Magnitudes		Catalogue Numbers	
		Flamsteed	Visual	HR	HD
29	e	5	4.14	1784	35369
30	Psi²	5	4.59	1811	35715
31		6	4.71	1834	36167
32	A	5	4.20	1839	36267
33	n¹	6	5.46	1842	36351
34	Delta	2	2.46c	1851/2	36485/6
35		6	5.64	1864	36653
36	Upsilon	4	4.62	1855	36512
37	Phi¹	5	4.41	1876	36822
38	n²	6	5.36	1872	36777
39	Lambda	4	3.39c	1879/80	36861/2
40	Phi²	5	4.09	1907	37160
41	Theta¹	6	4.85c	1893/4/5/6	37020/1/2/3
42	c	5	4.59	1892	37018
43+	Theta²	4	6.39	1897	37041
44	Iota	3½	2.77	1899	37043
45		5	5.26	1901	37077
46	Epsilon	2	1.70	1903	37128
47	Omega	5	4.57	1934	37490
48	Sigma	4	3.81	1931	37468
49	d	5	4.80	1937	37507
50	Zeta	2	1.76c	1948/9	37742/3
51	b	5	4.91	1963	37984
52		6	5.27	1999	38710
53	Kappa	3	2.06	2004	38771
54	Chi¹	5	4.41	2047	39587
55		6	5.35	2031	39291
56		6	4.78	2037	39400
57		5	5.92	2052	39698
58	Alpha	1	0.50	2061	39801
59		6	5.90	2100	40372
60		6	5.22	2103	40446
61	Mu	4	4.12	2124	40932
62	Chi²	6	4.63	2135	41117
63		6	5.67	2144	41361
64		6	5.14	2130	41040
65+		5½			
66		6	5.63	2145	41380
67	Nu	4½	4.42	2159	41753
68		6	5.75	2193	42509
69	f¹	6	4.95	2198	42545
70	Xi	4½	4.48	2199	42560
71		6	5.20	2220	43042
72	f²	6	5.30	2223	43153

(table continued on next page)

Table 131. The numbered stars of Orion *(continued)*

Flamsteed Number	Letter	Magnitudes		Catalogue Numbers	
		Flamsteed	Visual	HR	HD
73		6	5.33	2229	43247
74	k	6	5.04	2241	43386
75	l	6	5.39	2247	43525
76+		6			
77+		6	5.20	2334	45416
78+		6	5.55	2335	45433

The Lost, Missing, or Troublesome Stars of Orion

12

This star does not exist. In one of his observations of 13 Ori, Flamsteed erred in both right ascension and declination in reducing its coordinates to 1690, the epoch of his catalogue. This led him to believe he had seen a new star, 12 Ori. See BF 660 and Table B, p. 645.

26

This star does not exist. In one of his observations of 25 Ori, Flamsteed erred by 20' in noting its declination. As in the case of 12 Ori, this produced a new but erroneous set of coordinates that led him to believe he had seen a new star, 26 Ori. See BF 709 and Table B, p. 645.

43, Theta^2, HR 1897, 6.39V, *SA, HRP, BS,* HF 36, BF 750*.

In his revised edition of Flamsteed's catalogue, Baily noted with concern the large discrepancy between this star's actual magnitude, dimmer than 6, and that recorded in Flamsteed's "Catalogus Britannicus", 4. He stressed that he searched Flamsteed's notes but found nothing to explain the discrepancy. Indeed, in Halley's 1712 pirated edition of the Astronomer Royal's catalogue, this star's magnitude is listed as 3½. A possible explanation may lie in the fact that Theta^2, 43, is part of a multiple-star system that also includes Theta^1, 41. Perhaps on one occasion Flamsteed confused the combined light of the entire system, about magnitude 4.2, with its dimmer components. See Theta Ori in Part One.

65

This star does not exist. In one of his observations of 62 Ori, Flamsteed erred 3^m in noting its time of transit across his telescope's lens. This miscalculation led him to generate incorrect coordinates, which, in turn, led him to believe he had seen a new star, 65 Ori. See BF 838 and Table B, p. 645.

76, BF 890* and Table D, p. 646.

This star does not exist. Baily found no record — either in Flamsteed's printed works or in his manuscripts at the Royal Observatory at Greenwich — that he had ever observed this star. He guessed that its coordinates arose from errors in computation of the positions of 8 Mon and 63 Ori.

77, in Monoceros, HR 2334, 5.20V, *SA, HRP, BS,* GC 8355*.

78, in Monoceros, HR 2335, 5.55V, *SA, HRP, BS,* GC 8356*.

Pegasus, Peg

Pegasus, Flying Horse

Pegasus (Figure 94) is a Ptolemaic constellation that Flamsteed numbered from 1 to 89.

Table 132. The numbered stars of Pegasus

		Magnitudes		Catalogue Numbers	
Flamsteed Number	**Letter**	**Flamsteed**	**Visual**	**HR**	**HD**
1		4	4.08	8173	203504
2		4½	4.57	8225	204724
3		6	6.18	8265	205811
4		6	5.67	8270	205924
5		6½	5.45	8267	205852
6+	d Aqr	6	5.10	8277	206067
7		6	5.30	8289	206487
8	Epsilon	3	2.39	8308	206778
9		4½	4.34	8313	206859
10	Kappa	4	4.13	8315	206901
11+		6	5.64	8328	207203
12		6	5.29	8321	207089
13		6	5.29	8344	207652
14		6	5.04	8343	207650
15		6	5.53	8354	207978
16		6	5.08	8356	208057
17		6	5.54	8373	208565
18		5	6.00	8385	209008
19		6	5.65	8393	209167
20		6	5.60	8392	209166
21		5	5.80	8404	209459
22	Nu	5	4.84	8413	209747
23		6	5.63	8419	209833
24	Iota	4	3.76	8430	210027
25		6½	5.78	8438	210129
26	Theta	4	3.53	8450	210418
27	Pi^1	5	5.58	8449	210354
28		6½	6.46	8459	210516
29	Pi^2	4½	4.29	8454	210459

(table continued on next page)

Table 132. The numbered stars of Pegasus *(continued)*

		MAGNITUDES		CATALOGUE NUMBERS	
Flamsteed Number	Letter	Flamsteed	Visual	HR	HD
30		6	5.37	8513	211924
31		4½	5.01	8520	212076
32		6	4.81	8522	212097
33		6½	6.04	8532	212395
34		6	5.75	8548	212754
35		6	4.79	8551	212943
36		6½	5.58	8562	213119
37		6	5.48	8566	213235
38		6	5.63	8574	213323
39		6½	6.24	8586	213617
40		6	5.82	8618	214567
41		6½	6.21	8624	214698
42	Zeta	3	3.40	8634	214923
43	Omicron	5	4.79	8641	214994
44	Eta	3	2.94	8650	215182
45		6½	6.25	8660	215510
46	Xi	5	4.19	8665	215648
47	Lambda	4	3.95	8667	215665
48	Mu	4	3.48	8684	216131
49	Sigma	6	5.16	8697	216385
50	Rho	6	4.90	8717	216735
51		6	5.49	8729	217014
52		6	5.75	8739	217232
53	Beta	2	2.42	8775	217906
54	Alpha	2	2.49	8781	218045
55		5	4.52	8795	218329
56		5½	4.76	8796	218356
57		6	5.12	8815	218634
58		6	5.39	8821	218700
59		5¾	5.16	8826	218918
60		6	6.17	8827	218935
61		6	6.27	8842	219477
62	Tau	6	4.60	8880	220061
63		6	5.59	8882	220088
64		6	5.32	8887	220222
65		6	6.24	8891	220318
66		6	5.08	8893	220363
67		6½	5.57	8903	220599
68	Upsilon	6	4.40	8905	220657
69		6	5.98	8915	220933
70		5½	4.55	8923	221115
71		6	5.32	8940	221615
72		6	4.98	8943	221673
73		6	5.63	8948	221758
74		7	6.26	8960	222098

(table continued on next page)

Table 132. The numbered stars of Pegasus *(continued)*

Flamsteed Number	Letter	Magnitudes		Catalogue Numbers	
		Flamsteed	Visual	HR	HD
75		6	5.40	8963	222133
76+		6	6.29		222683
77		6	5.06	8991	222764
78		5½	4.93	8997	222842
79		6	5.96	9025	223461
80		6	5.79	9030	223637
81	Phi	6	5.08	9036	223768
82		6	5.30	9039	223781
83+		6	6.7		223792
84	Psi	6	4.66	9064	224427
85		6	5.75	9088	224930
86		5½	5.51	4	87
87		6	5.53	22	448
88	Gamma	2	2.83	39	886
89	Chi	6	4.80	45	1013

The Lost, Missing, or Troublesome Stars of Pegasus

6, in Aquarius as d Aqr, HR 8277, 5.10V, *HRP,* BF 2949 and Table A, p. 645.

This star is the same as 25 Aqr. It is one of Flamsteed's twenty-two duplicate stars.

11, HR 8328, 5.64V, *SA, HRP, BS,* BF 2979 and Table A, p. 645.

This star is the same as 27 Aqr. It is one of Flamsteed's twenty-two duplicate stars.

76, HD 222683, 6.29V, *SA, BS Suppl.,* GC 32932*.

83, HD 223792, 6.7V, BF 3271*, GC 33126*.

Tirion's *SA* and his *Uranometria 2000.0* erroneously assign 83 Peg to 23 Psc, HR 9035. These two stars are almost adjacent. 83 Peg is the dimmer star, 6.7V, to the northeast, while 23 Psc, the brighter one at 6.11V, is to the southwest. The mix-up between these two stars is also present in the atlases of Becvar and Norton. The second edition of *SA* leaves 23 Psc, HR 9035, unlabeled but properly identifies HD 223792 as 83 Peg.

Perseus, Per

Perseus, Champion

Perseus (Figure 81) is a Ptolemaic constellation that Flamsteed numbered from 1 to 59.

Table 133. The numbered stars of Perseus

		MAGNITUDES		CATALOGUE NUMBERS	
Flamsteed Number	**Letter**	**Flamsteed**	**Visual**	**HR**	**HD**
1		6	5.52	533	11241
2		6	5.79	536	11291
3		6½	5.69	568	11949
4	g	6	5.04	590	12303
5		6	6.36	627	13267
6+		6	5.31	645	13530
7		6¾	5.98	662	13994
8		7	5.75	661	13982
9	i	6	5.17	685	14489
10		7	6.25	696	14818
11		7	5.77	785	16727
12		6	4.91	788	16739
13	Theta	4	4.12	799	16895
14		6	5.43	800	16901
15+	Eta	6	3.76	834	17506
16		4	4.23	840	17584
17		5½	4.53	843	17709
18	Tau	5	3.95	854	17878
19+		6			
20		6	5.33	855	17904
21		4½	5.11	873	18296
22	Pi	4	4.70	879	18411
23	Gamma	3	2.93	915	18925
24		6	4.93	882	18449
25	Rho	4	3.39	921	19058
26	Beta	2½	2.12	936	19356
27	Kappa	4¾	3.80	941	19476
28	Omega	5	4.63	947	19656
29		6	5.15	987	20365

(table continued on next page)

Lost Stars

Table 133. The numbered stars of Perseus (continued)

Flamsteed Number	Letter	Magnitudes		Catalogue Numbers	
		Flamsteed	Visual	HR	HD
30		6	5.47	982	20315
31		5½	5.03	989	20418
32	l	6	4.95	1002	20677
33	Alpha	2½	1.79	1017	20902
34		6	4.67	1044	21428
35	Sigma	5	4.36	1052	21552
36		6½	5.31	1069	21770
37	Psi	5	4.23	1087	22192
38+	Omicron	6	3.83	1131	23180
39	Delta	3	3.01	1122	22928
40	o	6	4.97	1123	22951
41	Nu	4	3.77	1135	23230
42	n	6	5.11	1177	23848
43	A	5	5.28	1210	24546
44	Zeta	3	2.85	1203	24398
45	Epsilon	3	2.89	1220	24760
46	Xi	5	4.04	1228	24912
47	Lambda	4	4.29	1261	25642
48	c	5	4.04	1273	25940
49		6½	6.09	1277	25975
50		6½	5.51	1278	25998
51	Mu	4	4.14	1303	26630
52	f	5	4.71	1306	26673
53	d	6	4.85	1350	27396
54		6	4.93	1343	27348
55		6	5.73	1377	27777
56		7	5.76	1379	27786
57	m	6	6.09	1434	28704
58	e	5	4.25	1454	29094
59		6	5.29	1494	29722

The Lost, Missing, or Troublesome Stars of Perseus

6, on Perseus-Andromeda border, HR 645, 5.31V, *SA, HRP, BS,* GC 2653*.

15, Theta, HR 834, 3.76V, *SA, HRP, BS,* HF 17, BF 348*.

    Flamsteed's "Catalogus Britannicus" lists this star's magnitude as 6, an obvious error since his notes list it as 4, as does Halley's 1712 edition of his catalogue.

19

    This star does not exist. In one of his observations of 18 Per, Flamsteed erred in noting its time of transit, which resulted in incorrect coordinates. This led him to believe he had observed a new star, 19 Per. See BF 359 and Table B, p. 645.

38, Omicron, HR 1131, 3.83V, *SA, HRP, BS,* HF 44, BF 454*.

Flamsteed's 1725 "Catalogus Britannicus" lists this star as magnitude 6 although Halley's 1712 edition correctly lists it as 3½. The error in the "Catalogus Britannicus" may have arisen from the confusion between this star, Omicron, and 40, o (oh), 4.97V, which Flamsteed also listed as magnitude 6. See Omicron Per in Part One.

Pisces, Psc

Fishes

Pisces (Figure 101) is a Ptolemaic constellation that Flamsteed numbered from 1 to 113.

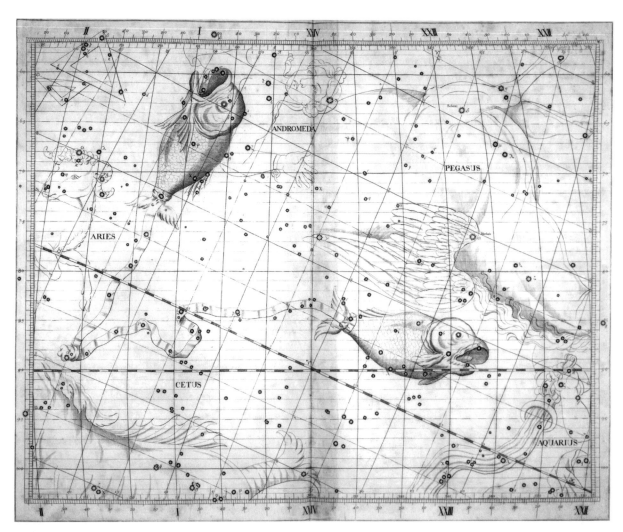

Figure 101. Flamsteed's Pisces. The vernal equinox, the First Point of Aries, ♈, is marked on the chart where the horizontal double line, the celestial equator, crosses the ecliptic, the slightly curved double line. The following stars in Pisces are nonexistent: **50 Psc,** 3°51' r.a., 71°25' n.p.d.; **56 Psc,** 5°56' r.a., 70°17' n.p.d.; **108 Psc,** 22°1' r.a., 68°28' n.p.d.

Table 134. The numbered stars of Pisces

		MAGNITUDES		CATALOGUE NUMBERS	
Flamsteed Number	Letter	Flamsteed	Visual	HR	HD
1		7	6.11	8715	216701
2		6	5.43	8742	217264
3		6	6.21	8750	217428
4	Beta	5	4.53	8773	217891
5	A	6	5.40	8807	218527
6	Gamma	4	3.69	8852	219615
7	b	5½	5.05	8878	220009
8	Kappa	5	4.94	8911	220825
9		6¾	6.25	8912	220858
10	Theta	5	4.28	8916	220954
11+		6	6.5		221147
12+		6	6.89		221146
13		6	6.38	8934	221409
14		6	5.87	8944	221675
15+		6	6.48		221833
16		6	5.68	8954	221950
17	Iota	6	4.13	8969	222368
18	Lambda	5	4.50	8984	222603
19		5	5.04	9004	223075
20		5½	5.49	9012	223252
21		6	5.77	9022	223438
22		6	5.55	9033	223719
23+		6	6.11	9035	223755
24		6	5.93	9041	223825
25		6	6.28	9042	223855
26		6	6.21	9048	224103
27		5	4.86	9067	224533
28	Omega	5	4.01	9072	224617
29		5	5.10	9087	224926
30		5	4.41	9089	224935
31		6	6.32	9092	224995
32	c	5½	5.63	9093	225003
33		4	4.61	3	28
34		6	5.51	26	560
35		6	5.79	50	1061
36		6	6.11	59	1227
37+		6	7.5		1243
38+		7	6.66		1317
39+		6	7.22		1352
40+		6	6.61		1563
41	d	6	5.37	80	1635
42		6	6.23	86	1796
43+		6	6.7		2035
44		6	5.77	97	2114
45+		6	6.78		2140

(table continued on next page)

Table 134. The numbered stars of Pisces (continued)

		MAGNITUDES		CATALOGUE NUMBERS	
Flamsteed Number	Letter	Flamsteed	Visual	HR	HD
46+		6	6.6		2410
47		6	5.06	103	2411
48		6	6.06	106	2436
49+		6	7.0		2714
50+		6			
51		6	5.67	132	2913
52		6	5.38	131	2910
53		7	5.89	155	3379
54		6	5.87	166	3651
55		6	5.36	167	3690
56+		6			
57		6	5.38	211	4408
58		7	5.50	213	4482
59		6	6.13	214	4490
60		6	5.99	216	4526
61		7	6.54	217	4568
62		6	5.93	221	4627
63	Delta	4	4.43	224	4656
64		6	5.07	225	4676
65	i	6	5.54c	230/1	4757/8
66		6	5.74	254	5267
67	k	6	6.00	262	5382
68	h	6	5.42	274	5575
69	Sigma	5	5.50	291	6118
70+		6	7.8		6077
71	Epsilon	4	4.28	294	6186
72		6	5.68	308	6397
73		6	6.00	307	6386
74	Psi[1]	5	4.92c	310/1	6456/7
75		6	6.12	319	6557
76+		5	6.27		6476
77		6	6.35c	313/4	6479/80
78		6	6.25	327	6680
79	Psi[2]	6	5.55	328	6695
80	e	5	5.52	330	6763
81	Psi[3]	6	5.55	339	6903
82	g	6	5.16	349	7034
83	Tau	5	4.51	352	7106
84	Chi	5	4.66	351	7087
85	Phi	5	4.65	360	7318
86	Zeta	4	5.18c	361/2	7344/5
87		7	5.98	364	7374
88		6½	6.03	367	7446
89	f	6	5.16	378	7804
90	Upsilon	5	4.76	383	7964

(table continued on next page)

Table 134. The numbered stars of Pisces *(continued)*

Flamsteed Number	Letter	Magnitudes Flamsteed	Visual	Catalogue Numbers HR	HD
91	1	6	5.23	389	8126
92+		7	6.8		8442
93	Rho	5	5.38	413	8723
94		5	5.50	414	8763
95+		7	7.00		8875
96+		6½	6.7		9024
97		6½	6.02	432	9100
98	Mu	5	4.84	434	9138
99	Eta	4	3.62	437	9270
100+		6	7.0		9656
101		6	6.22	455	9766
102	Pi	5	5.57	463	9919
103+		7¾	6.8		10113
104+		6½	6.76		10135
105		6½	5.97	475	10164
106	Nu	5	4.44	489	10380
107+		6½	5.24	493	10476
108+		6			
109		8	6.27	508	10697
110	Omicron	5	4.26	510	10761
111	Xi	6	4.62	549	11559
112		6½	5.88	582	12235
113	Alpha	3	3.82c	595/6	12446/7

The Lost, Missing, or Troublesome Stars of Pisces

11, HD 221147, 6.5V, *SA,* GC 32671*.

12, HD 221146, 6.89V, *SA,* GC 32672*.

15, HD 221833, 6.48V, *SA,* GC 32798*.

23, in Pegasus, HR 9035, 6.11V, *HRP,* BF 3268*.

    This star was erroneously labeled 83 Peg in *SA* and *Uranometria 2000.0*. It is unlabeled in the second edition of *SA*. See 83 Peg.

37, HD 1243, 7.5V, *SA,* GC 325*.

38, HD 1317, 6.66V, *SA, BS Suppl.,* GC 340*.

39, HD 1352, 7.22V, *SA,* GC 347*.

40, HD 1563, 6.61V, *SA, BS Suppl.,* GC 398*.

43, HD 2035, 6.7V, *SA,* GC 483*.

45, HD 2140, 6.78V, *SA, BS Suppl.,* GC 502*.

46, HD 2410, 6.6V, *SA,* GC 542*.

49, HD 2714, 7.0, SA, GC 598*.

50

    This star does not exist. When Bailey examined Flamsteed's manuscripts, he discovered that Flamsteed had marked the minute of this star's transit on November 30, 1697, as doubtful. If 1^m were to be added to the time of transit, the new coordinates would be the same as those of 52 Psc. The 1^m error resulted in the generation of incorrect coordinates that led Flamsteed to believe he had seen a new star, 50 Psc. See BF 36 and Table B, p. 645.

56

    This star does not exist. Flamsteed twice observed 56 but the observations were actually of 55. In a manuscript fragment of his catalogue, Flamsteed erased the coordinates for 55 but retained those for 56. Baily believed that they applied to one and the same star and that Flamsteed intended to find the mean between the two observations to obtain the correct coordinates, but somehow the two different sets of coordinates found their way into his "Catalogus Britannicus" of 1725. See BF 54 and Table B, p. 645. In Halley's 1712 edition of Flamsteed's catalogue, only one star (HF 52) is shown and it is listed with the mean coordinates of 55 and 56:

        55.........right ascension 5°54'10", declination +19°45'5"
        56.........right ascension 5°55'45", declination +19°42'45"
        HF 52...right ascension 5°55'00", declination +19°43'55"

70, HD 6077, 7.8V, *SA,* GC 1242*.

76, HD 6476, 6.27V, *SA,* GC 1320*.

92, HD 8442, 6.8V, BF 164*, BAC 413*, B 245*, P I, 63*, T 442*, H 100*.

    The star is properly identified in the second edition of *SA*.

95, HD 8875, 7.00V, *SA, BS Suppl.,* GC 1766*.

96, HD 9024, 6.7V, *SA,* GC 1795*.

100, HD 9656, 7.0V, *SA,* GC 1904*.

103, HD 10113, 6.8V, *SA,* GC 1998*.

104, HD 10135, 6.76V, *SA, BS Suppl.,* GC 2000*.

107, HR 493, 5.24V, *SA, HRP, BS,* BF 204 and Table A, p. 645.

    This star is the same as 2 Ari. It is one of Flamsteed's twenty-two duplicate stars.

108

    This star does not exist. In one of his observations of 109 Psc, Flamsteed erred 3° in computing its declination. This produced a new but erroneous set of coordinates that led him to believe he had seen a new star, 108 Psc. See BF 205 and Table B, p. 645.

Piscis Austrinus, PsA

Southern Fish

Piscis Austrinus (Figure 82) is a Ptolemaic constellation that Flamsteed numbered from 1 to 24.

Table 135. The numbered stars of Piscis Austrinus

		MAGNITUDES		CATALOGUE NUMBERS	
Flamsteed Number	Letter	Flamsteed	Visual	HR	HD
1+	Gamma Mic	5	4.67	8039	199951
2+		6	5.18	8076	200763
3+		6	5.42	8110	201901
4+	Epsilon Mic	4½	4.71	8135	202627
5		6	6.50	8214	204394
6		6	5.97	8230	204854
7		6	6.11	8256	205529
8		4½	5.73	8253	205471
9	Iota	4	4.34	8305	206742
10	Theta	4	5.01	8326	207155
11+		6	7.50		208851
12	Eta	5	5.42	8386	209014
13		6	6.47	8405	209476
14	Mu	4	4.50	8431	210049
15	Tau	5½	4.92	8447	210302
16	Lambda	4½	5.43	8478	210934
17	Beta	3	4.29	8576	213398
18	Epsilon	3½	4.17	8628	214748
19		5	6.17	8637	214966
20+		6	6.9		215452
21		6	5.97	8693	216210
22	Gamma	5	4.46	8695	216336
23	Delta	5	4.21	8720	216763
24	Alpha	1	1.16	8728	216956

*Lost Stars*_____

The Lost, Missing, or Troublesome Stars of Piscis Austrinus

1, in Microscopium as Gamma Mic, HR 8039, 4.67V, *SA, HRP,* CA 1701 in Microscopium.
2, in Microscopium, HR 8076, 5.18V, *SA, HRP,* CA 1713 in Microscopium, GC 29465*.
3, in Microscopium, HR 8110, 5.42V, *SA, HRP,* CA 1714 in Capricornus, G 58 in Microscopium, GC 29652*.
4, in Microscopium as Epsilon Mic, HR 8135, 4.71V, *SA, HRP,* CA 1722 in Microscopium.

A quarter of a century after Flamsteed's "Catalogus Britannicus" was published, Lacaille formed Microscopium from the stars between Piscis Austrinus and Sagittarius and assigned Greek letters to its brighter stars. These four were Flamsteed's numbered stars from the Southern Fish. Lacaille switched 1, 2, and 4 to the Microscope and 3 to the Sea Goat. Gould relocated 3 to the Microscope, where it remains today.

11, HD 208851, 7.50V, *SA,* BF 2994*, BAC 7653*, CAP 4399*.

20, HD 215452, 6.9V, *SA,* G 54*, GC 31751*.

Puppis, Pup

Ship's Stern

Puppis (Figure 88) is part of the Ptolemaic constellation Argo Navis that Lacaille divided into four parts: Carina, Puppis, Pyxis, and Vela (see Carina in Part One). From the latitude of the Royal Observatory at Greenwich, Flamsteed could see only the upper half of the Ship, the part Lacaille would later describe as Puppis. Flamsteed called the constellation, simply, Navis, the Ship, and numbered the stars in the upper portion of the Ship's Stern or Poop Deck from 1 to 22.

Table 136. The numbered stars of Puppis

Flamsteed Number	Letter	Magnitudes		Catalogue Numbers	
		Flamsteed	Visual	HR	HD
1		6	4.59	2993	62576
2		6	5.62c	3009/10	62863/4
3	l	4½	3.96	2996	62623
4		6	5.04	3015	62952
5		6	5.48	3029	63336
6		5	5.18	3044	63697
7	Xi	3½	3.34	3045	63700
8		5½	6.36	3063	64077
9		4	5.17	3064	64096
10		6	5.69	3073	64238
11	j	4	4.20	3102	65228
12		6	5.11	3123	65699
13+		4	4.39	3145	66141
14		6	6.13	3168	66834
15	Rho	3	2.81	3185	67523
16		5	4.40	3192	67797
17+		6			
18		6	5.54	3202	68146
19		4½	4.72	3211	68290
20		5½	4.99	3229	68752
21		6	6.16	3257	69665
22		6	6.11	3289	70673

The Lost, Missing, or Troublesome Stars of Puppis

13, in Canis Minor, HR 3145, 4.39V, *HRP,* BF 1125*, BAC 2673*, B 266*, B 58* in Canis Minor.

This star's Flamsteed number is often omitted from atlases and catalogues because it is located nowhere near the Ship. For some inexplicable reason, Flamsteed included it in the Ship even though it is more than a dozen degrees north of Puppis's border, past Monoceros, and into Canis Minor. This location has caused considerable consternation to generations of stellar cartographers who have tried all sorts of intricate ploys to somehow connect 13 with the other stars in the Ship. Bode found a more practical solution. He included 13 both in the Ship and also in the Little Dog, where he lettered it Lambda. In both constellations, Bode did acknowledge that the star was originally Flamsteed's 13 Nav.

17, BF 1155* and Table D, p. 646.

This star does not exist. When Baily examined Flamsteed's manuscripts, he found no record that the star had ever been observed. He suggested that it may have found its way into the "Catalogus Britannicus" as the result of a computation error for 20 Pup, which, in turn, generated erroneous coordinates that led Flamsteed to believe he had seen a new star, 17 Pup.

Sagitta, Sge

Arrow

Sagitta (figures 83 and 93) is a Ptolemaic constellation that Flamsteed numbered from 1 to 18.

Table 137. The numbered stars of Sagitta

Flamsteed Number	Letter	Magnitudes		Catalogue Numbers	
		Flamsteed	Visual	HR	HD
1		6	5.64	7301	180317
2		6	6.25	7369	182490
3+		6	6.82		182571
4	Epsilon	5	5.66	7463	185194
5	Alpha	4	4.37	7479	185758
6	Beta	4	4.37	7488	185958
7	Delta	4½	3.82	7536	187076
8	Zeta	6	5.00	7546	187362
9		6	6.23	7574	188001
10		6	5.36	7609	188727
11		6	5.33	7622	189090
12	Gamma	4	3.47	7635	189319
13		6	5.37	7645	189577
14+		6	5.67	7664	190229
15		6	5.80	7672	190406
16	Eta	6	5.10	7679	190608
17	Theta	6	6.48	7705	191570
18		6	6.13	7746	192836

The Lost, Missing, or Troublesome Stars of Sagitta

3, HD 182571, 6.82V, *SA, BS Suppl.,* GC 26802*.

14, in Aquila, HR 7664, 5.67V, *SA, HRP, BS,* GC 27812*.

Sagittarius, Sgr

Archer

Sagittarius (Figure 102) is a Ptolemaic constellation that Flamsteed numbered from 1 to 65.

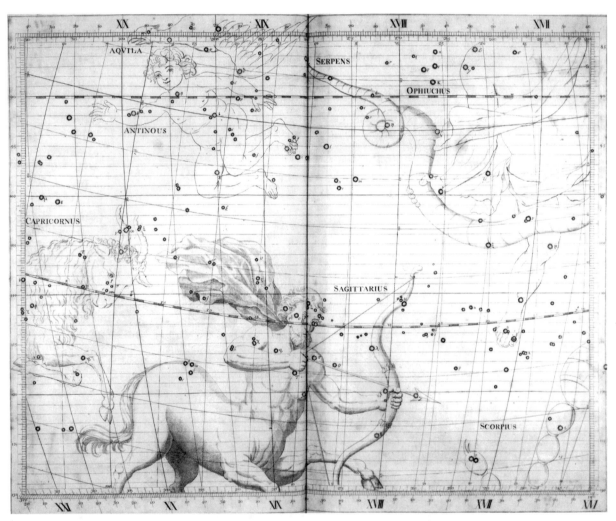

Figure 102. Flamsteed's Sagittarius. Included in this chart is the asterism Antinous although it is omitted from the chart of Aquila (Figure 83). The lower portion of the Archer is in Figure 107. The star 8 Sgr at 266°14' r.a. and 109°18' n.p.d. does not exist.

Table 138. The numbered stars of Sagittarius

		MAGNITUDES		CATALOGUE NUMBERS	
Flamsteed Number	Letter	Flamsteed	Visual	HR	HD
1+		6	4.98	6801	166464
2+		6	6.8		160042
3		6	4.54	6616	161592
4		6½	4.76	6700	163955
5+		7	6.7		164031
6		7	6.28	6715	164358
7		6	5.34	6724	164584
8+		7			
9		7	5.97	6736	164794
10	Gamma²	3	2.99	6746	165135
11+		7	7.28		165921
12+		7	6.66		166767
13	Mu	4	3.86	6812	166937
14		7	5.44	6816	167036
15		6	5.38	6822	167264
16		7	5.95	6823	167263
17+		7	7.0		167570
18		7	5.60	6888	169233
19	Delta	3	2.70	6859	168454
20	Epsilon	3	1.85	6879	169022
21		6	4.81	6896	169420
22	Lambda	4	2.81	6913	169916
23+		7	6.9		170457
24		7	5.49	6961	171115
25		7	6.51	6965	171237
26		6	6.23	7011	172546
27	Phi	5	3.17	7039	173300
28		7	5.37	7046	173460
29		6	5.24	7078	174116
30		6	6.61	7088	174309
31+		6	6.8		174596
32	Nu¹	5	4.83	7116	174974
33		6	5.69	7114	174947
34	Sigma	3¾	2.02	7121	175191
35	Nu²	5	4.99	7120	175190
36+	Xi¹		5.08	7145	175687
37+	Xi²	6	3.51	7150	175775
38	Zeta	3	2.60	7194	176687
39	Omicron	4	3.77	7217	177241
40	Tau	4	3.32	7234	177716
41	Pi	4	2.89	7264	178524
42	Psi	5	4.85	7292	179950
43	d	6	4.96	7304	180540
44	Rho¹	5	3.93	7340	181577
45	Rho²	6	5.87	7344	181645

(table continued on next page)

Table 138. The numbered stars of Sagittarius *(continued)*

		MAGNITUDES		CATALOGUE NUMBERS	
Flamsteed Number	Letter	Flamsteed	Visual	HR	HD
46	Upsilon	6	4.61	7342	181615
47	Chi^1	5	5.03	7362	182369
48+	Chi^2	5	7.1		182391
49	Chi^3	6	5.43	7363	182416
50		6	5.59	7375	182629
51	h^1	6	5.65	7431	184552
52	h^2	6	4.60	7440	184707
53		6	6.34	7470	185404
54	e^1	6	6.2	7476	185644
55	e^2	6	5.06	7489	186005
56	f	6	4.86	7515	186648
57		6	5.92	7561	187739
58	Omega	5	4.70	7597	188376
59	b	5	4.52	7604	188603
60	A	5	4.83	7618	189005
61	g	6	5.02	7614	188899
62	c	6	4.58	7650	189763
63		6	5.71	7649	189741
64+		6	6.34	7671	190390
65		6	6.55	7675	190454

The Lost, Missing, or Troublesome Stars of Sagittarius

1, HR 6801, 4.98V, *SA* erroneously as 11, *HRP, BS* erroneously as 11, BF 2470*, BAC 6161, G 40, BKH 30*, W 8*.

Considerable confusion surrounds 1 Sgr primarily because no star can be found at the coordinates listed for it in Flamsteed's catalogue. Baily discovered that the Astronomer Royal had made two mistakes in reducing this star to 1690 — a 10° error in right ascension and a 2° error in declination. Baily corrected these mistakes in his revised edition of Flamsteed's catalogue. Pickering took account of Flamsteed's mistakes as corrected by Baily and properly identified HR 6801 as 1 Sgr in the *HRP*. Later astronomers, though, have somehow confused 1 Sgr with 11 Sgr. Antonin Becvar, Dorrit Hoffleit, and Wil Tirion have all mistakenly labeled 1 Sgr as 11 Sgr. 1 Sgr is HR 6801; 11 Sgr is HD 165921, about ½° to the southwest (see 11 Sgr). Two of the most recent catalogues, however, properly identify 1 Sgr — Hirshfeld and Sinnott's *Sky Cat.* and Werner and Schmeidler's *Synopsis of the Nomenclature of the Fixed Stars,* pp. 424-25. In the second edition of *SA*, this star, HR 6801, is left unlabeled.

2, in Ophiuchus, HD 160042, 6.8V, BF 2408*, BAC 5954 in Ophiuchus, B 4*, P XVII, 160*, BKH 121* in Ophiuchus, G 158 in Ophiuchus.

5, HD 164031, 6.7V, *SA*, GC 24492*.

8

This star does not exist. In one of his observations of 9 Sgr, Flamsteed made a 5° error in declination while reducing it to the year 1690. This generated a new set of coordinates that led him to believe he had seen a new star, 8 Sgr. See BF 2448 and Table B, p. 645.

11, HD 165921, 7.28V, BF 2466*, BAC 6141, P XVII, 366, CAP 3480.

This star's Flamsteed number is often omitted from atlases and catalogues because no star exists at the coordinates listed in the "Catalogus Britannicus." When Baily examined Flamsteed's manuscripts, he discovered that Flamsteed observed this star twice, once on May 26, 1711, and again on the next day. When reducing the star to 1690, the epoch of his catalogue, he made two errors. He inadvertently calculated its right ascension as 266°35'15" when it should have been exactly 1° more and he mistakenly ascribed to 11 Sgr the declination of a neighboring star, 12 Sgr, -23°8'10", instead of its own declination of -23°59'. Baily made the necessary corrections in his revised edition of Flamsteed's catalogue. This is not the star designated as 11 Sgr in *SA* or *Uranometria 2000.0*. The second edition of *SA* leaves this star unnumbered. See 1 Sgr.

12, HD 166767, 6.66V, BF 2473*, BAC 6165*, CAP 3496*, the variable AP Sgr.

This star's Flamsteed number is omitted from most atlases and catalogues because no star exists in the sky at the coordinates listed in Flamsteed's catalogue. The Astronomer Royal observed this star on May 27, 1711, but in the process of reducing it to the year 1690 he made errors both in right ascension (1'30" too great) and declination (52' too great). Baily made the necessary corrections in his revised edition of Flamsteed's catalogue. Please note that the right ascension figure listed for this star in the *BAC* should be increased by 30^s. The star is properly identified in the second edition of *SA*.

17, HD 167570, 7.0V, *SA*, GC 24918*.

23, HD 170457, 6.9V, BF 2501*, BAC 6286*, CAP 3585*.

This star's Flamsteed number is often omitted from atlases and catalogues because there is no star in the sky at the coordinates listed in his "Catalogus Britannicus." In one of his observations of this star, Flamsteed erred in both right ascension (1' too great) and declination (40' too great) in reducing it to 1690. Baily corrected the mistakes in his revised edition of Flamsteed's catalogue. The star is properly identified in the second edition of *SA*.

31, HD 174596, 6.8V, *SA*, GC 25867*.

36, Xi^1, HR 7145, 5.08V, *SA, HRP, BS*, HF 21, BF 2548*.
37, Xi^2, HR 7150, 3.51V, *SA, HRP, BS*, HF 22, BF 2550*.

In Halley's 1712 edition of Flamsteed's catalogue, the magnitudes for these two stars are given as 5 and 4, respectively. Baily noted in his revised edition of Flamsteed's catalogue that the "Catalogus Britannicus" lists the magnitudes of 36 and 37 as 5 and 6, respectively, but in the copy at the New York Public Library, only a tiny dot is shown for 36's magnitude, probably a printing flaw of some kind. See also 126 Tau.

48, Chi^2, HD 182391, 7.1V, BF 2612*, BAC 6634*, P XIX, 94*, CAP 3827*, T 8888*.

Tirion's *SA* and *Uranometria 2000.0* erroneously label 49 as Chi^2 when it should be Chi^3. The second edition of *SA* properly designates 49 as Chi^3, but omits lettering 48 as Chi^2. See Chi Sgr in Part One. Baily discovered that Flamsteed's manuscripts describe this star as a telescopic sight *(telescopica)* with a magnitude of 8.

64, in Aquila, HR 7671, 6.34V, *SA*, GC 27850*.

This star is erroneously labeled in *HRP* as 64 Aql; it is unlabeled in the *BS*.

Scorpius, Sco

Scorpion

Scorpius (Figure 99) is a Ptolemaic constellation that Flamsteed numbered from 1 to 35.

Table 139. The numbered stars of Scorpius

		MAGNITUDES		CATALOGUE NUMBERS	
Flamsteed Number	**Letter**	**Flamsteed**	**Visual**	**HR**	**HD**
1	b	6	4.64	5885	141637
2	A	5	4.59	5904	142114
3		7	5.87	5912	142301
4		6	5.62	5917	142445
5	Rho	4	3.88	5928	142669
6	Pi	3	2.89	5944	143018
7	Delta	3	2.32	5953	143275
8	Beta[1,2]	2	2.76c	5984/5	144217/8
9	Omega[1]	5	3.96	5993	144470
10	Omega[2]	5	4.32	5997	144608
11		6	5.78	6002	144708
12	c[1]	6	5.67	6029	145483
13	c[2]	6	4.59	6028	145482
14	Nu	4	4.16c	6026/7	145501/2
15	Psi	5	4.94	6031	145570
16		6	5.43	6033	145607
17	Chi	6	5.22	6048	145897
18		4	5.50	6060	146233
19	o	6	4.55	6081	147084
20+	Sigma	5	2.89	6084	147165
21	Alpha	1	0.96	6134	148478
22	i	5½	4.79	6141	148605
23	Tau	4	2.82	6165	149438
24+		6	4.96	6196	150416
25		6	6.71	6225	151179
26	Epsilon	3	2.29	6241	151680
27		6	5.48	6288	152820
28+		6	6.30	6350	154418

(table continued on next page)

Table 139. The numbered stars of Scorpius *(continued)*

Flamsteed Number	Letter	Magnitudes		Catalogue Numbers	
		Flamsteed	Visual	HR	HD
29+		6	6.64		155685
30+		6	6.34		156026
31+		6½	6.81		156252
32+		6	6.6		156992
33+		7	6.19	6474	157588
34	Upsilon	4	2.69	6508	158408
35	Lambda	3	1.63	6527	158926

The Lost, Missing, or Troublesome Stars of Scorpius

20, Sigma, HR 6084, 2.89V, *SA, HRP, BS,* HF 20*, BF 2235*.

Flamsteed's "Catalogus Britannicus" of 1725 lists this star as magnitude 5, whereas Halley's 1712 edition describes it as magnitude 4.

24, in Ophiuchus, HR 6196, 4.96V, *SA, HRP,* BAC 5579, GC 22449*.

28, in Ophiuchus, HR 6350, 6.30V, *HRP,* BF 2346*, BAC 5788.

29, in Ophiuchus, HD 155685, 6.64V, BF 2360*, BAC 5800, B 177*, P XVII, 6*, T(P) 7963*.

In an accompanying note, the *BAC* mistakenly states that Flamsteed called this star 29 Oph.

30, in Ophiuchus, HD 156026, 6.34V, BF 2363*, BAC 5813*, B 181*, P XVII, 21*, T(P) 7979*.

31, in Ophiuchus, HD 156252, 6.81V, BF 2366* and Table A, p. 645.

This star is the same as 38 Oph. It is one of Flamsteed's twenty-two duplicate stars.

32, in Ophiuchus, HD 156992, 6.6V, BF 2373*, BAC 5846, B 188*, P XVII, 51*, T 7994.

33, in Ophiuchus, HR 6474, 6.19V, BF 2380*, BAC 5868, B 190*, P XVII, 77*, T(P) 8049*.

This star, 33 Sco, and the other stars listed above except for 20, are all included by the *BAC* in Ophiuchus since most are more than 5° past the border of Scorpius. Most atlases and catalogues omit their Flamsteed numbers to avoid confusion with Ophiuchus's numbered stars.

Serpens, Ser

Serpent

Serpens (Figure 100) is a Ptolemaic constellation that Flamsteed numbered from 1 to 64. In Bayer's *Uranometria*, the Serpent Bearer, Ophiuchus, holds Serpens facing away from the reader. As a result, Bayer lettered only the head and tail of the Serpent since its middle portion is hidden by Ophiuchus's torso. Although Flamsteed's atlas shows Ophiuchus facing the reader, most of the stars that the Astronomer Royal numbered in Serpens are concentrated either in the Serpent's head or tail. Gould decided, therefore, to split Serpens into two sections: (1) Pars Anterior, Caput (the Front Part, the Head); and (2) Pars Posterior, Cauda (the Rear Part, the Tail). Gould assigned Flamsteed's stars 1 to 50 to the Head and stars 53 to 64 to the Tail. For the fate of 51 and 52, see below.

Table 140. The numbered stars of Serpens

Flamsteed Number	Letter	Magnitudes		Catalogue Numbers	
		Flamsteed	Visual	HR	HD
1+	M Vir	7	5.53	5573	132132
2+		7	5.71	5594	132933
3		6½	5.33	5675	135482
4		6	5.63	5679	135559
5		6	5.06	5694	136202
6		6	5.35	5710	136514
7		7	6.28	5717	136831
8		7	6.12	5721	137006
9	Tau^1	6	5.17	5739	137471
10		6	5.17	5746	137898
11	A^1	6	5.51	5772	138562
12	Tau^2	7	6.22	5770	138527
13	Delta	3	3.8c	5788/9	138917/8
14		6	6.51	5799	139137
15	Tau^3	6	6.12	5795	139074
16		7	5.26	5802	139195
17+	Tau^4	6½	6.65		139216
18	Tau^5	6	5.93	5804	139225
19	Tau^6	6	6.01	5840	140027
20	Chi	6	5.33	5843	140160
21	Iota	5	4.52	5842	140159
22	Tau^7	6	5.81	5845	140232

(table continued on next page)

Table 140. The numbered stars of Serpens *(continued)*

		MAGNITUDES		CATALOGUE NUMBERS	
Flamsteed Number	Letter	Flamsteed	Visual	HR	HD
23	Psi	6	5.88	5853	140538
24	Alpha	2	2.65	5854	140573
25	A²	6	5.40	5863	140873
26	Tau⁸	6	6.14	5858	140729
27	Lambda	4	4.43	5868	141004
28	Beta	3	3.67	5867	141003
29+		5½	6.8		141040
30+		6	5.53	5875	141378
31	Upsilon	6	5.71	5870	141187
32	Mu	4	3.53	5881	141513
33+		6			
34	Omega	6	5.23	5888	141680
35	Kappa	4	4.09	5879	141477
36	b	6	5.11	5895	141851
37	Epsilon	3	3.71	5892	141795
38	Rho	3¾	4.76	5899	141992
39		6	6.10	5911	142267
40		7	6.29	5919	142500
41	Gamma	3	3.85	5933	142860
42+		6			
43		6	6.08	5976	144046
44	Pi	4	4.83	5972	143894
45		6	5.63	6004	144874
46+		6	6.74		144937
47		6	5.73	6010	145002
48+	q Her	6	6.08	6035	145647
49+		6	6.68		145958
50	Sigma	5	4.82	6093	147449
51+	Omega Her	6	4.57	6117	148112
52+		6			
53	Nu	4	4.33	6446	156928
54+		6			
55	Xi	4	3.54	6561	159876
56	Omicron	5	4.26	6581	160613
57	Zeta	3	4.62	6710	164259
58	Eta	3	3.26	6869	168723
59	d	6	5.21	6918	169985
60	c	6	5.39	6935	170474
61		6	5.94	6957	170920
62+		6	5.57	7135	175515
63	Theta¹,²	3	4.10c	7141/2	175638/9
64		6	5.57	7158	175869

The Lost, Missing, or Troublesome Stars of Serpens

1, in Virgo as M Vir, HR 5573, 5.53V, *SA, HRP, BS,* GC 20122*.

2, in Virgo, HR 5594, 5.71V, *SA, HRP,* GC 20212*.

17, Tau⁴, HD 139216, 6.65V, *SA, BS Suppl.,* GC 20983*.

29, HD 141040, 6.8V, BF 2154*, BAC 5219*, B 71*, P XV, 171*, T 7330*.

This star's Flamsteed number is often omitted from atlases and catalogues because there is no star in the heavens at the coordinates listed in the "Catalogus Britannicus." When Baily examined Flamsteed's manuscripts, he discovered that the Astronomer Royal had erred 2'30" in determining 29's declination. Baily corrected the error in his revised edition of Flamsteed's catalogue. The star is properly identified in the second edition of *SA*.

30, in Libra, HR 5875, 5.53V, *SA, HRP,* GC 21251*.

33

This star does not exist. In one of his observations of 34 Ser, Flamsteed erred 20° in reading its zenith distance. He corrected the error in the margin of his manuscript but somehow used the incorrect figure in computing the star's declination. This produced a new set of coordinates that led him to believe he had observed a new star, 33 Ser. See BF 2158 and Table B, p. 645.

42, BF 2191* and Table D, p. 646.

This star does not exist. Baily noted that there is no record that Flamsteed ever observed this star, and he was at a loss to explain how it found its way into the "Catalogus Britannicus." He assumed it might have arisen from a 2^h error in computing the right ascension of 69 Oph.

46, HD 144937, 6.74V, *SA, BS Suppl.,* GC 21692*.

48, in Hercules as q Her, HR 6035, 6.08V, *HRP,* BF 2218*.

See q Her in Part One.

49, in Hercules, HD 145958, 6.68V, *SA, BS Suppl.,* GC 21821*.

51, in Hercules as Omega Her, HR 6117, 4.57V, *HRP,* BF 2257 and Table A, p. 645.

This star is the same as 24 Her. It is one of Flamsteed's twenty-two duplicate stars.

52

This star does not exist. In one of his observations of 53 Ser, Flamsteed erred in determining its right ascension. This error led him to calculate a new set of coordinates and to believe he had seen a new star, 52 Ser. See BF 2375 and Table B, p. 645.

54

This star does not exist. In one of his observations of 56 Ser, Flamsteed erred in determining its right ascension. This generated a new but erroneous set of coordinates that led him to believe he had seen a new star, 54 Ser. See BF 2413 and Table B, p. 645.

62, in Aquila, HR 7135, 5.57V, *SA, HRP,* GC 25964*.

Sextans, Sex

Astronomical Sextant

Hevelius devised the constellation Sextans (figures 92 and 92a). Flamsteed acknowledged that it was a Hevelian constellation and included it in his atlas and catalogue, numbering its stars from 1 to 41.

Table 141. The numbered stars of Sextans

Flamsteed Number	Letter	Magnitudes		Catalogue Numbers	
		Flamsteed	Visual	HR	HD
1+		5	5.00	3827	83240
2+	H Hya	5	4.68	3834	83425
3+		6	7.1		84882
4		6	6.24	3893	85217
5+		6	7.0		85294
6		6	6.01	3899	85364
7		6	6.02	3906	85504
8	Gamma	6	5.05	3909	85558
9+		6	6.65		85762
10+		6	5.85	3926	86080
11+		5½	6.04	3938	86369
12		6	6.70	3945	86611
13		6	6.45	3961	87301
14		6	6.21	3973	87682
15	Alpha	4	4.49	3981	87887
16+		6	6.8		88048
17		6	5.91	3989	88195
18		6	5.65	3996	88333
19		6	5.77	4004	88547
20+		6	7.22		88697
21+		6	6.97		88764
22	Epsilon	6	5.24	4042	89254
23		5	6.66	4064	89688
24+		6	6.6		90043
25		6	5.97	4082	90044
26+		6	6.31		90473
27+		6	6.54		90485
28+		5			

(table continued on next page)

Table 141. The numbered stars of Sextans (continued)

Flamsteed Number	Letter	Magnitudes Flamsteed	Visual	Catalogue Numbers HR	HD
29	Delta	5	5.21	4116	90882
30	Beta	5	5.09	4119	90994
31+		6	6.99		91011
32+		6	7.1		91256
33		6	6.26	4182	92588
34+		6	6.6		92749
35		6	5.79	4193	92841
36		6	6.28	4201	93102
37+		6	6.37	4207	93244
38+		6	7.0		93431
39+		7	7.1		93704
40		6	6.61	4229	93742
41		6	5.79	4237	93903

The Lost, Missing, or Troublesome Stars of Sextans

1, In Leo, HR 3827, 5.00V, *SA, HRP,* BF 1357 and Table A, p. 645.

This star is the same as 10 Leo. It is one of Flamsteed's twenty-two duplicate stars.

2, in Hydra as H Hya, HR 3834, 4.68V, *SA, HRP,* GC 13316*.

3, HD 84882, 7.1V, *SA,* GC 13504*.

5, HD 85294, 7.0V, BF 1394*, BAC 3363*, B 19*, P IX, 191*, CAP 1777*, T 4326*.

This star's Flamsteed number is often omitted from atlases and catalogues because no star exists in the sky at the coordinates listed in his "Catalogus Britannicus." When observing this star on February 28, 1702, he erred 4' in determining its right ascension. Flamsteed never reobserved this star. Baily alerted astronomers to the error in his revised edition of Flamsteed's catalogue, and later in the *BAC* he equated it with Piazzi's IX, 191. The star is properly identified in the second edition of *SA*.

9, HD 85762, 6.65V, *SA, BS Suppl.,* GC 13617*.

10, in Leo, HR 3926, 5.85V, *HRP,* BF 1408*.

11, in Leo, HR 3938, 6.04V, *HRP,* BF 1411*.

16, HD 88048, 6.8V, *SA,* GC 13943*.

20, HD 88697, 7.22V, *SA,* GC 14043*.

21, HD 88764, 6.97V, *SA, BS Suppl.,* GC 14053*.

24, HD 90043, 6.6V, *SA,* GC 14267*.

26, HD 90473, 6.31V, *SA, BS Suppl.,* GC 14333*.

27, HD 90485, 6.54V, *SA, BS Suppl.,* GC 14338*.

28, BF 1486* and Table C, p. 645.

This star does not exist. Flamsteed observed this star on February 22, 1702, at $10^h48^m36^s$, but there is no star in the heavens located at the coordinates listed in his "Catalogus Britannicus." When Baily examined Flamsteed's manuscripts, he noted that the figure 48^m had originally read 58^m. He believed that Flamsteed erred 2^m in recording this star's time of transit and that the original entry should have read 50^m. Baily pointed out that if this assumption were correct, then the star Flamsteed observed would be 29 Sex. Peters agreed with Baily that 28 is the same as 29 ("Flamsteed's Stars," pp. 78-79).

31, HD 91011, 6.99V, *SA, BS Suppl.,* GC 14434*.

32, HD 91256, 7.1V, *SA,* GC 14476*.

34, HD 92749, 6.6V, *SA,* GC 14723*.

37, in Leo, HR 4207, 6.37V, *SA, HRP, BS,* GC 14805*.

38, in Leo, HD 93431, 7.0V, *SA,* GC 14846*.

39, HD 93704, 7.1V, *SA,* GC 14883*.

Taurus, Tau

Bull

Taurus (Figure 103) is a Ptolemaic constellation that Flamsteed numbered from 1 to 141. It is the constellation with the most stars in Flamsteed's catalogue. But in Halley's pirated edition of 1712, Ursa Major is the largest constellation, with 215 stars. This unauthorized catalogue contains stars from only forty-seven constellations. In his "Catalogus Brittanicus" of 1725, however, Flamsteed included eight new constellations: six Hevelian and two from Petrus Plancius. After making these additions to his catalogue, he shifted many of Ursa Major's stars to two of the new constellations, Camelopardalis and Lynx, thus reducing the Great Bear's stars to eighty-seven. See Andromeda for additional details about the constellations in the two catalogues.

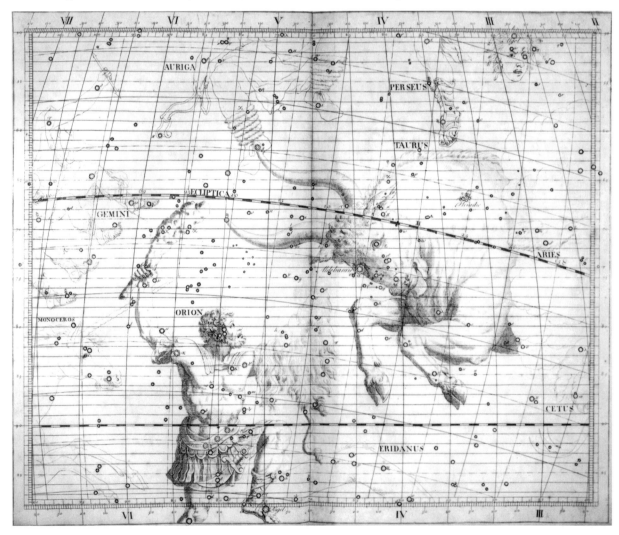

Table 142. The numbered stars of Taurus

		MAGNITUDES		CATALOGUE NUMBERS	
Flamsteed Number	Letter	Flamsteed	Visual	HR	HD
1	Omicron	4	3.60	1030	21120
2	Xi	4	3.74	1038	21364
3+		6			
4	s	6	5.14	1061	21686
5	f	5	4.11	1066	21754
6	t	6	5.77	1079	21933
7		6	5.92	1086	22091
8+		6			
9+		6	6.65		22374
10		4½	4.28	1101	22484
11		6	6.11	1118	22805
12		6	5.57	1115	22796
13		6	5.69	1126	23016
14		6	6.14	1132	23183
15+		6			
16		7	5.46	1140	23288
17		5	3.70	1142	23302
18		7	5.64	1144	23324
19	q	5	4.30	1145	23338
20		6	3.87	1149	23408
21		6½	5.76	1151	23432
22		7	6.43	1152	23441
23		5	4.18	1156	23480
24+		7	6.29		23629
25	Eta	3	2.87	1165	23630
26+		7½	6.47		23822
27		6	3.63	1178	23850
28		7½	5.09	1180	23862
29	u	6	5.35	1153	23466
30	e	5	5.07	1174	23793
31		6	5.67	1199	24263
32		6	5.63	1218	24740
33		7	6.06	1221	24769
34+		7			
35	Lambda	4	3.47	1239	25204

(table continued on next page)

Figure 103 (at left). Flamsteed's Taurus and Orion. The tiny 7th-magnitude star in Taurus between A and ω (Omega) and slightly to the right on the ecliptic at ♉ 28° is Uranus, which Flamsteed, unaware of its significance, labeled 34 Tau. One hundred years would have to elapse before 34 Tau was recognized as a planet by William Herschel. The following stars in Taurus do not exist: **3 Tau**, 48°16' r.a., 68°56' n.p.d.; **8 Tau**, 49°23' r.a., 71°51' n.p.d.; **15 Tau**, 51°37' r.a., 68° 2' n.p.d.; **34 Tau**, 55°48' r.a., 70°25' n.p.d.; **82 Tau**, 63°15' r.a., 75°37' n.p.d.; **100 Tau**, 70° 2' r.a., 74°10' n.p.d.; **124 Tau**, 80°3' r.a., 66°57' n.p.d.; **138 Tau**, 83°41' r.a., 76°10' n.p.d.

Table 142. The numbered stars of Taurus (continued)

		MAGNITUDES		CATALOGUE NUMBERS	
Flamsteed Number	Letter	Flamsteed	Visual	HR	HD
36		7	5.47	1252	25555
37	A[1]	5	4.36	1256	25604
38	Nu	4	3.91	1251	25490
39	A[2]	6	5.90	1262	25680
40		7	5.33	1253	25558
41		6	5.20	1268	25823
42	Psi	5	5.23	1269	25867
43	Omega[1]	6	5.50	1283	26162
44	p	6	5.41	1287	26322
45		7	5.72	1292	26462
46		7	5.29	1309	26690
47+		7	4.84	1311	26722
48		7	6.32	1319	26911
49	Mu	4	4.29	1320	26912
50	Omega[2]	6	4.94	1329	27045
51		7	5.65	1331	27176
52	Phi	5	4.95	1348	27382
53		7	5.35	1339	27295
54	Gamma	3	3.65	1346	27371
55+		7	6.89		27383
56		7	5.38	1341	27309
57	h	6½	5.59	1351	27397
58		7	5.26	1356	27459
59	Chi	5	5.37	1369	27638
60		7	5.72	1368	27628
61	Delta[1]	4	3.76	1373	27697
62		7	6.36	1378	27778
63		6	5.64	1376	27749
64	Delta[2]	4	4.80	1380	27819
65	Kappa[1]	5	4.22	1387	27934
66	r	5	5.12	1381	27820
67	Kappa[2]	5	5.28	1388	27946
68	Delta[3]	6	4.29	1389	27962
69	Upsilon	5	4.28	1392	28024
70		7	6.46	1391	27991
71		7	4.49	1394	28052
72		6	5.53	1399	28149
73	Pi	5	4.69	1396	28100
74	Epsilon	3½	3.53	1409	28305
75		7	4.97	1407	28292
76		7	5.90	1408	28294
77	Theta[1]	5	3.84	1411	28307
78	Theta[2]	5	3.40	1412	28319
79	b	5	5.03	1414	28355
80		7	5.58	1422	28485

(table continued on next page)

Table 142. The numbered stars of Taurus *(continued)*

		MAGNITUDES		CATALOGUE NUMBERS	
Flamsteed Number	Letter	Flamsteed	Visual	HR	HD
81		7	5.48	1428	28546
82+		7			
83		7	5.40	1430	28556
84+		7	6.34		28595
85		7	6.02	1432	28677
86	Rho	5	4.65	1444	28910
87	Alpha	1	0.85	1457	29139
88	d	5	4.25	1458	29140
89		7	5.79	1472	29375
90	c¹	5	4.27	1473	29388
91	Sigma¹	6	5.07	1478	29479
92	Sigma²	6	4.69	1479	29488
93	c²	6	5.46	1484	29589
94	Tau	5	4.28	1497	29763
95		6½	6.13	1499	29589
96		6	6.08	1537	30605
97	i	6	5.10	1547	30780
98	k	6	5.81	1590	31592
99		6	5.79	1586	31553
100+		6			
101+		6	6.76		31845
102	Iota	4	4.64	1620	32301
103		6	5.50	1659	32990
104	m	6	5.00	1656	32923
105		6	5.89	1660	32991
106	l	6	5.29	1658	32977
107+		6	6.5		33121
108		7	6.27	1711	34053
109	n	6	4.94	1739	34559
110		7	6.08	1774	35189
111		6½	4.99	1780	35296
112	Beta	2	1.65	1791	35497
113		6	6.25	1798	35532
114	o	5	4.88	1810	35708
115+		7½	5.42	1808	35671
116		6½	5.50	1814	35770
117		7	5.77	1816	35802
118		6	5.80	1821	35943
119+		7	4.38	1845	36389
120		7	5.69	1858	36576
121		6	5.38	1875	36819
122		7	5.54	1905	37147
123	Zeta	3	3.00	1910	37202
124+		6½			
125+		3	5.18	1928	37438

(table continued on next page)

Table 142. The numbered stars of Taurus *(continued)*

Flamsteed Number	Letter	Magnitudes		Catalogue Numbers	
		Flamsteed	Visual	HR	HD
126+			4.86	1946	37711
127+		6	6.7		37940
128+		6	6.8		38219
129		6	6.00	1985	38478
130		6	5.49	1990	38558
131		6	5.72	1989	38545
132		4	4.86	2002	38751
133		6	5.29	1993	38622
134		6	4.91	2010	38899
135		6	5.52	2016	39019
136		5	4.58	2034	39357
137		5	5.59	2033	39317
138+		6			
139		6	4.82	2084	40111
140+		6	6.91		40545
141+		6	6.37	2116	40724

The Lost, Missing, or Troublesome Stars of Taurus

3

This star does not exist. After one of his observations of 102 Tau, Flamsteed erred by 1½ hours in reducing its right ascension to 1690, the epoch of his catalogue. The mistake produced a new but erroneous set of coordinates that led Flamsteed to believe he had seen a new star, 3 Tau. See BF 645 and Table B, p. 645.

8

This star does not exist. After one of his observations of 104 Tau, Flamsteed erred by 1½ hours in reducing its right ascension. The mistake produced a new but erroneous set of coordinates that led him to believe he had seen a new star, 8 Tau. This mistake, like the one involving 3 Tau, above, resulted from observations he had made on October 1, 1704. See BF 654 and Table B, p. 645.

9, HD 22374, 6.65V, *SA, BS Suppl.,* GC 4307*.

15

This star does not exist. After observing 16 Tau on January 10, 1690, Flamsteed mistakenly copied the zenith distance of another star he had observed on the same day, Alpha Ari. This misstep produced a new set of coordinates that led him to believe he had seen a new star, 15 Tau. See BF 457 and Table B, p. 645.

24, HD 23629, 6.29V, *SA, BS Suppl.,* GC 4536*, in the Pleiades.

This is component B of the multiple-star system Struve I, 8. Component A is 25, Eta Tau (HR 1165, 2.87V). Flamsteed observed and numbered these two components. *BS Suppl.* designates 24 as well as 25 as Eta. See q Tau in Part One for a listing of the 14 Pleiades observed by Flamsteed.

26, HD 23822, 6.47V, *BS Suppl.,* BF 472*, BAC 1173*, in the Pleiades.

The star is properly identified in the second edition of *SA*.

34, Uranus.

Thinking it was just another dim 7th-magnitude star, Flamsteed observed Uranus on several occasions. He observed it first on December 13, 1690, in Taurus (34 Tau) and again almost a quarter of a century later on December 3, 1714, in Leo (BF 1647). Both of these sightings have been confirmed by Professor William Blitzstein of the Department of Astronomy and Astrophysics of the University of Pennsylvania, who compared Uranus's position on those dates with Flamsteed's manuscript notebooks at the Royal Greenwich Observatory (letter to author, September 12, 1994). Baily believed that Flamsteed may have observed Uranus on at least five additional occasions: March 22, 1712; February 21, 22, 27, and April 18, 1715. In an unpublished paper ("A Study by Modern Methods of Seven Alleged Pre-Discovery Observations of Uranus by John Flamsteed"), Professor Blitzstein argued that all of these sightings were indeed of Uranus. BF 1647, seen by Flamsteed in Leo and cited above, is one of 458 stellar objects that Flamsteed observed but that were inadvertently omitted from his "Catalogus Britannicus." See BF 1647 and *BF*, pp. 377, 392-93, and Table B, p. 645; Peters, "Flamsteed Stars," p. 79; *Sky & Telescope,* January 1978, p. 52.

47, HR 1311, 4.84V, *SA, HRP, BS,* HF 44, BF 523*.

Baily noted that this star's magnitude is described as 5½ in Flamsteed's manuscripts although it is later listed as 7 in both Halley's edition of Flamsteed's manuscript and the "Catalogus Britannicus."

55, HD 27383, 6.89V, *SA, BS Suppl.*

82

This star does not exist. After observing 81 Tau on February 4, 1691, Flamsteed copied the wrong declination. This mistake produced a new set of coordinates that led him to believe he had seen a new star, 82 Tau. See BF 571 and Table B, p. 645.

84, HD 28595, 6.34V, *SA, BS Suppl.,* GC 5495*.

100, BF 639* and Table C, p. 646.

This star does not exist. Flamsteed observed it on January 1, 1700, but there is no star in the sky at the coordinates listed in his "Catalogus Britannicus." Flamsteed noted in his original manuscript entry that this star's zenith distance — from which its declination was derived — was doubtful. On the same day, he observed two other stars, 96 and 101 Tau, and his observations of them also contain errors of declination. Peters, in "Flamsteed's Stars," pp. 71-73, proposed that 100 Tau was Bradley's 686. If Peters' suggestion is accepted, then this star is the same as Piazzi's IV, 246, BAC 1526, and HR 1585, 5.48V.

101, HD 31845, 6.76V, *SA, BS Suppl.,* GC 6085*.

107, HD 33121, 6.5V, *SA,* GC 6279*.

115, HR 1808, 5.42V, *SA, HRP, BS,* BF 714*.

In Flamsteed's "Catalogus Britannicus" of 1725, this star's magnitude is listed as 7½. Flamsteed's observational notes, however, refer to this star's magnitude as 6.

119, HR 1845, 4.38V, *SA, HRP, BS,* BF 724*.

In Flamsteed's "Catalogus Britannicus," this star's magnitude is listed as 7 although his observational notes describe it several times as magnitude 6 and once as 5½.

124

This star does not exist. After observing 44 Gem on February 22, 1705, Flamsteed made an error of 21° in reducing its right ascension to 1690, the epoch of his catalogue. This misstep produced a new but erroneous set of coordinates that led him to believe he had seen a new star, 124 Tau. See BF 988 and Table B, p. 645.

125, HR 1928, 5.18V, *SA, HRP, BS,* BF 754*.

In Flamsteed's "Catalogus Britannicus," this star's magnitude is listed as 3 although his observational notes describe it on one occasion as magnitude 5 and on another occasion as magnitude 6.

126, HR 1946, 4.86V, *SA, HRP, BS,* BF 762*, HF 117.

The copy of Flamsteed's "Catalogus Britannicus" in the New York Public Library gives no magnitude for this star although Halley's 1712 pirated edition lists this star as 6th magnitude. It is similarly listed in Baily's revised edition of Flamsteed's catalogue, which leads one to suspect that in some copies of the "Catalogus Britannicus" the magnitude for this star may not have been properly printed. See 36 and 37 Sgr.

127, HD 37940, 6.7V, *SA,* GC 7137*.

128, HD 38219, 6.8V, *SA,* GC 7181*.

138

This star does not exist. After observing 137 Tau on February 8, 1691, Flamsteed erred 13' in reducing its declination to 1690. This mistake produced a new set of coordinates that led him to believe he had seen a new star, 138 Tau. See BF 800 and Table B, p. 645.

140, HD 40545, 6.91V, *SA,* BF 818*, BAC 1916*.

141, in Gemini, HR 2116, 6.37V, *SA, HRP, BS,* GC 7610*.

TRIANGULUM, TRI

Triangle

Triangulum (Figure 81) is a Ptolemaic constellation that Flamsteed numbered from 1 to 16.

Table 143. The numbered stars of Triangulum

		MAGNITUDES		CATALOGUE NUMBERS	
Flamsteed Number	Letter	Flamsteed	Visual	HR	HD
1+		6	7.4		10407
2	Alpha	4	3.41	544	11443
3	Epsilon	6	5.50	599	12471
4	Beta	4	3.00	622	13161
5		7	6.23	634	13372
6	Iota	6	4.94	642	13480
7		6	5.28	655	13869
8	Delta	5	4.87	660	13974
9	Gamma	4	4.01	664	14055
10		6	5.03	675	14252
11		7	5.54	712	15176
12		6	5.29	717	15257
13		7	5.89	720	15335
14		6	5.15	736	15656
15		7	5.35	750	16058
16+		7	5.86	830	17471

The Lost, Missing, or Troublesome Stars of Triangulum

1, HD 10407, 7.4V, BF 200*, BAC 519*, B 7*, P I, 148*, T 552*.

This star's Flamsteed number is often omitted from atlases and catalogues because the coordinates listed in the "Catalogus Britannicus" do not correspond to any star in the heavens. Baily explained that Flamsteed erred 32^s in computing this star's right ascension. Taking this mistake into account, he noted in his revised edition of Flamsteed's catalogue that the star the Astronomer Royal observed was Piazzi's I, 148. The star is properly identified in the second edition of the *SA*.

16, in Aries, HR 830, 5.86V, *HRP,* BF 351*, BAC 866.

Since this star is more than 5° distant from the border of Triangulum, Baily switched it to Aries.

Ursa Major, UMa

Greater, Larger, or Big Bear

Ursa Major (Figure 104) is a Ptolemaic constellation that Flamsteed numbered from 1 to 87. In Halley's 1712 edition of Flamsteed's catalogue, 215 stars are listed in the Great Bear — the most in any constellation. For an explanation of the difference, see Taurus.

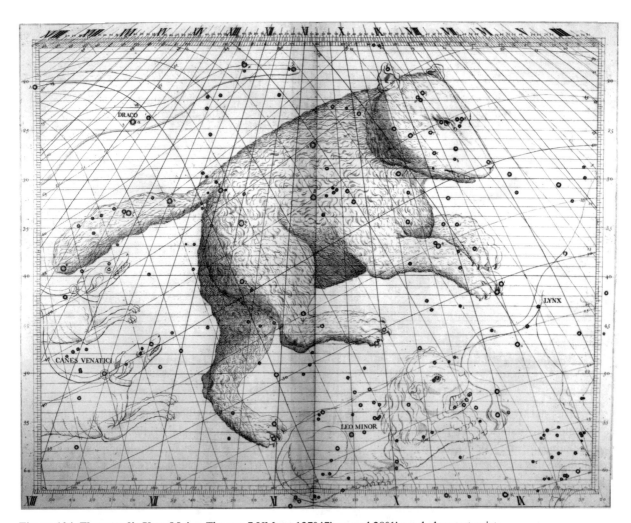

Figure 104. Flamsteed's Ursa Major. The star 7 UMa at 127°47' r.a. and 28°1' n.p.d. does not exist.

Table 144. The numbered stars of Ursa Major

		MAGNITUDES		CATALOGUE NUMBERS	
Flamsteed Number	Letter	Flamsteed	Visual	HR	HD
1	Omicron	4½	3.36	3323	71369
2	A	5	5.47	3354	72037
3	Pi¹	5	5.64	3391	72905
4	Pi²	6	4.60	3403	73108
5	b	5	5.72	3505	75486
6		5	5.58	3531	75958
7+		6			
8	Rho	5	4.76	3576	76827
9	Iota	4	3.14	3569	76644
10+		4	3.97	3579	76943
11	Sigma¹	5	5.14	3609	77800
12	Kappa	4	3.60	3594	77327
13	Sigma²	5	4.80	3616	78154
14	Tau	5	4.67	3624	78362
15	f	5	4.48	3619	78209
16	c	5	5.13	3648	79028
17		5	5.27	3660	79354
18	e	5	4.83	3662	79439
19+		6	5.97	3664	79452
20+		7	7.4		80130
21+		6	7.8		81104
22		7	5.72	3768	82189
23	h	4	3.67	3757	81937
24	d	4½	4.56	3771	82210
25	Theta	3½	3.17	3775	82328
26		5½	4.50	3799	82621
27		6	5.17	3839	83506
28		5	6.34	3865	84179
29	Upsilon	4	3.80	3888	84999
30	Phi	5	4.59	3894	85235
31		6	5.27	3917	85795
32		5	5.75	4026	88983
33	Lambda	3½	3.45	4033	89021
34	Mu	3	3.05	4069	89758
35		6	6.32	4106	90633
36		5	4.84	4112	90839
37		5	5.16	4141	91480
38		5	5.12	4178	92424
39		6	5.80	4187	92728
40+		6	7.11		93075
41		6½	6.34	4202	93132
42		5½	5.58	4236	93875
43		6	5.67	4235	93859
44		6	5.10	4246	94247
45	Omega	4½	4.71	4248	94334

(table continued on next page)

Lost Stars

Table 144. The numbered stars of Ursa Major *(continued)*

		MAGNITUDES		CATALOGUE NUMBERS	
Flamsteed Number	Letter	Flamsteed	Visual	HR	HD
46		6	5.03	4258	94600
47		6	5.05	4277	95128
48	Beta	2	2.37	4295	95418
49		6	5.08	4288	95310
50	Alpha	1½	1.79	4301	95689
51		7	6.00	4309	95934
52	Psi	3½	3.01	4335	96833
53	Xi	4	3.79c	4374/5	98230/1
54	Nu	4	3.48	4377	89262
55		5	4.78	4380	98353
56		6	4.99	4392	98839
57		6	5.31	4422	99787
58		6	5.94	4431	99984
59		6	5.59	4477	101107
60		6	6.10	4480	101133
61		6	5.33	4496	101501
62		6	5.73	4501	101606
63	Chi	4	3.71	4518	102224
64	Gamma	2	2.44	4554	103287
65+		7	5.87c	4560/1	103483/498
66		6	5.84	4566	103605
67		6	5.21	4594	104513
68		7	6.43	4641	106002
69	Delta	2½	3.31	4660	106591
70		6	5.55	4701	107465
71		7	5.81	4726	108135
72+		7	7.04		108346
73		6	5.70	4745	108502
74		6	5.35	4760	108844
75		6	6.08	4762	108861
76		6	6.02	4833	110462
77	Epsilon	3	1.77	4905	112185
78		6	4.93	4931	113139
79	Zeta	3	2.17c	5054/5	116656/7
80	g	5	4.01	5062	116842
81		5½	5.60	5109	118214
82		6	5.46	5142	119024
83		6	4.66	5154	119228
84		6	5.70	5187	120198
85	Eta	3	1.86	5191	120315
86		6	5.70	5238	121409
87+	i Dra	5	4.65	5226	121130

The Lost, Missing, or Troublesome Stars of Ursa Major

7, BF 1232* and Table C, p. 645.

This star apparently does not exist. Flamsteed observed it on January 23, 1696, but omitted its time of transit. In computing its right ascension, therefore, he based his calculations on what he thought was an earlier observation of this star, which was actually an observation of 5 UMa. Flamsteed may have suspected something was amiss since he marked this star as uncertain. Peters, in "Flamsteed's Stars," p. 78, believed it was BD+61 1070 (HD 73745, 7.4V). See b UMa in Part One.

10, in Lynx, HR 3579, 3.97V, *SA, HRP, BS,* GC 12434*.

19, in Lynx, HR 3664, 5.97V, *HRP,* BF 1302*, BAC 3144.

Since this star is more than 7° distant from Ursa Major, Baily switched it to Lynx.

20, HD 80130, 7.4V, *SA,* GC 12851*.

21, HD 81104, 7.8V, *SA,* GC 12980*.

40, HD 93075, 7.11V, *SA,* BF 1523*, BAC 3678*.

65, HR 4560/1, 6.54V, 7.03V, *SA, HRP, BS,* BF 1663*, BAC 4026*, BAC 4028, H 175*, ADS 8347.

This is a multiple-star system with a combined magnitude of 5.87V. Components A, B, and C, with magnitudes of 6.7V, 8.5V, and 8.3V, respectively, make up HR 4560, while component D is HR 4561. Astronomers question which star Flamsteed actually observed. Did he see the combined light of the entire system or did he succeed in separating the group? A century after Flamsteed, Giuseppe Piazzi (P XI, 83/4) saw both HR 4560 and 4561 and equated them with Flamsteed's 65. Baily, in his revised edition of Flamsteed's catalogue, noted that 65 had an "anonymous star [HR 4561] immediately following." Several years later, he included both HR 4560 and 4561 in the *BAC* but assigned 65 only to HR 4560 (BAC 4026) and left HR 4561 (BAC 4028) without a Flamsteed number, but with the comment that "it is the companion of 65...." On the other hand, Argelander (*UN,* p. 24) and Heis agreed with Piazzi's original proposal and equated both stars with Flamsteed's 65, as have most modern astronomers and cartographers.

72, HD 108346, 7.04V, *SA, BS Suppl.*

87, in Draco as i Dra, HR 5226, 4.65V, *HRP,* BF 1918 and Table A, p. 645.

This star is the same as 10 Dra. It is one of Flamsteed's twenty-two duplicate stars.

URSA MINOR, UMi

Lesser, Smaller, or Little Bear

Ursa Minor (Figure 89) is a Ptolemaic constellation that Flamsteed numbered from 1 to 24.

Table 145. The numbered stars of Ursa Minor

		MAGNITUDES		CATALOGUE NUMBERS	
Flamsteed Number	Letter	Flamsteed	Visual	HR	HD
1+	Alpha	3	2.02	424	8890
2+		6	4.25	285	5848
3		6	6.38	5305	124063
4		5	4.82	5321	124547
5		4	4.25	5430	127700
6+		7	7.4		130834
7	Beta	3	2.08	5563	131873
8+		6	6.84		133086
9+		7	6.66		133621
10+		7	7.2		134584
11		5	5.02	5714	136726
12+		7	7.2		136727
13	Gamma	3	3.05	5735	137422
14+		7	7.4		137686
15	Theta	5	4.96	5826	139669
16	Zeta	4	4.32	5903	142105
17+		7	6.9		143803
18+	Pi²	6	6.9		141652
19		5	5.48	6079	146926
20		6	6.39	6082	147142
21	Eta	5	4.95	6116	148048
22	Epsilon	4	4.23	6322	153751
23	Delta	3	4.36	6789	166205
24		6½	5.79	6811	166926

The Lost, Missing, or Troublesome Stars of Ursa Minor

1, Alpha, HR 424, 2.02V, *SA, HRP, BS,* HF 3, Polaris, Stella Maris.

Although Halley's 1712 edition properly lists 1's magnitude as 2, the "Catalogus Britannicus" lists it as magnitude 3.

2, in Cepheus, HR 285, 4.25V, *SA, HRP,* GC 1288*.

6, HD 130834, 7.4V, *SA,* GC 19918*.

8, HD 133086, 6.84V, *SA, BS Suppl.*

9, HD 133621, 6.66V, *SA, BS Suppl.,* GC 20236*.

10, HD 134584, 7.2V, *SA,* BF 2094*, BAC 5013*, P XV, 27*.

This star is usually omitted from atlases and catalogues because there is no star located in the heavens at the coordinates listed in Flamsteed's catalogue. Baily discovered that Flamsteed had erred 10' in calculating 10's right ascension. Taking this into account, Baily suggested in his revised edition of Flamsteed's catalogue that the star the Astronomer Royal observed was Piazzi's XV, 27.

12, HD 136727, 7.2V, BF 2107*, BAC 5078*, B 46*.

This star's Flamsteed number is usually omitted from atlases and catalogues because its right ascension is missing from the "Catalogus Britannicus." But based on Flamsteed's manuscripts, Baily reconstructed its approximate position and included it in his revised edition of Flamsteed's catalogue. The star is properly identified in the second edition of *SA*.

14, HD 137686, 7.4V, *SA,* BF 2133*, BAC 5102*.

Like 12 UMi, this star's Flamsteed number is usually omitted from atlases and catalogues because its right ascension is missing from the "Catalogus Britannicus." But Baily again succeeded in reconstructing this star's approximate right ascension and included it in his revised edition of Flamsteed's catalogue.

17, HD 143803, 6.9V, *SA,* GC 21454*.

18, Pi^2, HD 141652, 6.9V, *SA,* BF 2220*, BAC 5274*.

Like 12 and 14 above, this star's right ascension is not included in Flamsteed's "Catalogus Britannicus" of 1725. Moreover, its declination in the catalogue is 1' too great. Based on the data in Flamsteed's notebooks, Baily reconstructed 18's coordinates for his revised edition of Flamsteed's catalogue.

VIRGO, VIR

Virgin

Virgo (Figure 105) is a Ptolemaic constellation that Flamsteed numbered from 1 to 110.

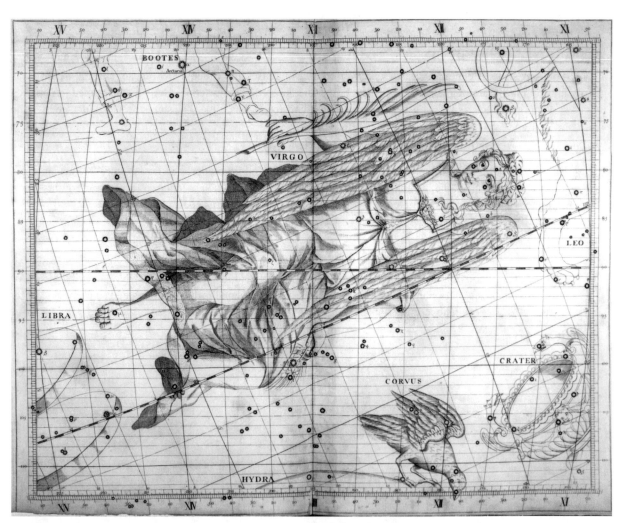

Figure 105. Flamsteed's Virgo. The autumnal equinox, the First Point of Libra, ♎, is where the horizontal double line, the celestial equator, crosses the ecliptic, the slightly curved double line. The following stars in Virgo are nonexistent: **18 Vir,** 183°17' r.a., 74°34' n.p.d.; **19 Vir,** 183°26' r.a., 77°34' n.p.d.; **22 Vir,** 185°15' r.a., 80° 6' n.p.d.; **23 Vir,** 185°24' r.a., 80°20' n.p.d.; **24 Vir,** 185°26' r.a., 82°44' n.p.d.; **42 Vir,** 189°49' r.a., 80°30' n.p.d.; **45 Vir,** 190°57' r.a., 92°58' n.p.d.; **52 Vir,** 193°53' r.a., 98°31' n.p.d.; **91 Vir,** 204°43' r.a., 87°25' n.p.d.

Table 146. The numbered stars of Virgo

		MAGNITUDES		CATALOGUE NUMBERS	
Flamsteed Number	Letter	Flamsteed	Visual	HR	HD
1	Omega	6	5.36	4483	101153
2	Xi	5	4.85	4515	102124
3	Nu	5	4.03	4517	102212
4	A^1	6	5.32	4528	102510
5	Beta	3	3.61	4540	102870
6	A^2	6	5.58	4559	103484
7	b	5½	5.37	4585	104181
8	Pi	5	4.66	4589	104321
9	Omicron	5	4.12	4608	104979
10		6	5.95	4626	105639
11		6	5.72	4629	105702
12		6½	5.85	4650	106251
13		6	5.90	4681	107070
14+		6	6.84		107161
15	Eta	3	3.89	4689	107259
16	c	3¾	4.96	4695	107328
17		6	6.40	4708	107705
18+		6			
19+		6			
20		6	6.26	4777	109217
21	q	6	5.48	4781	109309
22+		6			
23+		6			
24+		6			
25	f	6	5.87	4799	109704
26	Chi	5	4.66	4813	110014
27		6	6.19	4824	110377
28+		6	6.78		110418
29	Gamma	3	2.75c	4825/6	110379/80
30	Rho	5	4.88	4828	110411
31	d^1	6	5.59	4829	110423
32	d^2	6	5.22	4847	110951
33		6½	5.67	4849	111028
34		6	6.07	4855	111164
35		6	6.41	4858	111239
36+		6	5.70	4865	111397
37		6	6.02	4878	111765
38		6	6.11	4891	111998
39+		6	7.8		112036
40	Psi	5	4.79	4902	112142
41		6	6.25	4900	112097
42+		6			
43	Delta	3	3.38	4910	112300
44	k	6	5.79	4921	112846
45+		6			

(table continued on next page)

Table 146. The numbered stars of Virgo (continued)

Flamsteed Number	Letter	Magnitudes Flamsteed	Visual	Catalogue Numbers HR	HD
46		6	5.99	4925	112992
47	Epsilon	3	2.83	4932	113226
48		6	6.59	4937	113459
49		5	5.19	4955	114038
50		6	5.94	4961	114287
51	Theta	4	4.38	4963	114330
52+		6			
53		4½	5.04	4981	114642
54		6	6.28	4990	114846
55		6	5.33	4995	114946
56+		6	6.94		115062
57		6	5.22	5001	115202
58+		6	7.2		115466
59	e	6½	5.22	5011	115383
60	Sigma	5	4.80	5015	115521
61		4½	4.74	5019	115617
62+		6	6.73		115903
63		6	5.37	5044	116292
64		6	5.87	5040	116235
65		6	5.89	5047	116365
66		6½	5.75	5050	116568
67	Alpha	1	0.98	5056	116658
68	i	4	5.25	5064	116870
69		5½	4.76	5068	116976
70		6	4.98	5072	117176
71		6	5.65	5081	117304
72	l^1	6	6.09	5088	117436
73		6	6.01	5094	117661
74	l^2	6	4.69	5095	117675
75		6	5.55	5099	117789
76	h	6	5.21	5100	117818
77+		7	7.04		117878
78	o	6	4.94	5101	118022
79+	Zeta	6	3.37	5107	118098
80		6	5.73	5111	118219
81+		6	7.0		118511
82	m	6	5.01	5150	119149
83		6	5.60	5165	119605
84		6	5.36	5159	119425
85		6	6.19	5170	119786
86		6	5.51	5173	119853
87		6	5.43	5181	120052
88+		6	6.5		120235
89		5½	4.97	5196	120452
90	p	6	5.15	5232	121299

(table continued on next page)

Table 146. The numbered stars of Virgo (continued)

Flamsteed Number	Letter	Magnitudes		Catalogue Numbers	
		Flamsteed	Visual	HR	HD
91+		6			
92		6	5.91	5244	121607
93	Tau	5	4.26	5264	122408
94+		6	6.53		123177
95		6	5.46	5290	123255
96		5	6.47	5298	123630
97+		6	7.3		124248
98	Kappa	4	4.19	5315	124294
99	Iota	4	4.08	5338	124850
100	Lambda	4	4.52	5359	125337
101+		6	5.80	5352	125180
102	Upsilon	5	5.14	5366	125454
103+		5	6.8		125817
104		6	6.17	5406	126722
105	Phi	4	4.81	5409	126868
106		6	5.42	5410	126927
107	Mu	4	3.88	5487	129502
108		6	5.69	5501	129956
109		4	3.72	5511	130109
110		6	4.40	5601	133165

The Lost, Missing, or Troublesome Stars of Virgo

14, HD 107161, 6.84V, *SA,* GC 16798*.

18
19

These two stars do not exist. Flamsteed observed 70 and 71 Vir on several occasions, but on May 4, 1701 he erred 1^h in reducing their right ascension to 1690, the epoch of his catalogue. This mistake produced new sets of coordinates that led him to believe he had seen two new stars, 18 and 19 Vir. See BF 1851, 1853, and Table B, p. 645.

22, BF 1748* and Table D, p. 646.
23, BF 1751* and Table D, p. 646.
24, BF 1752* and Table D, p. 646.

These three stars do not exist. There is no record that Flamsteed ever observed them. Baily suggested that he may have observed three other stars in Bootes, and somehow confused their coordinates. The three he suggested Flamsteed observed are:

      BF 1968, BAC 4766, HR 5385/6, 6.86V, 5.12V, comb. mag. 4.87V.
      BF 1970, BAC 4771, HR 5388, 5.95V.
      BF 1971, BAC 4773, HD 126201, 7.20V.

These three are among the 458 stars that were observed by Flamsteed but inadvertently omitted from his printed catalogue (*BF*, p. 392).

28, HD 110418, 6.78V, *SA, BS Suppl.,* GC 17277*.

36, in Coma, HR 4865, 5.70V, *HRP,* BF 1770* and Table A, p. 645.

This star is the same as 29 Com. It is one of Flamsteed's twenty-two duplicate stars.

39, HD 112036, 7.8V, BF 1781*, BAC 4236*, B 363*, P XII, 210*, CAP 2299*, T 5916*.

The star is properly identified in the second edition of *SA*.

42, BF 1785*, *BAC*, p. 77.

This star does not exist. There is no record that Flamsteed ever observed it, and there is no star in the sky at the coordinates listed in his catalogue. There are, in all, eleven such mysterious objects in Flamsteed's catalogue.

45

This star does not exist. After Flamsteed observed 44 Vir on April 9, 1701, he made somehow an error of 50' in declination when copying figures to his catalogue. This error produced a new set of coordinates that led to the belief that a new star had been observed, 45 Vir. See BF 1791 and Table B, p. 645.

52, BF 1814* and Table D, p. 646.

This star does not exist. Like 42, there is no record that Flamsteed ever observed this star, and there is no star in the sky at the coordinates listed in his catalogue.

56, HD 115062, 6.94V, *SA, BS Suppl.,* GC 17935*.

58, HD 115466, 7.2V, *SA,* GC 17990*.

62, HD 115903, 6.73V, *SA, BS Suppl.,* GC 18037*.

77, HD 117878, 7.04v, *SA, BS Suppl.,* GC 18323*.

79, Zeta, HR 5107, 3.37V, *SA, HRP, BS,* HF 65, BF 1864*.

In Halley's 1712 edition of Flamsteed's catalogue, this star's magnitude is listed as 3 and in Flamsteed's manuscripts it is described as either 3rd or 4th magnitude, but in his "Catalogus Britannicus" it is erroneously listed as a 6th-magnitude star.

81, HD 118511, 7.0V, *SA,* BF 1867*, GC 18413*.

This is a triple-star system, ADS 8972 (Struve 1763). Component A is magnitude 7.0V; component B, barely 3" away, is 7.9V; and component C, 14" distant, is 11.0V. Flamsteed's manuscript catalogue lists 81's magnitude as 7½, although his published "Catalogus Britannicus" lists it as 6. Baily felt that the manuscript figure was correct and consequently, in his revised edition of Flamsteed's catalogue, he assigned 81 a magnitude of 7½. Since 81 is a multiple-star system, is it possible that Flamsteed saw components A and B on different occasions thus explaining the difference in magnitude between the manuscript and the printed catalogue?

88, HD 120235, 6.5V, *SA,* GC 18645*.

91, BF 1910* and Table C, p. 646.

This star does not exist. Flamsteed observed it on May 13, 1703, but there is no star in the heavens at the coordinates listed in his catalogue. Baily suggested that Flamsteed might have erred 2^m in recording its time of transit. If this supposition is accepted, the star he observed would be 92 Vir. Peters, in "Flamsteed's Stars," p. 80, also believed that this was the star Flamsteed observed.

94, HD 123177, 6.53V, *SA, BS Suppl.,* GC 19032*.

97, HD 124248, 7.3V, *SA,* GC 19161*.

101, in Bootes, HR 5352, 5.80V, *SA, HRP.*

103, HD 125817, 6.8V, *SA,* GC 19368*.

Vulpecula, Vul

Little Fox

Hevelius devised the constellation Vulpecula (figures 83 and 93). Flamsteed acknowledged that it was a Hevelian constellation and included it in his atlas and catalogue as the Little Fox and the Goose. He numbered its stars from 1 to 35.

Table 147. The numbered stars of Vulpecula

		Magnitudes		Catalogue Numbers	
Flamsteed Number	Letter	Flamsteed	Visual	HR	HD
1		5	4.77	7306	180554
2		6	5.43	7318	180968
3		6	5.18	7358	182255
4		6	5.16	7385	182762
5		6	5.63	7390	182919
6	Alpha	4	4.44	7405	183439
7		5	6.33	7409	183537
8		6	5.81	7406	183491
9		6	5.00	7437	184606
10		6	5.49	7506	186486
11+				7539	
12		5	4.95	7565	187811
13		6	4.58	7592	188260
14		5	5.67	7641	189410
15		4½	4.64	7653	189849
16		5	5.22	7657	190004
17		4½	5.07	7688	190993
18		5¾	5.52	7711	191747
19		6	5.49	7718	192004
20		5½	5.92	7719	192044
21		5½	5.18	7731	192518
22		5	5.15	7741	192713
23		4½	4.52	7744	192806
24		5	5.32	7753	192944
25		6	5.54	7789	193911
26		6	6.41	7874	196362

(table continued on next page)

Table 147. The numbered stars of Vulpecula *(continued)*

		MAGNITUDES		CATALOGUE NUMBERS	
Flamsteed Number	Letter	Flamsteed	Visual	HR	HD
27		5	5.59	7880	196504
28		6	5.04	7894	196740
29		5	4.82	7891	196724
30		6	4.91	7939	197752
31		6	4.59	7995	198809
32		5	5.01	8008	199169
33		6	5.31	8032	199697
34+		6	5.57	8165	203344
35		6	5.41	8217	204414

The Lost, Missing, or Troublesome Stars of Vulpecula

11, HR 7539, *HRP,* BF 2688*, HEV 1540*, the variable CK, the Nova of 1670.

Although Flamsteed did not actually observe the Nova, he included it in his catalogue, with the notation that Hevelius had seen it. Hevelius declared that he first noticed it in July 1670 when it reached magnitude 2½ or 3 before gradually disappearing in August 1671. It suddenly reappeared in March 1672 with a maximum magnitude of about 3 before again vanishing in September. It remained unobserved for the next 300 years until 1982, when a remnant of the Nova, together with some wisps of nebulosity, was discovered. See *Sky & Telescope,* January 1983, p. 12 and July 1986, p. 26.

34, in Pegasus, HR 8165, 5.57V, *SA, HRP, BS,* GC 29884*.

Figure 106 (at right). The planisphere of the northern sky from Flamsteed's *Atlas Coelestis.* This chart was prepared by Flamsteed's friend and associate Abraham Sharp after the Astronomer Royal's death. Unlike the other charts in the atlas, this one depicts the constellation figures "backward" as they would appear on a celestial globe. There are also several other differences, including the insertion of Hevelius's asterism Cerberus.

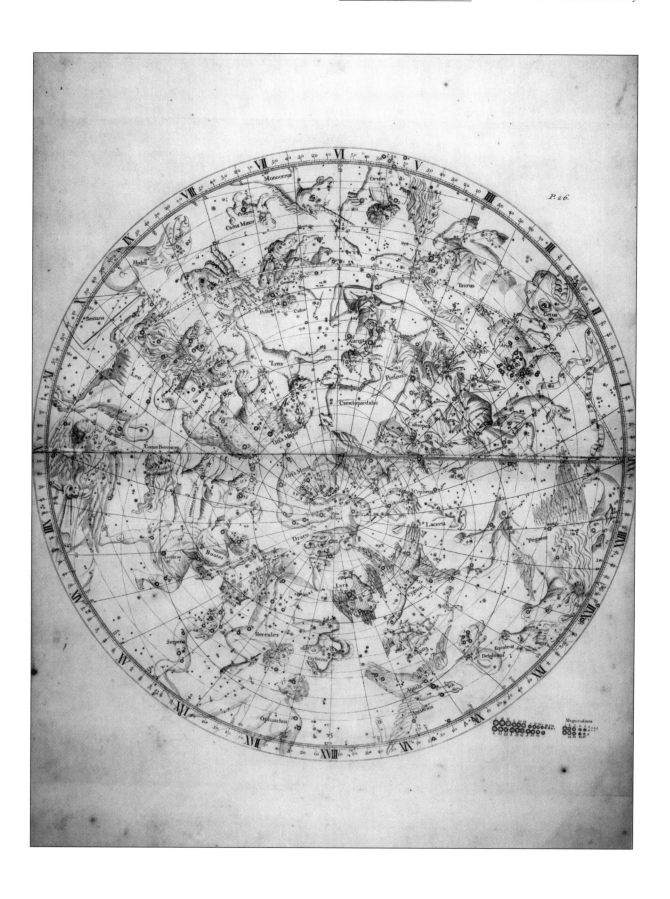

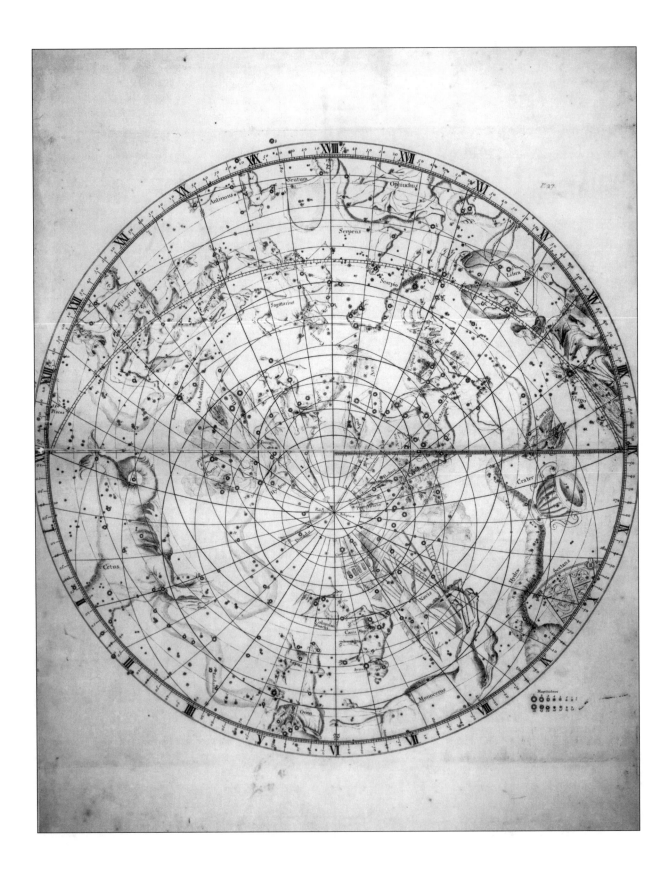

GLOSSARY

Asterism: A group of stars that is not one of the eighty-eight officially recognized constellations.

Binary: A double star, two stars that are close to each other. An optical or visual double are two stars that appear close to each other as seen from an observer on Earth although they may be a great distance apart. A true binary is a pair of stars that are close to each other in space.

Celestial Equator: The great circle around the celestial sphere midway between the celestial north and south poles. Where it crosses the ecliptic in Pisces, it marks the vernal equinox; and where it crosses the ecliptic on the other side of the celestial sphere in Virgo, it marks the autumnal equinox.

Celestial Poles: The positions in the heavens toward which the Earth's axis points. The star Polaris in Ursa Minor marks the approximate position of the celestial north pole while the star Sigma in Octans marks the approximate position of the celestial south pole.

Declination: The celestial equivalent of latitude measured in degrees, minutes, and seconds from the celestial equator, 0°, north or south to the celestial poles at 90°.

Ecliptic: The sun's apparent path across the heavens as seen from Earth. It is also the approximate path of the moon and planets.

Ecliptic Coordinates: Ecliptic latitude is measured north or south of the ecliptic 90° to the ecliptic north or south pole, and ecliptic longitude is measured eastward in degrees along the ecliptic from the vernal equinox, 0°. This system is no longer used in modern astronomy.

Ecliptic Poles: The points 90° north or south of the ecliptic. The north ecliptic pole is in Draco; the south ecliptic pole is in Dorado.

Epoch: Since stars' positions shift slightly each day because of the wobble of Earth's axis, stellar atlases and catalogues set the position of their stars according to a fixed date or epoch. The epoch of Bayer's atlas was 1600; Flamsteed's catalogue and atlas was 1690; Lacaille's was 1750. Most atlases and catalogues printed in recent years set their epoch at 2000.0, that is, January 1, 2000.

Figure 107 (at left). The planisphere of the southern sky from Flamsteed's *Atlas Coelestis*. Like the planisphere of the northern sky, this chart was also prepared by Abraham Sharp, but with the constellations drawn as they actually appear in the heavens, not backward as in Figure 106. Included in this chart are the southern Ptolemaic constellations that were omitted or partially omitted from the other charts: Ara, Centaurus, Lupus, Corona Australis, and Navis. Next to Navis is the asterism Robus Carolinum, King Charles's Oak (see figs. 31 and 65). Also included are Hevelius's Scutum, the Southern Cross, and the twelve newly devised constellations of the Dutch explorers and seafarers Keyzer and Houtman. They depict their constellation Dorado as guarding the south ecliptic pole, but instead of a Goldfish as shown by Bayer (Figure 34) or a Swordfish, Xiphias, as shown by Hevelius (Figure 35), Sharp depicts a *Sawfish*!

Equatorial Coordinates: Equatorial coordinates are measured on the celestial globe. Equatorial latitude or declination is measured from the celestial equator north or south 90° to the celestial poles. Celestial longitude or right ascension is measured in hours, minutes, and seconds, 0^h to 24^h — or occasionally in degrees, 0° to 360° — eastward around the celestial equator from the vernal equinox, 0^h right ascension. This is the system currently in use in modern astronomy.

Equinoctial Colure: An imaginary great circle — 0^h and 12^h right ascension — in the sky that runs down the celestial sphere from the north celestial pole to the south celestial pole. It intersects the celestial equator at the First Point of Aries, 0^h right ascension, in the constellation Pisces on one side of the sphere and the First Point of Libra, 12^h right ascension, in Virgo on the other. When the sun reaches these points in the sky, it is either the vernal (March 21-22) or autumnal (September 22-23) equinox.

Equinox: A point in the sky on either side of the heavenly sphere where the celestial equator, the equinoctial colure, and the ecliptic intersect. The vernal equinox, ♈, is currently in the constellation Pisces, and the autumnal equinox, ♎, is currently in Virgo. See **Equinoctial Colure**, above.

Light Year: The distance light travels in one year at the rate of approximately 186,000 miles a second. Light from the sun reaches Earth in about 8 minutes. Light from the sun to Pluto, the farthest planet in our solar system, takes about 5.5 hours. Light from Earth's nearest star, Alpha Centauri, takes about 4.2 years to reach Earth. Some objects in outer space are over 12 billion light years away.

Lucida: The brightest star in a constellation.

Magnitude: A star or planet's apparent brightness.

Meridian: An imaginary line in the sky extending from the celestial north pole over an observer's zenith and continuing on to the celestial south pole.

Modern Constellations: Constellations developed since Claudius Ptolemy's first century catalogue, which lists the forty-eight classic constellations. The forty modern constellations were primarily devised beginning in the sixteenth century by Plancius, Keyzer and Houtman, Hevelius, and Lacaille. Altogether there are eighty-eight constellations recognized in modern astronomy.

Nebula: A cloudy or hazy patch in the heavens surrounding a star or group of stars.

North Polar Distance (n.p.d.): A system of measuring declination from the north celestial pole, 0°, to the south celestial pole, 180°. It was employed by Flamsteed and some nineteenth century astronomers in their catalogues and atlases, but it is no longer in use.

Nova: A new star. Novae are usually variable stars that suddenly explode and produce an enormous amount of light. When a massive supergiant star explodes, it usually produces a supernova and marks the end of its life cycle.

Precession: The daily slight shift in the position of the stars caused by the wobble of Earth's axis. One complete wobble takes about 25,800 years.

Ptolemaic Constellation: One of the forty-eight constellations in the star catalogue of the Greek-Egyptian astronomer Claudius Ptolemy (A.D. 100? to A.D. 175?). His catalogue, the oldest surviving star catalogue, is found in Book 7 of his treatise on astronomy, the *Almagest*. Constellations devised after that time are usually referred to as "modern."

Radio Astronomy: Astral objects emit radio radiation or radio waves that can be picked up by receivers on Earth. Radio astronomy, using radio telescopes, can detect objects invisible to optical telescopes or obscured by cosmic dust. They can reveal the existence of stars and the remnants of stars, like the Supernova of 1653 in Cassiopeia. Radio objects are usually identified with upper-case Roman letters. Cassiopeia A, a remnant of the Supernova of 1653, is one of the strongest radio sources. The first radio telescope was at Jodrell Bank in England; one of the most powerful is at Socorro, New Mexico.

Glossary

Reduction: The term used to describe the process by which a star's position at a given day is "reduced" or adjusted to the position it would be at a given epoch. This is necessary because of the gradual daily shift of stellar positions caused by precession.

Right Ascension (r.a.): The celestial equivalent of longitude, measured either in hours (h), minutes (m), and seconds (s) of time or in degrees (°), minutes ('), and seconds ("). Distances are measured from 0^h or $0°$ at the First Point of Aries, ♈, in the constellation Pisces, eastward around the celestial sphere.

Sidereal Time: Also known as star time, sidereal time is based on the rotation of the Earth in relation to a specific point on the celestial sphere, the vernal equinox, ♈, 0^h right ascension. A sidereal day, 24 hours of right ascension, occurs when the stars return to the position in the sky they occupied the day before, or, more precisely, the time it takes one of the equinoxes to make two transits across the meridian. This transit takes 23 hours and 56 minutes, about 4 minutes shorter than the standard or solar day. See **Right Ascension**, above.

Solstice: The summer solstice (June 21-22 in the northern hemisphere) is when the sun reaches its highest point in the sky. The winter solstice (December 21-22 in the northern hemisphere) occurs when the sun reaches its lowest point in the sky.

Solstitial Colure: An imaginary great circle — 6^h and 18^h right ascension — that runs from the north celestial pole to the north ecliptic pole in Draco and that then continues to the south celestial pole and around the other side of the celestial globe to the south ecliptic pole in Dorado and back to the north celestial pole. The two points on either side of the celestial globe where the circle intersects the ecliptic are called the solstitial points. When the sun reaches one of these points on or about June 21, it is the time of the summer solstice, the highest point of the sun in the sky at noon. When the sun reaches the other of these points on or about December 21, it is the time of the winter solstice, the lowest point of the sun in the sky at noon.

Spectroscope: An instrument that breaks down a star's light into a rainbow band of colors, its spectrum, which reveals, among other things, a star's chemical elements as well as its temperature.

Spectroscopic Binary: Two stars that are so close to each other that their duplicity can be detected only with a spectroscope.

Supernova: See Nova.

Variable: A star whose magnitude changes. There are several types of variables. One famous variable, Algol, Beta Persei, changes in brightness every sixty-nine hours from magnitude 2.1 to magnitude 3.4. Algol is an eclipsing binary, two stars of different magnitudes circling each other. When the dimmer star crosses in front of the brighter one, it blocks out the latter's light, thus reducing its magnitude. The eclipse lasts about ten hours from start to finish. Another famous variable is Mira, Omicron Ceti. A long-period variable, it changes from a magnitude of 2.6 to 10.1 in a period of 332 days.

Visual Magnitude: A star's apparent visual magnitude, its **V** magnitude, is its brightness as seen by the human eye as measured by methods of modern photometry.

Zenith: An imaginary point in the sky directly over an observer. **Zenith Distance** is the distance of a celestial object from the zenith. Flamsteed added a star's zenith distance to the latitude of his observatory to determine the star's declination. For example, he determined the star H or 1 Gem's (HR 2134) declination on February 12, 1696, as follows: Its zenith distance was 28°14'10". He added to this the latitude of his observatory at Greenwich, 38°31' north polar distance, for a total of 66°45'10" n.p.d., or in modern terminology, 23°14'50" declination.

Zodiac: A band or zone, usually extending about 8° north and south of the ecliptic containing the twelve zodiacal constellations: Aries, Taurus, Gemini, Cancer, Leo, Virgo, Libra, Scorpius, Sagitarius, Capricorn, Aquarius, and Pisces. It is also the apparent path of the sun, moon, and planets across the sky.

APPENDIX I
LOCATING BAYER'S STARS ON HIS CHARTS

In processing Bayer's charts for this work, many star letters became so small that they have become difficult to discern. Consequently, this section lists all the lettered stars in the forty-eight Ptolemaic constellations of Bayer's *Uranometria* and describes where the stars are located on Bayer's charts. It should be noted that many of the descriptions are not those of Bayer but of the current author.

ANDROMEDA *(Figure 1)*

Alpha, α	the bright star on her head.
Beta, β	on the left side of the band about her waist.
Gamma, γ	the bright star on her left leg below her knee.
Delta, δ	of the two stars on her left shoulder, the brighter one above and to the left.
Epsilon, ε	of the two stars on her left shoulder, the smaller one below and to the right.
Zeta, ζ	of the two stars on her left elbow, the brighter one above and to the right.
Eta, η	of the two stars on her left elbow, the smaller one below and to the left.
Theta, θ	of the three stars on her right arm, the uppermost.
Iota, ι	of the two stars on her right wrist, the lower one to the right.
Kappa, κ	of the two stars on her right wrist, the upper one to the left.
Lambda, λ	at the top of her right hand.
Mu, μ	on the right side of the band about her waist.
Nu, ν	on the large link above Mu joining the band about her waist to the chain.
Xi, ξ	of the four stars on her right leg, the bright one under her knee.
Omicron, o	at the end of the chain.
Pi, π	above her left breast.
Rho, ρ	of the three stars on her right arm, the middle one.
Sigma, σ	of the three stars on her right arm, the lowest one.
Tau, τ	on her left thigh.
Upsilon, υ	on her left knee.
Phi, φ	above her right knee.
Chi, χ	of the four stars on her right leg, the lowest one.
Psi, ψ	on her right thumb.
Omega, ω	of the four stars on her right leg, the small one between Xi and Chi.
A	of the four stars on her right leg, the small one above and to the left of Xi and Omega.
b	the small star near the bottom of her skirt above her left leg.
c	the small star at the bottom of her right foot.

AQUARIUS *(Figure 4)*

Alpha, α	of the two stars on his left shoulder, the brighter one to the left.
Beta, β	the bright star on his right shoulder.
Gamma, γ	on his left forearm.
Delta, δ	the bright star on his left leg.
Epsilon, ε	of the two stars near the end of the cloth in his right hand, the brighter one to the right.
Zeta, ζ	of the three stars on his left hand, the middle one.
Eta, η	of the three stars on his left hand, the lowest one.
Theta, θ	of the two stars on his lower back on his left side, the brighter one to the right.
Iota, ι	on his right hip.
Kappa, κ	on the base of the vase's neck below his left forearm.
Lambda, λ	on the mouth of the vase.
Mu, μ	of the two stars near the end of the cloth in his right hand, the smaller one to the left.
Nu, ν	on his right hand.
Xi, ξ	near his right armpit.
Omicron, o	of the two stars on his left shoulder, the smaller one to the right.
Pi, π	the small star at the base of his left thumb.
Rho, ρ	of the two stars on his lower back, the smaller one to the left.
Sigma, σ	on his left thigh below the ecliptic.
Tau, τ	the small star behind his left knee.
Upsilon, υ	behind his right knee.
Phi, φ	of the first three stars on the flow of water, the upper one to the left.
Chi, χ	of the first three stars on the flow of water, the lowest one.
Psi, ψ	the group of three stars below Chi on the flow of water.

501

Omega, ω	the group of two stars below Psi on the flow of water.
A	of the three stars below Omega on the flow of water, the uppermost and brightest.
b	the group of three stars below A and to the right on the flow of water.
c	another group of three stars to the right of b on the flow of water.
d	the small star on his cap.
e	the small star above the ecliptic on the lower right side of his back.
f	the small star below his left thigh at the edge of the cloth.
g	the small star above his right knee.
h	of the first three stars on the flow of water, the upper one to the right.
i	of the three stars below Omega on the flow of water, the two to the left.

Aquila *(Figure 5)*

Alpha, α	of the four stars on the Eagle's neck, the brightest and uppermost.
Beta, β	of the four stars on his neck, the bottom one to the left.
Gamma, γ	of the four stars toward the front of his right wing scapular or shoulder feathers, the brightest.
Delta, δ	of the two stars near the side edge of his left wing covert feathers, the brighter one to the right.
Epsilon, ε	of the two bright stars on his tail feathers, the one at its tip.
Zeta, ζ	of the two bright stars on his tail feathers, the one to the south.
Eta, η	near Ganymede's right ear.
Theta, θ	on Ganymede's right hand.
Iota, ι	the bright star on Ganymede's chest.
Kappa, κ	behind Ganymede's right knee.
Lambda, λ	of the three stars on Ganymede's left foot, the brightest.
Mu, μ	at the front of the Eagle's left wing on his covert feathers near his body.
Nu, ν	of the two stars on Ganymede's left side, the upper one below the Eagle's wing.
Xi, ξ	of the four stars on the Eagle's neck, the small star next to Alpha.
Omicron, o	of the four stars near the base and toward the front of the Eagle's right wing scapular feathers, the one to the south and left of the bright star.
Pi, π	of the four stars near the base and toward the front of the Eagle's right wing scapular feathers, the one to the north and left of the bright star.
Rho, ρ	on the front edge of the Eagle's right wing.
Sigma, σ	on the Eagle's leg.
Tau, τ	on the Eagle's forehead.
Upsilon, υ	of the four stars on the Eagle's neck, the small star below and to the right.
Phi, φ	of the four stars near the base and toward the front of the Eagle's right wing scapular feathers, the one farthest to the left.
Chi, χ	the small star on the Eagle's back near its right wing.
Psi, ψ	on the Eagle's right wing among its secondary feathers.
Omega, ω	of the three small stars forming a triangle near the Eagle's tail, the one on the base to the right.
A	of the three small stars forming a triangle near the Eagle's tail, the one on the apex.
b	of the three small stars forming a triangle near the Eagle's tail, the one on the base to the left.
c	of the two stars near the side edge of his left wing covert feathers, the smaller one to the left.
d	of the two stars on Ganymede's left side, the lower and smaller one. The letter may not be visible on some charts.
e	the small star on Ganymede's belly.
f	the small star on Ganymede's left thigh. The letter may not be visible on some charts.
g	of the three stars on Ganymede's left foot, the small one on his heel.
h	of the three stars on Ganymede's left foot, the small one in the middle.

Ara *(Figure 6)*

Alpha, α	the bright star in the center of the lip of the Altar.
Beta, β	of the two stars on the left side of the fire, the brighter one near the tip of the flame.
Gamma, γ	at the base of the Altar near the middle leg.
Delta, δ	near the middle of the Altar.
Epsilon, ε	of the two stars at the left edge of the lip of the Altar, the northern one.
Zeta, ζ	of the two stars at the left edge of the lip of the Altar, the southern one.
Eta, η	of the two stars on the left side of the fire, the smaller star to the north.
Theta, θ	in the middle of the fire.

Aries *(Figure 7)*

Alpha, α	the bright star on his right horn.
Beta, β	the bright star on his left horn.
Gamma, γ	on his left ear.
Delta, δ	of the three stars on his tail, the lowest one to the right.
Epsilon, ε	at the base of his tail.
Zeta, ζ	of the three stars on his tail, the middle one.
Eta, η	near his left eye.
Theta, θ	near his nose.
Iota, ι	on his neck.
Kappa, κ	the small star below Alpha on his forehead.
Lambda, λ	the small star at the top of his head between his two horns.
Mu, μ	the small star in the middle of his body.

Nu, ν	the small star at the top of his back.
Xi, ξ	the small star on his right foreleg.
Omicron, o	of the two stars on his left hind leg, the one to the right.
Pi, π	of the two stars on his right thigh, the one to the right.
Rho, ρ	of the two stars on his right thigh, the one to the left.
Sigma, σ	of the two stars on his left hind leg, the one to the left.
Tau, τ	of the three stars on his tail, the small star near the tip.

Auriga *(Figure 10)*

Alpha, α	the bright star on his back.
Beta, β	of the two stars on his upper left arm, the brighter one.
Gamma, γ	on his left ankle.
Delta, δ	of the two stars on his cap, the lower and brighter one.
Epsilon, ε	on his right side directly above the two small kids.
Zeta, ζ	of the two stars on the two small kids, the one to the right.
Eta, η	of the two stars on the two small kids, the one to the left.
Theta, θ	of the two stars on his left thigh, the brighter one to the left.
Iota, ι	on his right leg.
Kappa, κ	at the very bottom of the harness in the zone of the zodiac.
Lambda, λ	of the four stars on his lower back, the middle one.
Mu, μ	of the four stars on his lower back, the bottom one to the right.
Nu, ν	of the two stars near his left elbow, the one to the left.
Xi, ξ	of the two stars on his cap, the smaller one at the top.
Omicron, o	on his neck.
Pi, π	of the two stars on his upper left arm, the smaller one.
Rho, ρ	of the four stars on his lower back, the uppermost.
Sigma, σ	of the four stars on his lower back, the bottom one to the left.
Tau, τ	of the two stars near his left elbow, the one to the right.
Upsilon, υ	of the two stars on his left thigh, the smaller one to the right.
Phi, φ	below his left thigh.
Chi, χ	on his right knee.
Psi, ψ	the ten small stars on his whip. The letter is omitted on the chart.

Boötes *(Figure 11)*

Alpha, α	the bright star above his left knee.
Beta, β	on his left ear.
Gamma, γ	on his left shoulder near his armpit.
Delta, δ	on his right shoulder.
Epsilon, ε	on his right side next to his right forearm.
Zeta, ζ	of the three stars on his right leg, the lowest.
Eta, η	of the three stars on his left leg, the uppermost.
Theta, θ	of the four stars on his left hand, the uppermost.
Iota, ι	of the four stars on his left hand, the middle one below Theta.
Kappa, κ	of the four stars on his left hand, the one to the right.
Lambda, λ	on his left forearm.
Mu, μ	on his staff near his right eye.
Nu, ν	of the two stars at the top of his staff, the brighter one to the right.
Xi, ξ	behind his right knee.
Omicron, o	of the three stars on his right leg, the uppermost.
Pi, π	of the three stars on his right leg, the middle one.
Rho, ρ	of the two stars on his waist, the upper one to the right.
Sigma, σ	of the two stars on his waist, the lower one to the left.
Tau, τ	of the three stars on his left leg, the middle one.
Upsilon, υ	of the three stars on his left leg, the lowest one.
Phi, φ	of the two stars on the top of his staff, the smaller one to the left.
Chi, χ	on his staff to the right of his elbow.
Psi, ψ	of the four stars on his right hand, the upper one to the right.
Omega, ω	of the four stars on his right hand, the lower one to the right.
A	below his left armpit.
b	of the four stars on his right hand, the upper one to the left.
c	of the four stars on his right hand, the lower one to the left.
d	on his upper left thigh.
e	the small star at the bottom of his tunic to the right of Alpha.
f	the small star at the bottom of his tunic to the left of Alpha.
g	of the four stars on his left hand, the one to the left.
h	in the middle of the sickle's blade.
i	of the two stars near the tip of the sickle's blade, the lower one to the right.
k	of the two stars near the tip of the sickle's blade, the upper one to the left.

Cancer *(Figure 13)*

Alpha, α	the bright star on the left claw.
Beta, β	on the tip of the fourth leg on the left side.
Gamma, γ	on his back just below the line marked K.
Delta, δ	on his back on the ecliptic.
Epsilon, ε	the fuzzy object on his back above and to the right of Delta, barely visible on some charts.
Zeta, ζ	on the edge of the shell near his hind quarters.
Eta, η	on its back to the right of Epsilon.
Theta, θ	on his back below Eta and south of the ecliptic.

Iota, ι	of the three stars at the bottom edge of the right claw, the one to the right.
Kappa, κ	at the bottom edge of the left claw to the left of Alpha.
Lambda, λ	of the two stars on the second leg on the right side, the one near the base of the leg.
Mu, μ	of the two stars on the fourth leg on the right side, the one near the base of the leg.
Nu, ν	on his right horn or feeler.
Xi, ξ	on his left horn or feeler.
Omicron, o	at the edge of the shell just above the base of the left claw.
Pi, π	near the mouth of the left claw.
Rho, ρ	of the three stars at the bottom edge of the right claw, the two stars to the left.
Sigma, σ	the three stars on the upper edge of the right claw.
Tau, τ	on the teeth of the right claw.
Upsilon, υ	the two stars on his back directly below the base of the right claw.
Phi, φ	the two stars on his first leg on the right side.
Chi, χ	of the two stars on his second leg on the right side, the one in the middle of the leg.
Psi, ψ	on his third leg on the right side.
Omega, ω	of the two stars on the fourth leg on the right side, the one in the middle of the leg.
A	of the three stars on the first leg on the left side, the two near the base of the leg.
b	of the three stars on the first leg on the left side, the one in the middle of the leg.
c	on the second leg on the left side.
d	the small pair on his back below the ecliptic.

Canis Major *(Figure 16)*

Alpha, α	the bright star on his mouth.
Beta, β	the star dangling from his collar.
Gamma, γ	on his forehead near his right ear.
Delta, δ	of the three stars in the center of his body, the brightest one below the other two.
Epsilon, ε	on the upper thigh of his right hind leg.
Zeta, ζ	on the knee or wrist of his right foreleg.
Eta, η	of the two stars on his lower back, the upper one. The smaller star below Eta is in Navis.
Theta, θ	on his left ear.
Iota, ι	near his right eye.
Kappa, κ	on the lower thigh of his right leg.
Lambda, λ	on the paw of his right hind leg.
Mu, μ	on his forehead between his eyes.
Nu, ν	the three small stars on his collar.
Xi, ξ	the two stars on the shoulder of his right leg.
Omicron, o	of the three stars on the center of his body, the two along the line marked φ.

Canis Minor *(Figure 17)*

Alpha, α	the bright star on his belly.
Beta, β	of the two stars on his neck, the brighter and lower one.
Gamma, γ	of the two stars on his neck, the smaller and higher one.
Delta, δ	the two stars on the paw of his left hind leg.
Epsilon, ε	the small star at the top of his head.
Zeta, ζ	on his right hind leg.
Eta, η	the small star on his chest.

Capricornus *(Figure 18)*

Alpha, α	the bright star, a visible double, on his head between his two horns.
Beta, β	the bright star above his eye.
Gamma, γ	of the two stars on the coil below the ecliptic, the one to the right.
Delta, δ	of the two stars on the coil below the ecliptic, the one to the left.
Epsilon, ε	of the two stars on his belly near the coil, the one to the right.
Zeta, ζ	of the two stars above his right hoof, the one to the right.
Eta, η	of three stars on his right shoulder, the one above the others.
Theta, θ	of the two stars on his chest just below the ecliptic, the one to the right.
Iota, ι	of the two stars on his chest just below the ecliptic, the one to the left. The letter may be missing on some charts.
Kappa, κ	of the two stars on his belly near the coil, the star to the left.
Lambda, λ	of the four stars on the upper section of the coil, the brightest and the one below the others.
Mu, μ	the star on the coil bisected by the ecliptic.
Nu, ν	the small star to the left of Alpha.
Xi, ξ	the small star to the right of Alpha.
Omicron, o	of the three small stars above his mouth, the lowest one. This star is barely visible on some charts.
Pi, π	of the three small stars above his mouth, the one in the middle. This star is barely visible on some charts.
Rho, ρ	of the three small stars above his mouth, the uppermost one.
Sigma, σ	above his nose. This star is barely visible on some charts.
Tau, τ	of the two stars on his neck, the upper one.
Upsilon, υ	of the two stars on his neck, the lower one.
Phi, φ	of the three stars on his right shoulder, the bottom one to the left.
Chi, χ	of the three stars on his right shoulder, the bottom one to the right.
Psi, ψ	on his left leg.
Omega, ω	on his right knee.
A	on his right thigh to the left of his knee.
b	of the two stars above his right hoof, the one to the left.
c	of the four stars on the upper section of the coil, the upper three small ones.

Cassiopeia *(Figure 20)*

Alpha, α	the star on her heart directly below her chin.
Beta, β	on the chair next to her right armpit.
Gamma, γ	the bright star below her waist.
Delta, δ	on her right thigh above her knee.
Epsilon, ε	on her left thigh.
Zeta, ζ	on her nose.
Eta, η	between her breasts.
Theta, θ	of the two stars on her left upper arm, the one to the left.
Iota, ι	on her left heel.
Kappa, κ	on the chair next to her waist. The bright object next to Kappa is the Supernova of 1572, Tycho's star.
Lambda, λ	below her right eye.
Mu, μ	of the two stars on her left upper arm, the one to the right.
Nu, ν	the first star on the frond.
Xi, ξ	the second star on the frond.
Omicron, o	the third star on the frond.
Pi, π	the fourth and last star near the top of the frond.
Rho, ρ	on her right elbow.
Sigma, σ	on her right forearm.
Tau, τ	at the top of the chair near her right elbow.
Upsilon, υ	the small star in the center of her waist.
Phi, φ	on the top of her right knee.
Chi, χ	on the side of her right knee.
Psi, ψ	on the side of the seat of the chair on the curved line marked D.
Omega, ω	on the front of the seat of the chair next to the curved line marked D.
A	near the bottom back corner of the chair.

Centaurus *(Figure 21)*

Alpha, α	the bright star on the hoof of his left foreleg.
Beta, β	the bright star on his left thigh.
Gamma, γ	on the hoof of his right foreleg.
Delta, δ	of the three stars on his belly, the bright one in the middle.
Epsilon, ε	of three stars on his left hind leg, the bright one in the middle.
Zeta, ζ	of the two stars on the hoof of his right hind leg, the brighter one above and to the left.
Eta, η	of the two stars on his right thigh, the brighter one above and to the left..
Theta, θ	of the three stars on his upper left arm, the one to the right near his shoulder.
Iota, ι	on his right shoulder.
Kappa, κ	on his left elbow.
Lambda, λ	on his pelvis.
Mu, μ	of the three stars on his rump, the one to the left.
Nu, ν	of three stars on his left hind leg, the lower one to the right.
Xi, ξ	of three stars on his left hind leg, the lower one to the left.
Omicron, o	of the three stars on his upper left arm, the one to the left.
Pi, π	of the two stars on the middle of the stalk of leaves, the one above and to the right. The stalk is actually a thyrsus, a wand symbolic of Bacchus.
Rho, ρ	of the two stars in the middle of the stalk of leaves, the one below and to the left. The letter is barely visible on some charts.
Sigma, σ	on his left wrist
Tau, τ	of the three stars to the right of his left armpit, the one above and to the right.
Upsilon, υ	of the three stars to the right of his left armpit, the lowest one.
Phi, φ	of the three stars to the right of his left armpit, the one above and to the left.
Chi, χ	of the two stars on the left side of his chest, the brighter one above.
Psi, ψ	on his back directly below his head.
Omega, o	on his lower back.
A	on the elbow of his left foreleg.
b	of the three stars on his belly, the small star below and to the left.
c	of the three stars on his rump, the one in the middle.
d	of the two stars on his right thigh, the smaller one to the right.
e	the small star right next to Beta on his left thigh, difficult to see because it is drawn inside Beta.
f	of the two stars on the hoof of his right hind leg, the smaller one below and to the right.
g	on his mouth.
h	on his forehead.
i	on his ear.
k	below his eye.
l	of the three stars on his upper left arm, the one in the middle.
m	on his left armpit.
n	of the two stars on the left side of his chest, the smaller one below.
o	the small star on his lower back to the left of his rump.
p	of the three stars on his rump, the one to the right.
q	of the three stars on his belly, the one to the right.

Cepheus *(Figure 22)*

Alpha, α	the star on his left shoulder.
Beta, β	on the left side of his belt.
Gamma, γ	on his right thigh.
Delta, δ	on his forehead.
Epsilon, ε	toward the front of his head-covering.
Zeta, ζ	of the two stars on his head-covering over his ear, the one above.
Eta, η	on his left wrist.
Theta, θ	on his left forearm.
Iota, ι	on his right upper arm.
Kappa, κ	on his left thigh above the hem of his tunic.
Lambda, λ	of the two stars on his head-covering over his ear, the one below.

Mu, μ	on the back of his head-covering.
Nu, ν	on the back of his neck.
Xi, ξ	on his upper back near the clasp of his cape.
Omicron, ο	on his right upper arm near his elbow.
Pi, π	directly below his right hand.
Rho, ρ	on the back of his right thigh.

CETUS *(Figure 23)*

Alpha, α	the bright star below his eye.
Beta, β	the bright star at the bottom of the coil.
Gamma, γ	of the three stars on his neck, the middle one.
Delta, δ	of the three stars on his neck, the lowest one.
Epsilon, ε	on his chest near his right flipper.
Zeta, ζ	of the five stars in the middle of his body, the uppermost of the three to the left. The letter may not be visible on some charts.
Eta, η	of the five stars in the middle of his body, the one to the right near the base of his tail.
Theta, θ	of the five stars in the middle of his body, the one above the others.
Iota, ι	the last letter on his tail under the band of the zodiac and by the line marked C. The letter Iota is not marked on some charts.
Kappa, κ	on his lower jaw.
Lambda, λ	above his eye.
Mu, μ	on his forehead.
Nu, ν	of the three stars on his neck, the uppermost.
Xi, ξ	the two stars on his mane with the letter Xi between them.
Omicron, ο	on the base of his neck near his right shoulder.
Pi, π	on the wrist of his left flipper.
Rho, ρ	on his chest above his left flipper.
Sigma, σ	on the forearm of his left flipper.
Tau, τ	of the five stars in the middle of his body, the lowest of the three to the left.
Upsilon, υ	on the elbow of his left flipper.
Phi, φ	the group of four stars on the coil.
Chi, χ	of the five stars in the middle of his body, the middle one of the three to the left.
Psi, ψ	the small star on his mane above Xi.

CORONA AUSTRALIS *(Figure 27)*

Alpha, α	the star at the bottom right. The letter is barely visible on some charts.
Beta, β	the third star to the left of Alpha.
Gamma, γ	the second star above the bright star marked ξ (Xi), on the left side of the constellation. Although Bayer included Xi in his chart of Corona Australis, he did not include it among its stars. He placed it instead in Sagittarius, where it is properly lettered Alpha Sgr.
Delta, δ	the third star above Xi.
Epsilon, ε	the first star to the left of Alpha.
Zeta, ζ	the second star to the left of Alpha.
Eta, η	the fourth star to the left of Alpha.
Theta, θ	the first star above Xi.
Iota, ι	the third star to the right above Alpha.

Kappa, κ	the second star to the right above Alpha.
Lambda, λ	the first star to the right above Alpha.
Mu, μ	of the two small stars to the right of Delta, the lower one.
Nu, ν	of the two small stars to the right of Delta, the upper one.

CORONA BOREALIS *(Figure 28)*

Alpha, α	the bright star at the bottom of the wreath.
Beta, β	the first star above Alpha to the right.
Gamma, γ	the first star to the left of Alpha.
Delta, δ	the second star to the left of Alpha.
Epsilon, ε	the third star to the left of Alpha.
Zeta, ζ	of the four stars on the right ribbon, the one at the bottom right.
Eta, η	of the two closely placed stars to the right of Beta, the upper one to the left.
Theta, θ	the second star above Alpha to the right.
Iota, ι	the fourth star to the left of Alpha.
Kappa, κ	of the four stars on the right ribbon, the one at the bottom left.
Lambda, λ	of the four stars on the right ribbon, the one at the top left.
Mu, μ	of the four stars on the right ribbon, the one at the top right.
Nu, ν	of the five stars on the left ribbon, the third from the bottom.
Xi, ξ	of the five stars on the left ribbon, the second from the bottom.
Omicron, ο	of the two closely placed stars to the right of Beta, the lower one to the right.
Pi, π	the third star above Alpha to the right.
Rho, ρ	on the knot joining the two ribbons to the wreath.
Sigma, σ	of the five stars on the left ribbon, the fifth from the bottom.
Tau, τ	of the five stars on the left ribbon, the fourth from the bottom.
Upsilon, υ	of the five stars on the left ribbon, the bottom one.

CORVUS *(Figure 29)*

Alpha, α	the star above his beak.
Beta, β	on his breast.
Gamma, γ	on his left wing.
Delta, δ	of the two stars on his right wing, the brighter one to the right.
Epsilon, ε	above his eye.
Zeta, ζ	on his neck.
Eta, η	of the two stars on his right wing, the smaller one to the left.

CRATER *(Figure 30)*

Alpha, α	of the two stars on the base, the one above and to the right.
Beta, β	of the two stars on the base, the lower one to the left.

Gamma, γ	of the three stars on the lower section above the base, the middle one.
Delta, δ	of the three stars on the lower section above the base, the one to the right.
Epsilon, ε	of the four stars on the upper section, the one to the right.
Zeta, ζ	of the four stars on the upper section, the one to the left.
Eta, η	next to the left handle. The letter is barely visible on some charts.
Theta, θ	on the right handle.
Iota, ι	of the four stars on the upper section, the upper of the two small stars in the middle.
Kappa, κ	of the four stars on the upper section, the lower of the two small stars in the middle.
Lambda, λ	of the three stars on the lower section above the base, the small star to the left.

CYGNUS (Figure 32)

Alpha, α	the bright star on the base of his right leg.
Beta, β	near his eye.
Gamma, γ	in the center of his breast.
Delta, δ	on the front edge of his right wing among his covert feathers.
Epsilon, ε	of the four stars on the front edge of his left wing, the third from the tip among his covert feathers.
Zeta, ζ	of the four stars on the front edge of his left wing, the second from the tip.
Eta, η	of the three stars on his neck, the left one.
Theta, θ	of the four stars in the middle of his right wing, the brighter of the two bottom ones.
Iota, ι	of the four stars in the middle of his right wing, the second from the tip.
Kappa, κ	of the four stars in the middle of his right wing, the uppermost near the tip.
Lambda, λ	of the four stars on the front edge of his left wing, the fourth from the tip among his scapular or shoulder feathers.
Mu, μ	of the four stars on the front edge of his left wing, the first one near the tip among his primary feathers.
Nu, ν	on his left talon.
Xi, ξ	on his left knee.
Omicron, o	the two stars on his right talon.
Pi, π	the two stars at the tip of his tail.
Rho, ρ	of the three stars on the left side of his tail, the middle one.
Sigma, σ	of the three stars on the back and middle feathers of his left wing, the uppermost toward his body.
Tau, τ	of the three stars on the back and middle feathers of his left wing, the middle one.
Upsilon, υ	of the three stars on the back and middle feathers of his left wing, the lowest one.
Phi, φ	of the three stars on his neck, the one to the right.
Chi, χ	of the three stars on his neck, the middle one.
Psi, ψ	of the three stars near the back of his right wing, the middle one.
Omega, ω	the two closely placed stars on his right knee, barely visible on some charts.
A	of the three stars on the left side of his tail, the bottom one to the right.
b	the three small stars at the base of its neck.
c	of the four stars in the middle of his right wing, the smaller of the two bottom ones.
d	of the three stars near the back of his right wing, the uppermost.
e	of the three stars near the back of his right wing, the lowest one.
f	the two small stars near the base of his tail on the right side.
g	of the three stars on the left side of his tail, the uppermost.

DELPHINUS (Figure 33)

Alpha, α	of the four bright stars forming a quadrilateral on his head, the one at the upper right.
Beta, β	of the four bright stars forming a quadrilateral on his head, the one at the bottom right.
Gamma, γ	of the four bright stars forming a quadrilateral on his head, the one at the upper left.
Delta, δ	of the four bright stars forming a quadrilateral on his head, the one at the bottom left.
Epsilon, ε	of the three stars on his tail, the bright one to the right.
Zeta, ζ	of the three small stars on his neck, the upper one next to Beta.
Eta, η	of the three small stars on his neck, the one below and to the right.
Theta, θ	of the three small stars on his neck, the one below and to the left.
Iota, ι	of the three stars on his tail, the small one to the north.
Kappa, κ	of the three stars on his tail, the small one to the south.

DRACO (Figure 36)

Alpha, α	the third bright star from the bottom of his tail.
Beta, β	to the left of his eye.
Gamma, γ	on his forehead.
Delta, δ	of the two stars on the second coil and on the curved line marked N, the one to the left..
Epsilon, ε	of the two stars on the second coil and on or next to the straight line marked ♉ (on the side of the chart), the one to the right.
Zeta, ζ	of the two stars on the third coil and on the curved line marked N, the brighter one to the right.
Eta, η	of the three stars between the third and fourth coils, the middle one.
Theta, θ	of the three stars between the third and fourth coils, the one to the left.
Iota, ι	on the fourth coil and marked i instead of Iota.

Kappa, κ	the next to last star on his tail.
Lambda, λ	the last star on his tail.
Mu, μ	on his tongue.
Nu, ν	in the back of his mouth.
Xi, ξ	on his cheek.
Omicron, o	the single star below the first coil. The letter is not marked on some charts.
Pi, π	in the center of the second coil and above the straight line marked ♈ (on the side of the chart).
Rho, ρ	of the two stars on the second coil and on the curved line marked N, the one to the right.
Sigma, σ	of the two stars on the second coil on or next to the straight line marked ♉ (on the side of the chart), the one to the left.
Tau, τ	of the four stars forming a quadrilateral below the second coil, the one below and to the right.
Upsilon, υ	of the four stars forming a quadrilateral below the second coil, the one above and to the right.
Phi, φ	of the four stars forming a quadrilateral below the second coil, the upper one to the left.
Chi, χ	of the four stars forming a quadrilateral below the second coil, the one below and to the left.
Psi, ψ	the single star between the second and third coils below the point marked L.
Omega, ω	of the two stars between the second and third coils below and to the left of the point marked L, the brighter one to the right.
A	on the bottom of the third coil on the straight line marked ♍ (on the side of the chart).
b	of the three stars in the beginning of the first coil, the lowest one.
c	of the three stars in the beginning of the first coil, the uppermost.
d	of the three stars in the beginning of the first coil, the middle one.
e	the small star on the second coil and on the straight line marked ♈ (on the side of the chart).
f	of the two stars between the second and third coils below and to the left of the point marked L, the smaller one to the left.
g	of the three stars between the third and fourth coils, the one to the right.
h	of the two stars on the third coil and on the curved line marked N, the smaller one to the left.
i	the small star on his tail next to Alpha. The letter is marked Iota on the chart.

Equuleus *(Figure 37)*

Alpha, α	the star between his ears.
Beta, β	to the left of his left ear. The letter is difficult to see on some charts.
Gamma, γ	on his nose.
Delta, δ	on his mouth.

Eridanus *(Figure 38)*

Alpha, α	the bright star at the river's end at the bottom of the chart.
Beta, β	of the three stars at the beginning of the river, the uppermost and brightest. Bayer considered Beta the first star in the river.
Gamma, γ	below the first large clump of reeds and the fourteenth star from the beginning of the river.
Delta, δ	below the first large clump of reeds and the sixteenth star from the beginning.
Epsilon, ε	below the first large clump of reeds and the seventeenth star from the beginning.
Zeta, ζ	the eighteenth star from the beginning.
Eta, η	the twentieth star from the beginning.
Theta, θ	below the third large clump of reeds near the bottom of the river and the fifth star above Alpha.
Iota, ι	the fourth star above Alpha.
Kappa, κ	the third star above Alpha.
Lambda, λ	of the three stars at the beginning of the river, the lowest one. Bayer considered Lambda the second star in the river.
Mu, μ	the sixth star from the beginning.
Nu, ν	the eighth star from the beginning.
Xi, ξ	the ninth star from the beginning.
Omicron, o	the two bright stars to the right of Xi. Bayer considered them the tenth and eleventh from the beginning.
Pi, π	below the first large clump of reeds and the fifteenth star from the beginning.
Rho, ρ	in the first bend of the river and the nineteenth star from the beginning.
Sigma, σ	in the first bend of the river and the twenty-first star from the beginning.
Tau, τ	the group of nine stars between the first and second bends of the river ending under the second large clump of reeds.
Upsilon, υ	the group of seven stars in the second bend of the river ending before the third large clump of reeds.
Phi, φ	the second star above Alpha.
Chi, χ	the first star above Alpha.
Psi, ψ	of the three stars at the beginning of the river, the middle one. Bayer considered Psi the third star in the river.
Omega, ω	the fifth star from the beginning.
A	the thirteen star from the beginning.
b	the fourth star from the beginning.
c	the seventh star from the beginning.
d	the twelfth star from the beginning next to the Omicron pair.

Gemini *(Figure 39)*

Alpha, α	the bright star on the forehead of Castor, the twin on the right.
Beta, β	the bright star on the neck of Pollux, the twin on the left.

Gamma, γ	on Pollux's left foot.		Lambda, λ	of the three stars on his upper right arm, the middle one.
Delta, δ	on Castor's right hand.			
Epsilon, ε	on Castor's left knee.		Mu, μ	of the three stars on his upper right arm, the one to the left.
Zeta, ζ	on Pollux's right knee.			
Eta, η	on Castor's left instep.		Nu, ν	of the two stars on the right side of the lion's paw, the one to the north.
Theta, θ	on Castor's left hand.			
Iota, ι	of the three stars on Castor's right shoulder, the brightest one to the right.		Xi, ξ	of the two stars on the right side of the lion's paw, the one to the south.
Kappa, κ	of the two stars on Pollux's right breast, the brighter one below.		Omicron, o	of the three stars on the left side of the lion's paw, the one to the south.
Lambda, λ	on Pollux's right thigh.		Pi, π	of the three stars on his right thigh, the one to the right.
Mu, μ	on Castor's left leg.			
Nu, ν	on Castor's right foot.		Rho, ρ	of the three stars on his right thigh, the one to the left.
Xi, ξ	on Pollux's right foot near his toes.			
Omicron, o	at the top of Castor's head-covering.		Sigma, σ	of the two stars on his left thigh just below the lion-skin, the brighter one to the north.
Pi, π	at the top of Pollux's head-covering.			
Rho, ρ	on Castor's left eye.		Tau, τ	on the side of his left knee.
Sigma, σ	the small star on Pollux's right cheek next to Beta.		Upsilon, υ	of the two closely placed stars below his left knee, the one to the right.
Tau, τ	on Castor's left shoulder.		Phi, φ	of the two closely placed stars below his left knee, the one to the left.
Upsilon, υ	of the three stars on Castor's right shoulder, the one to the left.		Chi, χ	on the calf of his left leg.
Phi, φ	on Pollux's right armpit.		Psi, ψ	above his left heel.
Chi, χ	on Pollux's upper right arm.		Omega, ω	on the club above his thumb
Psi, ψ	on Pollux's right forearm.		A	of the three stars on the left side of the lion's paw, the one to the north.
Omega, ω	the small star on Castor's belly under Pollux's left hand.		b	of the three stars on the left side of the lion's paw, the middle one.
A	on Castor's right side near his armpit.			
b	of the three stars on Castor's right shoulder, the small middle one.		c	of the two stars on his right hip to the left of the 30° line, the one to the north.
c	of the two stars on Pollux's right breast, the smaller one above.		d	of the two stars on his right hip to the left of the 30° line, the smaller one to the south.
d	the small star on Castor's right knee.		e	of the three stars on his right thigh, the middle one.
e	the small star on Pollux's right heel.			
f	of the two small stars near the end of the cloak by Pollux's right side, the one below and to the right.		f	on his right calf.
			g	of the two stars on his left thigh just below the lion-skin, the smaller one to the south.
g	of the two small stars near the end of the cloak by Pollux's right side, the upper one to the left.		h	the second star on the club above his thumb.
			i	the third star on the club above his thumb.
			k	the fourth star on the club above his thumb.
			l	the fifth star on the club above his thumb.
	HERCULES *(Figure 41)*		m	the sixth star on the club above his thumb.
			n	the seventh star on the club above his thumb.
Alpha, α	the bright star near his eye.		o	the eighth star on the club above his thumb.
Beta, β	of the three stars on his left shoulder, the bright one to the north.		p	the ninth star on the club above his thumb.
Gamma, γ	of the three stars on his left shoulder, the bright one to the south.		q	of the three stars on his left hand holding the club, the small one to the south.
Delta, δ	of the three stars on his upper right arm, the one to the right near his shoulder.		r	of the three stars on his left hand holding the club, the small one to the north.
Epsilon, ε	on the right side of his back at the edge of the lion-skin.		s	of the three stars on his left shoulder, the small one in the middle.
Zeta, ζ	on the lion-skin in the middle of his back.		t	on his right wrist.
Eta, η	the bright star at the bottom of the lion-skin at the base of its tail.		u	of the two small stars on the right side of his belly, the one to the right.
Theta, θ	at the back of his right knee.		w	of the two small stars on the right side of his belly, the one to the left.
Iota, ι	the bright star on his right leg above his ankle.			
Kappa, κ	of the three stars on his left hand holding the club, the middle one.		x	of the three stars at the bottom of his right foot, the one to the right.

Lost Stars

y   of the three stars at the bottom of his right foot, the one in the middle.

z   of the three stars at the bottom of his right foot, the one to the left.

HYDRA *(Figure 43)*

Alpha, α   the bright star between the first and second coils.
Beta, β   of the three stars on the third coil, the upper one to the left.
Gamma, γ   the second star from the end of the tail.
Delta, δ   near the tip of her nose.
Epsilon, ε   of the three stars over her left eye, the one to the right.
Zeta, ζ   of the three stars over her left eye, the one to the left.
Eta, η   of the two stars in front of her right eye, the one to the left.
Theta, θ   behind her right eye.
Iota, ι   of the three stars near the line marked K between the first and second coils, the upper one to the left.
Kappa, κ   of the five stars on the second coil, the one to the right.
Lambda, λ   of the five stars on the second coil, the uppermost of the middle three.
Mu, μ   of the five stars on the second coil, the one to the left.
Nu, ν   of the five stars between the second and third coils, the second one from the left.
Xi, ξ   of the three stars on the third coil, the upper one to the right.
Omicron, o   of the three stars on the third coil, the middle one below the other two.
Pi, π   the last star on the tail.
Rho, ρ   of the three stars over her left eye, the small lower one in the middle.
Sigma, σ   of the two stars in front of her right eye, the one to the right.
Tau, τ   of the three stars near the line marked K between the first and second coils, the two small ones to the right.
Upsilon, υ   of the five stars in the second coil, the two lower ones of the middle three.
Phi, φ   of the five stars between the second and third coils, the one to the right. Bayer considered the two dark unmarked stars (upper left) in this section of Hydra's body, as well as several others near the tail, part of the constellations Corvus and Crater not Hydra.
Chi, χ   of the five stars between the second and third coils, the one below and to the left
Psi, ψ   the third star from the end of the tail.
Omega, ω   the small star at the top of her head.
A   the small star above Alpha.
b   of the five stars between the second and third coils, the two small ones.

LEO *(Figure 48)*

Alpha, α   of the five stars on his chest, the bright one on the ecliptic.
Beta, β   of the two stars on his tail, the brighter one at the tip.
Gamma, γ   of the two stars on his neck, the lower one.
Delta, δ   of the three stars on the top of his back, the bright one to the left.
Epsilon, ε   of the two stars to the left of his mouth, the brighter one to the right.
Zeta, ζ   of the two stars on his neck, the upper one.
Eta, η   of the five stars on his chest, the uppermost.
Theta, θ   of the three stars on his right thigh, the uppermost.
Iota, ι   of the three stars on his right thigh, the lowest one.
Kappa, κ   on his nose.
Lambda, λ   on his mouth.
Mu, μ   between his right eye and ear.
Nu, ν   of the five stars on his chest, the one to the right of Alpha on the ecliptic.
Xi, ξ   of the two stars on his left front paw, the lower one.
Omicron, o   of the three stars on his right front paw, the uppermost one to the left.
Pi, π   on his right foreleg.
Rho, ρ   near his right armpit on the ecliptic. The letter may be difficult to see on some charts.
Sigma, σ   above the knee of his right hind leg.
Tau, τ   on the knee of his right hind leg below the ecliptic.
Upsilon, υ   of the three stars on the lower part of his right hind leg, the uppermost.
Phi, φ   of the three stars on the lower part of his right hind leg, the lowest one on his paw.
Chi, χ   of the two stars on his left thigh, the upper one.
Psi, ψ   of the two stars on his left front paw, the upper one on the ecliptic.
Omega, ω   of the three stars on his right front paw, the lowest one to the right.
A   of the five stars on his chest, the lowest one beneath Alpha.
b   of the three stars on the top of his back, the middle one.
c   of the two stars on his left thigh, the lower one to the right.
d   near his left hind knee.
e   of the three stars on the lower part of his right hind leg, the middle one.
f   on his left ear.
g   of the two stars to the left of his mouth, the smaller one to the left.
h   of the three stars on his right front paw, the middle one.
i   of the five stars on his chest, the small one above Alpha.
k   the small star in the center of his body behind his mane.

l	the small star on his belly. The letter is barely visible on some charts.
m	of the three stars on the top of his back, the small one to the right.
n	of the three stars on his right thigh, the small one in the middle.
o	of the two stars on his tail, the smaller one to the left.
p	the group of five small stars on his lower left hind leg.

Lepus *(Figure 50)*

Alpha, α	the star on his right shoulder.
Beta, β	on his right foreleg.
Gamma, γ	on his right hind leg near his paw.
Delta, δ	on his right hind leg at his hock or heel.
Epsilon, ε	on his left front paw.
Zeta, ζ	of the three stars in the center of his body, the one to the right.
Eta, η	of the three stars in the center of his body, the one in the middle.
Theta, θ	of the three stars in the center of his body, the one to the left.
Iota, ι	on his left ear.
Kappa, κ	at the top of his head between his ears.
Lambda, λ	of the two stars on his right ear, the lower one.
Mu, μ	on his check.
Nu, ν	of the two stars on his right ear, the upper one.

Libra *(Figure 51)*

Alpha, α	of the four stars in the right pan, the brightest.
Beta, β	near the bottom of the support piece.
Gamma, γ	of the four stars forming a quadrilateral in the left pan, the upper one to the right.
Delta, δ	on the beam to the right of the support piece.
Epsilon, ε	on the beam to the left of the support piece.
Zeta, ζ	of the four stars forming a quadrilateral in the left pan, the lower one to the left.
Eta, η	of the four stars forming a quadrilateral in the left pan, the upper one to the left.
Theta, θ	on the left end of the beam.
Iota, ι	below the front of the left pan.
Kappa, κ	below the side of the left pan on the ecliptic.
Lambda, λ	below the left end of the beam on the ecliptic.
Mu, μ	of the four stars in the right pan, the small one to the right.
Nu, ν	of the four stars in the right pan, the one to the left on the side of the pan.
Xi, ξ	of the four stars in the right plan, the uppermost.
Omicron, o	of the four stars forming a quadrilateral in the left pan, the one to the lower right.

Lupus *(Figure 52)*

Alpha, α	the star on his hind hock or heel.
Beta, β	of the two stars on his belly, the lower one.
Gamma, γ	below his jaw.

Delta, δ	of the two stars on his right front paw, the upper one to the right.
Epsilon, ε	of the two stars on his right front paw, the lower one to the left.
Zeta, ζ	of the three stars on his forearm, the one to the right.
Eta, η	of the three stars on his forearm, the middle one.
Theta, θ	at his armpit.
Iota, ι	of the three stars on his tail, the middle one.
Kappa, κ	of the three stars on his tail, the uppermost.
Lambda, λ	in the back of his mouth.
Mu, μ	of the two stars on his neck, the upper one.
Nu, ν	of the two stars on his neck, the lower one.
Xi, ξ	on his right side.
Omicron, o	on the paw of his hind leg.
Pi, π	of the two stars on his belly, the upper one.
Rho, ρ	on his haunch.
Sigma, σ	on his lower back.
Tau, τ	of the three stars on his tail, the lowest one.
Upsilon, υ	of the three stars on his forearm, the small faint one to the left.

Lyra *(Figure 54)*

Alpha, α	the bright star on the eagle's beak.
Beta, β	of the four stars near the base of the Lyre, the bright one to the right.
Gamma, γ	of the four stars near the base of the Lyre, the bright one to the left.
Delta, δ	of the two stars on the upper part of the Lyre, the one to the left.
Epsilon, ε	on the eagle's neck.
Zeta, ζ	of the two stars on the upper part of the Lyre, the one to the right.
Eta, η	of the three stars on the eagle's right wing, the uppermost.
Theta, θ	of the three stars on the eagle's right wing, the middle one.
Iota, ι	of the three stars on the eagle's right wing, the lowest to the right.
Kappa, κ	of the two stars on the eagle's left wing, the lower one.
Lambda, λ	of the four stars on the base of the Lyre, the small one to the left.
Mu, μ	of the two stars on the eagle's left wing, the upper one.
Nu, ν	of the four stars on the base of the Lyre, the small one to the right.

Navis *(Figure 19)*

Alpha, α	of the two stars on the left steering board, the bright one at the bottom.
Beta, β	the bright star at the top of the railing to the right of the backstays or braces at the stern.
Gamma, γ	the bright star on the stern below Beta.
Delta, δ	of the five stars below and to the right of the mast, the middle and brightest one.

Epsilon, ε at the top of the railing to the left of the ratlines.

Zeta, ζ on the hull between the sixth and seventh oars.

Eta, η above the fifth oar.

Theta, θ the bright star near the top of the fourth oar.

Iota, ι of the two uppermost stars on the stern, the one to the left on the curved line marked X.

Kappa, κ the brightest star in the middle of the stern surrounded by five somewhat smaller stars counterclockwise: Pi, Xi, Omicron, Sigma, and Rho.

Lambda, λ on the stern to the right of the left steering board.

Mu, μ the bright star at the crack in the middle of the fifth oar.

Nu, ν the two bright stars, one in the middle of what remains of the third oar and the other at the bottom of the fourth oar.

Xi, ξ the uppermost of the five stars surrounding Kappa on the stern below the curved line marked X.

Omicron, ο of the five stars surrounding Kappa on the stern, the one below Xi.

Pi, π of the five stars surrounding Kappa on the stern, the one to the upper right of Kappa below the line marked X.

Rho, ρ of the five stars surrounding Kappa on the stern, the one directly below Kappa.

Sigma, σ of the five stars surrounding Kappa on the stern, the one below and to the left.

Tau, τ on the stern to the left of Sigma.

Upsilon, υ next to the keel on the stern below and to the left of Rho.

Phi, φ the two closely placed stars at the top and to the left of the steering board.

Chi, χ At the top of the railing at the stern and to the right of Beta. There are two stars marked Chi on the chart, this star and the one below Gamma. The latter should be Psi (see Psi, below).

Psi, ψ Although no star is marked Psi on the chart, the star that should be Psi is the one erroneously marked Chi on the stern below Gamma. Bayer noted in his star list that there are fourteen stars in a series near the stern starting with Tau. The second is Chi (near Beta), the third Beta, the fourth Gamma, the fifth Psi but erroneous marked Chi on his chart (see Chi, above).

Omega, ω of the five stars below and to the right of the mast, the two lowest.

A of the five stars below and to the right of the mast, the two uppermost.

b on the hull between the first and second oars.

c on the rocks between the first two oars.

d six stars, five scattered on the rocks and one below in the water.

e of the two uppermost stars on the stern, the one to the right above the curved line marked X.

f at the stern above the steering board.

g at the stern on the keel at the water line.

h of the two stars on the left steering board, the small one above Alpha.

i on the hull between the last oar and the steering board.

k the small star on the hull between the fifth and sixth oars.

l the two small stars, one in the middle of the second oar and the other above Nu in the middle of the third oar.

m the small star on the hull between the fourth and fifth oars.

n four stars scattered on the rocks.

o the four stars on the mast.

p the three stars on the ship's railing between the ratlines and the mast.

q above Theta between the third and fourth oars.

r the small star near the top of the ninth oar.

s three small stars in and around the stub of the third oar.

Ophiuchus *(Figure 57)*

Alpha, α of the three stars on his head, the uppermost and brightest.

Beta, β of the two bright stars on his left shoulder, the one above and to the right.

Gamma, γ of the two bright stars on his left shoulder, the one below and to the left.

Delta, δ of the two bright stars on his right hand, the one above and to the right.

Epsilon, ε of the two bright stars on his right hand, the one below and to the left.

Zeta, ζ the bright star on his right thigh.

Eta, η on his upper left thigh below his cloak.

Theta, θ of the three stars on his lower left thigh, the upper one to the left.

Iota, ι of the two stars on his right shoulder, the one above and to the right.

Kappa, κ of the two stars on his right shoulder, the one below and to the left.

Lambda, λ on his right elbow.

Mu, μ on his left forearm.

Nu, ν on his left wrist.

Xi, ξ of the three stars on his lower left thigh, the upper one to the right.

Omicron, ο of the three stars on his lower left thigh, the one below the other two.

Pi, π of the four stars behind his left knee, the upper of the two on the right.

Rho, ρ of the three stars near his right ankle, the middle one.

Sigma, σ in the middle of his back.

Tau, τ on the left thumb.

Upsilon, υ the small star above his right knee.

Phi, φ on the calf of his right leg.

Chi, χ below his right calf.

Psi, ψ of the three stars near his right ankle, the one above and to the right.

Omega, ω	of the three stars near his right ankle, the one to the left.
A	of the four stars behind his left knee, the lower of the two on the right.
b	of the four stars behind his left knee, the second from the left.
c	of the four stars behind his left knee, the one farthest to the left.
d	near his left ankle.
e	of the three stars on his head, the small one below and to the right.
f	of the three stars on his head, the small one below and to the left.

ORION *(Figure 58)*

Alpha, α	the bright star on his left shoulder.
Beta, β	of the three stars on his right foot, the bright middle one. Bayer considered the star below Beta part of Eridanus not Orion.
Gamma, γ	of the two stars on his right shoulder, the bright one to the right. The letter may be difficult to see on some charts.
Delta, δ	of the three bright stars on his belt, the uppermost to the right.
Epsilon, ε	of the three bright stars on his belt, the middle one.
Zeta, ζ	of the three bright stars on his belt, the one to the left.
Eta, η	of the three stars below his belt on his right side, the lowest.
Theta, θ	of the three stars in the middle of his sword, the middle one.
Iota, ι	of the three stars in the middle of his sword, the lowest one.
Kappa, κ	on his left knee.
Lambda, λ	of the three stars on his head, the brightest one above the other two.
Mu, μ	on his left upper arm.
Nu, ν	of the two stars on his left hand, the one to the right on his thumb.
Xi, ξ	of the two stars on his left hand, the one to the left.
Omicron, o	of the three stars near the top of the lion-skin in his right hand, the two uppermost.
Pi, π	the group of six similar-sized stars on the lion-skin dangling below his right hand.
Rho, ρ	at the edge of the lion-skin dangling under his right shoulder.
Sigma, σ	the small star beneath the hilt of his sword.
Tau, τ	of the three stars on his right foot, the small one above.
Upsilon, υ	of the two stars near the top of his right boot, the one to the left near his sword.
Phi, φ	of the three stars on his head, the two smaller ones below.
Chi, χ	the two stars in the middle of his club.
Psi, ψ	of the five stars in a row on his back, the second from the right.
Omega, ω	of the five stars in a row on his back, the one to the left.
A	of the two stars on his right shoulder, the smaller one to the left.
b	at the top of his sword.
c	of the three stars in the middle of his sword, the uppermost one.
d	at the left side of his sword toward the tip.
e	of the two stars near the top of his right boot, the one to the right.
f	the two small stars near the base of his club.
g	of the three stars near the top of the lion-skin in his right hand, the small one below the other two.
h	of the two small stars on his upper right arm, the one above.
i	of the two small stars on his upper right arm, the one below.
k	on his left wrist.
l	on his left elbow.
m	of the five stars in a row on his back, the one on the right, barely visible on some charts.
n	of the five stars in a row on his back, the two small ones, the third and fourth from the right..
o	of the three stars below his belt on his right side, the uppermost.
p	of the three stars below his belt on his right side, the middle one.

PEGASUS *(Figure 60)*

Alpha, α	the bright star on the front edge of his wing.
Beta, β	the bright star on his left shoulder.
Gamma, γ	the bright star near the side of his wing among his primary feathers.
Delta, δ	the bright star where the clouds meet his body.
Epsilon, ε	on his nose.
Zeta, ζ	of the four stars on his neck, the brightest one to the right.
Eta, η	of the two stars near his left knee, the brighter one above and to the left.
Theta, θ	of the two stars near his eye, the brighter one above and to the left.
Iota, ι	behind his right leg.
Kappa, κ	above his right hoof.
Lambda, λ	of the two stars on his chest, the one below and to the right.
Mu, μ	of the two stars on his chest, the one above and to the left.
Nu, ν	of the two stars near his eye, the smaller one below and to the right.
Xi, ξ	of the four stars on his neck, the uppermost one to the left.
Omicron, o	of the two stars near his left knee, the smaller one below and to the right.
Pi, π	near his left hoof.
Rho, ρ	of the four stars on his neck, the bottom one.
Sigma, σ	of the four stars on his neck, the second from the bottom.

Tau, τ	of the four small stars at the tip of his feathers, the uppermost.
Upsilon, υ	of the four small stars at the tip of his feathers, the second from the top.
Phi, φ	the small star in the middle of his wing.
Chi, χ	of the four small stars at the tip of his feathers, the fourth from the top.
Psi, ψ	of the four small stars at the tip of his feathers, the third from the top.

PERSEUS *(Figure 61)*

Alpha, α	of the two stars on his back, the brighter one to the left.
Beta, β	the bright star by Medusa's right eye.
Gamma, γ	on his left shoulder.
Delta, δ	of the three stars near his waist, the one at the bottom and to the left. The letter is not clearly marked some charts.
Epsilon, ε	at back of his right knee.
Zeta, ζ	the bright star on his right instep.
Eta, η	on his left forearm.
Theta, θ	at the top of his shield.
Iota, ι	of the two stars on his back, the smaller one above and to the right.
Kappa, κ	on his right side near the edge of his shield.
Lambda, λ	of the two stars on his left thigh, the one to the left.
Mu, μ	of the three stars at his left knee, the middle one.
Nu, ν	on his right upper thigh below his pouch.
Xi, ξ	in the middle of his right leg.
Omicron, o	at his right heel.
Pi, π	on Medusa's left eye.
Rho, ρ	on Medusa's nose.
Sigma, σ	of the three stars near his waist, the uppermost.
Tau, τ	on his helmet near his ear.
Upsilon, υ	of the three stars on the blade of his sword, the one to the right.
Phi, φ	of the three stars on the blade of his sword, the one in the middle.
Chi, χ	of the three stars on the sword's hilt, the bright one to the left.
Psi, ψ	of the three stars near his waist, the middle one.
Omega, ω	on Medusa's right check.
A	of the two stars on his left thigh, the one to the right.
b	of the three stars at his left knee, the uppermost and to the left.
c	of the three stars at his left knee, the one to the right behind his knee.
d	in the middle of his left leg.
e	at the bottom of his left foot.
f	on the wing on his left foot.
g	of the three stars on the blade of his sword, the one to the left.
h	of the three stars on the sword's hilt, the small one to the right.
i	of the three stars on the sword's hilt, the one below the other two.
k	at his left elbow.
l	the small star on the right side of his pouch.
m	near the bottom of his left leg.
n	above his right ankle.
o	on the upper wing of his right foot.

PISCES *(Figure 63)*

Alpha, α	the bright star on the knot of the ribbon binding the two Fish together.
Beta, β	on the eye of the fish on the right, the southern fish.
Gamma, γ	on the gills of the southern fish.
Delta, δ	the second star on the ribbon from the tail of the southern fish.
Epsilon, ε	of the three stars on the loop of the ribbon binding the southern fish, the upper one to the right.
Zeta, ζ	of the three stars on the loop of the ribbon binding the southern fish, the upper one to the left.
Eta, η	the third star on the ribbon from Alpha toward the fish on the left, the northern fish.
Theta, θ	of the three stars on the back of the southern fish, the middle one.
Iota, ι	of the three stars on the back of the southern fish, the one to the left.
Kappa, κ	of the two stars near the belly of the southern fish, the one to the right.
Lambda, λ	of the two stars near the belly of the southern fish, the one to the left.
Mu, μ	the third star on the ribbon from Alpha toward the southern fish.
Nu, ν	the second star on the ribbon from Alpha toward the southern fish.
Xi, ξ	the first star on the ribbon from Alpha toward the southern fish.
Omicron, o	the first star on the ribbon from Alpha toward the northern fish.
Pi, π	the second star on the ribbon from Alpha toward the northern fish.
Rho, ρ	the fourth star on the ribbon from Alpha toward the northern fish.
Sigma, σ	near the eye of the northern fish.
Tau, τ	of the two stars below the mouth of the northern fish, the lower one.
Upsilon, υ	of the three stars on the belly of the northern fish, the middle one.
Phi, φ	of the three stars on the belly of the northern fish, the bottom one.
Chi, χ	of the four stars toward the tail of the northern fish, the left star of the two in the middle.
Psi, ψ	of the four stars toward the tail of the northern fish, the three stars in a line along its lower spine.
Omega, ω	of the two stars near the tail of the southern fish, the lower one to the right.
A	the small star below the eye of the southern fish.

b	of the three stars on the back of the southern fish, the one to the right.
c	of the two stars near the tail of the southern fish, the upper one to the left.
d	the first star on the ribbon from the tail of the southern fish.
e	of the three stars on the loop of the ribbon binding the southern fish, the one at the bottom below the ecliptic.
f	the fourth star on the ribbon from Alpha toward the southern fish.
g	of the two stars below the mouth of the northern fish, the upper one.
h	below the eye and near the gills of the northern fish.
i	of the two stars near the upper fins of the northern fish, the one above and to the right.
k	of the two stars near the upper fins of the northern fish, the one below and to the left.
l	of the three stars on the belly of the northern fish, the uppermost.

Piscis Austrinus *(Figure 64)*

Alpha, α	the bright star in his mouth.
Beta, β	of the three stars on the fin under his eye, the one to the left.
Gamma, γ	of the two small stars near his mouth, the one to the right.
Delta, δ	of the two small stars near his mouth, the one to the left next to Alpha.
Epsilon, ε	on the tip of the fin under his jaw.
Zeta, ζ	of the three stars on a horizontal line across his body, the one to the left.
Eta, η	of the three stars on a horizontal line across his body, the one to the right.
Theta, θ	of the two stars on the coil, the one above and to the left.
Iota, ι	of the two stars on the coil, the one below and to the right.
Kappa, κ	of the three stars on the fin under his eye, the one to the right.
Lambda, λ	of the three stars on a horizontal line across his body, the middle one.
Mu, μ	of the three stars on the fin under his eye, the one in the middle.

Sagitta *(Figure 66)*

Alpha, α	of the three stars on the fin, the uppermost.
Beta, β	of the three stars on the fin, the middle one.
Gamma, γ	of the three stars near the front of the shaft, the one to the right.
Delta, δ	of the two stars on the spike or barb, the lower one to the right.
Epsilon, ε	of the three stars on the fin, the lowest one.
Zeta, ζ	of the two stars on the spike or barb, the upper one to the left.
Eta, η	of the three stars near the front of the shaft, the middle one.
Theta, θ	of the three stars near the front of the shaft, the one to the left near the point.

Sagittarius *(Figure 67)*

Alpha, α	the bright star above his knee on his right foreleg.
Beta, β	the bright star on the hoof of his right foreleg.
Gamma, γ	on the point of the arrow.
Delta, δ	on his hand.
Epsilon, ε	on the bow below his hand.
Zeta, ζ	on his upper arm.
Eta, η	above the hoof of his left foreleg.
Theta, θ	on his right thigh.
Iota, ι	above the fetlock of his right hind leg. The letter is barely visible on some charts.
Kappa, κ	above the fetlock of his left hind leg.
Lambda, λ	on the upper part of his bow below the ecliptic.
Mu, μ	at the very top of his bow.
Nu, ν	the faint double star on his chin.
Xi, ξ	on his ear.
Omicron, o	of the two stars behind his neck, the one to the right.
Pi, π	of the two stars behind his neck, the one to the left.
Rho, ρ	of the three stars behind his head on his cape, the one in the middle.
Sigma, σ	of the two stars on his right shoulder, the one to the left.
Tau, τ	behind and below his right shoulder.
Upsilon, υ	of the three stars behind his head on his cape, the uppermost one to the left.
Phi, φ	of the two stars on his right shoulder, the one to the right.
Chi, χ	on his left shoulder.
Psi, ψ	between his shoulders.
Omega, ω	of the four stars on his rump, the upper one to the right.
A	of the four stars on his rump, the upper one to the left.
b	of the four stars on his rump, the one between the upper two and the lowest one.
c	of the four stars on his rump, the lowest one.
d	of the three stars behind his head on his cape, the one to the right near his head.
e	the small star in the middle of his cape.
f	of the two small stars near the bottom edge of his cape, the one below and to the right.
g	of the two small stars near the bottom edge of his cape, the one above and to the left.
h	the small star near his left side under the curved line marked M.

Scorpius *(Figure 68)*

Alpha, α	of the three stars in the center of his body, the bright one in the middle.
Beta, β	of the three stars at the base of his left claw, the bright one in the middle.
Gamma, γ	at the top of his right claw.

Lost Stars

Delta, δ	at the very top of his body near the base of his right claw.
Epsilon, ε	at the very bottom of his body near the base of his tail.
Zeta, ζ	of the two stars on the second joint of his tail, the upper one.
Eta, η	on the third joint of his tail.
Theta, θ	between the fourth and fifth joints of his tail.
Iota, ι	on the fifth joint of his tail.
Kappa, κ	between the fifth and sixth joints of his tail.
Lambda, λ	of the two stars on the last joint of his tail, the brighter one to the left.
Mu, μ	the two stars each marked Mu on his tail: one on the first joint, the other on the second joint below Zeta.
Nu, ν	of the three stars at the base of his left claw, the small one above and to the left.
Xi, ξ	of the two stars on his left pincer, the one below and to the right.
Omicron, o	of the three stars on his first right leg, the one toward the tip.
Pi, π	on his second right leg.
Rho, ρ	on his third right leg.
Sigma, σ	of the three stars in the center of his body, the small one above and to the right..
Tau, τ	of the three stars in the center of his body, the small one below and to the left.
Upsilon, υ	of the two stars on the last joint of his tail, the smaller one to the right.
Phi, φ	between the first and second joints of his left claw.
Chi, χ	on the upper part of the second joint of his left claw.
Psi, ψ	of the two stars on his left pincer, the one above and to the left.
Omega, ω	of the three stars at the base of his left claw, the small one below and to the left.
A	of the three stars on his first right leg, the uppermost one near his body.
b	of the three stars on his first right leg, the middle one.
c	the two stars on his body near the base of his fourth right leg.

SERPENS *(Figure 70)*

Alpha, α	of the three stars on the second bend, the bright one to the right.
Beta, β	of the four stars forming a small quadrilateral on his neck, the bright one above and to the right.
Gamma, γ	of the four stars forming a small quadrilateral on his neck, the bright one above and to the left.
Delta, δ	of the two stars on the first bend, the lower one to the right.
Epsilon, ε	of the three stars on the second bend, the one below and to the left of Alpha.
Zeta, ζ	the sixth star from his tail.
Eta, η	the fifth star from his tail.
Theta, θ	the star at the end of his tail.
Iota, ι	to the right of his eye.
Kappa, κ	to the left of his eye.
Lambda, λ	of the three stars on the second bend, the one above of Alpha.
Mu, μ	of the four stars on the third bend near the line marked L on the chart, the brightest and lowest one.
Nu, ν	of the three stars before the second coil, the one to the right. It is barely visible on some charts.
Xi, ξ	of the three stars before the second coil, the one in the middle.
Omicron, o	of the three stars before the second coil, the one to the left.
Pi, π	on his crest.
Rho, ρ	on his forehead above his eye.
Sigma, σ	at the top of the first coil.
Tau, τ	eight small stars in and around his mouth.
Upsilon, υ	of the four stars forming a small quadrilateral on his neck, the small star below and to the right.
Phi, φ	of the four stars forming a small quadrilateral on his neck, the small star below and to the left.
Chi, χ	of the two stars on the first bend, the upper one to the left.
Psi, ψ	of the two stars at the beginning of the third bend, the one to the right.
Omega, ω	of the two stars at the beginning of the third bend, the one to the left.
A	of the four stars on the third bend near the line marked L, the two upper ones on each side of the line.
b	of the four stars on the third bend near the line marked L, the small star next to Mu.
c	the fourth star from his tail.
d	the second star from his tail.
e	the third star from his tail.

Note: The unlettered stars in the Serpent belong to Ophiuchus.

TAURUS *(Figure 72)*

Alpha, α	the bright star on his right eye.
Beta, β	on the tip of his left horn.
Gamma, γ	of the four stars forming a quadrilateral on his snout, the one below and to the right.
Delta, δ	of the four stars forming a quadrilateral on his snout, the one above and to the right.
Epsilon, ε	next to his left eye.
Zeta, ζ	on the tip of his right horn.
Eta, η	the bright star in the Pleiades, the small group of six closely placed stars on his shoulder..
Theta, θ	of the four stars forming a quadrilateral on his snout, the one above and to the left.
Iota, ι	of the three stars at the base of his right horn, the bright one above and to the right.

Kappa, κ	of the two stars above his left eye, the lower one.
Lambda, λ	on his chest.
Mu, μ	of the two stars behind his right knee, the brighter one to the right.
Nu, ν	on his left knee.
Xi, ξ	of the five stars at the top of his left arm, the second from the bottom.
Omicron, o	of the five stars at the top of his left arm, the bottom one.
Pi, π	of the four stars forming a quadrilateral on his snout, the small one below and to the left.
Rho, ρ	of the two small stars to the left of his right eye, the one to the right.
Sigma, σ	of the two small stars to the left of his right eye, the one to the left.
Tau, τ	on his forehead above the ecliptic.
Upsilon, υ	of the two stars above his left eye, the upper one.
Phi, φ	of the two stars on his left ear, the upper one.
Chi, χ	of the two stars on his left ear, the lower one.
Psi, ψ	of the three stars on his neck, the uppermost.
Omega, ω	on his left check.
A	of the three stars on his neck, the bottom one.
b	above his right knee.
c	on his right knee.
d	on the side of his right knee.
e	in the middle of his left arm between his knee and his elbow.
f	of the five stars at the top of his left arm, the uppermost.
g	on his left hoof.
h	the small star at the tip of his snout.
i	the small star on his forehead above his right ear.
k	the small star at the top of his head between his two horns.
l	of the three stars at the base of his right horn, the small one in the middle.
m	of the three stars at the base of his right horn, the small one below the others.
n	of the two small stars in the middle of his right horn, the one to the right.
o	of the two small stars in the middle of his right horn, the one to the left.
p	of the three stars on his neck, the middle one.
q	the least bright star in the Pleiades.
r	of the two stars behind his right knee, the small one to the left.
s	of the five stars at the top of his left arm, the third from the bottom.
t	of the five stars at the top of his left arm, the small one to the left.
u	the small star below his knee in the middle of his left leg or shank.

TRIANGULUM *(Figure 73)*

Alpha, α	the star at its apex.
Beta, β	on the upper corner of its base.
Gamma, γ	on the lower corner of its base.
Delta, δ	on the base between the two corners.
Epsilon, ε	on the leg between the apex and the upper corner.

URSA MAJOR *(Figure 76)*

Alpha, α	of the four stars forming a quadrilateral on his back, the upper one to the right.
Beta, β	of the four stars forming a quadrilateral on his back, the lower one to the right.
Gamma, γ	of the four stars forming a quadrilateral on his back, the lower one to the left.
Delta, δ	of the four stars forming a quadrilateral on his back, the upper one to the left near the base of his tail.
Epsilon, ε	the first star on his tail.
Zeta, ζ	the second star on his tail.
Eta, η	the last star at the tip of his tail.
Theta, θ	of the two stars on his right foreleg, the one to the right.
Iota, ι	of the two stars on his left front paw, the upper one.
Kappa, κ	of the two stars on his left front paw, the lower one.
Lambda, λ	of the two stars on his right hind paw, the upper one.
Mu, μ	of the two stars on his right hind paw, the lower one. The letter may be difficult to see on some charts.
Nu, ν	of the two stars on his left hind paw, the upper one.
Xi, ξ	of the two stars on his left hind paw, the lower one.
Omicron, o	on his snout.
Pi, π	of the two stars on his eye, the one to the left.
Rho, ρ	of the two stars on his forehead, the one to the right.
Sigma, σ	of the two stars on his forehead, the one to the left.
Tau, τ	of the three stars around the back of his mouth, the uppermost.
Upsilon, υ	on his shoulder next to the line marked L.
Phi, φ	of the two stars on his right foreleg, the one to the left.
Chi, χ	on his right thigh.
Psi, ψ	of the two stars on his right hind leg, the upper one.
Omega, ω	of the two stars on his right hind leg, the lower one.
A	of the two stars on his eye, the one to the right.
b	of the three stars around the back of his mouth, the one at the bottom right.
c	of the three stars around the back of his mouth, the one at the bottom left.
d	on his right ear.
e	of the two stars on his left foreleg, the upper one.
f	of the two stars on his left foreleg, the lower one.

g	the barely visible small star next to the middle star on his tail.	b	below her left ear.
h	the small star on his neck.	c	over her left breast.

Ursa Minor *(Figure 77)*

Alpha, α	the bright star at the tip of his tail.
Beta, β	on his upper back to the right of his head.
Gamma, γ	above his right foreleg.
Delta, δ	the middle star on his tail.
Epsilon, ε	the third star from the tip of his tail.
Zeta, ζ	of the two stars on his right hip, the brighter one to the right.
Eta, η	above his right hind leg.
Theta, θ	of the two stars on his right hip, the smaller one to the left.

Virgo *(Figure 78)*

Alpha, α	the bright star on the spike of grain in her left hand.
Beta, β	the bright star on the upper part of her left wing.
Gamma, γ	of the two stars on her belly, the one on her left side.
Delta, δ	of the two stars on her belly, the one on her right side.
Epsilon, ε	of the two stars on her right wing, the brighter one to the right.
Zeta, ζ	the bright star above her right knee. The letter is difficult to see on some charts.
Eta, η	on her left shoulder.
Theta, θ	of the two stars on her left thigh, the brighter one to the left.
Iota, ι	on her right leg near her foot.
Kappa, κ	on her left leg near her foot.
Lambda, λ	on her left foot.
Mu, μ	on the tip of her right foot.
Nu, ν	of the five stars on her head, the one over her left eye.
Xi, ξ	of the five stars on her head, the second from the top of her head.
Omicron, o	to the right of her right ear.
Pi, π	on her nose.
Rho, ρ	on her right forearm near her wrist.
Sigma, σ	of the three stars along her right side, the one to the right next to the bottom strands of her hair.
Tau, τ	of the three stars along her right side, the one to the left.
Upsilon, υ	of the two stars near the bottom of her stole on her right side, the one to the right.
Phi, φ	of the two stars near the bottom of her stole on her right side, the one to the left.
Chi, χ	of the three stars near her left elbow, the middle one.
Psi, ψ	near her left wrist.
Omega, ω	of the five stars on her head, the one at the top.
A	of the five stars on her head, the two over and above her right eye.
b	below her left ear.
c	over her left breast.
d	the two small stars on her right elbow.
e	of the two stars on her right wing, the smaller one to the left.
f	of the three stars near her left elbow, the uppermost near the ecliptic.
g	on her left hand.
h	the small star on the ecliptic above and to the left of Alpha.
i	the small star below and to the left of Alpha.
k	of the two stars on her left thigh, the smaller one to the right.
l	the small star above her left knee.
m	the small star behind her left knee.
n	the small star on her leg below her left knee.
o	of the three stars along her right side, the one in the middle.
p	near her right knee.
q	of the three stars near her left elbow, the one at the bottom on her left wing.

APPENDIX II
LOST CONSTELLATIONS

Over the past several centuries, European astronomers have devised numerous constellations and asterisms to fill in otherwise unoccupied spaces in the sky and to honor certain objects or individuals, usually their sovereigns or patrons. Except for the eighty-eight constellatioins officially recognized by the International Astronomical Union included in Part One of this work, all of the others have fallen by the wayside. Listed below are some of these discarded and no longer recognized constellations. It is by no means a complete or definitive list, but it does include the more prominent obsolete constellations that have found their way into the astronomical literature. Their survival can be attributed to the fact that all of them are included in Bode's atlas of 1801, a large and beautifully crafted work that was a favorite of nineteenth-century astronomers. The primary sources for these out-dated constellations were Richard H. Allen's *Star Names* and Deborah J. Warner's *The Sky Explored*.

ANTINOUS

Devised by the Emperor Hadrian (AD 76-138) to commemorate his homosexual lover, Antinous of Bythnia. See Aquila in Part One.

CERBERUS

Hevelius devised Cerberus, the mythological three-headed canine guardian of the Underworld. See Canes Venatici, Vulpecula, and Figure 42 in Part One.

COR CAROLI

Devised by Charles Scarborough, the Heart of Charles I first appeared on Francis Lamb's planisphere of 1673. See Canes Venatici in Part One.

CUSTOS MESSIUM

The Harvest-Keeper or literally Guardian of the Harvest was devised by J. J. Lalande in 1775 to honor his friend and associate Charles Messier. The title is a play on words. "Messis" is Latin for harvest and *Messi*er was known to "harvest" or search out comets and nebulous objects in the heavens. The constellation was made up of a group of relatively small stars between Camelopardalis, Cassiopeia, and Cepheus.

FELIS

The Cat was devised by Lalande between Antlia and Hydra to honor his feline friends. Allen (p. 221) quoted him as saying: "I am very fond of cats. I will let this figure scratch on the chart." Bode included it in his atlas of 1782.

FREDERICI HONORES

Frederick's Glory or Friedrich's Ehre was devised by Bode in 1782 in honor of Frederick the Great (1712-1786) of Prussia. It consisted of a large group of small stars between Andromeda, Cassiopeia, Cepheus, and Cygnus.

GLOBUS AEROSTATICUS

The Aerial Balloon honored the pioneer balloonists, the Montgolfier brothers. Allen attributed the constellation to Lalande. Bode included it in his atlas of 1782 between Capricornus, Microscopium, and Piscis Austrinus.

LOCHIUM FUNIS

The nautical Log Line was used to measure a ship's speed and hence its daily distance. Bode devised it near Pyxis and included it in his atlas of 1801.

Machina Electrica

Devised by Bode and included in his atlas 1801, the Electric Machine represented an "electrostatic generator and prime conductor" (Warner, p. xii). It was made up of a group of stars below Cetus.

Mons Maenalus

Mount Maenalus, the home of Bootes in Arcadia, was devised by Hevelius. See Canes Venatici and Figure 26 in Part One.

Musca Borealis

The Northern Fly was probably devised by Petrus Plancius in 1613. See c Ari in Part One.

Officina Typographica

Formed by Bode, the Printing Office commemorated the 350th anniversary of the development of the printing press by Johann Gutenberg. It was located east of Sirius in Bode's atlas of 1801.

Psalterium Georgianum

George's Harp was devised by Maximillian Hell in 1781 to honor George III. Also called Harpa Georgii, it was formed from a group of stars between Taurus and Eridanus.

Rangifer

Pierre-Charles Le Monnier drew the Reindeer on his planisphere of the northern heavens in 1743 from the stars between Camelopardalis and Cassiopeia. It was meant to commemorate a trip he had made to Lapland.

Robur Carolinum

Charles's Oak was devised in 1679 by Halley in honor of his sovereign and patron Charles II. See figures 31 and 65.

Sceptrum Brandenburgicum

Devised by Gottfried Kirch in 1688 to honor the royal house of Prussia, the Brandenburg Scepter consisted of four stars in a vertical line between Lepus and Eridanus.

Solitaire

Formed by Pierre-Charles Le Monnier just above Hydra's tail, it first appeared in a plate in the *Mémoires, Académie Royale des Sciences* of 1776. Bode referred to it as Vogel Einsiedler, "Solitaire," in his atlas of 1782 and as Turdus Solitarius, the Solitary Thrush, in his atlas of 1801. Based on the plate in the *Mémoires,* it was probably distantly related to Townsend's Solitaire (*Myadestes townsendi*), certainly not to the extinct dodo, the Rodrigues Solitaire (*Pezophaps solitaria*), as some have claimed.

Taurus Poniatovii

Poniatowski's Bull was devised by Martin Odlanicky Poczobut in 1773 to honor King Stanislas II Poniatowski of Poland. Lalande placed it in his celestial globe of 1779 to the east of Ophiuchus's shoulder.

Triangulum Minus

Hevelius devised the Little or Lesser Triangle. See Canes Venatici and Figure 9 in Part One.

Tubus Herschelii Major, Tubus Herchelii Minor

Maximilian Hell devised these two constellations to honor William Herschel's discovery (1781) of Uranus. He placed them between Gemini and Taurus in his map of 1789 in the area of the zodiac where Uranus was first sighted. Tubus Major represented Herschel's large twenty-foot reflector and Tubus Minor his smaller seven-foot one. Bode, on the other hand, included only one telescope in his atlas of 1782, calling it simply Herschelsche Teleskop. In his atlas of 1801, he translated this into Latin as Telescopium Herschelii.

Select Bibliography

1. Atlases and Catalogues

Aitken, Robert Grant. *New General Catalogue of Double Stars within 120° of the North Pole.* 2 vols. Washington, D.C., 1932.

Aitken's catalogue lists and analyzes over 17,000 double stars reduced to 1900 and 1950.

Apianus, Peter. *Astronomicum Caesareum.* Ingolstadt, Germany, 1540.

Apian's work is especially interesting since it includes a sky chart depicting Bootes with several hounds, a possible prototype for the constellation Canes Venatici. This sky chart is the first by a European astronomer to show Alcor, g, 80 UMa, the faint 4th-magnitude companion of Mizar, Zeta, 79 UMa, in the handle of the Big Dipper. Islamic astronomers had noted Alcor's presence centuries earlier (Warner, *Sky Explored,* pp. 8-10).

Argelander, F. W. A. *Uranometria Nova: Catalogus Stellarum.* Berlin, 1843.

Argelander published both an atlas and catalogue of stars visible from Central Europe with the naked eye. His atlas contains seventeen charts: sixteen charts of the constellations he himself was able to observe from Germany and one chart of the southern skies. His catalogue is arranged by constellation, and, like Flamsteed's "Catalogus Britannicus" of 1725, individual stars are unnumbered. In both atlas and catalogue, right ascension is given in degrees, not in time, and the stars are reduced to the epoch 1840.

Backhouse, Thomas William. *Catalogue of 9842 Stars, or all Stars Very Conspicuous to the Naked Eye, for the Epoch of 1900.* Sunderland, England, 1911.

Backhouse's catalogue is especially valuable because it contains several pages of corrections for Heis's *Atlas Coelestis Novus.*

Baily, Francis. *An Account of the Revd. John Flamsteed, the First Astronomer-Royal.* London, 1835.

This biography of Flamsteed includes a revised and annotated version of Flamsteed's catalogue with the stars arranged by right ascension, not by constellation as in Flamsteed's original work. Also included are 458 additional stars that Flamsteed observed but that were inadvertently omitted from his printed catalogue. Two years later after the publication of this biography, Baily issued a *Supplement to the Account of the Revd. John Flamsteed, the First-Astronomer Royal,* which contains primarily papers and documents relating to the feud between Flamsteed and Newton and Halley. In 1966, the biography was reprinted in London with the supplement, but without the catalogue.

———. *The Catalogue of Stars of the British Association for the Advancement of Science.* London, 1845.

The *BAC,* as it has been called by generations of astronomers, was one of the most accurate, comprehensive, and advanced catalogues of its day. It lists the coordinates of over 8,000 stars for the epoch 1850 together with each star's proper motion, annual variation, and secular variation.

———. "A Catalogue of the Positions (in 1690) of 564 Stars Observed by Flamsteed, but not Inserted in his British Catalogue." *Memoirs of the Astronomical Society of London* [later the Royal Astronomical Society], Vol. IV, Part I (1830), pp. 129-64.

    The stars in this catalogue were later revised by Baily and the results included in his biography of Flamsteed.

———. "The Catalogues of Ptolemy, Ulugh Beigh, Tycho Brahe, Halley, Hevelius." *Memoirs of the Royal Astronomical Society,* XIII (1843).

    The stars in each of the catalogues are arranged in the order their authors originally intended. Anticipating the needs of the researcher, Baily numbered the stars and sought to identify them with their Bayer letters and Flamsteed numbers whenever possible.

———. "General Catalogue of the Principal Stars." *Memoirs of the Astronomical Society of London,* II (1826), Appendix.

    With just 2,870 stars reduced to 1830, this work is an early predecessor to the *BAC*.

Bartsch, Jacob. *Usus Astronomicus, Planispherii Stellati.* Strasbourg, France, 1624.

    Bartsch's work contains several maps of the heavens and lists the constellations, including the twelve new ones devised for the southern skies by Keyzer and Houtman. Like several other astronomers during the religious upheavals and fervor of the sixteenth and early seventeenth centuries, Bartch sought to "depaganize" the heavens by referring to the constellations with allusions to Biblical personages and events.

Bayer, Johannes. *Uranometria.* Augsburg, Germany, 1603.

    This is the first edition of Bayer's famous work. Unlike several later editions that contain only maps, the original includes a map and star list for each of the forty-eight Ptolemaic constellations. The original also includes a map of the southern skies showing the twelve new constellations devised by Keyzer and Houtman but does not include an accompanying star list. There are, in addition, two planispheres of the northern and southern hemispheres.

Becvar, Antonin. *Atlas Australis.* 2nd ed. Cambridge, Mass., 1976.

———. *Atlas Borealis.* 2nd ed. Cambridge, Mass., 1978.

———. *Atlas Eclipticalis.* 2nd ed. Cambridge, Mass., 1974.

    No longer in print, these three volumes provide a comprehensive portrait of the heavens for almost all stars down to magnitude 9, and sometimes beyond, for the epoch 1950. Each volume contains over 100,000 stars.

———. *Atlas of the Heavens: Atlas Coeli 1950.0.* Cambridge, Mass., 1962.

    Containing over 32,000 stars down to magnitude 7.75, this atlas is a much smaller, popular version of his three-volume work.

———. *Atlas of the Heavens - II: Catalogus 1950.0.* 4th ed. Cambridge, Mass., 1964.

    This catalogue was meant to accompany his *Atlas of the Heavens.*

Bevis, John. *Atlas Celeste [Uranographia Britannica].* Reprint. Alburg, England., 1987.

    This is a replica of one of the few nearly complete atlases prepared by Bevis. It contains both maps and star lists arranged by constellation with individual stars — over 3,500 — identified with their Bayer letters and Flamsteed numbers. Prepared about 1750, the *Uranographia* includes charts of the traditional Ptolemaic constellations and one of the southern skies depicting the twelve new constellations devised by Keyzer and Houtman. Bevis intended to call his work *Uranographia Britannica,* but the title pages of several incomplete sets that were published in 1786, fifteen years after his death, read *Atlas Celeste.*

Bode, Johann Elert. *Allgemeine Beschreibung und Nachweisung der Gestirne nebst Verzeichniss.* Berlin, 1801.

    This is Bode's catalogue of over 17,000 stars that was meant to accompany his *Uranographia,* his stellar atlas. Covering the entire sky, it is arranged by constellation and is based on the works of, among others, Hevelius, Flamsteed, Lacaille, Lalande, and Bode himself. The stars are reduced to 1801 and are identified with their Flamsteed numbers and Bayer and Lacaille letters.

———. *Uranographia.* Berlin, 1801.

    An exceptionally large, carefully drawn atlas of the entire sky, Bode's *Uranographia* was meant to be used by working astronomers in the field. Like his catalogue, it includes several constellations that would be recognized only by antiquarians and star-lore buffs: Herschel's Telescope, Electric Machine, Aerial Balloon, King George III's Harp, Printing Office, Solitary Thrush, Poniatowski's Bull, Brandenburg Scepter, and other long-forgotten asterisms.

———. *Vorstellung der Gestirne...nach der Pariser Ausgabe des Flamsteadschen Himmelsatlas.* Berlin, 1782. Reprint. Düsseldorf, 1973.

This book, with maps and star lists, is based on a French edition of Flamsteed's work, the 1776 atlas published by J. Fortin. See Lalande and Méchain.

Boss, Benjamin. *General Catalogue of 33342 Stars for the Epoch 1950.* 5 vols. Washington, D.C., 1937.

Many of the stars in the *General Catalogue* are identified with their Bayer letters and Flamsteed numbers.

Brahe, Tycho. *Tychonis Brahe Dani, Opera Omnia.* 15 vols. Edited by J. L. E. Dreyer *et al.* Copenhagen, 1913-29.

Tycho's first catalogue of 777 stars can be found in volume 2; his enlarged catalogue of over 1,000 stars is in volume 3.

Brisbane, Thomas Makdougall. *A Catalogue of 7385 Stars Chiefly in the Southern Hemisphere.* London, 1835.

As governor of New South Wales, Brisbane erected at Parramatta, just south of Sydney, the first permanent observatory in the southern hemisphere and was responsible for sponsoring this comprehensive catalogue of the southern skies, the first since Lacaille's work of a century before. The catalogue's stars are reduced to epoch 1825. It is especially helpful in identifying Lacaille's stars. See Kenneth Weitzenhoffer, "General Thomas Brisbane's Astronomical Adventures," *Sky & Telescope* (December 1992), pp. 620-22.

Cannon, Annie J., and Edward C. Pickering. *The Henry Draper Catalogue.* Vols. 91-99 of *Annals of the Astronomical Observatory of Harvard College* (1918-24).

These nine volumes catalogue a total of over a quarter of a million stars to epoch 1900. Many of the stars are identified in the notes with their Bayer or Lacaille letters but not with their Flamsteed numbers. Two additional volumes, 100 (1925-36) and 112 (1949), were added to the series, but they contain no notes on stellar nomenclature.

Delporte, Eugène. *Atlas Céleste.* Cambridge, England, 1930.

The charts in this work were prepared to illustrate the boundaries of the eighty-eight officially recognized constellations that were fixed, once and for all, by the International Astronomical Union in 1930. The boundary lines were drawn to epoch 1875. With one or two rare exceptions, the boundaries of the southern constellations were based on Gould's *Uranometria Argentina.*

Dreyer, J. L. E. *New General Catalogue,* 1888; *Index Catalogue,* 1895; *Second Index Catalogue,* 1908. Reprint (3 vols. in 1). London, 1971.

This reprint contains the *NGC* and *IC* catalogues of over 13,000 nonstellar objects reduced to epoch 1860. Dreyer prepared it to "revise, correct, and enlarge" John Herschel's *General Catalogue of Nebulae* published in 1864. The *NGC* should be used in conjunction with Dreyer's "Corrections to the New General Catalogue" that appeared in the *Monthly Notices, Royal Astronomical Society,* LXXIII (November 1912), pp. 37-40. See also Dorothy Carlson, "Some Corrections to Dreyer's Catalogues of Nebulae and Clusters," *Publications of the Lick Observatory,* XIII (1918), pp. 350-59.

Roger W. Sinnott's *NGC 2000.0* (Cambridge, Mass., 1988) merges the objects in Dreyer's three catalogues by right ascension and reduces them to epoch 2000, while Jack W. Sulentic and William G. Tifft's *The Revised New General Catalogue of Nonstellar Astronomical Objects* (Tucson, 1973) revises Dreyer's descriptions of the objects in the NGC and updates them to epoch 1975.

Flamsteed, John. *Atlas Coelestis.* London, 1729.

Flamsteed's atlas contains twenty-seven charts. Based on his "Catalogus Britannicus," the first twenty-five charts are of the constellations he himself was able to observe. The last two charts are planispheres of the northern and southern skies.

———. *Historia Coelestis Britannica.* 3 vols. London, 1725.

Volume 3 contains Flamsteed's catalogue of stars, "Catalogus Britannicus." It also contains, among others, the catalogues of Ptolemy, Tycho Brahe, and Hevelius with their stars rearranged by right ascension within each constellation, the same way Flamsteed arranged his own catalogue. This is important to bear in mind since some astronomers, like Bevis, refer to Hevelian stars using the order in Flamsteed's version of Hevelius's catalogue, not the order in Hevelius's original catalogue. Four years after the publication of his *Historia,* Flamsteed's star atlas, *Atlas Coelestis,* appeared, containing all the stars listed in his catalogue.

———. *Historiae Coelestis Libri Duo.* 2 vols. Edited by Edmund Halley. London, 1712.

The first sixty pages of the first volume contain Halley's edited, unauthorized, edition of Flamsteed's catalogue that was published against the expressed wishes of the Astronomer Royal. Flamsteed referred to it as the "Corrupted Catalogue." Unlike the authorized catalogue of 1725 which contains 2,935 stars, the pirated edition contains only 2,682 stellar objects.

Gould, Benjamin A. *Uranometria Argentina.* Vol. 1 of *Resultados del Observatorio Nacional Argentino en Córdoba* (1879).

The catalogue, arranged by constellation, lists all stars visible to the naked eye — 7,756 — within 100° of the south celestial pole reduced to the epoch 1875. Because he was high in the Andes and far from light pollution, Gould was able to observe stars as faint as 7th magnitude and sometimes even beyond. He also catalogued 981 nonstellar objects and published an atlas to accompany the catalogue. Gould's work is more than just another catalogue of southern stars, for he included detailed notes about magnitude, nomenclature, and stellar identification.

Grotius, Hugo. *Syntagma Arateorum.* Leiden, Netherlands, 1600.

Included in this work is a star list based exclusively on Ptolemy's catalogue. Most interesting, however, are Grotius's comments on stellar nomenclature that include references to non-European and especially Islamic sources, an indication of the new horizons that were opening up to the astronomers and scientists of Western Europe.

Heis, Eduard. *Atlas Coelestis Novus: Catalogus Stellarum.* Cologne, 1872.

Under the title *Atlas Coelestis Novus,* Heis published both an atlas and catalogue with the stars reduced to the epoch 1855. Like Argelander, Heis arranged his catalogue by constellation and noted right ascension in degrees, not in time. *Atlas Coelestis* lists 5,421 stars visible with the naked eye from Central Europe, 2,153 more stars than Argelander's *Uranometria Nova.*

Helmer, Karl, ed. *Johannes Bayer, Sternzeichen und Sternbilder.* Dortmund, Germany, 1981.

This small paperback book contains reduced but excellent reproductions of all of Bayer's star charts. Like some of the later editions of the *Uranometria,* it omits Bayer's star lists and commentaries. There is a short account of the history of the *Uranometria* at the end of the book.

Hevelius, Johannes. *Prodromus Astronomiae.* Danzig, 1690.

Like Flamsteed's *Historia,* Hevelius's magnum opus, the *Prodromus,* is a posthumous work. It is divided into three parts. The first part is titled "Prodromus Astronomiae" and discusses various astronomical topics, including a detailed explanation and analysis of the twelve new constellations that Hevelius himself devised. The second part is titled "Catalogus Stellarum Fixarum" and includes not only his own catalogue of 1,564 stars reduced to the epoch 1660 and arranged by constellation but also the catalogues of several other astronomers, including Ptolemy and Tycho Brahe. The third part is titled "Firmamentum Sobiescianum, sive Uranographia" and contains two planispheres and fifty-four charts, including one depicting all the new southern constellations devised by Keyzer and Houtman, although their stars are not listed in his catalogue. His star charts were reprinted in the former Soviet Union under the title *Jan Hevelius: The Star Atlas,* ed. V. P. Shcheglov, 3rd ed. (Tashkent, 1979).

Because of two drawbacks, Hevelius's atlas never became as popular as Bayer's, although it is more accurate. It was printed with the constellation figures "backward;" that is, as they would appear on a celestial globe, thus precluding its use by working astronomers. Secondly, like the stars in his catalogue, the stars in his atlas were unlabeled. Although Bayer's atlas had been available for over three-quarters of a century, Hevelius failed to make use of Bayer's letters, thus making it difficult for astronomers to reference his stars.

Hirshfeld, Alan, and Roger W. Sinnott. *Sky Catalogue 2000.0.* 2 vols. Cambridge, Mass., 1982-91.

Volume 1, originally published in 1982, lists over 45,000 stars down to magnitude 8.05V. With the assistance of François Ochsenbein, a revised edition was issued in 1991 with over 50,000 stars. Volume 2, published in 1985, contains double stars, variable stars, and nonstellar objects.

Hoffleit. Dorrit. *The Bright Star Catalogue.* 4th ed. New Haven, 1982.

    The *BS* lists over 9,000 stars down to magnitude 6.5 reduced to epochs 1900 and 2000. It is a revised and updated version of Pickering's *Harvard Revised Photometry* of 1908. Although Pickering identified stars with their proper Greek and Roman letters, Professor Hoffleit has eliminated all references to Roman letters. See also Hoffleit's "Additions and Corrections to the Bright Star Catalogue," *Bulletin d'Information du Centre de Données Stellaires de Strasboug,* XXVII (1984), pp. 161-95.

Hoffleit, Dorrit *et al. A Supplement to the Bright Star Catalogue.* New Haven, 1983.

    A continuation of *The Bright Star Catalogue,* this supplement catalogues over 2,500 additional stars, some of which Hoffleit inadvertently omitted from the earlier work although they were brighter than the limiting magnitude of 6.50V.

Houtman, Frederick de. *Spraeckende Woordboeck Inde Maleysche ende Madagaskarche Talen met Vele Arabische ende Turksche Woorden.* Amsterdam, 1603.

    The appendix to this dictionary consists of a thirteen-page catalogue of southern stars. In addition to several southern Ptolemaic constellations, it includes the Southern Cross and the twelve new constellations that Houtman and Keyzer devised. All in all, Houtman's catalogue contains just over 300 stars arranged by constellation with their magnitudes and coordinates.

Kepler, Johannes. *Gesammalte Werke.* 18 vols. to date. Edited by Walther von Dyck *et al.* Munich, 1937-.

    Kepler's Rudolphine Tables of 1627 can be found in volume 10. The tables are arranged into three groups, or classes, of stars. The first class is a revised version of Tycho's expanded catalogue of 1,000 stars. The second consists of over 400 stars drawn primarily from Ptolemy's catalogue that were not included in Tycho's work. The last class includes southern stars mostly drawn from the catalogue of Frederick de Houtman. Altogether, Kepler's Rudolphine Tables list over 1,700 stars reduced to 1600, and like all the other astronomical works of that time, stellar coordinates are based on ecliptical longitude and latitude, not right ascension and declination.

    Volume 1 contains a detailed account of the Nova of 1604 in the leg of Ophiuchus or Serpentarius, the Serpent Bearer. In an effort to correct the errors found in Tycho's catalogue of 1,000 stars and in Bayer's *Uranometria,* Kepler also included a detailed map of the area together with a catalogue of forty-seven neighboring stars down to the 6th magnitude.

Kholopov, P. N. *General Catalogue of Variable Stars.* 4th ed. 3 vols. Moscow, 1985-87.

    The fourth edition lists more than 28,000 variables by constellation reduced to 1950.

Klepesta, Josef. *Uranometria.* Prague, 1938.

    Prepared by a prominent Czech astronomer, this is a small, updated, paperback version of Bayer's masterpiece with drawings somewhat similar to those in the original. Meant primarily for popular consumption, it does not contain Bayer's star lists.

Knobel, Edward Ball. "On Frederick de Houtman's Catalogue of Southern Stars, and the Origin of the Southern Constellations." *Monthly Notices of the Royal Astronomical Society,* LXXVII (March 1917), pp. 414-32.

    Included in this essay on the travels and travails of Keyzer and Houtman is an English translation of Houtman's 1603 catalogue of southern stars that includes the twelve new constellations that he and Keyzer devised.

———, ed. *Ulugh Beg's Catalogue of Stars.* Washington, D.C., 1917.

    This book provides important information about Ulugh Beg's catalogue. Ulugh Beg, the grandson of Tamerlane, established an observatory at Samarkand and recalculated the positions of the stars in Ptolemy's catalogue reduced to A.H. (Anno Hegirae, in the year of the Hegira) 841 or A.D. 1437.5 (June 1437).

Lacaille, Nicholas-Louis de. *A Catalogue of 9766 Stars in the Southern Hemisphere.* Edited by Francis Baily. London, 1847.

    This is Lacaille's complete catalogue of the southern skies that was reduced to 1750 and finally published almost a century after the original observations were made at Cape Town.

———. *Coelum Australe Stelliferum.* Paris, 1763.

    This work contains Lacaille's catalogue of 1,942 stars reduced to 1750. He selected these stars, the brighter ones, from his earlier observations of the southern skies. The catalogue's format is almost modern. The stars are all numbered and arranged by right ascension, not by constellation, although constellation names and individual stellar letters are supplied. It is a revised, corrected, and somewhat enlarged version of his earlier, preliminary catalogue of 1752.

———. "Table des Ascensions Droites et des Déclinaisons Apparentes." *Mémoires de l'Académie Royale des Sciences, 1752,* pp. 539-92.

This is Lacaille's first or preliminary catalogue, containing 1,935 southern stars. Written in French, the catalogue introduces the fourteen new constellations that Lacaille devised. Like his revised catalogue of 1763, this one too is arranged by right ascension with the stars individually numbered. A short essay following the catalogue explains why he chose the names for his new constellations and why he decided to break up Argo Navis, the Ship, into three separate constellations. Although the *Memoires* of the Royal Academy are dated 1752, they were actually published in 1756. For a brief account of Lacaille's life and work, see Davis S. Evans, "Nicholas de la Caille and the Southern Sky," *Sky & Telescope* (July 1980), pp. 4-7; and Owen Gingerich's extensive essay on Lacaille in the *Dictionary of Scientific Biography*.

Lalande, Joseph-Jérôme de. *A Catalogue of Those Stars in the Histoire Céleste Française of Jérôme DeLalande.* Edited by Francis Baily. London, 1847.

Lalande's catalogue, based on observations made in the last decade of the eighteenth century at Paris, lists over 47,000 stars from the north celestial pole to the Tropic of Capricorn down to the 9th magnitude. They were edited, reduced to 1800, and arranged according to right ascension by Baily.

———. *Histoire Céleste Française.* Paris, 1801.

This is a detailed, day-by-day account of the observations made by Lalande and his associates. The information contained in these observations provided Baily with the data he needed to edit the catalogue cited above.

Lalande, Joseph-Jérôme de, and P. F. A. Méchain. *Atlas Céleste de Flamstéed, Publié en 1776, Par J. Fortin.* 3rd ed. Paris, 1795.

This is a revised and updated version of Flamsteed's atlas, with a planisphere of the southern skies based on Lacaille's observations. It also contains a catalogue of almost 1,000 stars. As an indication that science is never totally divorced from politics, the title page of this work notes that it was published in *"L'an IIIe. de la République François* (the third year of the French Republic)," and that it was reviewed, corrected, and updated by *Citoyens* (citizens) Lalande and Méchain. Luckily Lalande lost only his title to the revolution, not his head.

Maclear, Thomas. *Catalogue of 4,810 Stars for the Epoch 1850.* London, 1884.

Popularly known as the *Cape Catalogue of 1850,* this work is especially helpful in identifying stars in the southern skies since it is cross-referenced to the works of Lacaille, Piazzi, Brisbane, Taylor, Baily, and others.

Mallas, John H., and Evered Kreimer, ed. *The Messier Album.* Cambridge, Mass., 1978.

This work includes a facsimile reproduction of Charles Messier's list of 103 nebulous objects and star clusters that was originally published in 1784. It also includes a detailed account, with photographs, of all 110 objects that Messier and his associate Pierre Francois André Méchain eventually observed.

Norton, Arthur P. *Norton's Star Atlas and Reference Book.* 16th ed. Cambridge, Mass., 1973.

An old standby, *Norton's* contains references to over 8,400 stars down to magnitude 6.35, reduced to 1950.

Peters, C. H. F. "Flamsteed's Stars 'Observed, but Not Existing.'" *Memoirs of the National Academy of Sciences,* III (1886), pp. 69-83.

This article provides a detailed analysis of the twenty-two stars in Flamsteed's catalogue that Baily, in his revised edition of Flamsteed's catalogue, had pronounced as nonexistent.

Peters, C. H. F., and Edward Ball Knobel. *Ptolemy's Catalogue of Stars: A Revision of the Almagest.* Washington, D.C., 1915.

Although outdated by more recent works, such as Toomer's *Ptolemy's Almagest,* this catalogue is still valuable for its copious notes and references to various editions of the *Almagest.*

Piazzi, Joseph [Giuseppe]. *Praecipuarum Stellarum Inerrantium Positiones Mediae Ineunte Saeculo XIX.* Palermo, 1814.

Arranged by hour of right ascension, Piazzi's catalogue identifies many of Bayer and Flamsteed stars. It lists over 7,600 stars reduced to 1800.

Piccolomini, Alessandro. *De le Stelle Fisse Libro Uno.* Venice, 1579.

This work, containing both charts and star lists for the forty-eight Ptolemaic constellations, identifies stars with Roman letters. Compared to Bayer's *Uranometria,* however, Piccolomini's atlas appears primitive in quality and style, although barely twenty-four years separate the two. Surprisingly, it was written in the vernacular, an indication perhaps that it was meant for popular consumption rather than for scholarly work. One should bear in mind that as late as the twentieth century, astronomical studies, like Dreyer's voluminous edition of Tycho's works, were still being written in Latin.

Pickering, Edward. *Revised Harvard Photometry.* Vol. 50 of *Annals of the Astronomical Observatory of Harvard College* (1908).

The Harvard Revised Photometry, as it is more commonly known, is the forerunner of *The Bright Star Catalogue.*

Rumker, Charles. *Preliminary Catalogue of Fixed Stars Intended for a Prospective of a Catalogue of the Stars of the Southern Hemisphere.* Hamburg, 1832.

Rumker's catalogue of almost 500 stars reduced to 1827 was based on observations he made at the Parramatta Observatory in New South Wales under the direction of Governor Thomas Brisbane.

Schiller, Julius. *Coelum Stellatum Christianum Concavum.* Augsburg, Germany, 1627.

In a grandiose effort to depaganize the heavens, Schiller, caught up in the religious fervor of the Reformation and Counter Reformation, substituted Biblical figures for all the forty-eight Ptolemaic constellations as well as the twelve new southern constellations devised by Keyzer and Houtman. He even went so far as to change the twelve signs of the ecliptic to twelve religious symbols: the Key of St. Peter, the Cross of St. Andrew, the Staff of St. James the Elder, and the like.

Prepared with the assistance of Bayer, Schiller's atlas, as Deborah Warner has suggested (*Sky Explored,* p. 232), is an updated, revised, enlarged, and more accurate edition of the *Uranometria*. There are, however, some other major differences between the two. Most noticeably, Schiller's constellation drawings are reversed, as they would appear on a celestial globe and not as they would actually appear to an earth-bound observer. Secondly, stars are identified with numbers, not letters. Finally, two sets of charts are included for each constellation — one set depicting only stars, unadorned with any constellations figures; the other, decorated with Schiller's own religious version of the constellations, the *Coelum Christianum,* the Christian Heaven. Despite the differences, Schiller's work is especially helpful in identifying several of Bayer's "lost" stars.

Schjellerup, H. C. F. C. *Description des Étoiles...par...Abd-Al-Rahman Al-Sufi.* St. Petersburg, 1874.

Al-Sufi, or Azophi as he was known to Europeans, was a tenth-century Islamic astronomer who revised Ptolemy's catalogue, drew a series of constellation figures, and noted several objects not mentioned in the *Almagest,* such as the Andromeda Nebula or Galaxy and Alcor in the handle of the Big Dipper (see Apianus above). Schjellerup translated Al-Sufi's work and reproduced his original illustrations.

Schoener, Johann. *Opera Mathematica.* Nuremberg, 1561.

This work includes a section entitled "Globi Stelliferi," which contains Ptolemy's star catalogue with stellar positions reduced to 1550. The copy in the John Carter Brown Library at Brown University originally belonged to Bayer and contains, among other things, hand-written notations, probably Bayer's, reducing Schoener's coordinates to 1600 in preparation for his forthcoming *Uranometria*. Professor N. M. Swerdlow believes Bayer may have been copying the coordinates from Tycho's catalogue of 777 stars, which, like Bayer's *Uranometria,* was based on the epoch 1600. See Swerdlow's "A Star Catalogue Used by Johannes Bayer."

Scovil, Charles E. *The AAVSO Variable Star Atlas.* 2nd ed. Cambridge, Mass., 1990.

Sponsored by the American Association of Variable Star Observers, this work contains 177 charts of the heavens and lists all variables down to magnitude 9.5. The charts, based on the Smithsonian Astrophysical Observatory's *Star Atlas* of 1969, contain about 260,000 stars and nonstellar objects with coordinates set at epoch 1950.

Smithsonian Astrophysical Observatory. *Star Catalogue.* 4 vols. Washington, D.C., 1966. Reprint 1971.

The SAO catalogues almost 260,000 stars reduced to 1950. Although it contains a wealth of information about each star, it does not identify stars with either Bayer letters or Flamsteed numbers.

Stone, Edward James. *The Cape Catalogue of Stars.* Cape Town, 1878.

Stone's catalogue contains 2,892 stars in the southern sky reduced to 1840.

———. *The Cape Catalogue of 1159 Stars.* Cape Town, 1873.

The stars in this catalogue of the southern skies are reduced to 1860.

Taylor, Thomas Glanville. *A General Catalogue of the Principal Fixed Stars*. Madras, 1844.

> Often referred to as the *Madras General Catalogue*, this catalogue contains over 11,000 stars reduced to 1835. Taylor was the astronomer for the British East India Company, and he made his observations at the Company's observatory at Madras, India. This is the first or preliminary *Madras General Catalogue*.

———. *Taylor's General Catalogue of Stars*. Rev. ed. Edinburgh, 1901.

> This catalogue and the 1844 edition cited above are most helpful, each in its own way, in identifying southern stars. The first edition refers to many stars by their Flamsteed numbers, while the revised edition cross-references stars to the catalogues of Brisbane, Lacaille, and Piazzi. Used in conjunction with each other, they are invaluable in ferreting out lost or missing stars.

Tirion, Wil. *Sky Atlas 2000.0*. Cambridge, Mass., 1981.

> *Sky Atlas* contains twenty-six maps that chart 43,000 stars to magnitude 8. Individual stars are identified with both Bayer letters and Flamsteed numbers.
>
> A second edition by Tirion and Roger Sinnott appeared in 1998 with over 81,000 stars to magnitude 8.5. Unfortunately, the editors admitted in the Introduction that "this edition of the atlas contains fewer *anachronistic names* [emphasis added] than the first one did." In effect, what the editors did was to remove some number and letter designations, thus making the second edition less valuable for purposes of stellar identification. Consequently all references to *Sky Atlas (SA)* are to the first edition unless otherwise noted.

Tirion, Wil, et al. *Uranometria 2000.0*. 2 vols. Richmond, Va., 1987-88.

> A much more detailed atlas than his *Sky Atlas 2000.0*, Tirion's *Uranometria 2000.0* contains 473 charts showing the position of over 330,000 stars down to magnitude 9.5. Included in volume one is an illustrated introductory essay on stellar cartography.

Toomer, G. J. *Ptolemy's Almagest*. New York, 1984.

> Toomer's work contains the latest, most informative English edition of Ptolemy's catalogue.

Webb, T. W. *Celestial Objects for Common Telescopes*. 2 vols. New York, 1962.

> This is an updated reprint of the sixth edition of 1917. Webb's *Celestial Objects* was originally published in 1859. The current edition reduces all the objects cited by Webb to epoch 2000.

Werner, Helmut, and Felix Schmeidler. *Synopsis of the Nomenclature of the Fixed Stars*. Stuttgart, 1986.

> Stars are arranged by constellation and cross-referenced to the catalogues of many of the major astronomers, such as Ptolemy, Tycho, Bayer, Flamsteed, Gould, and others. Although a most important work, it is, unfortunately, limited to stars brighter than magnitude 5.5, thus greatly reducing its value in stellar identification. See my review in the *Journal for the History of Astronomy*, XIX (February 1988), pp. 59-61.

2. Other Works

Allen, Richard Hinckley. *Star Names: Their Lore and Meaning*. 1899. Reprint. New York, 1963.

> Originally published as *Star-Names and Their Meanings*, this is still the classic work on star lore although somewhat outdated by more recent research.

Argelander, F. W. A. *De Fide Uranometriae Bayeri Dissertatio*. Bonn, 1842.

> This brief essay by one of the giants of nineteenth century astronomy analyzes the accuracy of Bayer's *Uranometria* and finds that, despite some flaws in positions and magnitudes, *Uranometria* is sufficiently accurate for naked-eye observations.

Ashworth, William B., Jr. "John Bevis and his Uranographia (*ca.* 1750)." *Proceedings of the American Philosophical Society*, CXXV (February 1981), pp. 52-73.

> This is a detailed account of how Bevis sought to secure financial support for the publication of his ill-fated *Uranographia* or, as it is more popularly known, *Atlas Celeste*.

Barton, George A. "Traces of the Rhinoceros in Ancient Babylon." *Journal of the Society of Oriental Research*, X (1926), pp. 92-95.

Bates, William Nickerson. *Euripides, A Student of Human Nature*. 1930. Reprint. New York, 1969.

> This work includes all of Euripides' plays, including excerpts and summaries of his lost ones. Two of these,

Melanippe Sophe and *Melanippe Desmotis*, relate to the origin of the constellation Pegasus and possibly Equuleus.

Blitzstein, William. "A Study by Modern Methods of Seven Alleged Pre-Discovery Observations of Uranus by John Flamsteed." 1955. Unpublished paper in the possession of the author.

———. "The Seven Identified Observations of Uranus Made by John Flamsteed Using His Mural Arc." *The Observatory,* CXVIII (August 1998), pp. 219-22.

Boorstin, Daniel J. *The Discoverers.* New York, 1985.

This work is especially valuable for its sections devoted to the measurement of time, the development of astronomy, and the beginnings of modern science.

Burnham, Robert, Jr. *Burnham's Celestial Handbook.* 3 vols. New York, 1978.

An indispensable guide to astronomy, *Burnham's Handbook* meets the needs both of the interested amateur and the serious scholar.

Chapman, Allan. *The Preface to John Flamsteed's Historia Coelestis Britannica.* London, 1982.

This book, which is number 52 of the Maritime Monographs and Reports of the National Maritime Museum, includes a brief account of Flamsteed's life, his instruments, and his work. It also provides an English translation of the introduction to the third volume of the *Coelestis Britannica,* which contains Flamsteed's history of astronomy from classical times to the eighteenth century and an account of the founding of the Royal Greenwich Observatory.

Cirlot, J.E. *A Dictionary of Symbols.* 2nd ed. New York, 1972.

Condos, Theony. *Star Myths of the Greeks and Romans: A Source Book.* Grand Rapids, Mich., 1997.

This work is especially valuable in tracing the origin of many of the Greek and Roman astral myths and legends. It includes translations of Hyginus's *Poetic Astronomy,* and Eratosthenes or Pseudo-Eratosthenes' *Catasterismi,* or, as Condos refers to it, *The Constellations.* Condos selected those sections of these works that apply to astronomy and, in addition, added explanations and commentaries for each selection.

Contenau, Georges. *Everyday Life in Babylon and Assyria.* London, 1954.

This work provides valuable information on the early history of Mesopotamia.

Cook, S. A., F. E. Adcock, and M. P. Charlesworth. *The Cambridge Ancient History.* Vol. VII, *The Hellenistic Monarchies and the Rise of Rome.* Cambridge, England, 1954.

This volume provides valuable information on the constellation Coma Berenices and the Ptolemaic rulers of Hellenistic Egypt.

De Vaux, Roland. *The Early History of Israel.* Philadelphia, 1978.

Dvornik, Francis. *The Slavs in European History and Civilization.* New Brunswick, N.J., 1962.

This book supplies details on Hevelius's Poland and the military achievements of John III Sobieski, who inspired the creation of the constellation Scutum.

Epstein, I., ed. *The Babylonian Talmud.* 18 vols. London, 1938-52.

Finegan, Jack. *Light from the Ancient Past: The Archeological Background of Judaism and Christianity.* Princeton, 1959.

Fish, T. "The Zu Bird." *Bulletin of the John Rylands Library,* XXXI (1948), pp. 162-71.

In addition to describing the Zu Bird of Mesopotamian mythology, this essay casts light on the origin of the constellations Taurus and Monoceros.

Frankfort, Henri. *Kingship and the Gods: A Study of Ancient Near Eastern Religion as the Integration of Society & Nature.* Chicago, 1948.

This work provides an in depth analysis of the gods of ancient Mesopotamia and the importance of the *Akitu* or New Year Festival.

Frazer, James George. *The Golden Bough: A Study in Magic and Religion.* 3rd ed. 1912. Reprint. New York, 1966.

Part V, Volume I, *Spirits of the Corn and of the Wild,* describes the worldwide customs of planting grain that helps explain the significance of the constellation Virgo with her spike (Spica, Alpha Vir) of wheat.

Freedman, H., and Maurice Simon, eds. *Midrash Rabbah.* 10 vols. London, 1939.

Frymer-Kensky, Tikve. "Adad." In *Encyclopedia of Religion.* New York, 1987, I, 26-27.

    This essay, like the others in the *Encyclopedia of Religion,* is thoughtful, incisive, and well written by an outstanding scholar of ancient Near Eastern history and civilization.

Gantz, Timothy. *Early Greek Myth, A Guide to Literary and Artistic Sources.* Baltimore, 1993.

    This is a monumental work that is especially valuable in tracing the origin of many of the myths and legends of ancient Greece.

Gardner, John, and John Maier, eds. *Gilgamesh.* New York, 1984.

    This is the latest, annotated translation of the first known epic in human history.

Gaster, Theodor H. *Thespis: Ritual, Myth and Drama in the Ancient Near East.* New York, 1950.

    The work not only explores the religious beliefs and rituals of the peoples of the ancient Near East but provides complete translations of some of their epic poems that help to explain the origins of several astral legends.

Gayley, Charles Mills. *The Classic Myths in English Literature and in Art.* 1893. Reprint (n.p.). 1939.

    Although an old work and largely superceded by Gantz's *Early Greek Myth,* this book contains some valuable information on obscure mythological figures.

Gingerich, Owen. *The Great Copernicus Chase and Other Adventures in Astronomical Literature.* Cambridge, Mass., 1992.

    This book brings together in one volume some of the more interesting essays of one of the leading authorities on the history of science.

———. "Nicholas-Louis de Lacaille." In *Dictionary of Scientific Biography.* New York, 1973, VII, 542-45.

Gleadow, Rupert. *The Origin of the Zodiac.* New York, 1969.

    This book traces the concept of the zodiac as it emerged in various cultures throughout the world, but devotes only one chapter to ancient Mesopotamia, where the zodiac, as we know it today, first appeared.

Gordon, Cyrus H. "Belt-Wrestling in the Bible World." *Hebrew Union College Annual,* XXIII (1950-51), pp. 131-36.

    This, and the following two essays, first alerted the author to the significance of belt-wrestling in explaining the configuration of the constellation Hercules, the Kneeler.

———. "The Glyptic Art of Nazu." *Journal of Near Eastern Studies,* VII (October 1948), pp. 261-66.

———. "Western Asiatic Seals in the Walters Art Gallery." *Iraq,* VI (Spring 1939), pp. 3-34.

Hamilton, Edith. *Mythology.* Boston, 1942.

    This book, the work of a renowned Greek scholar, explains clearly and in great detail the classic legends and myths of the ancient world.

Hansen, Donald P. "New Votive Plaques from Nippur." *Journal of Near Eastern Studies,* XXII (July 1963), pp. 145-66.

    This article describes several plaques with fish motifs that may have significance for the origin of the constellation Pisces.

Hartner, Willy. "The Earliest History of the Constellations in the Near East and the Motif of the Lion-Bull Combat." *Journal of Near Eastern Studies,* XXIV (January-April 1965), pp. 1-16.

    Invaluable and thought-provoking, this essay on the meaning and origin of the constellations also contains sky charts depicting Mesopotamian stellar configurations.

Heimpel, Wolfgang. "A Catalogue of Near Eastern Venus Deities." *Syro-Mesopotamian Studies,* IV (December 1982), pp. 9-22.

    This article traces the course by which the Sumerian goddess Inanna, the queen of heaven, assumed the attributes of other Near Eastern deities, such as Ishtar and Astarte, and became associated with the planet Venus.

Heuter, Gwyneth. "Star Names — Origins and Misconceptions." *Vistas in Astronomy,* XXIX (1986), pp. 237-51.

    Although this work is primarily concerned with exploring the Arabic origin of stellar names, it traces several back to the ancient Near East.

Hoffleit, Dorrit. "Discordances in Star Designations." *Bulletin d'Information du Centre de Données Stellaires de Strasbourg,* XVII (1979), pp. 38-65.

    This article notes the discordances in Bayer letters and Flamsteed numbers among various stellar catalogues, but does not explain how or why these discordances arose.

Hooke, S. H., ed. *Myths, Ritual, and Kingship: Essays on the Theory and Practice of Kingship in the Ancient Near East and in Israel.* Oxford, 1958.

    Of special interest is the essay by Professor Sidney Smith on rituals and myths in Mesopotamia.

Hunger, Hermann, and David Pingree. *MUL.APIN: An Astronomical Compendium in Cuneiform.* Horn, Austria, 1989.

    MUL.APIN ("Plow Star," mul is Sumerian for star), are the first words of an astronomical text from Mesopotamia. Although the tablets on which it was written date from the seventh century B.C., the text itself is much earlier and may date from the third millennium. The text contains a catalogue of 60 stars, dates of heliacal risings and settings, the path of the moon through the heavens, an intercalation scheme, planetary theory, etc. The introduction and notes are most helpful in identifying stars and constellations.

Jacobsen, Thorkild. "Mesopotamian Religions." In *Encyclopedia of Religion.* New York, 1987, IX, 447-66.

———. *Toward the Image of Tammuz and Other Essays on Mesopotamian History and Culture.* Cambridge, Mass., 1970.

    Chapters one through six explore in depth Mesopotamian religious beliefs.

———. *The Treasures of Darkness: A History of Mesopotamian Religion.* New Haven, 1976.

    Like the preceding work, *Treasures of Darkness* is itself a treasure and an indispensable study of the religions and philosophic beliefs of the people of Mesopotamia by one of the foremost scholars of the ancient Near East.

King, Henry C. *The History of the Telescope.* 1955. Reprint. New York, 1979.

    This classic study of the telescope describes in the second chapter the developments leading up to the invention of the first successful telescope.

Kramer, Samuel Noah. *History Begins at Sumer.* 3rd ed. Philadelphia, 1981.

    This book is a popular introduction to the people and culture of ancient Sumer.

———. *Sumerian Mythology: A Study of Spiritual and Literary Achievement in the Third Millennium B.C.* Rev. ed. New York, 1961.

    An in depth study of Sumerian religious beliefs, this book is especially important for translating and analyzing the various Sumerian epics and other literary works.

———. *The Sumerians: Their History, Culture, and Character.* Chicago, 1963.

    Kramer's extensive translations of Sumerian literary works are most helpful in exploring the origin of Mesopotamian astral legends.

———, ed. *Mythologies of the Ancient World.* Garden City, N.Y. 1961.

    The chapters by Kramer on Sumerian and Akkadian mythology and by Cyrus Gordon on Canaanite mythology are brief summaries of the religious beliefs of these peoples.

Krupp, E. C. *Beyond the Blue Horizon: Myths and Legends of the Sun, Moon, Stars, and Planets.* New York, 1991.

    This is an excellent book that summarizes the latest available knowledge about astral myths from all over the globe and explains difficult astronomical concepts with clarity and ease.

———. "Night Gallery: The Function, Origin, and Evolution of Constellations." *Archaeoastronomy: The Journal of Astrology in Culture,* XV (2000), pp. 43-63.

Kunitzsch, Paul. "Peter Apian and 'Azophi': Arabic Constellations in Renaissance Astronomy." *Journal for the History of Astronomy,* XVIII (May 1987), pp. 117-24.

    This article describes how several Arabic names found their way into the star charts of Peter Apian in the sixteenth century.

Kunitzsch, Paul, and Tim Smart. *Short Guide to Modern Star Names and their Derivation.* Wiesbaden, Germany, 1986.

    Co-author Paul Kunitzsch is the recognized authority on the derivation of Arabic star names.

Langdon, Stephen Herbert. *Babylonian Menologies and the Semitic Calendars.* London, 1935.

    In describing the Babylonian calendar, this book explains the origin and meaning of the months, notes the reigning stars of each month, and describes activities that were to be allowed or prohibited on each day of the month.

———. *The Babylonian Epic of Creation [Enuma Elish].* Oxford, 1923.

    This work includes a translation of the Mesopotamian account of how Marduk defeated Chaos, ordered the universe, and became king of the gods. It also contains a lengthy introduction and extensive notes.

———. *The Legend of Etana and the Eagle, or the Epical Poem "The City They Hated."* Paris, 1932.

    Langdon relates the epic tale of man's first attempt to fly to heaven on the wings of an eagle.

———. *The Mythologies of All Races.* Vol. V, *Semitic.* 1931. Reprint. New York, 1964.

    Langdon's work on Mesopotamian mythology should be used with considerable caution. Since it was published, new material has been uncovered, new advances in cuneiform studies have been made, and a new generation of scholars trained to translate and evaluate the new material has emerged. In short, Langdon's work on Mesopotamian mythology has been largely outdated by the writings of Frankfort, Jacobsen, and Kramer.

Lombardo, Stanley, ed. and trans. *Sky Signs: Aratus' Phaenomena.* Berkeley, Calif., 1983.

    Lombardo's work provides a modern translation of the third-century B.C. Greek poem on astronomy.

Lum, Peter. *The Stars in Our Heaven: Myths and Fables.* New York, 1948.

    Often neglected by those interested in star lore, this book brings together astral myths and legends from around the world. It also includes charts of the heavens.

Marcus, David. "Enki." In *Encyclopedia of Religion.* New York, 1987, V, 106-7.

    This is a short, but insightful account of the Sumerian god of fresh water.

Masselman, George. *The Cradle of Colonialism.* New Haven, 1963.

    A detailed account of Dutch expansionism in the East Indies, this book provides the background for the experience of Keyzer and Houtman, the first Europeans to devise new constellations from the stars in the Southern Sky.

McNeill, William H. *The Rise of the West: A History of the Human Community.* New York, 1963.

    In a sweeping view of world history from ancient times to the twentieth century, this insightful book explores how Western European civilization, with its science and learning, its technology and industry, and its concepts of individual freedom and human values, became the dominant force in the modern world.

Mozel, Philip. "The Real Berenice's Hair." *Sky & Telescope* (May 1990), pp. 485-86.

Neugebauer, Otto. *The Exact Sciences in Antiquity.* 2nd ed. Providence, R.I., 1957.

———, ed. *Astronomical Cuneiform Texts: Babylonian Ephemerides of the Seleucid Period for the Motion of the Sun, the Moon, and the Planets.* 3 vols. London, 1955.

———. "The History of Ancient Astronomy: Problems and Methods." *Journal of Near Eastern Studies,* IV (January 1945), pp. 2-38.

———. "The Survival of Babylonian Methods in the Exact Sciences of Antiquity and Middle Ages." *Proceedings of the American Philosophical Society,* CVII (December 1963), pp. 528-35.

    Written by one of the truly great scholars of the history of science, these four works are invaluable as background material for anyone interested in studying the history of science in antiquity.

O'Neil, W. M. *Time and the Calendars.* Sydney, 1975.

    This book provides the reader with a detailed, scholarly study of how different civilizations developed different measures of time.

Oppenheim, A. Leo. "Mesopotamian Mythology II," *Orientalia,* XVII (1948), pp. 17-57.

    Like the work of Cyrus Gordon, this essay helped to untangle the mystery of the constellation Hercules, the Kneeler.

Pannekoek, A. *A History of Astronomy.* London, 1961.

Pasachoff, Jay M., and Donald H. Menzel. *A Field Guide to the Stars and Planets.* 3rd ed. Boston, 1992.

    With star charts drawn by Wil Tirion, the foremost modern stellar cartographer, this is one of the best astronomy handbooks for the average reader or amateur astronomer.

Pritchard, James B. *The Ancient Near East in Pictures Relating to the Old Testament.* 2nd ed. Princeton, 1969.

    This book is especially valuable since it contains many illustrations from walls, boundary stones, tablets, and temples that depict people, royalty, deities, and symbols from the ancient world.

——, ed. *Ancient Near Eastern Texts Relating to the Old Testament.* 3rd ed. Princeton, 1969.

    Like its companion volume cited above, this book is invaluable for anyone interested in studying the ancient world. It includes collections of documents and texts that help in exploring the life, times, and beliefs of the people of the Fertile Crescent. It is especially valuable for providing the latest translations of such works as "Etana" and "The Creation Epic."

Reiner, Erica, and David Pingree. *Babylonian Planetary Omens,* Part 1, "Enuma Anu Enlil," Tablet 63: The Venus Tablet of Ammisaduqa. *Bibliotheca Mesopotamica.* Edited by Giorgio Buccellati. Vol. 2, Facile 1. Malibu, Calif., 1975.

——. *Babylonian Planetary Omens,* Part 1, "Enuma Anu Enlil," Tablets 50-51. *Bibliotheca Mesopotamica.* Edited by Giorgio Buccellati. Vol. 2, Facile 2, Malibu, Calif., 1981.

    The introductions to this volume and the preceding one help to identify many of the stars and constellations of ancient Mesopotamia.

Reinhold, Meyer. *Essentials of Greek and Roman Classics, A Guide to the Humanities.* Great Neck, N.Y., 1946.

    As its title implies, this work provides a brief, but intelligent summary of the deities and legendary figures of Greek and Roman mythology by a perceptive and scholarly professor of the classics.

Ridpath, Ian. *Star Tales.* New York, 1988.

    This is one of the latest efforts to retell the mythological origins of the constellations. It is an excellent work in recounting astral legends. Like so many others, however, it emphasizes Greek and Roman sources rather than those from the Near East. It is beautifully illustrated with star charts from the atlases of Bayer, Hevelius, Flamsteed, and Bode.

Rogers, John H. "The Origins of the Ancient Constellations: I. The Mesopotamian Traditions; II. The Mediterranean Traditions." *Journal of the British Astronomical Association,* CVIII (February and April 1998), pp. 9-27, 79-89.

    In part I, the author believes that the Mesopotamians devised the zodiacal and "parazodiacal" animal constellations such as Hydra, Corvus, Aquila, and Piscis Austrinus. In part II, he attributes the northern Ptolomaic constellations to the Minoans and Greeks.

Roux, Georges. *Ancient Iraq.* Cleveland, 1964.

Roy, Archie E. "The Origin of the Constellations." *Vistas in Astronomy,* XXVII (1984), pp. 171-97.

    This article presents an interesting thesis that argues that many of the forty-eight Ptolemaic constellations were originally devised by the Minoans of ancient Crete.

Sachs, Abraham J., and Hermann Hunger. *Astronomical Diaries and Related Texts from Babylonia.* 2 vols. Vienna, 1988-89.

    The introduction explains in detail the calendar and the methods of time measurement in ancient Babylon.

Schaefer, Bradley E. "The Latitude and Epoch for the Formation of the Southern Greek Constellations." *Journal for the History of Astronomy,* XXXIII (November 2002), pp. 313-50.

    The author asserts that based on latitude and visibility the Mesopotamians most certainly devised the zodiacal and southern Ptolomaic constellations.

Schwab, Gustav. *Gods & Heroes, Myths & Epics of Ancient Greece.* 1847. Reprint. New York, 1965.

    This work was originally published as *Die Sagen des Klassischen Altertums* (The Myths of Classic Antiquity). Like many of the older works on classic mythology, this book is most helpful in ferreting out many obscure references. It also provides a detailed account of all of Hercules' Twelve Labors.

Sluiter, Engel. "The Telescope before Galileo." *Journal for the History of Astronomy,* XXVIII (August 1997), pp. 223-234.

    This article asserts that the claim made by Johan Sachariassen that his father had invented the telescope — possibly as early as 1590 — was an effort "to rob [Hans] Lipperhey of the honour...."

Staal, Julius D. W. *The New Patterns in the Sky: Myths and Legends of the Stars.* Blacksburg, Va., 1988.

In addition to recounting the legends and stories surrounding the constellations, this work also contains star charts illustrating the modern constellations as well as charts showing how the constellations appeared to other peoples and cultures of the world. Although helpful in reconstructing the classical Greek and Roman legends, it contains only a small amount of material on the ancient Near East.

Swerdlow, N. M. "A Star Catalogue Used by Johannes Bayer." *Journal for the History of Astronomy,* XVII (August 1986), pp. 189-97.

This essay provides a wealth of information about Bayer's methodology and the overall reliability of his *Uranometria.*

Tester, S. J. *A History of Western Astrology.* Bury Saint Edmunds, Eng., 1987.

Tester's work is a scholarly account of astrology from ancient times to the eighteenth century.

Van Buren, Elizabeth Douglas. *Symbols of the Gods in Mesopotamian Art.* Rome, 1945.

With its many illustrations and line drawings, Van Buren's book is especially valuable in interpreting the figures and symbols on Mesopotamian tablets, boundary stones, walls, cylinder seals, and plaques.

———. "Fish-Offerings in Ancient Mesopotamia." *Iraq,* X (Autumn 1948), pp. 101-21.

This essay is helpful in tracing the origin of the constellation Pisces.

Van der Waerden, B. L. "On Babylonian Astronomy I: The Venus Tablets of Ammisaduqa." *Jaar Bericht voor Aziatisch-Egyptisch Genootschap (Ex Oriente Lux),* X (1948), pp. 414-24.

———. "Babylonian Astronomy. II.: The Thirty-Six Stars." *Journal of Near Eastern Studies,* VIII (January 1949), pp. 6-26.

———. "Babylonian Astronomy. III.: The Earliest Computations." *Journal of Near Eastern Studies,* X (January 1951), pp. 20-34.

———. "History of the Zodiac." *Archiv für Orientforschung,* XVI (1953), pp. 216-30.

Professor van der Waerden's series of ground-breaking essays are significant contributions to the history of science since they discuss Near Eastern astronomy and trace the origin of many of the constellations to the peoples of Mesopotamia.

Wagman, Morton. "Flamsteed's Missing Stars." *Journal for the History of Astronomy,* XVIII (August 1987), pp. 209-23.

———. "Hercules, the Champion." *Journal for the History of Astronomy,* XXIII (May 1992), pp. 134-36.

———. "Who Numbered Flamsteed's Stars?" *Sky & Telescope,* (April 1991), pp. 380-81.

See also "Letters," *Sky & Telescope* (November 1991) for important comments on this article by Adam Perkins and Owen Gingerich.

Warner, Deborah J. *The Sky Explored: Celestial Cartography, 1500-1800.* New York, 1979.

An invaluable addition to the history of astronomy, Warner's book contains maps and informative commentary that were most helpful in the preparation of the present work.

Willmoth, Frances, ed. *Flamsteed's Stars: New Perspectives on the Life and Work of the First Astronomer Royal, 1646-1719.* Bury Saint Edmunds, England, 1997.

This work contains a collection of valuable essays on various aspects of the life and work of John Flamsteed. Especially relative to the current work are essays by Mordechai Feingold, Alan Cook, and Owen Gingerich on Flamsteed's feud with Hally and Newton and on the differences between the catalogues of 1712 and 1725.

Wolf, A. *A History of Science, Technology, and Philosophy in the 16th and 17th Centuries.* 2 vols. Reprint of 2nd ed., New York, 1959.

This work summarizes European scientific and technological achievements that ushered in the Age of Reason. The numerous diagrams and pictures of the various instruments and inventions that were developed during this period are especially interesting.

Zerubavel, Eviatar. *The Seven Day Circle: The History and Meaning of the Week.* New York, 1985.

Index

This index primarily includes names and objects; however, not all names or objects that appear in the body of the text, tables, synopses, appendixes, or charts have been included since this would cause unnecessary clutter. Only when the name or object has some significance or importance has it been cited in this index. Figures in bold indicate the page(s) upon which the main entry for a constellation appears in Part One: Lettered Stars and/or Part Two: Numbered Stars.

Abzu (Apsu), 171
Achilles, 93
Adad (Hadad), 320
Adam and Eve, 220
Adamath, 21
Aegean Sea, 116
Aegeus, 116
Aegis, 50
Aeolus, 142
Aeson, 82
Aethra, 293
Akitu (New Year) Festival, 45, 193, 292, 308
Alcides, 62
Alcmene, 62
Alcyone, 296, 298
Alexander the Great, 38
Alexandria, 112
Alphabet, 292-293
Alphonsine Tables, 4
Al-Sufi, 4
Amalthea, 50
Amphitrite, 89, 132
Amphitryon, 62
An, 33
Anat, 225
Androgeus, 116
Andromeda, **21-24,** 89, 100, 103, 237, **333-337**
Antiochus, 112
Antinous, 30, 39, 284, 519
Antlia, 7, **25-29**
Anunitum, 247
Apama II, 112
Aphrodite (Venus), 42, 116-117, 248, 251
Apian, Peter, 68
Apis, 7
Apollo, 42, 121, 132, 144, 204, 220, 226, 266, 287
Apostles, 155
Apus, 7, **30-32,** 60, 107, 231
Aqhat, 225

Aquarius, **33-37,** 78, 184, 248, 251, 275, 292, **338-342**
Aquila, 34, **38-41,** 284, 328, **343-346**
Ara, 6, 39, **42-44,** 93, 333
Aratus, 144
Arcadia, 54, 68
Arcas, 54, 308, 312
Arctophylax, 54, 68, 134, 308
Ares (Mars), 42, 116
Argelander, F. W .A., 4, 6, 8, and *passim*
Argonauts, 46, 82
Argo Navis, 6, 7, 82, 110, 255, 262, 303, 315, 459
Argus, 82
Ariadne, 116
Aries, 9, 25, **45-49,** 233, 292, 301, **347-350**
Arion, 132
Aristotle, 3
Artemis (Diana), 42, 142, 225-226, 233, 308
Asclepius, 93, 138, 220, 266, 287
Asterion, 68
Asterope (Sterope), 298
Astypalaea, 191
Athamas, 45
Athena (Minerva), 21, 42, 100, 103, 138, 220, 237
Athenaeus, 191
Athens, 40, 116
Atlas, 42, 54, 138, 237, 293, 298
Atlas Mountains, 138, 237
Augeas, 328
Augsburg, 241
Auriga, **50-53, 351-353**
Azimeth, 55, 112

Baal, (*see* Adad)
Babylon, 45, 193
Bacchantes, 204
Bacchus, (*see* Dionysus)
Baily, Francis, 6-8, 40, 112, 182, 189, 202, 211, 262, 268, 290, 299, 328, 432, and *passim*
Bartsch, Jacob, 30, 49, 60, 110, 124, 134, 211, 213

Bayer, Johannes, 3-10, 16 and *passim*
Bayer's duplicate stars, 23
Bayer's nebulous objects, 168
Beelzebub, 49
Beg, Ulugh, 4
Beirut, 155
Bellerophon, 233
Belt-Wrestling, 163
Berenice II, 112
Bering Strait, 308
Bevis, John, 5-6, 8, 40, and *passim*
Black Sea, 46, 82
Bleau, William Jansen, 213
Blitzstein, William, 479
Bode, Johann, 8, 58, 251, 312, and *passim*
Boeotia, 82
Bootes, 54-57, 68, 112, 308, **354-356**
Bow-Star (Sirius), 71, 225
Boyle, Robert, 25, 152
Bradley, James, 6
Brahe, Tycho, 3, 112, 223, 383, and *passim*

Cadmus, 116
Caelum, 7, **58-59**
Cain and Abel, 225
Calliope, 204
Callisto, 54, 308, 312
Camelopardalis, 8, **60-61**, 189, 333, **357-360**, 474
Cancer, **62-67**, **361-364**
Canes Venatici, 8, 54, 60, **68-70**, 189, 333, **365-366**
Canis Major, 6, **71-75**, 76, 191, 268, **367-368**
Canis Minor, 54, 71, **76-77**, 191, **369**
Cape of Good Hope, 7
Cape Town, 207
Capricornus, 33, **78-81**, 248, 284, **370-371**
Carina, 6, **82-88**, 255, 262, 315, 459
Cassini, Giovanni, 299
Cassiopeia, 21, **89-92**, 100, 103, 237, **372-374**
Castor and Pollux, 82, 127, 155-156, 157-158
Celaeno, 298
Celsus, 3
Centaurus, 6, 7, 8, **93-99**, 124, 197, 268, **375**, 432
Cephalus, 71
Cepheus, 21, 89, **100-102**, 103, 237, **376-377**
Cerberus, 68, 103, 164, 165, 328, 519
Ceres, (*see* Demeter)
Ceto, 103, 138
Cetus, 21, 89, 100, **103-106**, 138, 237, **378-382**
Chamaeleon, 7, **107-108**, 231
Chaos, 42
Chara, 68
Charles I, 69-70, 519
Charles II, 125, 256, 285, 520
Chimaera, 103, 233
Chiron, 93, 142, 197, 233
Cimmerians, 268
Circinus, 7, **109**, 303
Clymene, 144
Coelum Stellatum Christianu, (*see* Schiller)
Columba, 7, 60, **110-111**
Coma Berenices, 8, 30, 55, 69, **112-113**, 189, 333, **383-384**
Conon of Samos, 112
Copernicus, Nicholas, 3
Cor Caroli, 69-70, 519

Corona Australis, 6, **114-115**, 117, 299, 333
Corona Borealis, **116-118**, **385-386**
Coronis, 266
Corvus, **119-120**, 121, 171, **387-389**
Crater, **121-123**, 171, **390-391**, 415
Crateres Achaion, 121
Crete, 42, 50, 71, 116
Cronus (Saturn), 42, 50, 93
Crotus, 268
Crux, 6, 93, **124-126**, 127
Cupid, 248
Custos Messium, 519
Cyclopes, 220, 266
Cygnus, **127-131**, **392-395**
Cynosura, 312
Cyrene (Libya), 112

Daedalus, 116
Danae, 237
Danzig, 290
David's Harp, 49
Delphinus, **132-133**, **396**
Delporte, Eugène, 114, 275, 287
Demeter (Ceres), 42, 54, 321
Demetrius the Fair, 112
Demiphon, 121
Derceto, 251
Descartes, René, 109
Deucalion, 142
Diana, (*see* Artemis)
Dike, 321
Dilbat (planet Venus), 171
Diomedes, 328
Dionysus (Bacchus *or* Liber), 54, 62, 114, 116-117, 204, 293
Dodecatemoria, 309
Dorado, 7, 60, 107, **134-137**, 231
Double Cluster, 90, 164, 238, 240-241
Draco, **138-141**, **397-399**
Dragon, 171, 197
Dumuzi, (*see* Tammuz)

East Indies, 60
Ebro, 144
Echidna, 103, 138
Egypt, 112
El, 155
Electra, 296, 298
Eleusa, 121
Enki (Ea), 33, 71, 78, 144, 171, 247-248, 251, 320
Enkidu, 163, 220, 292
Enlil, 33
Equuleus, **142-143**, **400-401**
Eratosthenes, 163
Erech, 193
Erichthonius, 50
Eridanus, 6, **144-151**, **402-404**
Eridu (Abu Shahrein), 144
Erigone, 54
Erymanthian boar, 328
Esther, 171
Etana, 38
Ethiopia, 21
Euphrates, 45, 144, 247-248, 251
Euripides, 233

Europa, 71, 293
Eurydice, 204
Eurystheus, 62, 328
Ezekiel, 225
Ezinu, 320

Felis, 519
Fera, (*see* Lupus)
Fermat, Pierre de, 109
Figures
   1. Bayer's Andromeda, 22
   2. Lacaille's planisphere of the southern heavens, 26
   2a. Lacaille's planisphere from 153° to 265° (enlarged), 27
   2b. Lacaille's planisphere from 218° to 325° (enlarged), 27
   2c. Lacaille's planisphere from 323° to 0° and from 0° to 94° (enlarged), 28
   2d. Lacaille's planisphere from 25° to 152° (enlarged), 28
   3. Bayer's Tabula XLIX, the southern heavens, and an enlarged view of Apus from Tabula XLIX, 31
   4. Bayer's Aquarius, 34
   5. Bayer's Aquila, 38
   6. Bayer's Ara, 43
   7. Bayer's Aries, 46
   8. Hevelius's Aries, 47
   9. Hevelius's Triangulum Majus, Triangulum Minus, and Musca, 47
   10. Bayer's Auriga, 51
   11. Bayer's Bootes, 55
   12. Hevelius's Camelopardalis, 61
   13. Bayer's Cancer, 63
   14. Hevelius's Canes Venatici, 69
   15. Lelande and Méchain's Canes Venatici with Cor Caroli, 69
   16. Bayer's Canis Major, 72
   17. Bayer's Canis Minor, 77
   18. Bayer's Capricornus, 79
   19. Bayer's Navis, 83
   20. Bayer's Cassiopeia, 90
   21. Bayer's Centurus, 94
   22. Bayer's Cepheus, 101
   23. Bayer's Cetus, 104
   24. Bayer's Chamaeleon, 108
   25. Hevelius's Columba and Lepus, 111
   26. Hevelius's Coma Berenices, 113
   27. Bayer's Corona Australis, 115
   28. Bayer's Corona Borealis, 117
   29. Bayer's Corvus, 120
   30. Bayer's Crater, 122
   31. Hevelius's Crux and Centaur, 125
   32. Bayer's Cygnus, 128
   33. Bayer's Delphinus, 133
   34. Bayer's Dorado, 135
   35. Hevelius's Dorado, 136
   36. Bayer's Draco, 139
   37. Bayer's Equuleus, 143
   38. Bayer's Eridanus, 145
   39. Bayer's Gemini, 156
   40. Bayer's Grus, 161
   41. Bayer's Hercules, 164
   42. Hevelius's Hercules with Cerberus, 165
   43. Bayer's Hydra, 172
   44. Bayer's Hydrus, 177
   45. Bayer's Indus, 179
   46. Hevelius's Indus, 180
   47. Hevelius's Lacerta, 183
   48. Bayer's Leo, 185
   49. Hevelius's Leo Minor, 190
   50. Bayer's Lepus, 192
   51. Bayer's Libra, 194
   52. Bayer's Lupus, 198
   53. Hevelius's Lynx, 203
   54. Bayer's Lyra, 205
   55. Hevelius's Monoceros, 212
   56. Bayer's Musca, 213
   57. Bayer's Ophiuchus, 221
   58. Bayer's Orion, 226
   59. Bayer's Pavo, 231
   60. Bayer's Pegasus, 234
   61. Bayer's Perseus, 238
   62. Bayer's Phoenix, 243
   63. Bayer's Pisces, 249
   64. Bayer's Piscis Austrinus, 252
   65. Hevelius's Argo Navis, 256
   66. Bayer's Sagitta, 267
   67. Bayer's Sagittarius, 269
   68. Bayer's Scorpius, 276
   69. Hevelius's Scutum, 285
   70. Bayer's Serpens, 287
   71. Hevelius's Sextans, 291
   72. Bayer's Taurus, 294
   73. Bayer's Triangulum, 302
   74. Bayer's Triangulum Australe, 304
   75. Bayer's Tucana, 305
   76. Bayer's Ursa Major, 309
   77. Bayer's Ursa Minor, 313
   78. Bayer's Virgo, 321
   79. Bayer's Piscis Volans, 327
   80. Hevelius's Anser and Vulpecula, 329
   81. Flamsteed's Andromeda, Perseus, and Triangulum, 334
   82. Flamsteed's Aquarius, Capricornus, and Piscis Austrinus, 338
   83. Flamsteed's Aquila, Sagitta, Vulpecula with Anser, and Delphinus, 343
   84. Flamsteed's Aries, 347
   85. Flamsteed's Camelopardalis and Auriga, 351
   86. Flamsteed's Coma Berenices, Bootes, and Canes Venatici, 354
   87. Flamsteed's Cancer, 361
   88. Flamsteed's Monoceros, Canis Major, Canis Minor, Lepus, and Navis, 367
   89. Flamsteed's Cassiopeia, Cepheus, Ursa Minor, and Draco, 372
   90. Flamsteed's Cetus, 378
   91. Flamsteed's Hercules, Corona Borealis, and Lyra, 385
   92. Flamsteed's Hydra, Crater, Corvus, Sextans, and Virgo, 387
   92a. Flamsteed's Hydra, Crater, Corvus, Sextans, and Virgo from a defective copy, 388
   93. Flamsteed's Lyra, Cygnus, Lacerta, Vulpecula with Anser, and Sagitta, 392
   94. Flamsteed's Pegasus and Equuleus, 400
   95. Flamsteed's Eridanus, Orion, and Lepus, 402
   96. Flamsteed's Gemini, 405
   97. Flamsteed's Leo, 419
   98. Flamsteed's Lynx and Leo Minor, 424
   99. Flansteed's Libra and Scorpius, 429
   100. Flamsteed's Ophiuchus and Serpens, 438
   101. Flamsteed's Pisces, 452
   102. Flamsteed's Sagittarius, 462

103. Flamsteed's Taurus and Orion, 474
104. Flamsteed's Ursa Major, 482
105. Flamsteed's Virgo, 488
106. Flamsteed's planisphere of the northern sky, 495
107. Flamsteed's planisphere of the southern sky, 496

Flamsteed, John, 5-11, 15-17, 110, 121, 169, 182, 202, 284-285, 290, and *passim*
Flamsteed's coordinates, 334-335
Flamsteed's nebulous objects, 337
Flamsteed's nonexistent stars, 334-335
Flamsteed's use of clock time for right ascension, 350
Fornax, 7, **152-154**
Frederici Honores, 285, 519
Frederick the Great, 285, 519

Gaea (Ge), 42, 103, 138, 226, 275
Galilei, Galileo, 3, 169, 209, 285, 299
Ganymede, 33-34, 38-39
Garden of Eden, 220
Gemini, 9, **155-159, 405-409**
George III, 285, 520
Geryon, 328
Gilgamesh, 30, 119, 163, 220, 247, 275, 292
Glauber, Johann, 152
Globus Aerostaticus, 519
Golden Fleece, 46, 82
Gorgons, 103, 138, 233, 237
Gould, Benjamin, 6-8, 82, 171, 197, 217, 262, 268, 275, 285, 287, 290, 299, 468, and *passim*
Graiae, 103, 138, 237
Grus, 7, 60, **160-162**
Guericke, Otto von, 25
Gutenberg, Johann, 520

Habrecht, Isaac, 264
Hades, 42, 164, 204, 220, 266
Hadley, John, 217
Hadrian, 39
Haiia, 193
Halley, Edmund, 5, 110, 285, 290, 333, and *passim*
Harmonia, 116
Harpa Georgii, (*see* Psalterium Georgianum)
Heis, Eduard, 4, 6, 8, 188, and *passim*
Helen of Troy, 127
Helice, 312
Helle, 45-46
Hellespont (Dardanelles), 46, 121
Hephaestus (Vulcan), 42, 50, 62, 116
Hera (Juno), 42, 54, 62, 93, 100, 117, 138, 293, 308
Hercules, 30, 49, 62, 68, 82, 93, 138, **163-168**, 266, 328, **410-414**
Hercules' twelve tasks, 328
Hermes (Mercury), 42, 50, 82, 191, 204, 226, 237, 301
Herschel, Caroline, 6
Herschel, John, 25
Herschel, William, 6, 337, 475, 520
Hesperides, 138, 328
Hestia (Vesta), 42
Hevelius, Johannes, 5, 8, 15-16, 54, 68, 110, 164, 182, 189, 202, 284-285, 290, 328, 357, 365, 383, 418, 424, 433, 436, 471, 474, 493, and *passim*
Hipparchus, 30, 142
Hippe, 142, 233
Hippocrates, 220
Hippodamia, 50

Hippolyte, 328
Hondius, Jodocus, Sr., 124
Hooke, Robert, 290
Horologium, 7, **169-170**
Houtman, Frederick de, 3, 6, 15, 30, 60, 68, 107, 110, 134, 160, 176, 179, 213, 231, 243, 303, 305, 326
Hubble Space Telescope, 17
Huygens, Christiaan, 169
Hyades, 292, 293
Hyas, 293
Hydra, 6, 62, 103, 121, **171-175**, 328, 390, **415-417**
Hydrus, 7, **176-178**
Hyginus, 116-117, 121, 226, 248
Hyrieus, 226

Iapetus, 42
Ibex, 33, 78, 184, 248, 251, 275, 292
Icarius, 54, 76
Inanna (Ishtar), 119, 171, 247, 292-293, 320
Indus, 7, 107, **179-181**
Innes, R. T. A., 96
Ino, 45
International Astronomical Union (IAU), 9, 11, 29, 114, 158, 196, 210, 240, 275, 287, 364, 415
Io, 100, 293
Iolaus, 62
Ionian Sea, 100
Isaiah, 155
Ishkur, 320
Ishtar, (*see* Inanna)
Ixion, 93

Jaffa (Tel-Aviv), 21
Janssen, Sacharias, 209, 241
Jason, 46, 82, 93
Jericho, 284
Jerusalem, 155, 225
Jesus, 30, 131, 155, 284
Jonathan's Javelin, 49
Josephus, 21
Juno, (*see* Hera)
Jupiter, (*see* Zeus)

Kaerius, Petrus, 49, 60, 211
Kepler, Johannes, 3, 30, 49, 134, 213, 223
Keyser, Pieter Dircksen, 3, 6, 15, 30, 60, 68, 107, 134, 160, 176, 179, 213, 231, 243, 303, 305, 326
Klepesta, Josef, 4
Knossos, 116
Kuppuru, 45

Labyrinth, 116
Lacaille, Nicholas-Louis de, 6-10, 15, 25, 58, 82, 93, 107, 109, 110, 114, 124, 134, 144, 152, 160, 169, 171, 176, 179, 197, 207, 209, 213, 215, 217, 231, 243, 246, 251, 255, 262, 264, 268, 275, 282, 290, 303, 305, 315, 459, and *passim*
Lacerta, 8, 60, 68, **182-183**, 189, 333, **418**
Ladon, 103, 138, 328
Laelaps, 71
Lamb, Francis, 70
Leda, 127, 156
Leeuwenhoek, Anton van, 209
Leo, 9, 33, **184-188**, 251, 275, 292, **419-423**
Leo Minor, 8, 60, 68, **189-190**, 333, **424-427**

Lepus, 71, **191-192, 428**
Leros, 191
Leto, 225
Libra, 9, 25, **193-196,** 275, **429-431**
Lipperhey, Hans, 241
Lochum Funis, 519
Lupus, 6, 7, 93, **197-201, 432**
Lycaon, 54
Lynx, 8, 60, 68, 189, **202-203,** 333, **433-434,** 474
Lyra, 49, **204-206,** 328, **435**

Machina Electrica, 520
Madagascar, 107
Mad-Dog, 171, 197
Maera, 54, 76
Magas, 112
Magellanic Clouds, 135, 177, 207
Maia, 54, 298
Malpighi, Marcello, 209
Malus, 262
Marduk, 21, 71, 171, 197, 275
Mars, (see Ares)
Mastusius, 121
Medicean stars, 285
Medici, Cosmo de, 285
Medina, Pedro de, 124
Medusa, 233, 237
Megara, 62
Melanippe, 233
Mensa, 7, **207-208**
Mercury, 299, (also see Hermes)
Merope, 296, 298
Messier, Charles, 519
Microscopium, 7, **209-210**
Middelburg, 209, 241
Milky Way, 42, 130, 212, 267, 272, 299, 333, 337
Minerva, (see Athena)
Minos, 116, 293
Minotaur, 116
Monoceros, 8, 60, 189, **211-212,** 333, **436-437**
Mons Maenalus, 68-69, 113, 520
Montgolfier Brothers, 519
Mordecai, 171
Moses, 225
Mt. Sinai, 284
Mul-Apin, 45
Musca, 7, **213-214**
Musca Borealis, 49, 347, 520
Myrtilus, 50

Nabu, 45
Nanshi, 193
Navis, (see Argo Navis)
Nebiru (planet Jupiter), 171
Nemean lion, 103, 184, 328
Nemesis, 127
Nephele, 45
Neptune, (see Poseidon)
Nereids, 21, 89, 132
Newton, Isaac, 5, 333
Nidabe, 193
Nile, 144, 225
Ningizzida (Ningishzida), 220
Ninhursaga (Ninma), 33

Ninshubur, 119
Ninurta, 71
Nippur, 247
Noah, 33, 119, 142, 211, 220, 284
Noah's Ark, 110
Noah's Dove, 60, 110
Norma, 7, **215-216,** 303
Northern Cross, 127
Northern Fly, 49, (see Musca Borealis)
Nova of 1572, 90, 92
Nova of 1600, 128, 131, 330
Nova of 1604, 223
Nova of 1670, 334, 494
Nyx, 127

Oceanus, 42, 233, 293
Octans, 7, **217-219,** 290
Oenomaus, 50
OfficinaTypographica, 520
Ophiuchus, **220-224,** 266, 287, 299, **438-442,** 468
Oracle of Ammon, 100
Oracle of Delphi, 62
Orion, 9, 71, 76, 191, **225-230,** 268, 275, 293, **443-445**
Orpheus, 82, 204
Orthos, 103
Osiris, 225
Ovid, 226, 243

Pan, 78
Pasiphae, 116
Passover, 45, 284
Pavo, 7, 60, 107, **231-232**
Pegasus, 45, **233-236,** 301, **446-448**
Peloponnesus, 50
Pelops, 50
Pentecost, 155
Perses, 237
Perseus, 21, 62, 89, 100, 103, 233, **237-242, 449-451**
Phaethon, 144
Philip of Macedon, 50
Philomelus, 54
Philyra, 93
Phineus, 100
Phoenice, 312
Phoenix, 7, **243-245**
Phorcys, 103, 138
Phrixus, 45-46
Piccolomini, Alexandro, 5
Pictor, 7, **246**
Pisces, **247-250,** 251, 301, **452-456**
Piscis Austrinus, 6, 248, **251-254, 457-458**
Piscis Volans, (see Volans)
Plancius, Petrus, 7, 15, 30, 49, 60, 110, 124, 134, 160, 176, 211, 303, 305, 333, 357, 436, 474
Pleiades, 54, 292-293, 296, 298
Pleione, 293, 298
Po, 144
Polophylax, 134
Polydeuces, (see Castor and Pollux)
Pontus, 103
Poseidon (Neptune), 21, 42, 89, 100, 103, 132, 226, 233
Praesepe or Beehive Cluster, 62
Precession, 16, 249
Prometheus, 42, 266

Propus, 159
Proxima Centauri, 96
Psalterium Georgianum, 285, 520
Ptolemy I, 112
Ptolemy II, 60
Ptolemy III, 112
Ptolemy, Claudius, 3, 4, 6, 23, and *passim*
Puppis, 6, 7, 82, **255-261**, 262, 315, **459-460**
Pyxis, 6, 7, 82, 109, **262-263**, 459

Rangifer, 520
Ras Shamra, 155
Rebecca's Camel, 60
Re'em, 60, 211
Reticulum, 7, **264-265**
Rhea, 42, 50, 93
Rhine, 144
Rhombus, 264
Rhone, 144
Richaud, Jean, 96
Robor Carolinum, 125, 256, 520
Rome, 284
Rosh Hashanah, 193
Royer, Augustin, 5
Rudolphine Tables, 30, 134, 213

Sabbath, 284
Sagitta, **266-267, 461**
Sagittarius, 6, 114, **268-274**, 284, 299, **462-465**
Sampson, 49, 60
Saturn, 299, (*also see* Cronus)
Satyr, 62, 268
Scarborough, Charles, 70, 519
Sceptrum Brandenburgicum, 520
Schiller, Julius, 23, 49, 105, 223, 240, 253, 279
Scorpian-Man, 171, 197, 275
Scorpius, 6, 9, 33, 184, 193, 226, 251, **275-281**, 292, 299, **466-467**
Sculptor, 7, **282-283**
Scutum, 8, 68, **284-286**
Scythians, 268
Semele, 116-117
Senex, John, 5
Septem Triones, 308, 312
Septuagint, 60, 211
Serpens, **287-289, 468-470**
Seven, 284
Sextans, 8, 60, 68, 189, 217, **290-291**, 333, **471-473**
Shacher, 155
Shala, 320
Shalem, 155
Sharp, Abraham, 271, 494, 497
Shat-al-Arab, 144
Shavuot, 155, 284
Shemini Atzeres, 193
Sicily, 132, 301
Sidon, 71
Sobieski, John III, 284-285
Solitaire, 520
Sparta, 127
Sphinx, 103, 171, 197
Stymphalian birds, 328
Suculae, (*see* Hyades)
Sukkot, 193

Swammerdam, Jan, 209
Syria, 112, 320

Tammuz (Dumuzi), 45, 225
Taurus, 9, 33, 45, 184, 211, 225, 233, 251, 275, **292-298**, 301, **474-480**
Taurus Poniatovii, 520
Taygeta, 298
Telescope, 241, 290
Telescopium, 7, **299-300**
Telescopium Herschelii, (*see* Tubus Herschelii)
Thebes, 62, 71
Themis, 321
Therion, (*see* Lupus)
Theseus, 82, 116
Thetis, 142, 233
Tiamat, 21, 71, 171, 197, 275
Tigris, 247-248
Tisha B'Av, 225
Titans, 42, 50, 54, 62, 78, 93, 138, 237, 266, 293
Triangulum, 45, 68, 233, **301-302, 481**
Triangulum Australe, 7, 215, **303-304**
Triangulum Minus, 47, 68, 347, 520
Tros, 33
Tubus Herschelii Major, 520
Tubus Herschelii Minor, 520
Tucana, 7, 60, 107, 231, **305-307**
Turdus Solitarius, (*see* Solitaire)
Twelve, 30, 68
Tyndareus, 156
Typhon, 78, 248

Ugarit, 155, 225
Ur, 193
Uranus, 42
Uranus (planet), 475, 479, 520
Urion, 226
Ursa Major, 54, 68, 112, **308-311**, 312, 474, **482-485**
Ursa Minor, 308, **312-314, 486-487**
Uruk (Erech), 292
Utnapishtim, 33, 119, 220

Vela, 6, 7, 82, 255, 262, **315-319**, 459
Venus, 299, (*see* Aphrodite)
Vespucci, Amerigo, 124
Vesta, (*see* Hestia)
Vienna, 284
Viper, 171, 197
Virgo, 54, 112, **320-325, 488-492**
Volans, 7, 60, 107, 231, **326-327**
Vulcan, (*see* Hephaestus)
Vulgate of St. Jerome, 60, 211
Vulpecula, 8, 60, 68, 189, **328-330**, 333, **493-494**

Xiphias, 134

Yom Kippur, 193

Zeus (Jupiter), 34, 39, 42, 50, 54, 62, 71, 76, 78, 93, 100, 117, 121, 127, 138, 156, 184, 204, 220, 226, 233, 237, 266, 293, 301, 308, 312, 320-321